# Lecture Notes in Electrical Engineering

## Volume 937

The book series *Lecture Notes in Electrical Engineering* (LNEE) publishes the latest developments in Electrical Engineering—quickly, informally and in high quality. While original research reported in proceedings and monographs has traditionally formed the core of LNEE, we also encourage authors to submit books devoted to supporting student education and professional training in the various fields and applications areas of electrical engineering. The series cover classical and emerging topics concerning:

- Communication Engineering, Information Theory and Networks
- Electronics Engineering and Microelectronics
- Signal, Image and Speech Processing
- Wireless and Mobile Communication
- Circuits and Systems
- Energy Systems, Power Electronics and Electrical Machines
- Electro-optical Engineering
- Instrumentation Engineering
- Avionics Engineering
- Control Systems
- Internet-of-Things and Cybersecurity
- Biomedical Devices, MEMS and NEMS

For general information about this book series, comments or suggestions, please contact leontina.dicecco@springer.com.

To submit a proposal or request further information, please contact the Publishing Editor in your country:

**China**

Jasmine Dou, Editor (jasmine.dou@springer.com)

**India, Japan, Rest of Asia**

Swati Meherishi, Editorial Director (Swati.Meherishi@springer.com)

**Southeast Asia, Australia, New Zealand**

Ramesh Nath Premnath, Editor (ramesh.premnath@springernature.com)

**USA, Canada:**

Michael Luby, Senior Editor (michael.luby@springer.com)

**All other Countries:**

Leontina Di Cecco, Senior Editor (leontina.dicecco@springer.com)

**** This series is indexed by EI Compendex and Scopus databases. ****

Peter Han Joo Chong · Akhtar Kalam ·
Antonio Pascoal · Manas Kumar Bera
Editors

# Emerging Electronics and Automation

## Select Proceedings of E2A 2021

Springer

*Editors*
Peter Han Joo Chong
Auckland University of Technology
Auckland, New Zealand

Akhtar Kalam
School of Engineering and Science
Victoria University
Melbourne, VIC, Australia

Antonio Pascoal
Institute for Systems and Robotics
Lisbon, Portugal

Manas Kumar Bera
Department of Electronics
and Instrumentation Engineering
National Institute of Technology Silchar
Silchar, India

ISSN 1876-1100 ISSN 1876-1119 (electronic)
Lecture Notes in Electrical Engineering
ISBN 978-981-19-4299-0 ISBN 978-981-19-4300-3 (eBook)
https://doi.org/10.1007/978-981-19-4300-3



This Springer imprint is published by the registered company Springer Nature Singapore Pte Ltd.
The registered company address is: 152 Beach Road, #21-01/04 Gateway East, Singapore 189721, Singapore

# About the Conference

The International Conference on Emerging Electronics and Automation (E2A) 2021 is the first annual international conference organized by the Department of Electronics and Instrumentation Engineering at NIT Silchar, Assam, India. E2A 2021 is scheduled to be held online for three days between December 17 and 19, 2021.

This conference focuses on trending research and product outcomes in the domain of control and robotics, signal processing, communication systems, sensors and instrumentation, MEMS and VLSI, AI and soft computing. E2A 2021 will provide a common platform for the academicians, researchers, industrial players and budding scholars to discuss their results, innovations, challenges encountered during research and concerns with peers and update their knowledge by interacting with leading researchers of respective domains. It offers a great opportunity for them to network together, present research progress and addresses new challenges. This conference promises to provide knowledge enriching keynote lectures to the participants by world-renowned researchers in the relevant thematic areas.

# Message from the Director

On behalf of the academic community in NIT Silchar, I welcome you to the International Conference on Emerging Electronics & Automation (E2A 2021) from December 17 to 19, 2021.

Being a research conducive green campus, NIT Silchar encourages young minds to develop attributes and knowledge beyond the constraints of traditional classroom teaching and learning. We stand among the top engineering institutions of north-eastern part of India.

Responding to the challenges posed due to the pandemic in a rapidly advancing global environment for research, the online E2A 2021 conference will focus on the interdisciplinary fields of instrumentation, signals, systems and computing. The conference wishes to offer a conducive environment to explore avenues for creative thinking and scientific steps needed for sustainable engineering solutions for the challenges ahead of us.

I would like to congratulate the Department of Electronics and Instrumentation Engineering for hosting this international event and look forward for an exciting three days of ideas sharing and academic collaborations being established. I am sure the participants from diverse fields of science and technology would find this event fulfilling and enriching.

I wish you all a successful conference.

Director
NIT Silchar

# Message from the Honorary Chair

I would like to congratulate the faculty at NIT Silchar for their initiative in organizing and realizing this first International Conference on Emerging Electronics and Automation (E2A). Control, Communication and Computation (C^3) assisted by artificial intelligence (AI) and machine learning (ML) are emerging as the drivers of modern technology in all traditional engineering fields as well as in new ones such as driverless cars, unmanned aerial vehicles as well as renewable energy applications such as solar power, wind power and others. It is most appropriate that India, as an emerging power, and its universities and research organizations take a lead in and invest in these technologies of the future.

The conference program put together by the organizers is of high quality, and I am honored to be associated with this conference. I look forward to participating in this and future editions of this conference.

With best wishes.

Shankar Bhattacharyya
Texas A&M University, USA

# Message from the General Chair

It is my pleasure to co-organize the 1st International Conference on Emerging Electronics and Automation (E2A) 2021 organized by the Department of Electronics and Instrumentation Engineering at NIT Silchar, India.

As one of the General Chairs of E2A, I am glad to see that the conference covers various important topics in the areas of control and robotics, communications, instrumentation, MEMS and VLSI, AI and computing. It is an excellent opportunity and networking platform for us, academicians, researchers and industrial engineers and managers, to discuss the current research problems, present our research progress and address new challenges together. This year, ECA is well organized to cover various important topics in industrial electronics, AI & machine learning, power electronics and wireless sensors networks. In addition, the conference has invited a number of renowned researchers worldwide to share their research experience and thoughts in those relevant research areas. I am confident that the academicians, researchers and students participated in the conference will be benefited greatly by the variety and important research topics to be presented and discussed in the coming days.

I wish all success to E2A 2021 as well as all the various parties involved.
Look forward to seeing and meeting you in the conference.
Thank you.

Peter Han Joo Chong
Auckland University of Technology Auckland, New Zealand

# Message from Head of the Department

As the head of the department, I am pleased to welcome you all to the International Conference on Emerging Electronics & Automation (E2A 2021). It is the first international conference going to be organized by the Department of EIE, NIT Silchar. By the age, may be this conference is young, but the standard of the research articles received, the papers are selected for publication and the coverage of the thrust areas can meet international standards.

To keep up with the growth of technological revolution in the field of electronics, instrumentation, control and automation, the Department of EIE brings before you E2A 2021. This conference is a unique forum for exchange of innovative ideas, technical expertise for technological advancements in these fields. It includes keynote addresses from world-renowned academicians and paper presentation by the researchers from different parts of the country. It is a matter of joy for us to welcome the participants to this conference.

I thank Springer, which agreed to publish the peer-reviewed and selected papers of this E2A 2021 conference as proceedings with in their prestigious *Lecture Notes in Electrical Engineering* which is indexed by EI Compendex and Scopus databases.

In a nutshell, the conference promises to transcend to a new and unprecedented level of excellence.

Thank you.

Dr. Manas Kumar Bera
NIT Silchar

# Message from the Organizing Chairs

On behalf of the E2A 2021 organizing committee, we are honored and delighted to welcome you to the 1st International Conference on Emerging Electronics and Automation (E2A) 2021 to be held virtually from 17 to 19 December 2021 at the Department of Electronics and Instrumentation Engineering, NIT Silchar, Silchar, Assam, India. The conference focuses on trending research and product outcomes in the domain of control and robotics, signal processing, communication systems, sensors and instrumentation, MEMS and VLSI, energy and power systems, AI and soft computing. The conference is technically co-sponsored by Springer, whereas the other sponsors are Edutech, Allied Telesis, Novo Computing, Geons Logix, Promptech Instruments and Sushma Electronics. The technical program consists of 6 keynote speeches, 1 speech from the industry and around 42 technical papers. The peer-reviewed and selected papers will be published as proceedings with Springer in prestigious Lecture Notes in Electrical Engineering series.

The conference is a result of the hard work, support, dedication and effort from all the stakeholders. In particular, we are thankful to Director of the Institute and Patron of the Conference Prof. Sivaji Bandyopadhyay for his guidance and vision in organizing the conference. Our sincere thanks to Honorary Chair Prof. S. P. Bhattacharya, Texas A&M University, USA, for his valuable suggestion. The immense help and support received from General Chairs, namely Prof. Peter Han Joo Chong, Auckland University of Technology, New Zealand, Prof. Akhtar Kalam, College of Engineering and Science, Victoria University, Australia, Dr. Antonio Pascoal, Institute for System and Robotics, Lisbon, Portugal, and Dr. Manas Kumar Bera, Head, Department of Electronics and Instrumentation Engineering, NIT Silchar, have been instrumental in improving the quality of the conference. Our sincere gratitude to General Co-Chair, Dr. S. H. Laskar, Associate Professor, Department of Electronics and Instrumentation Engineering, NIT Silchar, for sharing his knowledge and wisdom. We are thankful to the various committees, namely Technical Program Chairs, Publication Chairs, Publicity Chairs, Hospitality Chairs, Session Chairs and Finance Chairs. We are extremely thankful to the Technical Program Committee for their thorough and timely reviewing of the papers and our sponsors who have helped us to organize E2A 2021 in a better way. The recognition should go to the Local Organizing Committee members, namely Registration Committee, Sponsorship Committee, Website and Online Platform Committee, who have all worked extremely hard for the minute details of the conference programs. Last but not least, we are thankful to all the participants without whose presence the program would not have been a success.

Thank you for your presence at the conference. Enjoy the three days.

Organizing Chairs

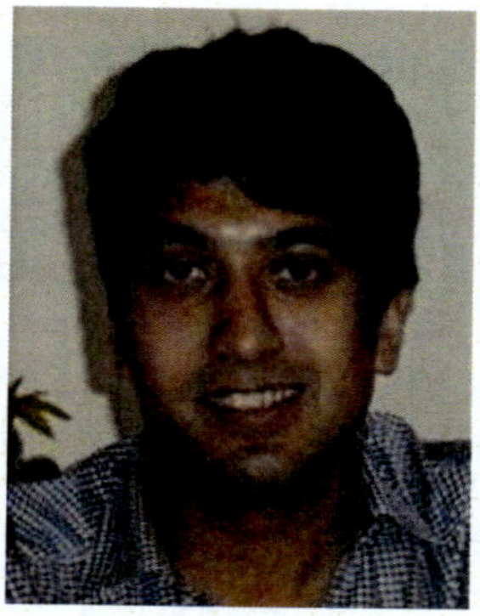

Dr. Ranjay Hazra

Dr. Koena Mukherjee

Dr. Sudipta Chakraborty

Dr. Anup Kumar Sharma

# Keynote Speeches

*Name*: **Professor S. P. Bhattacharyya**

*Affiliation*: Texas A&M University, USA

*About*: Dr. Shankar P. Bhattacharyya is Professor in the Dept. of Electrical Engineering, at the College of Engineering of Texas A&M University (TAMU). He is currently Robert M. Kennedy Professor of Electrical Engineering. He has been active in the field of automatic control systems, and his research interest also lies in multivariable control system analysis and design, computer-aided control system design (CACSD), robust control theory and applications. He is a graduate from IIT Bombay and has done his Masters and Ph.D. from Rice University. He started his career as Assistant Professor at the Federal University, Rio de Janeiro, Brazil, in 1971 and continued there till 1980. Later, he joined the Texas A&M University as Associate Professor and has been working there ever since. During this time, he has been awarded with numerous awards and recognitions for his contribution in the field of engineering. Some of the prestigious awards he has received in recent years are the Texas A&M University Outstanding ECEN Professor Award, International Federation of Automatic Control, IFAC Fellow Award for fundamental contributions to

Robust Control and Controller Synthesis, Kenneth L. Clinton Award for contributions to the Study Abroad program and many more. He has been Foreign Member of the National Academy of Engineering, Brazil, since 2015 and Foreign Member of Brazilian Academy of Sciences since 2011. He has authored several books in Control Theory and Automatic Control Systems.

*Title*: **Recent Developments on the Design of PID Controllers**

*Abstract*: The Proportional Integral Derivative (PID) controller is ubiquitous by its presence in all control systems, be they traditional ones such as motor control, power electronics, process control or modern ones such as driverless cars, UAVs, renewable energy systems such as solar panels, and windmills. Any progress on PID control thus has an enormous impact on many applications.

In this talk, we outline significant recent (last 20 years') developments in the theory and design of PID controllers. This account begins with the effective computation of the complete stabilizing set S for continuous time as well as discrete time systems using a nested linear programming algorithm. With this set in hand, it is possible to impose multiple design specifications such as prescribed gain and phase margins as well as H_infinity margins. An analytical approach to such multiobjective design problems is developed and illustrated via examples. Additionally, a novel extension of single-input single-output design methods to multivariable controllers is also developed and illustrated by an example. The results follow those described in the recent monograph [1].

[1] Ivan Rodriguez Diaz, Sangjin Han and S. P. Bhattacharyya Analytical Design of PID Controllers, Springer, 2019.

*Name*: **Dr. Shahar Kvatinsky**

*Affiliation*: Andrew and Erna Viterbi Faculty of Electrical and Computer Engineering, Technion—Israel Institute of Technology, Israel.

*About*: Shahar Kvatinsky is Associate Professor at the Andrew and Erna Viterbi Faculty of Electrical and Computer Engineering, Technion—Israel Institute of Technology. He received the Ph.D. degree in Electrical Engineering from the Technion—Israel Institute of Technology in 2014. From 2006 to 2009, he worked as a circuit designer at Intel. From 2014 and 2015, he was a post-doctoral research fellow at Stanford University. He is a member of the Israel Young Academy. He is the head of the Architecture and Circuits Research Center at the Technion, chair of the IEEE Circuits and Systems in Israel, and an editor of Microelectronics Journal and Array. He has been the recipient of numerous awards: 2021 Norman Seiden Prize for Academic Excellence, 2020 MDPI Electronics Young Investigator Award, 2019 Wolf Foundation's Krill Prize for Excellence in Scientific Research. His current research is focused on circuits and architectures with emerging memory technologies and design of energy efficient architectures.

*Title*: **Making Real Memristive Processing-in-Memory Faster and Reliable**

*Abstract*: Memristive technologies are attractive candidates to replace conventional memory technologies and storage class memories and can also be used to perform logic and arithmetic operations using a technique called "stateful logic." Combining data storage and computation in the memory array enables a novel non-von Neumann architecture, where both the operations are performed within a memristive memory processing unit (mMPU). mMPU relies on adding computing capabilities to the memristive memory cells without changing the basic memory array structure. The use of an mMPU alleviates the primary restriction on performance and energy in von Neumann machine, which is the data transfer between CPU and memory.

This talk focuses on the various aspects of mMPU. I will discuss its architecture and implications on the computing system and software, as well as examining the microarchitectural aspects. Then, I will present examples of applications that can benefit from processing within memristive memory and show how mMPU can be improved to accelerate different applications and how the poor reliability of memristors can be improved as part of the mMPU operation.

*Name*: **Professor Moncef Gabbouj**

*Affiliation*: Tampere University, Finland

*About*: Dr. Gabbouj is Professor of Information Technology at the Department of Computing Sciences, Tampere University, Tampere, Finland. He was Academy of Finland Professor during 2011–2015. His research interests include dig data analytics, multimedia content-based analysis, indexing and retrieval, artificial intelligence, machine learning, pattern recognition, nonlinear signal and image processing and analysis, voice conversion and video processing and coding. He is the past chairman of the IEEE CAS TC on DSP and committee member of the IEEE Fourier Award for Signal Processing. He is Finland Site Director of the USA NSF IUCRC-funded Center for Visual and Decision Informatics (CVDI) and leads the Artificial Intelligence Research Task Force of Finland's Ministry of Economic Affairs and Employment-funded Research Alliance on Autonomous Systems (RAAS).

*Title*: **Recent Progress on the Design of PID Controllers**

*Abstract*: Operational Neural Networks (ONNs) are new generation network models targeting to address two major drawbacks of conventional convolutional neural networks (CNNs): the homogenous network configuration and the "linear" neuron model that can only perform linear transformations over previous layer outputs. ONNs can perform any linear or nonlinear transformation with a proper combination of "nodal" and "pool" operators. This is a great leap toward expanding the neuron's learning capacity in ONNs, requiring the use of a single nodal operator for all synaptic connections of every neuron. This restriction has recently been lifted by introducing a superior neuron called the "generative neuron" where each nodal operator can be customized during the training in order to maximize learning. As a result, the network is able to self-organize the nodal operators of its neurons' connections. Self-organized ONNs (Self-ONNs) composed with superior generative neurons can achieve diversity even with a compact configuration. We shall explore several signal processing applications of neural network models equipped with the superior neuron.

*Name*: **Professor Peter Han Joo Chong**

*Affiliation*: Auckland University of Technology, Auckland, New Zealand

*About*: Professor Peter Han Joo Chong is currently an Associate Head of School (Research) in the School of Engineering, Computer and Mathematical Sciences at Auckland University of Technology (AUT), New Zealand. He was previously Head of Department of Electrical and Electronic Engineering (EEE) at AUT. He received the Ph.D. degree from the University of British Columbia, Canada, in 2000. He is Adjunct Professor in the Department of Information Engineering, Chinese University of Hong Kong, Hong Kong. He was previously Associate Professor (tenured) in the School of EEE at Nanyang Technological University, Singapore. Between 2013 and 2016, he was a Director of Infinitus, Centre for Infocomm Technology. His research interests are in the areas of wireless and mobile networks including MANETs/VANETs, V2X, Internet of Things/vehicles, AI for wireless networks and 5G networks.

*Title*: **Fuzzy Logic-based Resource Allocation Algorithm for 5G C-V2X Networks**

*Abstract*: Cellular V2X (C-V2X) communications in 5G enable more advanced services with requirements of ultra-low latency and ultra-high reliability. How to make full use of the limited physical-layer resources is a key determinant to guarantee the quality of service (QoS). Therefore, resource allocation plays an essential role in exchanging information between vehicles, infrastructure and other devices.

In this talk, we spotlight C-V2X in 5G cellular networks. In order to intelligently and reasonably allocate resources, we propose a self-adaptive fuzzy logic-based strategy. To evaluate the network performance for this adaptive strategy, a system model for C-V2X is built for urban areas, and typical safety and non-safety services are deployed in the network. Our results reveal that the proposed fuzzy logic-based algorithm can substantially improve resource utilization and satisfy the requirements of C-V2X, compared with prior counterparts, which cannot provide guaranteed services due to low resource utilization.

*Name*: **Professor Antonio Pascoal**

*Affiliation*: Institute for System and Robotics, Lisbon, Portugal

*About*: Professor Antonio Pascoal is Founder of the Dynamical Systems and Ocean Robotics Lab (DSORLab) of ISR, Portugal, and currently acting as Professor of Control and Robotics at IST, University of Lisbon, Portugal. His Ph.D. is in Control Science from the University of Minnesota, Minneapolis, MN, USA, 1987. He is Adjunct Scientist, National Institute of Oceanography (NIO), Goa, India. He is expertise in dynamical systems theory, marine robotics, navigation, guidance and control of autonomous vehicles and networked control and estimation with applications to air and underwater robots. He has coordinated and participated in a large number of international projects that have led to the design, development and field testing of single and multiple autonomous marine and air vehicles and systems in cooperation with partners in India (National Institute of Oceanography, Goa), USA (Naval Postgraduate School, Monterey, CA), Korea (KAIST, Daejeon) and Europe. He is a co-author of the monograph Time-Critical Cooperative Control of Autonomous Air Vehicles. He has received the IEEE OES AUV Distinguished Lifetime Technical Achievement Award in 2020.

*Title*: **Cooperative Marine Robots: From Theory to Scientific and Commercial Applications**

*Abstract*: The last decade has witnessed tremendous progress in the development of marine technologies that are steadily affording scientists advanced equipment and methods for ocean exploration and exploitation. Recent advances in marine robotics, sensors, computers, communications and information systems are being applied to the development of sophisticated technologies that will lead to safer, faster and far more efficient ways of exploring the ocean frontier, especially in hazardous conditions. As part of this trend, there has been a surge of interest worldwide in the development of autonomous marine robots capable of roaming the oceans freely and collecting data at the surface of the ocean and underwater on an unprecedented scale. Representative examples are autonomous surface craft (ASC) and autonomous underwater vehicles (AUVs). The mission scenarios envisioned call for the control of single or multiple AUVs acting in cooperation to execute challenging tasks without close supervision of human operators. This talk addresses the general topic of cooperative motion planning, navigation and control of marine vehicles, both from a theoretical and a practical perspective. The presentation builds upon practical developments and experiments. Examples of scientific and commercial missions with ASCs and AUVs, acting alone or in cooperation, set the stage for the main contents of the presentation. From a theoretical standpoint, a number of challenging problems are addressed in the general area of networked systems subjected to stringent communication constraints, namely i) cooperative motion control using event-driven control and communications and ii) range-based multiple target localization and tracking using tools from the areas of optimal motion planning and estimation theory. Some of the results obtained are illustrated with videos from actual field tests with multiple marine robots exchanging information over acoustic and optical networks. The core material presented in the talk was obtained in the scope of the following EU-funded projects: • MORPH (http://cordis.europa.eu/project/rcn/101726_en.html) • WiMUST (http://www.wimust.eu/) • H2020 EU Marine Robotics Research Infrastructure Network (https://www.eumarinerobots.eu/) • RAMONES (https://ramones-project.eu/).

*Name*: **Dr. Anamika Dubey**

*Affiliation*: WSU Pullman, USA

*About*: Dr. Anamika Dubey is Assistant Professor in the School of Electrical Engineering and Computer Science in Washington State University. She is a graduate from IIT Roorkee. She has done her M.S. and Ph.D. in the area of electrical power systems from the University of Texas, Austin. She has received two U.S. patents for "Locating Multi-Phase Faults in Ungrounded Power Distribution Systems," and "Decoupled Three-Phase Power Flow Analysis Method for Unbalanced Power Distribution Systems." She has bagged several awards and grants for her research contribution in the field of electrical power systems.

# Future Directions

We take this opportunity to inform you that we are going to organize E2A 2022 in December 2022.

# Committee

## Patron

Prof. Sivaji Bandyopadhyay, Director, NIT Silchar

## Honorary Chair

Prof. S. P. Bhattacharyya, Texas A&M University, USA

## General Chair

Prof. Peter Han Joo Chong, Auckland University of Technology, New Zealand
Dr. Antonio Pascoal, Institute for Systems and Robotics, Lisbon
Prof. Akhtar Kalam, College of Engineering and Science, Victoria University
Dr. Manas Kumar Bera, NIT Silchar

## General Co-chair

Dr. S. H. Laskar, NIT Silchar

## Organizing Chairs

Dr. Ranjay Hazra, NIT Silchar

Dr. Koena Mukherjee, NIT Silchar
Dr. Sudipta Chakraborty, NIT Silchar
Dr. Anup Kumar Sharma, NIT Silchar

## Technical Program Chairs

### Control and Robotics

Prof. Indra Narayan Kar, IIT Delhi
Dr. K. Mukherjee, NIT Silchar

### Signal Processing

Prof. R. Balasubramanian, IIT Roorkee
Dr. K. Shankar, NIT Silchar

### Communication System

Dr. Hari Krishna Garg, NUS Singapore
Dr. Ranjay Hazra, NIT Silchar

### Sensors and Instrumentation

Prof. N. C. Shivaprakash, IISc Bangalore
Dr. Rajdeep Dasgupta, NIT Silchar

### MEMS and VLSI

Prof. Roy Paily, IIT Guwahati
Dr. S. K. Pandey, NIT Silchar

## AI and Soft Computing

Prof. Rajib Bandyopadhyay, Jadavpur University
Dr. Munmun Khanra, NIT Silchar

## Publication Chairs

Dr. Rajdeep Dasgupta, NIT Silchar
Dr. Munmun Khanra, NIT Silchar
Dr. Arun Kumar Sunaniya, NIT Silchar

## Finance Chairs

Prof. A. K. Barbhuiya, NIT Silchar
Dr. Lalu Seban, NIT Silchar
Dr. Shivendra Kumar Pandey, NIT Silchar

## Advisory Committee

Prof. R. Balasubramanian, IIT Roorkee
Prof. Debasish Ghosh, IIT Roorkee
Dr. Hari Krishna Garg, NUS Singapore
Dr. A. Abudhahir, B.S. A.R. CIST, Chennai
Prof. Roy Paily, IIT Guwahati
Dr. Subhojit Ghosh, NIT Raipur

## Hospitality Chairs

Prof. R. D. Mishra, NIT Silchar
Dr. Sudarsan Sahoo, NIT Silchar
Dr. Vipin Chandra Pal, NIT Silchar

## Publicity Chairs

Prof. B. K. Roy, NIT Silchar
Dr. Rajeeb Dey, NIT Silchar
Dr. Wasim Arif, NIT Silchar
Dr. Koena Mukherjee, NIT Silchar

## Technical Program Committee

Dr. Shyam Kamal, IIT BHU
Dr. S. Mohamed Yacin, Majmaah University, Saudi Arabia
Dr. Smriti Aggarwal, MNNIT Allahabad
Dr. Chiranjib Koley, NIT Durgapur
Dr. Soumya Ranjan Sahoo, IIT Kanpur
Dr. Prabhat Kumar Sharma, VNIT Nagpur
Dr. Anshul Jaiswal, IIT Roorkee
Dr. Altaf Badar, NIT Warangal
Dr. Twinkle Tripathy, IIT Kanpur
Dr. Girraj Sharma, JECRC University, Jaipur
Dr. Prateek Dolas, IIT Roorkee
Dr. Surabhi Homchowdhury
Dr. Sneha Gajbhiye, IIT Palakkad
Dr. R. Periyasamy, NIT Trichy
Dr. Pravin Prajapati, AD Patel Institute of Technology, Gujarat
Dr. Abhishek Rajan, NIT Sikkim
Dr. Abhisek Kumar Behera, IIT Roorkee
Dr. Vinal Patel, IIITM Gwalior
Dr. OmPrakash Acharya, KIIT University
Dr. Arunangshu Ghosh, NIT Patna
Dr. Sohom Chakraborthy, IIT Roorkee
Dr. Varun Bajaj, IIITDM Jabalpur
Dr. Pratap Khuntia
Dr. D. S. Gurjar, NIT Silchar
Dr. R. Jeyasenthil, NIT Warangal
Manika Saha, Sister Nivedita University, West Bengal
Dr. Manoranjan Rai Bharti, NIT Hamirpur
Dr. Sisir Kumar Nayak, IIT Guwahati
Dr. Shailaja Kurode, COEP
Dr. Jayendra Kumar, NIT Jamshedpur
Dr. Hari Shankar Singh, Thapar University
Dr. Satish Kumar Dubey, IIT Delhi
Dr. Axay Mehta, IITRAM

Dr. Swagatadeb Sahoo, NIT Jamshedpur
Dr. Jagannath Mallick, Korea
Dr. R. Periyasamy, NIT Trichy
Dr. P. S. Lal Priya, College of Engineering, Trivandrum
Dr. Dinesh, Thapar University
V. Rama, NIT Warangal
Dr. Shweta Gautam, Netaji Subhas University of Technology
Dr. Quanxin Zhu, Hunan Normal University, China
Dr. Rahul Upadhyay, Thapar University
Prof. (Dr.) Madhurima Chattopadhyay, Heritage Inst. Technology, Kolkata
Prof. Asheesh K. Singh, MNNIT Allahabad
Dr. Sandip Ghosh, IIT BHU
Dr. Pankaj Pratap Singh, CIT Kokrajhar
Dr. A. Vimala Juliet, SRM, Chennai
Dr. Omhari Gupta, NIT Jamshedpur
Prof. Bidyadhar Subudhi, IIT Goa
Dr. Arvind Yadav, Parul University
Arindam Bit, NIT Raipur
Dr. Jitendra Kumar, NIT Jamshedpur
Dr. Subhendu Bhasin, IIT Delhi
Dr. Ribhu, IIT Guwahati
Dr. Surekha Bhanot, BITS Pilani
Dr. Nitin Singh, MNNIT Allahabad
Dr. K. Srinivasan, NIT Trichy
Dr. Sudarsan Mukherjee, IIT Guwahati
Dr. Dilbag Singh, NIT Jalandhar
Dr. Anshul Agarwal, NIT Delhi
Dr. Sandeep Kumar, NIT Delhi
Dr. Ekant Sharma, IIT Roorkee
Prof. Akhtar Kalam, Victoria University
Navneet Kumar Singh, MNNIT Allahabad
Dr. Deep Gupta, VNIT Nagpur
Dr. Anuradha Ravi, SMU Singapore
Dr. A. Abudhahir, B.S. A.R. CIST, Vandalur
Dr. Sheetla Prasad, Galgotias University, Greater Noida
Dr. Shitala Prasad, A STAR, Singapore
Dr. Amrit Mukherjee, Anhui University, PR, China
Dr. N. C. Shivaprakash, IISc Bangalore
Dr. Prashant Kumar Tiwari, MNNIT Allahabad
Dr. Roshan Kumar, Henan University, China
Dr. Sukwinder Singh, NIT Jalandhar
Dr. Tarikul Islam, Jamia Millia Islamia, Delhi
Dr. Vijay Pratap Singh, REC Sonbhadra, UP
Dr. Bhanu, NIT Kurukshetra
Dr. Sarthak Singhal, MNIT Jaipur

Dr. Ritu Sharma, MNIT Jaipur
Dr. Shashi Singh, NIT Kurukshetra
Dr. Asim Kumar Naskar, NIT Rourkela
Dr. Ravinder Kumar, Qualcomm, Bangalore
Dr. Nikhil Deep Gupta, VNIT Nagpur
Dr. Maloth Naresh, GNIT Telangana
Prof. B. K. Roy, NIT Silchar
Dr. Suneel Yadav, IIIT Allahabad
Dr. Shashikant Sharma, IIIT Ranchi
Dr. Sahaj Saxena, Thapar University
Dr. Avadh Pati, NIT Silchar
Dr. Himanshu Maurya, IIIT Allahabad
Dr. Ashish Kumar, JECRC University
Dr. S. K. Aggarwal, Thapar University
Dr. Souvik Ganguly, Thapar University
Dr. Radhika Gour, VNIT Nagpur
Dr. Satish M., NIT Warangal
Dr. S. Prakash, Thapar University
Dr. Vikram Chopra, Thapar University
Dr. Khadim Siddiqui, SRMU Lucknow
Dr. Shivam Verma, IIT BHU
Dr. Amrita Sinha, Thapar University
Dr. Swati Sondhi, Thapar University
Dr. Mukesh Singh, Thapar University
Dr. Mahendra Sakare, IIT Ropar
Dr. Manoj Badoni, Thapar University
Dr. Raja Rout, Manipal University
Dr. Shakti Singh, Thapar University
Dr. Tejendra Dixit, IIIT Kancheepuram
Dr. Shakti Singh, Thapar University
Prof. Richa Negi, MNNIT Allahabad
Dr. S. Prakash, Thapar University
Dr. Saurabh Kumar Pandey, IIT Patna
Dr. Amit Kumar, Thapar University
Dr. Manikant, C. V. Raman Global University.
Dr. Sunanda Sinha, MNIT Jaipur
Dr. Amitesh Kumar, NIT Patna
Dr. Pawan Kumar, Thapar University
Dr. L. Seban, NIT Silchar
Dr. Shashi Bhushan, NIT Kurukshetra
Dr. Pankaj Pal, NIT Uttarakhand
Dr. Rajan Kumar, NIT Hamirpur
Dr. Vipin Chandra Pal, NIT Silchar
Dr. Pradeep Kumar, NIT Kurukshetra
Dr. Menka, MNIT Jaipur

Dr. Janardan Kundu, Government College of Engineering and Technology.
Dr. Tripti Goel, NIT Silchar
Dr. Ravi Kumar Maddila, MNIT Jaipur
Dr. Ritabrata Bhattacharya, Cadence
Dr. Anish Ahmad, Tezpur University
Dr. R. K. Karsh, NIT Silchar
Dr. Brajesh Rawat, IIT Ropar
Dr. Abhishek Kumar Upadhyay, Technische Universitet Dresden
Dr. Suman Maiti, IIT Kharagpur
Prof. Naimur Rahman Kidwai, Integral University, Lucknow
Dr. Arjun Singh Yadav, NIT Rourkela
Dr. Ashis Kumar Dhara, NIT Durgapur
Dr. Prasenjit Basak, Thapar University
Dr. Shanky Saxena, Lovely University
Dr. Shalendra Kumar Saxena, Weizmann Institute of Science, Israel
Tanmoy Dam, University of New South Wales
Dr. Sulabh Sachan, M. J. P. Rohilkhand University
Dr. Satyasai Jagannath Nanda, MNIT Jaipur
Dr. Arvind Kumar, IGNTU
Dr. R. Murugan, NIT Silchar
Dr. Sudarsan Sahoo, NIT Silchar
Dr. Rajat Kumr Giri, IIIT Bhubaneswar
Dr. Amit Kr Yadav, NIT Sikkim
Dr. Badal Soni, NIT Silchar
Dr. M. D. Borah, NIT Silchar
Dr. Sabyasachi Patra, IIIT Bhubaneswar
Dr. Tarun Biswas, NIT Sikkim
Dr. Piyush Kant, Berkeley
Soumen Biswas, NIT Silchar
Dr. Bunil Kumar Balabantaray, NIT Meghalaya
Dr. Prasanjit Dey, NIT Sikkim
Dr. Tarikul Islam, Jamia Millia Islamia, Delhi
Dr. Tanmoy Malakar, NIT Silchar
Dr. Punit Kumar Jain, NIT Rourkela
Dr. Anulekha Saha, NIT Sikkim
Dr. Abhishek Midya, Memorial Sloan Kettering Cancer Center
Dr. Rumi Rajbongshi, Tezpur University
Dr. Sandeep Kumar Dash, NIT Mizoram
Dr. Narsimha Reddy Sura, KITS
Dr. Rinku Rabidas, Assam University
Asifa Yesmin, NIT Silchar
Dr. Ratnakar Dash, NIT Rourkela
Dr. Samyuktha Kannan, IFPRI
Dr. Jayasree Chakrabarty, Memorial Sloan Kettering Cancer Center
Dr. Rafik Ahmad, BBDITM Lucknow

Pravin Nambisan, NIT Silchar
Dr. Anjan Kr Ray, NIT Sikkim
Dr. Prabina Pattanayak, NIT Silchar
Dr. Shivendra Kumar Pandey, NIT Silchar
Dr. Arun P, St. Joseph, College of Engg. and Tech., Kerala
Dr. Munish Kumar, MRS, Punjab Technical University
Krishanu Nath, NIT, Silchar
Dr. Koena Mukherjee, NIT, Silchar
Dr. Rahul Antony, St. Gaits College of Engineering, Kerala
Dr. Piyush Samant, Chandigarh University
Dr. Shahedul Laskar, NIT Silchar
Dr. Sudipta Chakraborty, NIT Silchar
Mr. Vishnu Mohan, College of Engineering, Kidangoor, Kerala
Dr. Alok Kumar Dubey, NIT Silchar
Dr. Rajdeep Dasgupta, NIT Silchar
Praveen Nambisan, NIT Silchar
Ganesh Chiilakalapudi, NIT Silchar
Dr. K. Shankar, NIT Silchar
Jeetendra Prasad, Vidisha Engineering College
Priyam Kar, NIT Silchar
Dr. Ranjay Hazra, NIT Silchar
Pankaj Kumar Pal, NIT Uttarakhand
Subhra Sankha Sarma, NIT Silchar
Kamlesh Kumar Singh, Amity University
Dr. Manju Khurana, Thapar University
Vikas Singh, IIT Kanpur
Komal Agarwal, MNNIT Allahabad
Ahmad Faiz Minai, Integral University
Akshaya Pati, KIIT University

## Registration Committee

Dr. K. Shankar, NIT Silchar
Dr. Jupitara Hazarika, NIT Silchar

## Sponsorship Committee

Dr. Rajdeep Dasgupta, NIT Silchar
Dr. Munmun Khanra, NIT Silchar
Dr. S. K. Pandey, NIT Silchar
Dr. Bipul Das, NIT Silchar

## Website and Online Platform Committee

Dr. Ranjay Hazra, NIT Silchar
Dr. Lalu Seban, NIT Silchar
Dr. Sudipta Chakraborty, NIT Silchar

## Local Organizing Committee

All EIED Faculty, NIT Silchar

# Contents

## Signal Processing Track—2



## MEMS and VLSI Track—1



## MEMS and VLSI Track—2

# About the Editors

**Peter Han Joo Chong** is Associate Head of School (Research) at the School of Engineering, Computer and Mathematical Sciences, Auckland University of Technology, New Zealand. He received the B.Eng. (with distinction) in Electrical Engineering from the Technical University of Nova Scotia, Canada, in 1993, and the M.A.Sc. and Ph.D. degrees in Electrical and Computer Engineering from the University of British Columbia, Canada, in 1996 and 2000, respectively. His current research projects focus on machine learning techniques applied to software defined vehicular networks. He has been developing techniques of deep reinforcement learning (DRL)-based resource management for future 5G-V2X networks. His research interest areas are wireless/mobile communications systems including radio resource management, multiple access, MANETs/VANETs, green radio networks, and 5G-V2X networks. He has published over 200 journal and conference papers, 01 edited books, and 9 book chapters in the relevant areas.

**Akhtar Kalam** has been at Victoria University (VU) since 1985. He is the former Deputy Dean of the Faculty of Health, Engineering, and Science and Head of Engineering of the College of Engineering and Science. Currently, he is the Head of External Engagement. He is also current Chair of the Academic Board in the Engineering Institute of Technology, Perth, Australia, and Editor in Chief of the *Australian Journal of Electrical & Electronics Engineering*. Further, he has a Distinguished Professorship position at the University of New South Wales, Sydney, Australia, and three Indian and five Malaysian universities. He has also been appointed as Editor in Chief of the *Australian Journal of Electrical and Electronic Engineering*. He has wide experience in educational institutions and industries across four continents. He received his B.Sc. and B.Sc. Engineering from Calcutta University and Aligarh Muslim University, India. He completed his M.S. and Ph.D. at the University of Oklahoma, USA, and the University of Bath, UK. He has worked with Ingersoll Rand and other electrical manufacturers. Other teaching appointments include the University of Technology, Baghdad, Iraq, and Capricornia Institute of Advanced Education, Rockhampton, Queensland. He has conducted research, provided industrial consultancy, and published more than 542 publications in his area of expertise.

He has written 26-plus books in the area. More than 35 Ph.D. students have graduated under his supervision and he is an external examiner of many external doctoral students in Australia and overseas. He provides consultancy for major electrical utilities, manufacturers, and other industry bodies in his field of expertise. He is a fellow of EA, IET, AIE, a life member of IEEE, and a member of the CIGRE AP B5 Study Committee.

**Antonio Pascoal** received the Licenciatura degree in Electrical Engineering from the Instituto Superior Técnico (IST), Lisbon, Portugal, in 1975 and the M.S. degree in Electrical Engineering and the Ph.D. degree in Control Science from the University of Minnesota, Minneapolis, USA, in 1983 and 1987, respectively. His Ph.D. research work was conducted under the supervision of Prof. Pramod Khargonekar. His expertise includes dynamical systems theory, marine robotics, navigation, guidance, and control of autonomous vehicles, and networked control and estimation with applications to air and underwater robots. His long-term goal is to contribute to the development of advanced robotic systems for ocean resources exploration and exploitation. He is currently the Director of the FCT-IST Ph.D. program on Networked Interactive Cyber-Physical Systems (NETSyS). He has supervised or co-supervised 11 postdoctoral and 15 Ph.D. students. He has published 98 books, book chapters, peer-reviewed journal papers, and 250 conference papers.

**Manas Kumar Bera** is an Assistant Professor in the Department of Electronics and Instrumentation Engineering at the National Institute of Technology, Silchar. He completed his Ph.D. from the Indian Institute of Technology Bombay in 2015, M.E. from Jadavpur University in 2006, and B.E. from the University of Burdwan in 2003. His areas of expertise are control theory and its applications. bio-medical application of control. He has published several papers in national and international journals.

# Control and Robotics Track—1

# Performance Assessment of Unstable Systems Subjected to Ramp Input Changes

**K. Ghousiya Begum**

## 1 Introduction

A huge massive research effort has been articulated toward the controller performance assessment to avoid erroneous tuning and unsuitable control structures. To enhance the quality of the products and the safety factors, the tuning of proportional integral (PI) controllers has turned out to be one of the essential elements to rectify the performance issues. The designed controller has to meet the control target, and its performance has to be improved efficiently. In order to satisfy the above requirements, a benchmark has to be proposed to estimate the performance of the control loops. Many researchers proposed diverse methods to accomplish MVC as the standard benchmark for PID controllers [1–3]. Besides the MVC-based benchmark, an idle index has been presented to identify whether the controller is properly tuned [4]. The area index [5] was presented to provide the guidelines for retuning PI controllers to avoid the sluggish and oscillatory nature of the control loops. In [6], a ratio index was developed to disclose the damping nature of the control loop. The theoretical bound of error (IAE) for PI/PID control loops was developed for the system, whose reference signal is subjected to step set point changes [7]. Also, an IAE-based index was proposed from the closed-loop output responses subjected to step input changes using internal model control (IMC) tuning rules for stable and integral processes [8, 9]. In [10], an IMC-based IAE index was designed to measure the performance of the control scheme from the closed-loop outputs subjected to ramp, step, and general types of input changes. The behavior of the PI control loop developed based on the direct synthesis (DS) tuning rule was assessed in rejecting the load disturbances [11]. The linear industrial controllers incorporated in control loops were evaluated based on IAE–TV-based performance index [12]. A tool called entropy performance

K. Ghousiya Begum (✉)
School of Electrical and Electronics Engineering, SASTRA University, Thirumalaisamudram, Thanjavur 613401, India
e-mail: ghousiyabegum@eie.sastra.edu



P. H. J. Chong et al. (eds.), *Emerging Electronics and Automation*, Lecture Notes in Electrical Engineering 937, https://doi.org/10.1007/978-981-19-4300-3_1

assessment (EPA) was elaborated for the multi-loop cascade process where the load variables are subjected to Gaussian probability distribution in [13]. Following this, a novel assessment method has been proposed which relies on the diminution of the error to find optimum tuning settings [14]. Also, the control quality factor indices were proposed for real and model-based control loops in offline and online mode for HVAC systems by [15]. A new technique based on acquired data for instantaneous performance valuation and to retune the PID controllers was discussed in [16]. A comprehensive assessment of control loops and their assessment using recognized standards were also discussed in [17]. The assessment of the performance of the fractional controllers designed for gas turbine engine has been carried out [18]. Recently, the stability and performance of the controller designed for microgrid connected to unified power quality conditioner have been assessed [19]. In [20], a novel technique using fractal measures has been implemented to assess the performance of generalized predictive controller.

Although these methods have been applied to industries effectively, they are only applicable to stable processes. Many real-time systems exhibit unstable characteristics. The existence of an unstable system has been extensively studied in [21], and efforts have been taken throughout to control these processes [22, 23]. There are many enhanced controllers designed for stabilizing the unstable processes. Nevertheless, the performance assessment criterion for unstable processes is inadequate. A performance index based on the theoretically derived IAE obtained using the response of closed-loop system with step changes in input and load variable for unstable processes has been developed in [24]. However, TV is also important which emphasizes the smoothness of the controller output. Hence, by utilizing the theoretical bounds of manipulated variable in terms of total variation (TV) and integral absolute error (IAE) as reasonable standards, an IAE–TV index is developed to estimate the input tracking characteristics of PI controller loops. This satisfies the control target, and it acts as a criterion required to investigate whether the suggested control technique could assure the performance needs on reference input tracking.

Thus, in this article, to assess the performance of the PI controller deployed for time delayed unstable first-order process, an IAE–TV index has been proposed. The IAE–TV assessment index is developed by utilizing the theoretical bounds of TV and IAE as reasonable benchmarks, and the condition required to obtain the enhanced performance is also discussed. The assessment procedure validates the input tracking performance of PI control loops designed for unstable models subjected to ramp type of input changes. The performance PI control loop deployed for the unstable system will be the perfect one only when the derived IAE–TV index value (number) is close to one.

## 2 Problem Statement

Figure 1 depicts the feedback control system with filter (two degrees of freedom-2 DoF), where $G(s)$ is the model, $G_c(s)$ is the controller, $F_r(s)$ is the pre-filter, $R(t)$

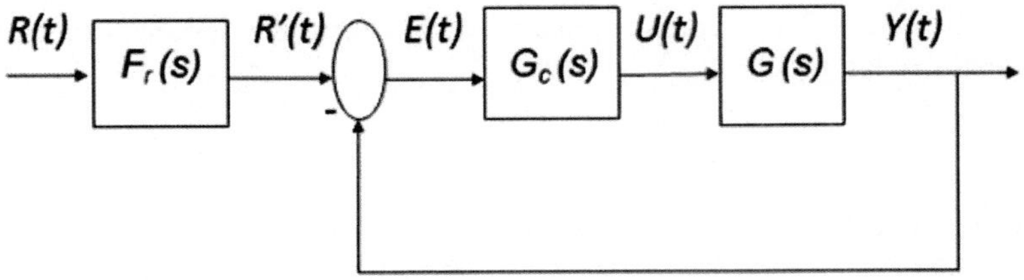


**Fig. 1** 2 DoF feedback control loop

is the ramp input, $E(t)$ is offset, $U(t)$ is the control variable, and $Y(t)$ is the output variable.

Here, the process is considered to be a time delayed unstable first-order model

$$G(s) = \frac{Ke^{-\theta s}}{\tau s - 1} \tag{1}$$

and the conventional controller $G_c(s)$ configuration is

$$G_c(s) = k_c\left(1 + \frac{1}{\tau_i s}\right) \tag{2}$$

The set point here is the ramp input which takes the form as follows:

$$R(t) = \begin{cases} \text{kt}, & 0 \le t < \frac{A}{k} \\ A, & \frac{A}{k} \le t < \infty \end{cases} \tag{3}$$

where $A$ and $k$ are the final values of the ramp input change and the slope of the ramp signal, which is commonly used in practice. For $G(s)$ in (1), the closed-loop transfer function for time delayed unstable first-order model is

$$G_{cl}(s) = \frac{Y(s)}{R(s)} = \frac{F_r(s)G(s)C(s)}{1 + G(s)C(s)} = \frac{1}{(\lambda s + 1)^2}e^{-\theta s} \tag{4}$$

where $R(s)$, $Y(s)$ denote the Laplace transformation of ramp input $R(t)$, process output $Y(t)$, correspondingly. $\lambda$ is the user-selected tuning parameter, and the robustness is calculated by the value of maximum sensitivity function (Ms).

$$\text{Ms} = \max\left|\frac{1}{1 + G(j\omega)G_c(j\omega)}\right| \tag{5}$$

The solution has to be found for the three problems using the above specifications. The first one is to derive a dimensionless index to assess the servo characteristics. The second one is to realize the condition required for the control scheme to attain better performances. The final one is to analyze the derived assessment index from the simulation examples.

## 3 IAE–TV-Based Assessment Index

This segment elucidates on establishing the theoretical bound of TV of control signal and also illustrates about IAE–TV-based assessment index to analyze the servo tracking characteristics.

### 3.1 *Standard Assessment Benchmarks*

Total variation (TV) measures the smoothness of the input control signal to the process. The theoretical bound of TV for closed-loop response in (4) is derived in this section. The ramp input is decomposed into two parts as

$$R(t) = R_1(t) + R_2(t) \tag{6}$$

$R_1(t) = \mathrm{kt}, 0 \le t < \infty$, and

$$R_2(t) = \begin{cases} 0, & 0 \le t < \frac{A}{k} \\ -k\left(t - \frac{A}{k}\right), & A/k \le t < \infty \end{cases}$$

The $s$ domain representation of the ramp signal $R(t)$ is

$$R(s) = R_1(s) + R_2(s) = \left(1 - e^{-\frac{A}{k}s}\right)\frac{k}{s^2} \tag{7}$$

Here, $U(s)$ can be derived using $G_{\mathrm{cl}}(s)$ in (4) and $R(s)$ in (7).

$$U(s) = \frac{G_{\mathrm{cl}}(s)R(s)}{G(s)} = \frac{k\left(1 - e^{-\frac{A}{ks}}\right)}{K}\left(\frac{\tau + 2\lambda}{s} - \frac{1}{s^2} - \frac{\tau + 2\lambda}{s + \frac{1}{\lambda}} - \frac{\tau + \lambda}{\lambda}\frac{1}{\left(s + \frac{1}{\lambda}\right)^2}\right) \tag{8}$$

The total variation of the manipulated variable is

$$\mathrm{TV}_0 = \int_0^\infty \frac{\mathrm{d}(U(t))}{\mathrm{d}t}\mathrm{d}t \tag{9}$$

After taking the inverse Laplace transform of controller output $U(s)$ in (8), its derivative $U(t)$ is

$$\frac{dU(t)}{dt} = \begin{cases} \frac{k}{K}\left[-1 + e^{-\frac{t}{\lambda}} + \frac{(\tau+\lambda)}{\lambda^2}te^{-\frac{t}{\lambda}}\right] & ;0 \le t < \frac{A}{k} \\ \frac{k}{K}\left[\begin{array}{c} \frac{(\tau+2\lambda)}{\lambda}\left(e^{\frac{A}{k\lambda}} - 1\right)e^{-\frac{t}{\lambda}} - \frac{A}{k}\frac{(\tau+\lambda)}{\lambda^2}e^{\frac{A}{k\lambda}}e^{-\frac{t}{\lambda}} \\ -\frac{(\tau+\lambda)}{\lambda}\left(e^{\frac{A}{k\lambda}} - 1\right)\left(-\frac{te^{-\frac{t}{\lambda}}}{\lambda} + e^{-\frac{t}{\lambda}}\right) \end{array}\right] & ;\frac{A}{k} \le t < \infty \end{cases} \tag{10}$$

Substituting (10) in (9), we obtain

$$TV_0 = \frac{k}{K}\left\{\begin{array}{l} \int_0^{\frac{A}{k}}\left[-1 + e^{-\frac{t}{\lambda}} + \frac{(\tau+\lambda)}{\lambda^2}te^{-\frac{t}{\lambda}}\right]dt \\ + \int_{A/k}^{\infty}\left[\begin{array}{c} \frac{(\tau+2\lambda)}{\lambda}\left(e^{\frac{A}{k\lambda}} - 1\right)e^{-\frac{t}{\lambda}} - \frac{A}{k}\frac{(\tau+\lambda)}{\lambda^2}e^{\frac{A}{k\lambda}}e^{-\frac{t}{\lambda}} \\ -\frac{(\tau+\lambda)}{\lambda}\left(e^{\frac{A}{k\lambda}} - 1\right)\left(-\frac{te^{-\frac{t}{\lambda}}}{\lambda} + e^{-\frac{t}{\lambda}}\right) \end{array}\right]dt \end{array}\right\}$$

And the theoretical minimum bound of TV for ramp input is

$$TV_0 = \begin{cases} \left|\frac{k}{K}\right| - \frac{A}{k} + 2(\tau + 2\lambda)\left(1 - e^{-\frac{A}{k\lambda}}\right) - 2\frac{(\tau+\lambda)}{\lambda}\frac{A}{k}e^{-\frac{A}{k\lambda}} & ; \quad 0 < \lambda \le \tau \\ \left|\frac{A}{k}\right| & ; \quad \tau < \lambda < \infty \end{cases} \tag{11}$$

The general performance metric IAE is considered as

$$IAE = \int_0^{\infty} |E(t)|dt = \int_0^{\infty} |(R(t) - Y(t))|dt \tag{12}$$

The theoretical minimum bound of IAE for ramp input change is denoted as $IAE_0$ [24]:

$$IAE_0 = |A|(2\lambda + \theta) \tag{13}$$

## 3.2 *Performance Assessment Index*

By considering the lower bound of $TV_0$ in (11) and lower bound of $IAE_0$ in (13), two dimensionless assessment indices are derived as

$$\eta_{TV} = \frac{\min(TV_0, TV_{actual})}{\max(TV_0, TV_{actual})} \tag{14}$$

$$\eta_{IAE} = \frac{\min(IAE_0, IAE_{actual})}{\max(IAE_0, IAE_{actual})} \tag{15}$$

And the actual values of TV and IAE are denoted as $TV_{actual}$ and $IAE_{actual}$, and they can be calculated using the sampling period $H$ as shown below

$$TV_{actual} = \sum_{i=1}^{\infty} |U_{i+1}(t) - U_i(t)|H \tag{16}$$

$$IAE_{actual} = \sum_{i=1}^{\infty} |R_i(t) - Y_i(t)|H \tag{17}$$

The range of both indices $\eta_{TV}$ in (14) and $\eta_{IAE}$ in (15) is [0, 1]. Using $\eta_{TV}$ and $\eta_{IAE}$, a general assessment index is derived as

$$\eta = \eta_{TV}.\eta_{IAE} \tag{18}$$

Hereafter, $\eta$ is mentioned as the IAE–TV index which describes the level of the reference tracking characteristics. The desired range of index $\eta$ is [0, 1] with the desired value equal to 1 so as to indicate that, if $\eta \to 1$, the servo response obtained and the deployed controller are reasonable. Both $IAE_{actual}$ and $TV_{actual}$ values are close to the desired benchmarks $IAE_0$ in (13) and $TV_0$ in (11), correspondingly. Both the performance indices $\eta_{TV}$ or $\eta_{IAE}$ should not move away from 1, so as to make their product $\eta$ close to 1. The suggested index is applicable to the controller design, provided the closed-loop response experiencing the ramp input change is the same as $G_{cl}(s)$ in (4) and the process model $G(s)$ is UFOPTD in (1).

## 4 Conditions for Better Performance

This section defines the control scheme and the condition required to provide satisfactory performance. The PI controller settings are [23]

$$k_c = \frac{\beta\tau}{K(\lambda^2 + \beta\theta)};\ \tau_i = \beta;\ \tau_d = 0 \tag{19}$$

where $\beta = \frac{2\lambda+\theta-\tau\lambda^2}{\tau\theta+1}$ and $\lambda$ is the tuning parameter, and it should be chosen carefully to achieve the desired level of robustness (Ms).

### 4.1 Single Feedback (SFB) Control Scheme

Let us adopt the simple SFB scheme as shown in Fig. 1. The controller $G_c(s)$ is a PI controller with the control parameters given in (19). The feedback response is

$$G_{\text{cl}}^{\text{SFB}}(s) = \frac{G(s)G_c(s)F_r(s)}{1+G(s)G_c(s)} = G_d(s)G_c(s)F_r(s)$$

$$G_{\text{cl}}^{\text{SFB}}(s) = \frac{\tau_i s+1}{(\beta s+1)} * \frac{e^{-\theta s}}{(\lambda s+1)^2} \tag{20}$$

Comparing (20) with (4), the condition required for $G_{\text{cl}}^{\text{SFB}}(s)$ to be the same as $G_{\text{cl}}(s)$ in (4) is $\beta = \tau_i$. The procedure of the evaluation method is outlined in the form of a flowchart in Fig. 2 for a desired level of robustness. If model dynamics are not available, one shall use the closed-loop identification techniques already available in the literature [25] to identify a model as

$$\tilde{G}(s) = \frac{\tilde{K}e^{-\tilde{\theta}s}}{\tilde{\tau}s-1} \tag{21}$$

### 4.2 Simulations Results

To affirm the validity of the theoretical bound of TV and the new evaluation method, several examples are reviewed.

**Example 1** Let us investigate the process $G(s) = \frac{e^{-0.6s}}{5s-1}$ [21] subjected to ramp type of set point change as in Fig. 1. The lower bound of TV can be estimated (13). The feedback controller takes the PI form [23] and the controller settings for $\lambda = 2$ and $\text{Ms} = 1.8009$ in (5), and the controller settings are $k_c = 3.9941$ and $\tau_i = 6.1364$. The ramp input is

$$R(t) = \begin{cases} 0, & 0 \le t < 5 \\ 0.2(t-5), & 5 \le t < 10 \\ 1, & 10 \le t < \infty \end{cases} \tag{22}$$

Figure 3 shows the closed-loop responses of SFB scheme for different $\lambda$ values. Figure 4 displays that the theoretically derived $\text{TV}_o$ and $\text{TV}_{\text{actual}}$ values are closer.

**Example 2** Let us examine another process $G(s) = \frac{4e^{-2s}}{4s-1}$ experiencing a ramp input change as

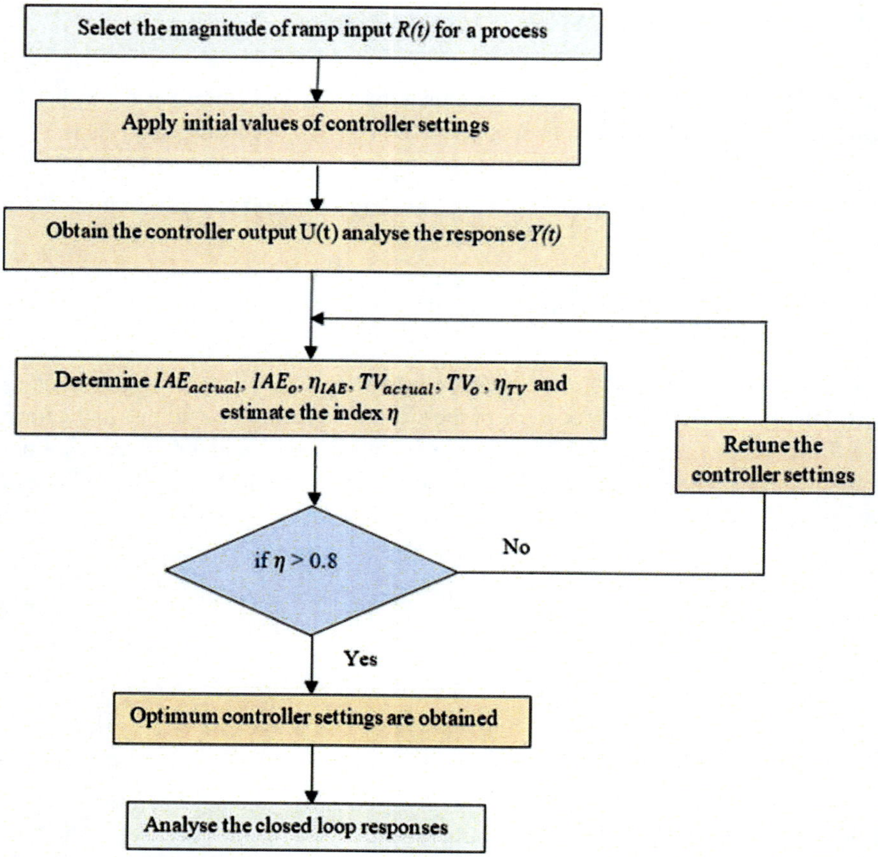


**Fig. 2** Flowchart of the proposed method

$$R(t) = \begin{cases} 0, & 0 \le t < 10 \\ 0.2(t-5), & 10 \le t < 15 \\ 2, & 15 \le t < \infty \end{cases} \tag{23}$$

In the first case of the SFB control scheme, an initial set of PI controller parameters (not using the DS rule) is used as mentioned in the first column of Table 1. The process output $Y(t)$ and controller output $U(t)$ are shown in Fig. 5 along with ramp input $R(t)$. The procedure of the suggested assessment method discussed in Sect. 4.1 is realized as follows. An estimated model is obtained with $\tilde{K} = 4.011, \tilde{\theta} = 2, \tilde{\tau} = 4.069$ as per (21). The performance assessment results of the SFB control scheme with initial controller settings are given in Table 1. The $\eta = 0.5288$ index value is not the expected value of 1, and the set point tracking characteristic is not satisfactory.

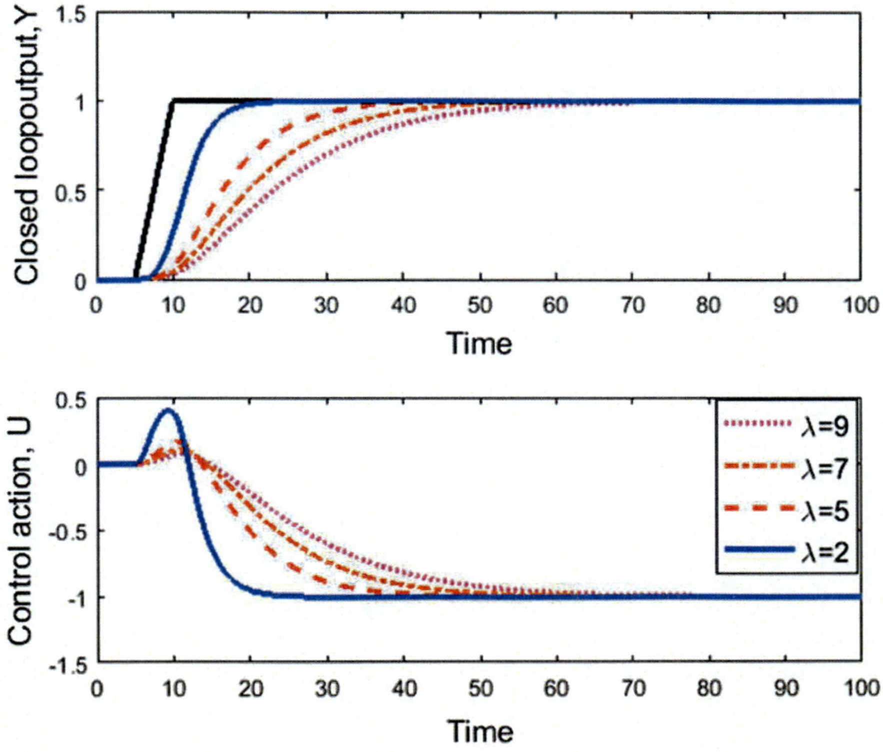


**Fig. 3** Responses of closed loop (SFB scheme) for ramp input for diverse λ values

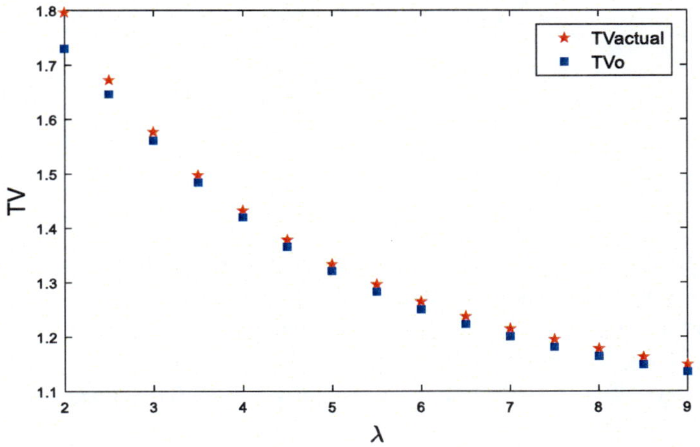


**Fig. 4** $TV_o$ and $TV_{actual}$ values for diverse λ values using SFB scheme

**Table 1** Performance assessment results

| | SFB (initial) | SFB (19) |
|---|---|---|
| $k_c$ | 0.3431 | 0.3853 |
| $\tau_i$<br>$IAE_{actual}$ | 29.412<br>25.7402 | 30.5970<br>21.6267 |
| $IAE_0$ | 21.60 | 21.60 |
| $\eta IAE$ | 0.8392 | 0.9988 |
| $TV_{actual}$ | 0.9925 | 0.6836 |
| $TV_0$ | 0.6255 | 0.6281 |
| $\eta_{TV}$ | 0.6302 | 0.9189 |
| $\eta$ | 0.5288 | 0.9177 |

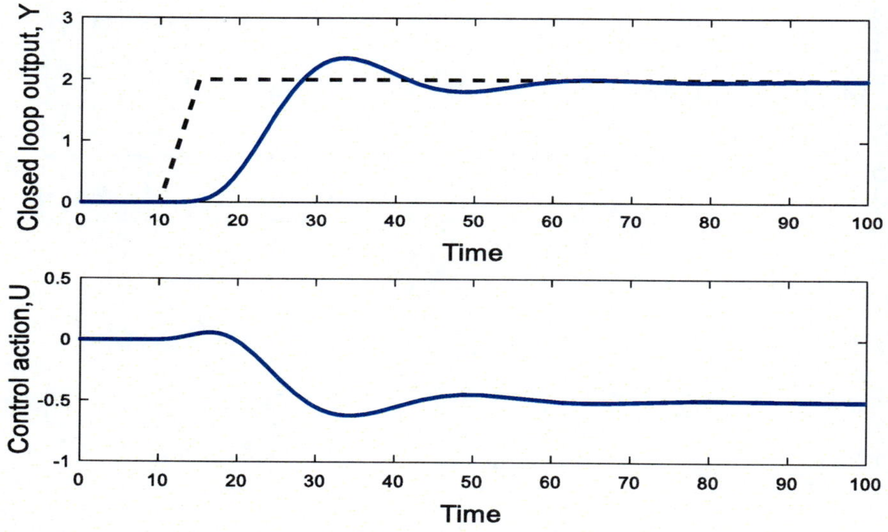


**Fig. 5** SFB control scheme with initial controller settings

The regulation parameter λ needs to be chosen in such a method so that it provides a reasonable compromise between robustness and performance.

It is carefully selected after analyzing the values of Ms. Hence, λ is chosen as 4.4 for Ms = 5.7, and the desired controller settings obtained as per (19) are given in the second column of Table 1. Table 1 also provides the performance assessment results. Figure 5 proves that the SFB scheme with initial controller settings does not provide good results, whereas the SFB with the desired controller settings in (19) yields satisfactory performance as shown in Fig. 6 with $\eta = 0.9177$ (close to 1), when compared to the initial controller settings of SFB control scheme.

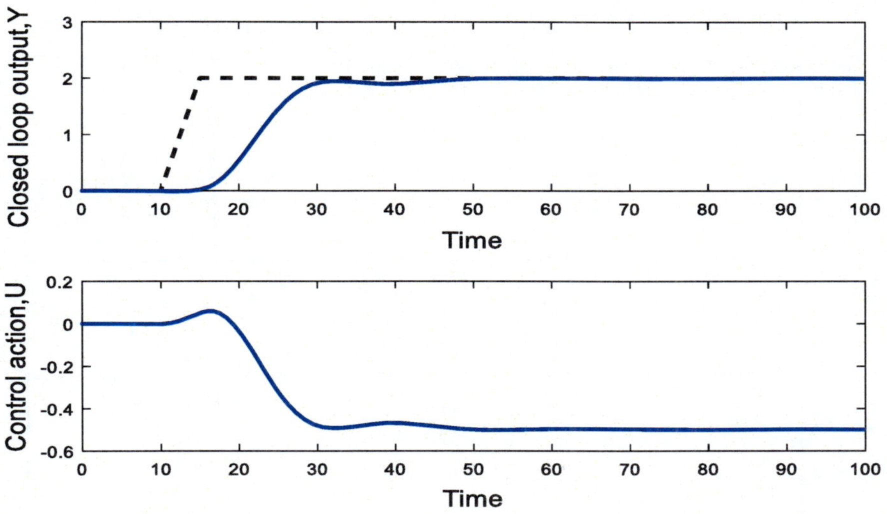


**Fig. 6** Responses for SFB control scheme with desired controller settings

**Example 3** Let us examine another process, a continuous bioreactor whose model transfer function is $G(s) = \frac{-5.89e^{-0.1s}}{5.86s-1}$. The system experiences a ramp input change as given below.

$$R(t) = \begin{cases} 0, & 0 \le t < 15 \\ 0.6(t-15), & 15 \le t < 20 \\ 3, & 20 \le t < \infty \end{cases} \tag{24}$$

The actual values of TV and IAE appear nearer to theoretically derived ones for varied values of λ. This is justified based on the closeness of $\text{TV}_{\text{actual}}$ and $TV_0$ given in Table 2 and displayed as a response plot as in Fig. 7. The IAE–TV-based performance index is $\eta > 0.8$, and therefore, retuning of the controller is not required.

**Table 2** Performance assessment results for Example 3

| λ | $\text{IAE}_{\text{actual}}$ | $\text{IAE}_0$ | $\text{TV}_{\text{actual}}$ | $\text{TV}_0$ | $\eta$ |
|---|---|---|---|---|---|
| 2 | 12.3 | 12.3 | 1.0108 | 1.006 | 0.9953 |
| 3.5 | 21.3 | 21.3 | 0.8343 | 0.8299 | 0.9948 |
| 5 | 30.3 | 30.3 | 0.7293 | 0.7192 | 0.9862 |
| 7.5 | 45.3 | 45.3 | 0.6397 | 0.627 | 0.98 |
| 9 | 54.3 | 54.3 | 0.6101 | 0.5977 | 0.9796 |
| 10 | 60.3 | 60.3 | 0.5957 | 0.5838 | 0.9799 |

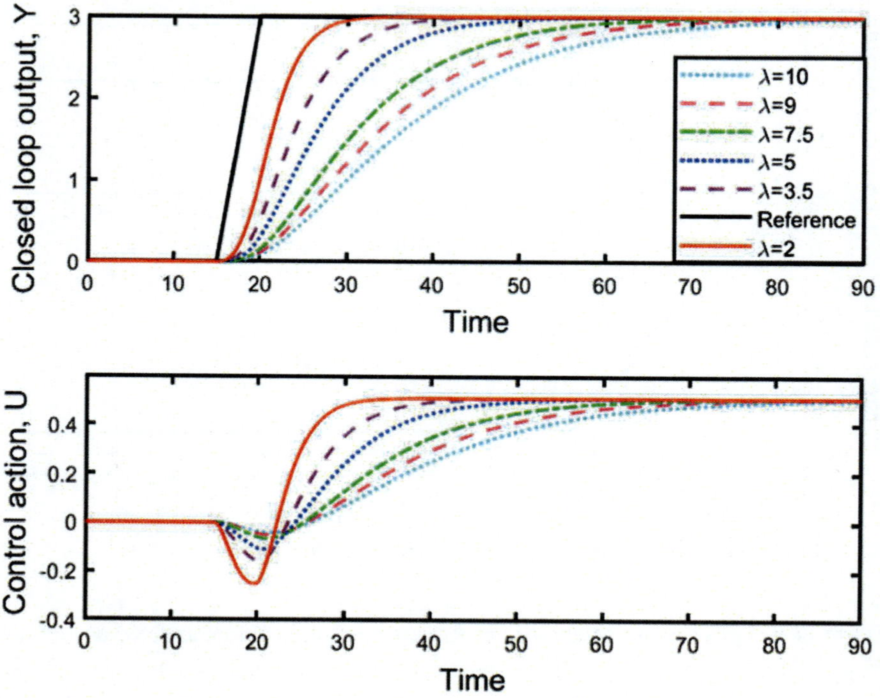


**Fig. 7** Responses of SFB scheme for diverse λ values for Example 3

# 5 Conclusions

The performance of the control scheme deployed for the unstable process has been discussed in this article. The process is subjected to ramp input, and the closed-loop output response experiences a ramp input variation. The behavior of the controller was assessed by considering the theoretical bounds of TV and IAE as standard benchmarks. Based on this, a combined performance index named IAE–TV index, $\eta$, was derived to assess the control behavior for the unstable systems. The control scheme was explored with the help of a feedback PI controller which uses the DS rule, and the condition to obtain the desired servo response was also discussed. The proposed performance index was investigated to evaluate whether the control objective is attained or not. The proposed assessment scheme was established via simulation results, and accordingly, the IAE–TV index estimates the controller action in terms of smooth control output and reference tracking characteristics.

## References


1. Hugo AJ (2006) Performance assessment of single-loop industrial controller. J Process Control 16(8):785–794
2. Sendjaja AY, Kariwala AY (2009) Achievable PID performance using sums of squares programming. J Process Control 19(6):1061–1065
3. Fu R, Xie L, Song Z, Cheng Y (2012) PID control performance assessment using iterative convex programming. J Process Control 22(9):1793–1799
4. Hägglund T (2005) Industrial implementation of on-line performance monitoring tools. Control Eng Pract 13(11):1383–1390
5. Visioli A (2006) Method for proportional-integral controller tuning assessment. Ind Eng Chem Res 45(8):2741–2747
6. Salsbury TI (2005) A practical method for assessing the performance of control loops subject to random load changes. J Process Control 15(4):393–405
7. Huang H-P, Jeng J-C (2002) Monitoring and assessment of control performance for single loop systems. Ind Eng Chem Res 41(5):1297–1309
8. Veronesi M, Visioli A (2010) An industrial application of a performance assessment and retuning technique for PI controllers. ISA Trans 49(2):244–248
9. Veronesi M, Visioli A (2010) Performance assessment and retuning of PID controllers for integral processes. J Process Control 20(3):261–269
10. Yu Z, Wang J, Huang B, Bi Z (2011) Performance assessment of PID control loops subject to setpoint changes. J Process Control 21(8):1164–1171
11. Yu Z, Wang J (2012) Assessment of proportional-integral control loop performance for input load disturbance rejection. Ind Eng Chem Res 51(36):11744–11752
12. Yu Z, Wang J, Huang B, Li J, Bi Z (2014) Performance assessment of industrial linear controllers in univariate control loops for both set point tracking and load disturbance rejection. Ind Eng Chem Res 53(27):11050–11060
13. Zhang J, Zhang L, Chen J, Xu J, Li K (2015) Performance assessment of cascade control loops with non-Gaussian disturbances using entropy information. Chem Eng Res Des 104:68–80
14. de Francisco DO, Trierweiler JO, Farenzena M (2019) A novel PID autotuning approach: how to correct bad tuning by closed-loop performance assessment. IFAC-PapersOnLine 52(1):184–189
15. Liu R, Li Y, O'Neill ZD, Zhou X (2019) Development of control quality factor for HVAC control loop performance assessment—II: field testing and results (ASHRAE RP-1587). Sci Technol Built Environ 25(7):873–888
16. Gao X, Yang F, Shang C, Huang D (2017) A novel data-driven method for simultaneous performance assessment and retuning of PID controllers. Ind Eng Chem Res 56(8):2127–2139
17. Domański PD (2020) Control performance assessment: theoretical analyses and industrial practice. Springer, Switzerland
18. Tsoutsanis E, Meskin N (2018) Performance assessment of classical and fractional controllers for transient operation of gas turbine engines. IFAC-PapersOnLine 51(4):687–692
19. Iqbal A, Waqar A, Elavarasan RM, Premkumar M, Ahmed T (2021) Stability assessment and performance analysis of new controller for power quality conditioning in microgrids. Int Trans Electr Energ Syst 31:e12891
20. Domański PD, Ławryńczuk M (2017) Assessment of predictive control performance using fractal measures. Nonlinear Dyn 89:773–790
21. Sree RP, Chidambaram M (2017) Control of unstable single and multi-variable systems. Narosa Publishers, New Delhi
22. Rao AS, Rao VSR, Chidambaram M (2007) Simple analytical design of modified smith predictor with improved performance for unstable first-order plus time delay (FOPTD) processes. Ind Eng Chem Res 46(13):4561–4571
23. Cho W, Lee J, Edgar TF (2014) Simple analytic proportional-integral-derivative (PID) controller tuning rules for unstable processes. Ind Eng Chem Res 53(13):5048–5054

24. Begum KG, Rao AS, Radhakrishnan TK (2018) Performance assessment of control loops involving unstable systems for set point tracking and disturbance rejection. J Taiwan Inst Chem Eng 85:1–17
25. Ananth I, Chidambaram M (1999) Closed-loop identification of transfer function model for unstable systems. J Franklin Inst 336(7):1055–1061

# Level Control of a Two-Tank System Using Quantitative Feedback Theory-Based Controller

**Sandipan Prasad Chakravarty, Aniket Roy, Aritra Sinha, and Prasanta Roy**

## 1 Introduction

Liquid level control in an interconnected tank system is an integrated part of most of the process industries such as chemical industries, oil refinery, power plant, water treatment, food processing, distillery plants. The main purpose of the liquid level control is to maintain a desired level of liquid in the tank that depends on the variation of the inflow and outflow quantities of liquid. Previous work done in this field show a wide use of proportional integral derivative (P-I-D) controllers where, the proposed controllers successfully met the desired specifications like reduced overshoot and minimal steady-state error. However, when it comes to ensuring the robustness of the controller to model uncertainties, P-I-D controller fails in producing a robust and satisfactory system performance [1]. This calls for a control strategy in the form of QFT which can ensure the robust stability criterion in non-linear uncertain systems.

### *1.1 Why Quantitative Feedback Theory (QFT)?*

Classical feedback controller like P-I, P-I-D, Cascade controller, etc. for an uncertain plant fails to give the expected performance in presence of parametric uncertainty. It demands for a robust control strategy which shall be efficient in achieving the desired specifications regardless of the changes encountered by the plant parameters in the procedure. Issac Horowitz developed a new form of frequency domain technique in

S. P. Chakravarty (✉) · A. Roy · A. Sinha · P. Roy
National Institute of Technology, Silchar, Assam 788010, India
e-mail: sandipan_rs@ee.nits.ac.in

P. Roy
e-mail: prasanta@ee.nits.ac.in



P. H. J. Chong et al. (eds.), *Emerging Electronics and Automation*, Lecture Notes in Electrical Engineering 937, https://doi.org/10.1007/978-981-19-4300-3_2

the 1960s that copes with the practicable designing of the feedback systems. It was termed as Quantitative Feedback Theory (QFT) that assists in obtaining a response from the system lying within the pre-determined bounds despite of output disturbance and known plant uncertainties [2]. QFT works in a platform of frequency domain analysis that assists in providing a deal between the attainable performance and the controller complication in the chosen frequency range. It relates the closed-loop specifications with the open-loop specifications using Nichols chart. Nichols plot is preferred over other methods of analysis because in root locus, the parametric uncertainty of the uncertain plant cannot be defined whereas, it is nicely depicted in frequency domain techniques like Bode, Nyquist or Nichols plot. But, in Bode plot the presence of plant uncertainty can be observed as a bold line irrespective of frequency which can be eliminated in Nyquist plot. But, it is difficult to analyse the variation of the plant due to gain uncertainty in Nyquist plot. So, in Nichols plot these parametric uncertainties can be plotted in the form of templates at a particular frequency. The vital advantage of this proposed controller strategy in the form of QFT is the reduction of cost of feedback, i.e. loop gain to obtain an optimum location of open-loop transfer function (OLTF) thereby, meeting the desired performance specifications [3].

## *1.2 Literature Review*

The design of a robust decentralized P-I controller is proposed by [4] for controlling the water level of a coupled tank system using a frequency domain model matching technique. The design of an adaptive fuzzy controller is proposed for the level control of a two-tank system in [5] with a comparative analysis against a P-I controller. A back-stepping sliding mode P-I controller is proposed for level control by [6] with a comparative analysis against other sliding mode robust controllers despite plant uncertainties. A comparative analysis between cascade fractional order P-I, P-D against integer order P-I and P-D is proposed for the level control of a two-tank system in [7]. The design of a sliding mode controller for the level control of a two-tank system is proposed and a comparative analysis is showed against P-I-D controller in [8]. The development of a chattering-free sliding mode controller strategy is shown for the level control of a coupled tank system for proving its robustness in [9]. A fractional order P-I controller is designed for controlling the water level in a coupled tank system in [10]. An intelligent control strategy in the form of fuzzy P-I-D controller is proposed for liquid level control in [11]. A comparative analysis is presented between model predictive controller and P-I controller based on different tuning methods for liquid level control in [12]. A comparative study is done by [13] between a artificial neural network-based controller and a model predictive controller in a level control system. A fuzzy logic controller is used by [14] for maintaining the level of the liquid in the presence of disturbances.

## *1.3 Research Contribution of the Paper*

The research contributions of this paper are as below:

- The two-tank system is modelled as a second-order linear system with interval type of parametric uncertainties.
- A novel robust controller using QFT-based approach is proposed.
- A comparative analysis is done for the proposed controller against a P-I-D, a Cascade controller and other reported controller strategies.

# 2 Modelling of Two-Tank System

The two-tank system under consideration is shown in Fig. 1. In this section, The transfer function between the water level of tank 2 and the control input to the pump is found. A combination of both physical laws and data driven approach is used to determine the parametric uncertainties in interval form.

## *2.1 Terminologies and Notations*

The terminologies and notations used throughout this section are listed below:

- $H_i$: Liquid level of the $i$th tank (cm),
- $A_i$: Cross sectional area of the $i$th tank (cm$^2$),
- $a_i$: Cross sectional area of the outlet $i$th tank (cm$^2$),

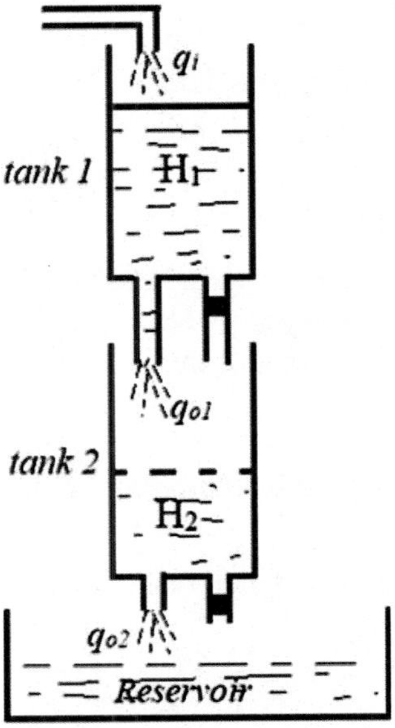


**Fig. 1** Two-tank system as a type-0 order-2 plant

- $u$: Input voltage to the submersible pump (V),
- $c$: Pump constant $cm^3s^{-1}V^{-1}$,
- $\bar{H}_i$: Steady state operating point of $H_i$,
- $\bar{U}$: Steady state operating point of $U$,
- Incremental level $h_i = H_i - \bar{H}_i$,
- Incremental pump voltage $u = U - \bar{U}$.

## 2.2 Mathematical Modelling

Using Bernoulli's equation and the continuity of flow, the dynamics of the system can be written as [15],

$$A_1 \frac{dH_1}{dt} + a_1\sqrt{2gH_1} = cU, \tag{1}$$

$$A_2 \frac{dH_2}{dt} + a_2\sqrt{2gH_2} = a_1\sqrt{2gH_1} \tag{2}$$

Equations (1) and (2) are linearized around the operating point $H_i = \bar{H}_i, U = \bar{U}$. The linearized equations in terms of incremental variables are given below [15].

$$\frac{dh_1}{dt} + \lambda_1 h_1 = \eta u, \tag{3}$$

$$\frac{dh_2}{dt} + \lambda_2 h_2 = \lambda_1 h_1 \tag{4}$$

$$\text{where, } \lambda_i = \frac{a_i}{A_i}\sqrt{\frac{g}{\bar{H}_i}} \text{ and } \eta = \frac{c}{A_1}$$

Applying Laplace Transform of (3) and upon simplification we get,

$$\frac{H_2(s)}{U(s)} = P(s) = \frac{\eta\lambda_1}{(s+\lambda_1)(s+\lambda_2)} \tag{5}$$

where $H_2(s)$ and $U(s)$ are Laplace transform of $u$ and $h_2$. The parameters $\eta$, $\lambda_1$ and $\lambda_2$ along with the uncertain intervals are found from the experiments as described in the next subsection.

## 2.3 Derivation of Interval Type of Parametric Uncertainty

If the outlet of tank 1 (refer to Fig. 1) is closed, then (1) reduces to

$$\frac{dh_1}{dt} = \eta u \tag{6}$$

For a fixed control input $u$, the rate of change of water level ($\frac{dh_1}{dt}$) in tank 1 is measured to find $\eta$. It is observed that the parameter $\eta$ varies in the interval as given below.

$$0.109 \leq \eta \leq 0.2408 \tag{7}$$

Now if the outlet of tank 2 is closed keeping the tank 1 outlet open, then (4) reduces to

$$\frac{dh_2}{dt} = \lambda_1 h_1 \tag{8}$$

The level in tank 1 is kept constant at different levels. For each case, the rate of change of water level in tank 2 ($\frac{h_2}{dt}$) is noted. A set of values of $\lambda_1$ may be found using (8). Thus, we evaluate $\lambda_1$ as given below.

$$0.0348 \leq \lambda_1 \leq 0.1145 \tag{9}$$

Now, both the outlets of tank 1 and tank 2 are kept open. The water is allowed to flow until steady-state is achieved. Under steady-state (4) reduces to

$$\lambda_1 h_1 - \lambda_2 h_2 = 0 \tag{10}$$

The knowledge of $\lambda_1$ is already known whilst obtaining (9). The measurement of $h_1$ and $h_2$ along with the knowledge of $\lambda_1$ helps to determine $\lambda_2$ as given below:

$$0.0378 \leq \lambda_2 \leq 0.080 \tag{11}$$

Equation (5) represents the system transfer function and (7), (9) and (11) shows the parametric uncertainty in interval format.

# 3 Design of QFT-Based Controller

## *3.1 Choice of Control Structure*

In this section a 2-DOF control architecture is chosen for the two-tank liquid level control plant $P(s)$ as shown in Fig. 2. The control algorithm involves the design of a robust controller $G_c(s)$ followed by the design of a pre-filter $F(s)$, to meet a set of performance specifications.

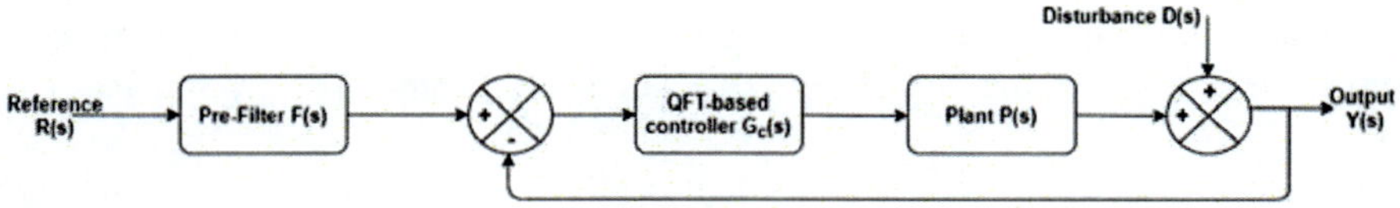


**Fig. 2** Schematic diagram of the proposed control structure

## *3.2 Performance Specifications*

The performance specifications as given in (12)–(14) for a single-input single-output control structure considering $G_c(s)$ and $F(s)$ to be the controller and pre-filter, respectively, are shown below:

1. Stability specification:

$$\left|\frac{P(j\omega)G_c(j\omega)}{1+P(j\omega)G_c(j\omega)}\right| \leq 1.5,\ \forall\,\omega \in [0.002\ 100]\ (\text{rad-s}^{-1}) \tag{12}$$

2. Specification for output disturbance rejection:

$$\left|\frac{1}{1+P(j\omega)G_c(j\omega)}\right| \leq 0.3,\ \forall\,\omega \in [0.002\ 0.02]\ (\text{rad-s}^{-1}) \tag{13}$$

3. Reference tracking specification:

$$\delta_{lm}(\omega) \leq \left|\frac{P(j\omega)G_c(j\omega)F(j\omega)}{1+P(j\omega)G_c(j\omega)}\right| \leq \delta_{um}(\omega)\ \forall\,\omega \in [0.002\ 0.42]\ (\text{rad-s}^{-1}) \tag{14}$$

$$\text{where, } \delta_{lm}(\omega) = \left|\frac{0.01}{3.33(j\omega)^3+1.667(j\omega)^2+0.233(j\omega)+0.01}\right|$$

$$\delta_{um}(\omega) = \left|\frac{0.033(j\omega)+0.01}{(j\omega)^2+0.1(j\omega)+0.01}\right|$$

## *3.3 Choice of Nominal Plant and Construction of Plant Templates*

The design of the controller starts with an arbitrarily chosen nominal plant ($N(s)$) from the set of $N = 200$ uncertain plants as below.

$$N(s) = \frac{0.00379}{s^2+0.1945s+0.00916} \tag{15}$$

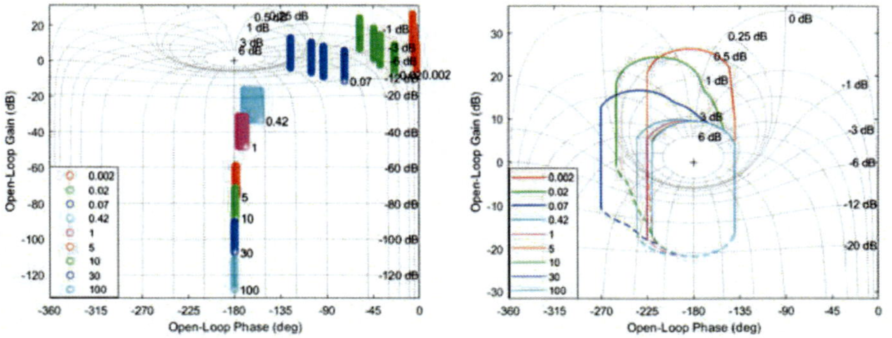


**Fig. 3** Templates for $P(j\omega)$ (left) and Stability bounds (right) at selected frequencies ($\omega$) in rad-s$^{-1}$

The plant templates of the uncertain plant $P(s)$ at some selected frequencies as shown in Fig. 3 are developed on the Nichols chart [16] by drawing the plot of open-loop magnitude-phase for all the plants, within the range of uncertainty.

## *3.4 Computation of Bounds*

The controller is designed for the uncertain plant $P(s)$ in such a way that it meets the performance specifications defined in (12)–(14). The design of the respective performance bounds along with the controller $G_c(s)$ is done by the use of QFT toolbox [2]. The plotting of the stability bounds on the Nichols chart at a specified frequency is done by the movement of the plant templates along the circumference of the closed-loop M-circle, considering the nominal plant $N(s)$ as the reference.

The stability bounds for the OLTF: $G_c(s)P(s)$ as shown in Fig. 3 are computed in such a way that the specification (12) is fulfilled despite the plant uncertainties. The output disturbance rejection bounds for the OLTF: $G_c(s)P(s)$ as shown in Fig. 4 at the chosen frequencies are computed in a similar manner thereby, satisfying the performance requirement in (13) regardless of the plant ($P(s)$) uncertainties. Then, the reference tracking bounds are computed for selected frequencies as shown in Fig. 4. After that, all the bounds are grouped together to obtain the group bounds as shown in Fig. 5 (left). Finally, a common permissible area of all specific performance bounds as shown in Fig. 5 is obtained in the form of composite (intersection) bounds. The designed transfer function of the controller $G_c(s)$ is:

$$G_c(s) = \frac{168.26(s + 0.291)(s + 0.1872)}{s(s + 2.281)} \tag{16}$$

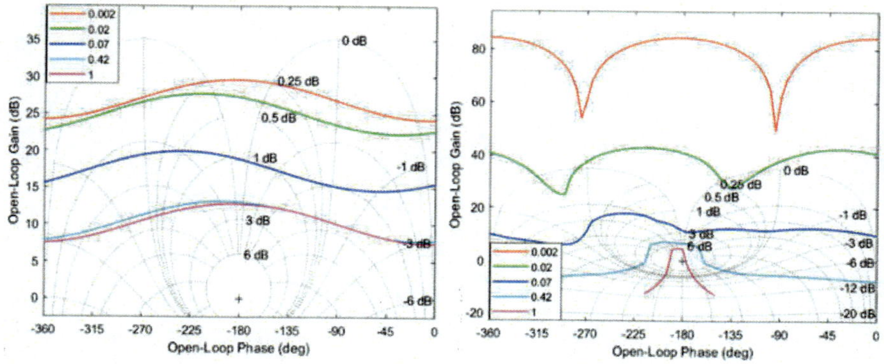


**Fig. 4** Sensitivity bounds (left) and Robust tracking bounds (right) at selected frequencies ($\omega$) in rad-s$^{-1}$

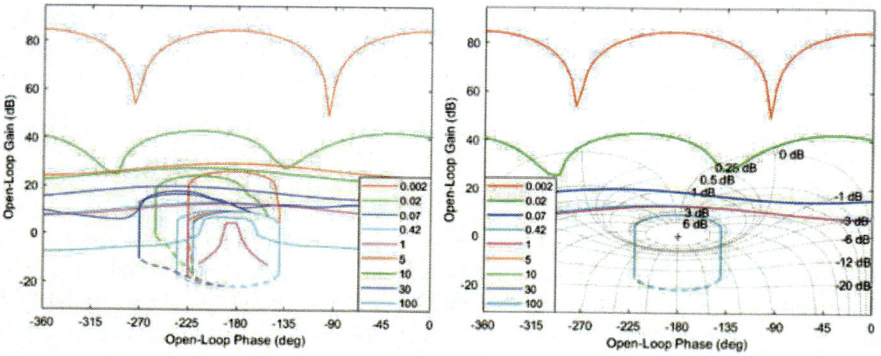


**Fig. 5** Group bounds (left) and Intersection bounds (right) at selected frequencies ($\omega$) in rad-s$^{-1}$

## 3.5 *Loop Shaping*

After the determination of all the respective bounds on the Nichol's Chart, the loop shaping technique is used using QFT Toolbox to design the controller $G_c(s)$ as shown in Fig. 6. The Nichol's plot of the compensated plant are reshaped to lie in the region defined by those convergence bounds thereby, meeting all the closed-loop specifications at each selected frequency.

## 3.6 *Design of Pre-filter*

After the design of the controller $G_c(s)$, the pre-filter $F(s)$ as shown in Fig. 6 is designed to ensure that the magnitude of the frequency response plots always stays within the upper and lower tracking boundary as designated by the tracking specification in (14). The transfer function of the pre-filter $F(s)$ is:

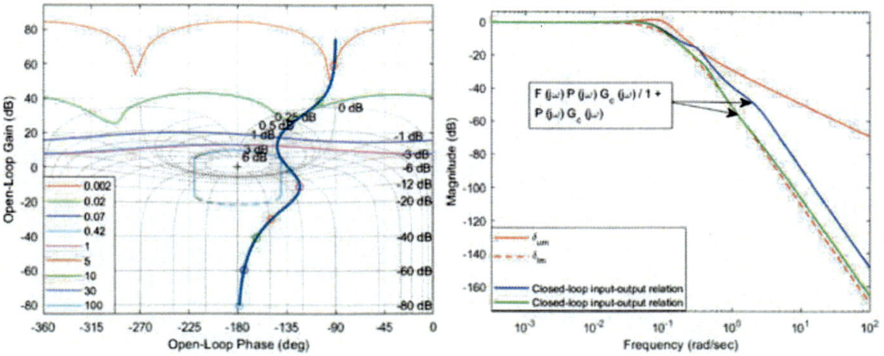


**Fig. 6** Synthesis of controller $G_c(s)$ by loop shaping(left) and Frequency magnitude response of pre-filter $F(s)$ compensated closed-loop system (right)

$$F(s) = \frac{0.005692s^2 + 0.006704s + 0.0003618}{s^3 + 0.1709s^2 + 0.01238s + 0.0003619} \tag{17}$$

## 4 Result and Discussion

The testing of the performance of the controlled two-tank system plant is verified by the use of numerical simulation of 200 uncertain plants in MATLAB. The two-tank system is allowed to track a multiple-step input at $t = 1$ and 2 s. The plant responses along with the output disturbances applied at 150 and 400 s for 200 uncertain plants and its corresponding control effort signals are shown in Fig. 7.

It is observed that the response of the nominal plant settles at 85 s with about 5% overshoot. The worst case overshoot and settling time are about 7% and 100 s, respectively. A satisfactory performance is achieved by the use of the proposed QFT-based robust controller predominantly.

It is seen that output disturbances are successfully rejected by the controlled system and a constant set point is achieved for all plants along with the rejection of disturbances at the output, in the presence of plant uncertainties. Also, it is determined that the control efforts are within the saturation limit of the actuator.

### *4.1 Comparative Analysis with Existing Controllers*

A comparative analysis as shown in Table 1 is performed against a P-I-D and a cascade controller for the uncertain two-tank system plant in this paper. The output plant responses using P-I-D and cascade controller as shown in Fig. 8 illustrates about the step responses for different plants including the nominal plant whose time

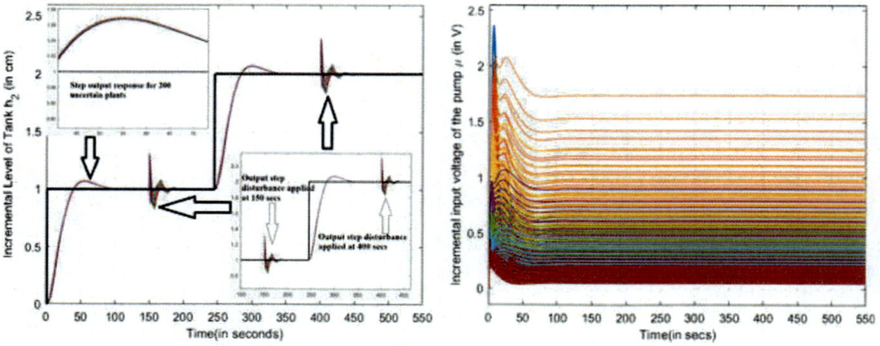


**Fig. 7** Plant output response with output disturbance rejection (left) and Control effort ($u(s)$) for 200 uncertain plants (right)

**Table 1** Comparative analysis

| Controllers | $\% M_p$ | IAE (units) | $t_s$ (secs) |
|---|---|---|---|
| P-I-D | 8.5 | 10.65 | 20 |
| Cascade | 26 | 17 | 40 |
| QFT | 5 | 0.75 | 85 |

domain specifications are described and compared with our proposed QFT-based robust controller in Table 1. A comparative analysis as shown in Table 2 is discussed with existing control strategies in terms of settling time ($t_s$), integral absolute error (IAE) and maximum peak overshoot ($M_p$). In [1], the use of P-I-D controller for liquid level control lacks in better performance with respect to plant uncertainty compared to our paper. However, in [4] where, a robust decentralized P-I controller is used, our paper performs better compared to it. The use of back-stepping sliding mode P-I controller approach in [6] consists of computational burden compared to our strategy. In [13] the controller design is complex as it is time consuming and in [14] it fully depends upon human expertise. In comparison to the above controller strategies, the proposed QFT-based robust controller used in this paper gives about 7.35% $M_p$ with a settling time ($t_s$) of about 85 secs and an IAE of about 0.75 units irrespective of model uncertainties and external disturbances. However, the QFT based controller $G_c(s)$ is designed despite the plant uncertainties at minimum cost of feedback thereby, proving its robustness in maintaining the stability of the uncertain plant as well as in rejecting the output disturbances up to 70%.

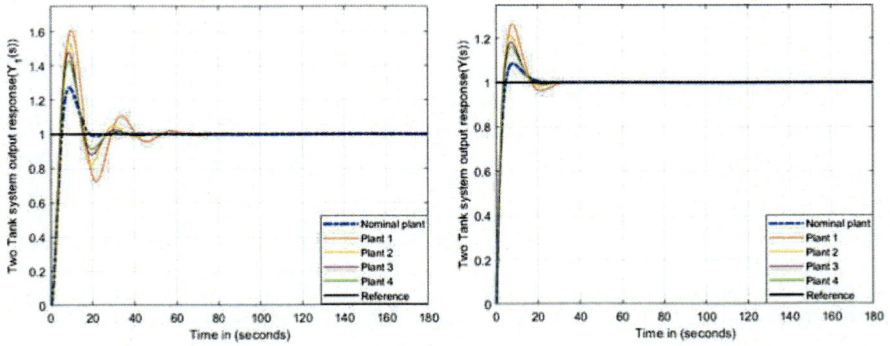


**Fig. 8** Two-Tank plant output response using Cascade controller (left) and using P-I-D controller (right)

**Table 2** Comparative analysis with reported works

| Paper | Controller strategy | % $M_p$ | IAE (units) | $t_s$ (secs) |
|---|---|---|---|---|
| This paper | QFT | 5% | 0.75 | 85 |
| [6] | Back-stepping sliding-mode P-I | 6.9% | – | 95 |
| [4] | Robust de-centralised P-I | – | 18.3 | 495 |
| [13] | Model predictive controller | 25.94% | – | 0.22 |
| [14] | Fuzzy logic controller | – | 12.6 | 62 |
| [1] | P-I-D | 10% | – | 12 |

## 5 Conclusions

In this paper, an uncertain model of a two-tank liquid level system is developed using both physical laws and data driven approach. The parametric uncertainty is quantified in interval form using data driven approach. The quantitative feedback theory (QFT) technique is used for the design of the controller $G_c(s)$ for the uncertain two-tank system followed by the design of the pre-filter $F(s)$. The simulation results show that the above control strategy successfully meets multiple performance specifications despite the plant uncertainty. A comparative analysis against reported controller strategies shows a satisfactory and improved performance for the QFT-based robust controller.

## References

1. Fellani MA, Gabaj AM (2015) PID controller design for two tanks liquid level control system using Matlab. Int J Electr Comput Eng (IJECE) 5(3):436–442
2. Borghesani C, Chait Y, Yaniv O (2003) The QFT frequency domain control design toolbox for use with MATLAB user's guide. Terasoft Inc., San Diego, pp 1–148

3. Sanz MG (2014) Encyclopedia of systems and control. Springer, Cleveland
4. Mahapatro SR, Subudhi B, Ghosh S (2019) Design and experimental realization of a robust decentralized PI controller for a coupled tank system. ISA Trans 89:158–168
5. Basçi A, Derdiyok A (2016) Implementation of an adaptive fuzzy compensator for coupled tank liquid level control system. Measurement 91:12–18
6. Aksu IO, Coban R (2018) Sliding mode PI control with back stepping approach for MIMO nonlinear cross-coupled tank systems. Int J Robust Nonlinear Control 29:1854–1871
7. Kar B, Roy P (2018) A comparative study between cascaded FOPI-FOPD and IOPI-IOPD controllers applied to a level control problem in a coupled tank system. J Control Autom Electr Syst 29:340–349
8. Abbas H, Asghar S, Qamar S (2012) Sliding mode control for coupled-tank liquid level control system. In: 2012 10th International conference on frontiers of information technology
9. Derdiyok A, Baçi A (2013) The application of chattering-free sliding mode controller in coupled tank liquid-level control system. Korean J Chem Eng 30(3):540–545
10. Roy P, Roy BK (2016) Fractional order PI control applied to level control in coupled two tank MIMO system with experimental validation. Control Eng Pract 48:119–135
11. Liang L (2011) The application of fuzzy PID controller in coupled-tank liquid-level control system. In: 2011 International conference on electronics, communications and control (ICECC)
12. Shehu IA, Wahab NA (2016) Applications of MPC and PI controls for liquid level control in coupled-tank systems. In: 2016 IEEE international conference on automatic control and intelligent systems (I2CACIS), Malaysia, 119–124
13. Jibril M, Tadese M, Tadese EA (2020) Tank liquid level control using NARMA-L2 and MPC controllers. Researcher 12(7):23–27
14. Al Tahtawi AR, Yahya S, Setiadi B, Marsya C (2020) The implementation of embedded fuzzy logic controller on liquid level control system. Adv Eng Res: Int Seminar Sci Appl Technol (ISSAT 2020) 198:161–166
15. Ogata K, Yang Y (2002) Modern control engineering, vol 4. Prentice-Hall
16. Houpis, Constantine H (1999) Quantitative feedback theory: fundamentals and applications. CRC Press

# Design of Continuous Time Nonlinear Control with Quadruple Tank System Application

**Sowmya Gupta and Chinmay Rajhans**

## 1 Introduction

Stability is one of the most crucial aspect while designing the controller. One of the most promising optimization-based controller is the Model Predictive Control (MPC) [1]. MPC finds applications in every field of science, engineering and technology [2–4].

Commonly used stabilizing constraints include (a) Terminal equality constraint, (b) Terminal penalty term, and (c) Terminal inequality constraint [5]. An enormous process has taken place in the field of nominal stability of linear MPC [6] and Nonlinear MPC (NMPC) [7–9]. Grimm et al. have presented systems when small changes in the parameters alters the stability characteristics of the controller [10]. Hence formally establishing the asymptotic stability result becomes necessary.

Executing the terminal equality constraint is relatively easy [11], however, main limitation is that it is highly conservative and often leads to infeasibility specifically when using constrained formulations. Michalska and Mayne conceptualized dual mode MPC scheme where in the idea of terminal region was introduced [12]. NMPC controller is expected to drive the plant trajectory into a region, termed as terminal region, around the set point using a set of feasible inputs in a finite horizon time. Subsequently localized linear controller will take the system trajectory to the set point. This idea was extended by Chen and Allgöwer where in NMPC control acted inside of the terminal region instead of a linear control law, which has resulted in the concept of Quasi Infinite Horizon—Nonlinear Model Predictive Control (QIH-NMPC) formulation [13].

Chen and Allgöwer presented one approach for terminal region characterization for the case of the continuous time MPC formulation [13]. The approach to characterize the terminal region using inequality-based conditions was presented. First

S. Gupta · C. Rajhans (✉)
VJTI, Mumbai, India
e-mail: rajhanschinmay2@gmail.com



P. H. J. Chong et al. (eds.), *Emerging Electronics and Automation*, Lecture Notes in Electrical Engineering 937, https://doi.org/10.1007/978-981-19-4300-3_3

major drawback of the approach is a tuning parameter which is nearly independent of the NMPC formulation stage weighting matrices. Second limitation of Chen and Allgöwer's approach is that it provides a scalar tuning parameter which restricts the design to a single scalar for increasing the terminal region.

Several researchers have presented approaches for the terminal region characterization for the case of discrete time NMPC formulations [14–17]. However, they cannot be applied to the continuous time NMPC formulations due to the involvement of sampling time [18].

Current work presents one linear quadratic regulator-based novel approach for the terminal region characterization for the case of continuous time MPC formulation. Approach provides two distinct degrees of freedom as additive weighting matrices (tuning parameters) for increasing the terminal region.

Novel LQR-based approach is simulated using a quadruple tank system. This system is used by various researchers for showcase their controller performance [19, 20]. Application of the NMPC formulation is very limited in the literature and is another contribution of the current work.

Second section presents the continuous time NMPC formulation in detail. In addition, approach by Chen and Allgöwer [13] is stated formally along with its limitation. Third section presents the proposed LQR-based approach for the terminal region characterization. Subsequently, asymptotic stability result is presented. Forth section gives numerical characterization steps using the novel approach. Fifth section presents the terminal region characterization using demonstration case study. Sixth section details case study results. Seventh section gives the conclusions.

## 2 Continuous Time NMPC Formulation

Consider nonlinear system (continuous time) given by

$$\frac{\mathrm{d}\mathbf{x}(t)}{\mathrm{d}t} = \mathbf{f}(\mathbf{x}(t), \mathbf{u}(t)) \tag{1}$$

$$\mathbf{x}(0) = \mathbf{x}_0 \tag{2}$$

where $\mathbf{x}(t) \in \mathcal{X} \subset \mathbb{R}^{n_x}$: the state vector and $\mathbf{u}(t) \in \mathcal{U} \subset \mathbb{R}^{n_u}$: the input vector. Assumptions are stated as follows:

**C1** System function $\mathbf{f} : \mathbb{R}^{n_x} \times \mathbb{R}^{n_u} \to \mathbb{R}^{n_x}$ is a twice continuously differentiable function.

**C2** The origin $\mathbf{0} \in \mathbb{R}^{n_x}$ is equilibrium point i.e. $\mathbf{f}(\mathbf{0}, \mathbf{0}) = \mathbf{0}$.

**C3** Vector $\mathbf{u}(t)$ are inside a closed and convex set $\mathcal{U} \subset \mathbb{R}^{n_u}$.

**C4** The system (1) has unique solution for any $\mathbf{x}_0 \in \mathcal{X}$ and any piece wise right continuous input signal.

**C5** $\mathbf{x}(t)$ is measured at any time $t$.

**C6** External disturbances do not affect the system dynamics.

## 2.1 NMPC Formulation

Nonlinear MPC formulation given for the nonlinear continuous system (1) is stated as:

$$\min_{\overline{\mathbf{u}}_{[l,l+T_p]}} J\left(\mathbf{x}(l), \overline{\mathbf{u}}_{[l,l+T_p]}\right) \tag{3}$$

with

$$J\left(\mathbf{x}(l), \overline{\mathbf{u}}_{[l,l+T_p]}\right) = \int_{l}^{l+T_p} \left\{\mathbf{z}(m)^T \mathbf{W}_x \mathbf{z}(m) + \overline{\mathbf{u}}(m)^T \mathbf{W}_u \overline{\mathbf{u}}(m)\right\} \mathrm{d}m + \mathbf{z}(l+T_p)^T \mathbf{P} \mathbf{z}(l+T_p) \tag{4}$$

$$\overline{\mathbf{u}}_{[l,l+T_p]} = \left\{\mathbf{u}(m) \in \mathcal{U} : m \in \left[l, l+T_p\right]\right\} \tag{5}$$

subject to

$$\frac{\mathrm{d}\mathbf{z}(m)}{\mathrm{d}m} = \mathbf{f}\left(\mathbf{z}(m), \overline{\mathbf{u}}(m)\right) \text{ for } m \in \left[l, l+T_p\right] \tag{6}$$

$$\mathbf{z}(l) = \mathbf{x}(l) \tag{7}$$

$$\mathbf{z}\left(l+T_p\right) \in \Omega \tag{8}$$

where $\mathbf{W}_x$ and $\mathbf{W}_u$ are weighting matrices for state and input, respectively, of dimensions $(n_x \times n_x)$, and $(n_u \times n_u)$ respectively. $\mathbf{P}$ defines the terminal penalty term and is a matrix of dimension $(n_x \times n_x)$. Matrices $\mathbf{W}_x, \mathbf{W}_u, \mathbf{P}$ are positive definite and symmetric matrices. $T_p$ is a prediction horizon time or control horizon time. $\mathbf{z}(m)$ is the predicted state vector of the MPC formulation and $\overline{\mathbf{u}}(m)$ is the future control moves. Note that time variables $l = t$, and $m = \tau$ may be also used.

## 2.2 Design of MPC Controller

The terminal region is designed as invariant set for system (1) under the linear control law having gain $\mathbf{K}$. Terminal penalty term is computed using the following condition:

$$\mathbf{z}(l+T_p)^T \mathbf{P} \mathbf{z}(l+T_p) \geq \int_{l+T_p}^{\infty} \left\{\mathbf{z}(m)^T \mathbf{W}_x \mathbf{z}(m) + \overline{\mathbf{u}}(m) \mathbf{W}_u \overline{\mathbf{u}}(m)\right\} \mathrm{d}m \tag{9}$$

with $\overline{\mathbf{u}}(m) = -\mathbf{K}\mathbf{z}(m) \in \mathcal{U}$ for all $m \geq (l+T_p)$ and for all $\mathbf{z}(l+T_p) \in \Omega$.

Solution to the standard optimal control problem (3) can be computed. Controller is designed using moving horizon framework. Accordingly, only the control move at time $t$ is implemented as,

$$\mathbf{u}(t) = \overline{\mathbf{u}}^{*}(t) \tag{10}$$

Whole process of optimization is repeated at $t + \delta$ with $\delta$ being an sufficiently small sampling period.

## 2.3 *Chen and Allgöwer's Approach*

Before proceeding to the proposed LQR based approach, a look at Chen and Allgöwer's approach is required. Consider, Jacobian linearization of the system around the origin as,

$$\frac{\mathrm{d}\mathbf{x}(t)}{\mathrm{d}t} = \mathbf{A}\mathbf{x}(t) + \mathbf{B}\mathbf{u}(t) \tag{11}$$

where

$$\mathbf{A} = \left[\frac{\partial \mathbf{f}}{\partial \mathbf{x}}\right]_{(\mathbf{0},\mathbf{0})} \text{ and } \mathbf{B} = \left[\frac{\partial \mathbf{f}}{\partial \mathbf{u}}\right]_{(\mathbf{0},\mathbf{0})}$$

One additional assumption is required at this stage.

**C7** The linearized system (11) is stabilizable.

Chen and Allgöwer presented terminal region as,

$$\Omega \equiv \left\{\mathbf{x} \in \mathbb{R}^n | \mathbf{x}^T \mathbf{P} \mathbf{x} \le \alpha, -\mathbf{K}\mathbf{x} \in \mathcal{U}\right\} \tag{12}$$

where linear gain $\mathbf{K}$ and the terminal penalty matrix $\mathbf{P}$ are the steady-state solutions of the modified Lyapunov equations given as follows:

$$(\mathbf{A}_K + \kappa\mathbf{I})^T \mathbf{P} + \mathbf{P}(\mathbf{A}_K + \kappa\mathbf{I}) = -\mathbf{Q}^* \tag{13}$$

where

$$\mathbf{Q}^* = \mathbf{W}_x + \mathbf{K}^T \mathbf{W}_u \mathbf{K} \tag{14}$$

where $\mathbf{A}_K = \mathbf{A} - \mathbf{B}\mathbf{K}$.

It can be noted that once stage cost weighting matrices $\mathbf{W}_x$, $\mathbf{W}_u$ are chosen, there is barely any degree of freedom left to the designer for increasing of the terminal region. This results in a very conservative terminal region.

## 3 Alternate Approach for the Terminal Region Characterization

Consider the LQR-based approach given as follows:

**Lemma 1** *Suppose that assumptions C1 to C7 are satisfied. Let* $\widetilde{\mathbf{W}}_x > \mathbf{W}_x$ *and* $\widetilde{\mathbf{W}}_u > \mathbf{W}_u$ *be any positive definite matrices.* $\mathbf{P}_{LQ}$*: steady-state solution of the following equations:*

$$\mathbf{A}_{K_{LQ}}^T \mathbf{P}_{LQ} + \mathbf{P}_{LQ}\mathbf{A}_{K_{LQ}} = -\left(\widetilde{\mathbf{W}}_x + \mathbf{K}_{LQ}^T \widetilde{\mathbf{W}}_u \mathbf{K}_{LQ}\right) \tag{15}$$

$$\mathbf{K}_{LQ} = \left(\widetilde{\mathbf{W}}_u\right)^{-1} \mathbf{B}^T \mathbf{P}_{LQ} \tag{16}$$

*where* $\mathbf{A}_{K_{LQ}} = \mathbf{A} - \mathbf{B}\mathbf{K}_{LQ}$*. Then there exists a constant* $\alpha > 0$ *which defines region given as,*

$$\Omega \equiv \left\{\mathbf{x} \in \mathbb{R}^{n_x} | \mathbf{x}^T \mathbf{P}_{LQ} \mathbf{x} \le \alpha, -\mathbf{K}_{LQ}\mathbf{x} \in \mathcal{U}\right\} \tag{17}$$

*For any terminal state* $\mathbf{x}(t + T_p) \in \Omega$*, the following equation holds true.*

$$\mathbf{z}(t + T_p)^T \mathbf{P}_{LQ} \mathbf{z}(t + T_p) \ge \int_{t+T_p}^{\infty} \left\{\mathbf{z}(\tau)^T \mathbf{W}_x \mathbf{z}(\tau) + \overline{\mathbf{u}}(\tau) \mathbf{W}_u \overline{\mathbf{u}}(\tau)\right\} d\tau \tag{18}$$

***Proof*** Consider a Lyapunov candidate chosen as

$$V(\mathbf{x}) = \mathbf{x}^T \mathbf{P}_{LQ} \mathbf{x} \tag{19}$$

Using (15), derivative of the function $V(\mathbf{x})$ is computed as,

$$\frac{\mathrm{d}V(\mathbf{x})}{\mathrm{d}t} = \mathbf{x}^T \left(\mathbf{A}_{K_{LQ}}^T \mathbf{P}_{LQ} + \mathbf{P}_{LQ}\mathbf{A}_{K_{LQ}}\right)\mathbf{x} + 2\mathbf{x}^T \mathbf{P}_{LQ} \mathbf{\Phi}_K(\mathbf{x}) \tag{20}$$

Defining matrices,

$$\Delta\mathbf{W}_x \equiv \widetilde{\mathbf{W}}_x - \mathbf{W}_x > 0 \text{ and } \Delta\mathbf{W}_u \equiv \widetilde{\mathbf{W}}_u - \mathbf{W}_u > 0 \tag{21}$$

one can write

$$\widetilde{\mathbf{W}}_x + \mathbf{K}_{LQ}^T \widetilde{\mathbf{W}}_u \mathbf{K}_{LQ} = \mathbf{Q}^* + \Delta\mathbf{Q} \tag{22}$$

$$\mathbf{Q}^* = \mathbf{W}_x + \mathbf{K}_{LQ}^T \mathbf{W}_u \mathbf{K}_{LQ} \tag{23}$$

$$\Delta\mathbf{Q} = \Delta\mathbf{W}_x + \mathbf{K}_{LQ}^T \Delta\mathbf{W}_u \mathbf{K}_{LQ} \tag{24}$$

and Eq. (15) can be re-written as follows

$$\mathbf{A}_{K_{LQ}}{}^{T}\mathbf{P}_{LQ} + \mathbf{P}_{LQ}\mathbf{A}_{K_{LQ}} = -\left(\mathbf{Q}^{*} + \Delta\mathbf{Q}\right) \tag{25}$$

Equations (20) and (25) are combined as follows:

$$\frac{\mathrm{d}V(\mathbf{x})}{\mathrm{d}t} = -\mathbf{x}^{T}\mathbf{Q}^{*}\mathbf{x} + \left(-\mathbf{x}^{T}\Delta\mathbf{Q}\mathbf{x} + 2\mathbf{x}^{T}\mathbf{P}_{LQ}\boldsymbol{\Phi}_{K}(\mathbf{x})\right) \tag{26}$$

Consider second term of the expression (26),

$$\boldsymbol{\Psi}(\mathbf{x}) := \left(\mathbf{x}^{T}\Delta\mathbf{Q}\mathbf{x} - 2\mathbf{x}^{T}\mathbf{P}\boldsymbol{\Phi}_{K}(\mathbf{x})\right) \tag{27}$$

Using (27) in (26),

$$\frac{\mathrm{d}V(\mathbf{x})}{\mathrm{d}t} = -\mathbf{x}^{T}\mathbf{Q}^{*}\mathbf{x} - \boldsymbol{\Psi}(\mathbf{x}) \tag{28}$$

If $\Omega$ is chosen such that

$$\boldsymbol{\Psi}(\mathbf{x}) = \left(\mathbf{x}^{T}\Delta\mathbf{Q}\mathbf{x} - 2\mathbf{x}^{T}\mathbf{P}\boldsymbol{\Phi}_{K}(\mathbf{x})\right) \geq 0 \tag{29}$$

then

$$\frac{\mathrm{d}V(\mathbf{x})}{\mathrm{d}t} \leq -\mathbf{x}^{T}\mathbf{Q}^{*}\mathbf{x} \tag{30}$$

Integrating (30) on the interval, $[t + T_p, \infty)$, gives

$$V(\mathbf{x}(t + T_p)) \geq \int_{t+T_p}^{\infty} \mathbf{x}(\tau)^{T}\mathbf{Q}^{*}\mathbf{x}(\tau)d\tau \tag{31}$$

which is identical to the equation (18) for any terminal state $\mathbf{x}(t + T_p) \in \Omega$.

Satisfaction of the terminal inequality constraint given by (8) implies that

$$V(\mathbf{x}(t + T_p)) = \mathbf{x}^{T}(t + T_p)\mathbf{P}\mathbf{x}(t + T_p) \leq \alpha \tag{32}$$

In addition, equation (30) implies that $\frac{dV(\mathbf{x})}{dt} < 0$. Subsequently,

$$V(\mathbf{x}(t + T_p + \delta)) = \mathbf{x}^{T}(t + T_p + \delta)\mathbf{P}\mathbf{x}(t + T_p + \delta) \leq \alpha \tag{33}$$

i.e. $\mathbf{x}(t + T_p + \delta) \in \Omega$ for any $\delta > 0$.

Consider the asymptotic stability result as follows:

**Theorem 1** *Let (a) the assumptions given by C1–C7 hold true and (b) the MPC problem be feasible at $t = 0$, then the nonlinear system (1) under MPC controller is asymptotically stable at the origin.*

***Proof*** From candidate Lyapunov equation (19) from Lemma 1,

$$V(\mathbf{x}) = \mathbf{x}^T \mathbf{P} \mathbf{x} \tag{34}$$

Consider three main properties [21]:

- $V(\mathbf{0})$ is 0 since $V(\mathbf{0}) = (\mathbf{0}^T)\mathbf{P}(\mathbf{0}) = 0$.
- $\mathbf{P}$ is positive definite matrix, hence, $V(\mathbf{x}) = \mathbf{x}^T\mathbf{P}\mathbf{x} > 0$ for all $\mathbf{x} \neq \mathbf{0}$.
- Using (30) and $\mathbf{Q}^* > 0$ implies

$$\frac{\mathrm{d}V(\mathbf{x})}{\mathrm{d}t} \leq -\mathbf{x}^T\mathbf{Q}^*\mathbf{x} < 0 \tag{35}$$

Thus, the candidate function $V(\mathbf{x})$ becomes the Lyapunov function for the nonlinear system under MPC controller. Hence, the closed loop system is asymptotically stable.

## 4 Terminal Region Characterization

Lemma 1 presents a condition for the terminal region characterization. In order to numerically compute the terminal region and subsequently implement the NMPC controller, following steps are used for the proposed approach:

**S1** Upper Bound Set:
Compute the largest number $\gamma$ such that inputs constraints remain satisfied inside set $\Omega_\gamma$.

$$\Omega_\gamma \equiv \left\{\mathbf{x} \in \mathbb{R}^{n_x} | \mathbf{x}^T\mathbf{P}\mathbf{x} \leq \gamma, -\mathbf{K}_{LQ}\mathbf{x} \in \mathcal{U}\right\} \tag{36}$$

**S2** Computation of the Terminal Region using inequality-based method:
Compute the largest number $\alpha \in (0, \gamma]$ such that

$$\left[\min_{\mathbf{x} \in \Omega} \mathbf{\Psi}(\mathbf{x})\right] = 0 \tag{37}$$

The necessary stability condition given by equation (37) ensures that $\mathbf{\Psi}(\mathbf{x}) \geq 0$ for all $\mathbf{x} \in \Omega$.

Hyper volume is computed for state dimension of 4 as

$$H_{V4} = \frac{\pi^2\alpha^2}{2\sqrt{det(\mathbf{P})}} \tag{38}$$

## 5 Quadruple Tank Case Study

Consider a benchmark four state systemm, i.e., quadruple tank nonlinear system given in [20], which is similar to system in [19].

System dynamics equations are given as follows [19]:

$$\dot{X}_1 = -\frac{a_1}{A_1}\sqrt{2gX_1} + \frac{a_3}{A_1}\sqrt{2gX_3} + \frac{\gamma_1}{A_1}U_1 \tag{39}$$

$$\dot{X}_2 = -\frac{a_2}{A_2}\sqrt{2gX_2} + \frac{a_4}{A_2}\sqrt{2gX_4} + \frac{\gamma_2}{A_2}U_2 \tag{40}$$

$$\dot{X}_3 = -\frac{a_3}{A_3}\sqrt{2gX_3} + \frac{(1-\gamma_2)}{A_3}U_2 \tag{41}$$

$$\dot{X}_4 = -\frac{a_4}{A_4}\sqrt{2gX_4} + \frac{(1-\gamma_1)}{A_4}U_1 \tag{42}$$

where $X_i$ are the heights (i.e., states) of water, $a_i$ are cross sectional areas of outflows, and $A_i$ are the internal areas of $i$th tank for $i = 1, 2, 3, 4$; $U_j$ are the flow rates (i.e. control inputs) for $j = 1, 2$; $g$ is the gravitational constant; $0 < \gamma_l < 1$ are the split ratios for $l = 1, 2$. Parameters for the system are kept identical to the one given in [20] (Table 1).

To compare the current approach with Chen and Allgöwer's approach, model parameters given by Raff et al. [20] are used. The terminal region is calculated at the following origin:

$$\mathbf{X}_s = \begin{bmatrix} 13.9883 \\ 14.0644 \\ 14.2562 \\ 21.4277 \end{bmatrix} \text{ cm and } \mathbf{U}_s = \begin{bmatrix} 43.4 \\ 35.4 \end{bmatrix} \text{ ml/s} \tag{43}$$

Thus, the model equations are transformed using change of variables as follows:

$$\mathbf{x}(t) = \mathbf{X}(t) - \mathbf{X}_s \text{ and } \mathbf{u}(t) = \mathbf{U}(t) - \mathbf{U}_s \tag{44}$$

**Table 1** Quadruple tank system parameter values

| Parameter | Value | Unit | Parameter | Value | Unit |
|---|---|---|---|---|---|
| $A_1$ | 50.27 | $cm^2$ | $a_1$ | 0.233 | $cm^2$ |
| $A_2$ | 5027 | $cm^2$ | $a_2$ | 0.242 | $cm^2$ |
| $A_3$ | 28.27 | $cm^2$ | $a_3$ | 0.127 | $cm^2$ |
| $A_4$ | 28.27 | $cm^2$ | $a_4$ | 0.127 | $cm^2$ |
| $\gamma_1$ | 0.4 | – | $\gamma_2$ | 0.4 | – |
| $g$ | 981 | $cm/s^2$ | – | – | – |

The control problem is formulated in terms of the perturbation variables $(\mathbf{x}, \mathbf{u})$. Stage cost weighting matrices are chosen as follows:

$$\mathbf{W}_x = \text{diag}\left(\left[\,1\ 1\ (1\times 10^{-5})\ (1\times 10^{-5})\,\right]\right) \text{ and } \mathbf{W}_u = \text{diag}\left(\left[\,0.01\ 0.01\,\right]\right) \quad (45)$$

The control input constraints are stated as follows:

$$\mathcal{U} = \left\{ \begin{bmatrix} U_1 \\ U_2 \end{bmatrix} \in \mathbb{R}^2 \mid \begin{bmatrix} 0 \\ 0 \end{bmatrix} \leq \begin{bmatrix} U_1 \\ U_2 \end{bmatrix} \leq \begin{bmatrix} 60 \\ 60 \end{bmatrix} \right\} \quad (46)$$

Raff et al. [20] have given $\mathbf{P}$ matrix obtained using the approach given by Chen and Allgöwer [13].

$$\mathbf{P} = \text{diag}\left(\left[\,6.55\ 6.55\ 7.92\ 31.70\,\right]\right) \quad (47)$$

and the terminal region computed as $\Omega = \left\{\mathbf{x} : \mathbf{x}^T\mathbf{P}\mathbf{x} \leq 792.4\right\}$ with hyper volume of 29856.

To simplify the numerical computation of the terminal region using the LQR-based approach, following parameterization is carried out:

$$\widetilde{\mathbf{W}}_x = \rho_x \mathbf{W}_x \text{ and } \widetilde{\mathbf{W}}_u = \rho_u \mathbf{W}_u \quad (48)$$

with $\rho_x > 1$ and $\rho_u > 1$ being the tuning parameters.

Table 2 presents a summary of the terminal regions for the quadruple tank system example by Raff et al. [20]. In the case study, terminal region computed using LQR-based approach was found significantly larger by hyper volume on comparison with the terminal region with approach by Chen and Allgöwer [13].

Maximum terminal region is obtained using LQR based approach for $\rho_x = 20$ and $\rho_u = 200$ resulting in linear gain matrix $\mathbf{K}_{LQ}$ and terminal penalty matrix $\mathbf{P}_{LQ}$ as follows:

$$\mathbf{K}_{LQ} = \begin{bmatrix} 1.0013\ 0.5331\ 0.1278\ 0.2613 \\ 0.6278\ 0.9700\ 0.3040\ 0.0923 \end{bmatrix} \quad (49)$$

$$\mathbf{P}_{LQ} = \begin{bmatrix} 313.4577 & -40.9009 & 74.4913 & -23.1609 \\ -40.9009 & 308.6622 & -24.3180 & 65.5670 \\ 74.4913 & -24.3180 & 37.7673 & -15.8857 \\ -23.1609 & 65.5670 & -15.8857 & 33.3047 \end{bmatrix} \quad (50)$$

**Table 2** Comparison of terminal regions for literature and LQR-based approaches

| Approach | Tuning Parameters | $\alpha$ | Hyper volume |
|---|---|---|---|
| Chen and Allgöwer [13] | $\kappa = 0.020425$ | 792.4 | 29856 |
| LQR based approach | $\rho_x = 20, \rho_u = 1$ | 4.0728 | 1.2081 |
| LQR based approach | $\rho_x = 20, \rho_u = 200$ | 6747.37 | 41538.36 |

Terminal region is given as $\Omega = \{\mathbf{x}|\mathbf{x}^T\mathbf{P}_{LQ}\mathbf{x} \leq 6747.37\}$ with a hyper volume of 41538, which is 1.3913 times larger than the terminal region presented by Raff et al. [20] using hyper volume as measurement criterion. This terminal region is used for carrying out the continuous time NMPC simulations.

To check the MPC results, the initial conditions $A,\ B,\ C$ are taken as:

$$\mathbf{X}_A = \begin{bmatrix} 9.25 \\ 10.00 \\ 6.84 \\ 17.00 \end{bmatrix}; \mathbf{X}_B = 1.9\mathbf{X}_s = \begin{bmatrix} 26.58 \\ 26.72 \\ 27.09 \\ 40.71 \end{bmatrix}; \mathbf{X}_C = 0.2\mathbf{X}_s = \begin{bmatrix} 2.80 \\ 2.81 \\ 2.85 \\ 4.29 \end{bmatrix} \text{cm} \quad (51)$$

Initial condition $A$ is very close to the initial condition specified by Raff et al. [20]. The nonlinear plant simulation and the model predictions in the NMPC formulations are carried out using the explicit Euler method. The sampling time is kept at $\delta = 5\ s$ and prediction and control horizon are kept at $T_p = 25\ s$.

Figure 1 presents states plot. System converges to the steady state operating point as expected, and is according to the asymptotic stability result.

Figure 2 shows control inputs plot. Initially one of the inputs is at saturation and eventually the inputs converge to the corresponding steady-state values.

Figure 3 gives plot of the terminal constraint value. Values remains negative for the entire time duration indicating that the terminal constraint was always satisfied. Also terminal constraint value approaches $-\alpha$, indicating that the predicted state trajectory has reached the origin in the deviation variables.

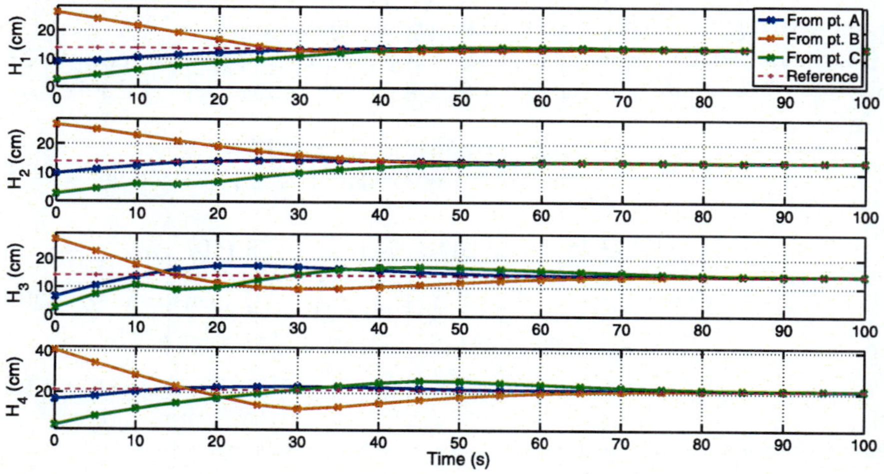


**Fig. 1** Plot of states in actual variables

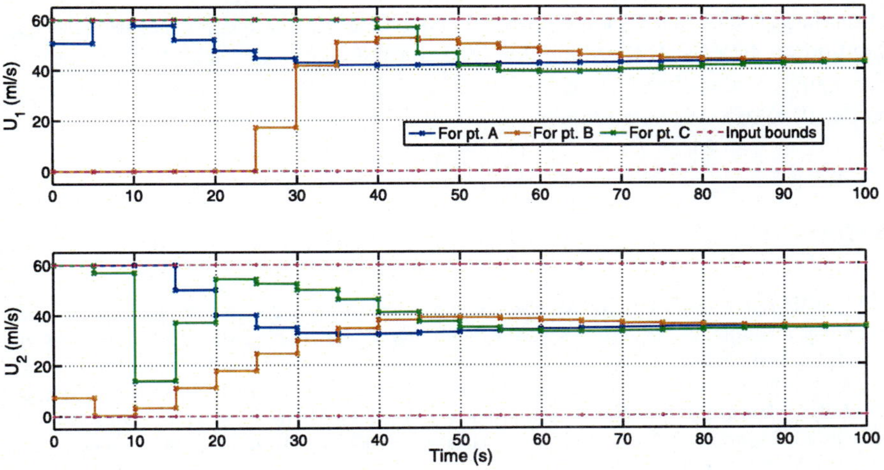


**Fig. 2** Plot of control inputs

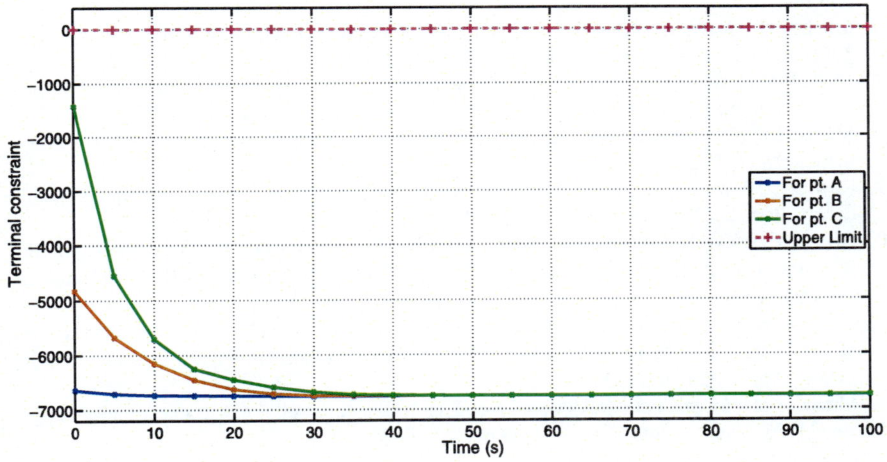


**Fig. 3** Plot of terminal constraint value

# 6 Conclusions

Approaches presented in the literature for the terminal region characterization for the MPC formulations (continuous time) provide limited degrees of freedom and often lead to a conservative terminal regions. A novel LQR-based approach is presented in the current work which exhibits a large degrees of freedom for increasing of the terminal regions.

Effectiveness of the approach is simulated using quadruple tank system example. In the case study, it is observed that terminal region computed using the LQR-based approach is approximately 1.39 times larger by hyper volume as compared to the largest terminal region computed using Chen and Allgöwer's approach [13]. Continuous time NMPC simulations validate the asymptotic stability property of the designed controller. Future work will involve extending the result for the output feedback NMPC formulation.

## References


1. Lee JH (2011) Model predictive control: review of the three decades of development. Int J Control Autom Syst 9(3):415–424. https://doi.org/10.1007/s12555-011-0300-6, http://www.springer.com/12555
2. Mayne D, Michalska H (1990) Receding horizon control of nonlinear systems. IEEE Trans Autom Control 35(7):814–824. https://doi.org/10.1109/9.57020
3. Qin SJ, Badgwell TA (2003) A survey of industrial model predictive control technology. Control Eng Pract 11(7):733–764
4. Camacho EF, Bordons CC (2007) Model predictive control. Springer, Berlin
5. Rawlings J, Mayne D (2009) Model predictive control: theory, computation, and design. Nob Hill Pub
6. Muske KR, Rawlings JB (1993) Model predictive control with linear models. AIChE J 39(2):262–287. https://doi.org/10.1002/AIC.690390208
7. Chen WH (2010) Stability analysis of classic finite horizon model predictive control. Int J Control Autom Syst 8(2):187–197. https://doi.org/10.1007/s12555-010-0202-z
8. Mayne DQ (2014) Model predictive control: recent developments and future promise. Automatica 50(12):2967–2986
9. Rawlings JB, Mayne DQ, Diehl M (2017) Model predictive control: theory, computation, and design, 2nd edn. Nob Hill Publishing, LLC
10. Grimm G, Messina MJ, Tuna SE, Teel AR (2004) Examples when nonlinear model predictive control is nonrobust. Automatica 40(10):1729–1738. https://doi.org/10.1016/j.automatica.2004.04.014
11. Keerthi SS, Gilbert EG (1988) Optimal infinite-horizon feedback laws for a general class of constrained discrete-time systems: stability and moving-horizon approximations. J Optim Theory Appl 57(2):265–293
12. Michalska H, Mayne D (1993) Robust receding horizon control of constrained nonlinear systems. IEEE Trans Autom Control 38(11):1623–1633
13. Chen H, Allgöwer F (1998) A quasi-infinite horizon nonlinear model predictive control scheme with guaranteed stability. Automatica 34(10):1205–1217
14. Johansen TA (2004) Approximate explicit receding horizon control of constrained nonlinear systems. Automatica 40(2):293–300
15. Yu S, Qu T, Xu F, Chen H, Hu Y (2017) Stability of finite horizon model predictive control with incremental input constraints. Automatica 79:265–272. https://doi.org/10.1016/J.AUTOMATICA.2017.01.040
16. Rajhans C, Griffith D, Patwardhan SC, Biegler LT, Pillai H (2018) Two approaches for terminal region characterization in discrete time quasi-infinite horizon NMPC
17. Rajhans C, Griffith DW, Patwardhan SC, Biegler LT, Pillai HK (2019) Terminal region characterization and stability analysis of discrete time quasi-infinite horizon nonlinear model predictive control. J Process Control 83:30–52. https://doi.org/10.1016/j.jprocont.2019.08.002
18. Astrom KJ, Wittenmark B (1997) Computer-controlled systems: theory and design. Prentice Hall
19. Johansson K (2000) The quadruple-tank process: a multivariable laboratory process with an adjustable zero. IEEE Trans Control Syst Technol 8(3):456–465. https://doi.org/10.1109/87.845876

20. Raff T, Huber S, Nagy ZK, Allgöwer F, (2006) Nonlinear model predictive control of a four tank system: an experimental stability study. In: IEEE conference on computer aided control system design, 2006 IEEE international conference on control applications, 2006 IEEE international symposium on intelligent control, pp 237–242. IEEE. https://doi.org/10.1109/CACSD-CCA-ISIC.2006.4776652
21. Khalil HK (2002) Nonlinear systems. Prentice Hall

# Continuous Time Nonlinear Controller Design with Increased Terminal Region

**Chinmay Rajhans and Sowmya Gupta**

## 1 Introduction

One of the most promising optimization-based controller is the Model Predictive Control (MPC) [1]. MPC finds applications in every field of science, engineering and technology [2–4]. Commonly used stabilizing constraints include (a) Terminal equality constraint, (b) Terminal penalty term, and (c) Terminal inequality constraint [5]. An enormous process has taken place in the field of nominal stability of linear MPC [6] and Nonlinear MPC (NMPC) [7–9].

Executing terminal equality constraint is convenient [10], however main limitation is that it is highly conservative and often leads to infeasibility specifically when using constrained formulations. Michalska and Mayne conceptualized dual mode MPC scheme where in the idea of terminal region was introduced [11]. NMPC controller is expected to drive the plant trajectory into a region, termed as terminal region, around the set point using a set of feasible inputs in a finite horizon time. Subsequently localized linear controller will take the system trajectory to the set point. This idea was extended by Chen and Allgöwer where in NMPC control acted inside of the terminal region instead of using a linear control law, which has resulted in the concept of Quasi Infinite Horizon-Nonlinear Model Predictive Control (QIH-NMPC) formulation [12].

Chen and Allgöwer presented one approach for terminal region characterization for the case of continuous time MPC formulation [12]. The approach to characterize the terminal region using an inequality-based conditions was presented. First major drawback of their approach is a tuning parameter which is nearly independent of the NMPC formulation stage weighting matrices. Second limitation of Chen and Allgöwer's approach is that, it provides a scalar tuning parameter which restricts the design to a single scalar for increasing the terminal region. Several researchers have

C. Rajhans (✉) · S. Gupta
VJTI, Mumbai, India
e-mail: rajhanschinmay2@gmail.com

 
P. H. J. Chong et al. (eds.), *Emerging Electronics and Automation*, Lecture Notes in Electrical Engineering 937, https://doi.org/10.1007/978-981-19-4300-3_4

developed terminal region characterization approaches for the MPC formulations (discrete time) [13–16]. However, they cannot be applied to the continuous time MPC formulations directly due to the involvement of sampling time [17].

Current work proposed an arbitrary controller-based approach for the terminal region characterization for the continuous time NMPC formulation. Approach provides two degrees of freedom in the form of additive weighting matrices as tuning parameters for shaping the terminal region.

Proposed arbitrary controller-based approach is demonstrated using a complex mathematical example [18]. It can be observed later that the terminal regions obtained using the proposed arbitrary controller-based approach is larger than the terminal region obtained using the approach from the literature. This results in larger number of initial conditions becoming feasible with prediction horizon time of one unit of sampling time which were otherwise infeasible.

Second section presents the continuous time NMPC formulation in detail. In addition, approach by Chen and Allgöwer [12] is stated formally along with its limitation. Third section presents the proposed arbitrary controller-based approach for the terminal region characterization. Forth section gives numerical characterization steps using the proposed approach. Fifth section presents the terminal region characterization using demonstration case study. Sixth section details case study results. Seventh section gives the conclusions.

## 2 Continuous Time NMPC Formulation

Consider nonlinear system (continuous time) given by

$$\frac{d\mathbf{x}(t)}{dt} = \mathbf{f}(\mathbf{x}(t), \mathbf{u}(t)) \tag{1}$$

$$\mathbf{x}(0) = \mathbf{x}_0 \tag{2}$$

where $\mathbf{x}(t) \in \mathcal{X} \subset \mathbb{R}^{n_x}$: the state and $\mathbf{u}(t) \in \mathcal{U} \subset \mathbb{R}^{n_u}$: the input. Assumptions are stated as follows:

**C1** System dynamics $\mathbf{f}: \mathbb{R}^{n_x} \times \mathbb{R}^{n_u} \rightarrow \mathbb{R}^{n_x}$ is a twice continuously differentiable function.

**C2** The origin $\mathbf{0} \in \mathbb{R}^{n_x}$ is equilibrium point i.e. $\mathbf{f}(\mathbf{0}, \mathbf{0}) = \mathbf{0}$.

**C3** The inputs $\mathbf{u}(t)$ are inside a closed and convex set $\mathcal{U} \subset \mathbb{R}^{n_u}$.

**C4** The system (1) has a unique solution for any $\mathbf{x}_0 \in \mathcal{X}$ and any piece wise right continuous input signal.

**C5** $\mathbf{x}(t)$ is measured at any time $t$.

**C6** External disturbances do not affect the system dynamics.

## 2.1 NMPC Formulation

NMPC formulation for the continuous time system (1) is stated as:

$$\min_{\overline{\mathbf{u}}_{[t,t+T_p]}} J\left(\mathbf{x}(t), \overline{\mathbf{u}}_{[t,t+T_p]}\right) \tag{3}$$

with

$$J\left(\mathbf{x}(t), \overline{\mathbf{u}}_{[t,t+T_p]}\right) = \int_t^{t+T_p} \left\{\mathbf{z}(\tau)^T \mathbf{W}_x \mathbf{z}(\tau) + \overline{\mathbf{u}}(\tau)^T \mathbf{W}_u \overline{\mathbf{u}}(\tau)\right\} \mathrm{d}\tau + \mathbf{z}(t+T_p)^T \mathbf{P} \mathbf{z}(t+T_p) \tag{4}$$

$$\overline{\mathbf{u}}_{[t,t+T_p]} = \left\{\mathbf{u}(\tau) \in \mathcal{U} : \tau \in \left[t, t+T_p\right]\right\} \tag{5}$$

subject to

$$\frac{\mathrm{d}\mathbf{z}(\tau)}{\mathrm{d}\tau} = \mathbf{f}\left(\mathbf{z}(\tau), \overline{\mathbf{u}}(\tau)\right) \text{ for } \tau \in \left[t, t+T_p\right] \tag{6}$$

$$\mathbf{z}(t) = \mathbf{x}(t) \tag{7}$$

$$\mathbf{z}\left(t+T_p\right) \in \Omega \tag{8}$$

where $\mathbf{W}_x$ and $\mathbf{W}_u$ are state and input weighting matrices of dimension $(n_x \times n_x)$, $(n_u \times n_u)$ respectively. $\mathbf{P}$ is the terminal penalty matrix of dimension $(n_x \times n_x)$. $\mathbf{W}_x, \mathbf{W}_u, \mathbf{P}$ are symmetric positive definite matrices. $T_p$ is a finite prediction horizon time and is identical to the control horizon time. $\mathbf{z}(\tau)$ denotes the predicted state of the MPC formulation and $\overline{\mathbf{u}}(\tau)$ denotes the future input moves.

## 2.2 Design of MPC Controller

The terminal region is designed as invariant set for the system (1) under the linear control law with gain $\mathbf{K}$. Terminal penalty term is computed using the following condition: The terminal penalty term is computed such that single term having larger value than that of the sum of all the predicted stage cost terms from end of horizon time to infinity.

$$\mathbf{z}(t+T_p)^T \mathbf{P} \mathbf{z}(t+T_p) \geq \int_{t+T_p}^{\infty} \left\{\mathbf{z}(\tau)^T \mathbf{W}_x \mathbf{z}(\tau) + \overline{\mathbf{u}}(\tau) \mathbf{W}_u \overline{\mathbf{u}}(\tau)\right\} \mathrm{d}\tau \tag{9}$$

with $\overline{\mathbf{u}}(\tau) = -\mathbf{K}\mathbf{z}(\tau) \in \mathcal{U}$ for all $\tau \geq (t + T_p)$ and for all $\mathbf{z}(t + T_p) \in \Omega$.

Solution to the standard optimal control problem (3) can be computed. Controller is designed using moving horizon framework. Accordingly, only the control move at time $t$ is implemented as,

$$\mathbf{u}(t) = \overline{\mathbf{u}}^*(t) \tag{10}$$

Whole process is repeated at next time point $t + \delta$ with $\delta$ being the sampling time.

## 2.3 Chen and Allgöwer's Approach

Before proceeding to the proposed arbitrary controller-based approach, a look at Chen and Allgöwer's approach is required. Jacobian linearization around the origin as,

$$\frac{\mathrm{d}\mathbf{x}(t)}{\mathrm{d}t} = \mathbf{A}\mathbf{x}(t) + \mathbf{B}\mathbf{u}(t) \tag{11}$$

where

$$\mathbf{A} = \left[\frac{\partial \mathbf{f}}{\partial \mathbf{x}}\right]_{(\mathbf{0},\mathbf{0})} \text{ and } \mathbf{B} = \left[\frac{\partial \mathbf{f}}{\partial \mathbf{u}}\right]_{(\mathbf{0},\mathbf{0})}$$

One additional assumption is required at this stage.

**C7** The linearized system (11) is stabilizable.

Terminal region characterization approach by Chen and Allgöwer is stated as follows:

$$\Omega \equiv \left\{\mathbf{x} \in \mathbb{R}^n | \mathbf{x}^T \mathbf{P}_{\mathrm{CA}} \mathbf{x} \leq \alpha,\ -\mathbf{K}_{\mathrm{CA}} \mathbf{x} \in \mathcal{U}\right\} \tag{12}$$

where linear gain $\mathbf{K}_{\mathrm{CA}}$ is the solution of the following modified Lyapunov equations:

$$\mathbf{A}_{K_{\mathrm{CA}}}^T \mathbf{P} + \mathbf{P}\mathbf{A}_{K_{\mathrm{CA}}} = -\left(\mathbf{W}_x + \mathbf{K}_{\mathrm{CA}}^T \mathbf{W}_u \mathbf{K}_{\mathrm{CA}}\right) \tag{13}$$

$$\mathbf{K}_{\mathrm{CA}} = (\mathbf{W}_u)^{-1} \mathbf{B}^T \mathbf{P} \tag{14}$$

After $\mathbf{K}_{\mathrm{CA}}$ is available, $\mathbf{A}_{K_{\mathrm{CA}}} = \mathbf{A} - \mathbf{B}\mathbf{K}_{\mathrm{CA}}$ is computed and subsequently the terminal penalty $\mathbf{P}_{\mathrm{CA}}$ becomes the steady-state solution of the modified Lyapunov equation given as follows:

$$\left(\mathbf{A}_{K_{\mathrm{CA}}} + \kappa\mathbf{I}\right)^T \mathbf{P}_{\mathrm{CA}} + \mathbf{P}_{\mathrm{CA}} \left(\mathbf{A}_{K_{\mathrm{CA}}} + \kappa\mathbf{I}\right) = -\mathbf{Q}^* \tag{15}$$

where

$$\mathbf{Q}^* = \mathbf{W}_x + \mathbf{K}_{\mathrm{CA}}^T \mathbf{W}_u \mathbf{K}_{\mathrm{CA}} \tag{16}$$

Parameter $\kappa > 0$ is chosen such that $\kappa < -\text{Re}[\lambda_{\max}(\mathbf{A} - \mathbf{BK}_{\text{CA}})]$. Note that the expression $Re[\lambda_{\max}(\mathbf{A} - \mathbf{BK}_{\text{CA}})]$ is the real part of the right most eigenvalue of $\mathbf{A}_{K_{\text{CA}}}$ i.e. eigen value having largest real part and it is negative due to the fact that linear matrix $\mathbf{A}_{K_{\text{CA}}}$ is stable by design. It can be noted that once stage cost weighting matrices $\mathbf{W}_x$, $\mathbf{W}_u$ are chosen, there is barely any degree of freedom left to the designer for shaping of the terminal region. This results in a very conservative terminal region.

## 3 Alternate Approach for the Terminal Region Characterization

Arbitrary controller-based approach makes use of an arbitrary linear controller, which is typically designed using one of the methods such as Linear Quadratic Gaussian Control (LQGC) [19, 20], pole placement [21]. We prove the following theorem for arbitrary controller-based approach:

**Theorem 1** *Let the assumptions C1–C7 hold true and a stable controller is designed, i.e., matrix* $\mathbf{A}_K = (\mathbf{A} - \mathbf{BK})$ *is stable. Let* $\Delta\mathbf{Q}$ *be a positive definite matrix. Let* $\mathbf{P}$ *is the steady-state solution of the following Lyapunov equation:*

$$\mathbf{A}_K^T\mathbf{P} + \mathbf{P}\mathbf{A}_K = -(\mathbf{Q}^* + \Delta\mathbf{Q}) \tag{17}$$

*where* $\mathbf{Q}^*$ *is given by (16). Then there exists constant* $\alpha > 0$ *defining an closed region of the form*

$$\Omega \equiv \left\{\mathbf{x} \in \mathbb{R}^{n_x} | \mathbf{x}^T\mathbf{P}\mathbf{x} \le \alpha, -\mathbf{K}\mathbf{x} \in \mathcal{U}\right\} \tag{18}$$

*such that the region* $\Omega$ *is invariant for the system with controller* $\mathbf{u}(t) = -\mathbf{K}\mathbf{x}(t)$. *Additionally, for any* $\mathbf{x}(t + T_p) \in \Omega$ *the inequality given by (19) holds true.*

$$\mathbf{z}(t + T_p)^T\mathbf{P}\mathbf{z}(t + T_p) \ge \int_{t+T_p}^{\infty} \left\{\mathbf{z}(\tau)^T\mathbf{W}_x\mathbf{z}(\tau) + \bar{\mathbf{u}}(\tau)\mathbf{W}_u\bar{\mathbf{u}}(\tau)\right\} d\tau \tag{19}$$

*Proof* Since $\mathbf{A}_K = (\mathbf{A} - \mathbf{BK})$ is stable, hence, the eigenvalues of $\mathbf{A}_K$ are having negative real part. Using the solvability necessary condition of the Lyapunov equation, unique solution $\mathbf{P} > 0$ can be calculated which solves (17). According to assumption C2, the origin $\mathbf{0} \in \mathbb{R}^{n_u}$ is inside the compact set $\mathcal{U}$. Subsequently, one can compute $\gamma$ which defines the upper bound set $\Omega_\gamma$ as

$$\Omega_\gamma \equiv \left\{\mathbf{x} \in \mathbb{R}^{n_x} | \mathbf{x}^T\mathbf{P}\mathbf{x} \le \gamma, -\mathbf{K}\mathbf{x} \in \mathcal{U}\right\} \tag{20}$$

Now, let $0 < \alpha \le \gamma$ specify a region of the form given by Eq. (21).

$$\Omega \equiv \left\{\mathbf{x} \in \mathbb{R}^{n_x} | \mathbf{x}^T\mathbf{P}\mathbf{x} \le \alpha\right\} \tag{21}$$

As the input constraints are satisfied in $\Omega_\gamma$ and $\Omega \subseteq \Omega_\gamma$ (by virtue of $0 < \alpha \leq \gamma$), the system dynamics can be equivalently viewed as an input unconstrained system in the set $\Omega$. Consider a vector $\mathbf{\Phi}_K(\mathbf{x})$ representing the nonlinearity in the system dynamics defined as

$$\mathbf{\Phi}_K(\mathbf{x}) = \mathbf{f}(\mathbf{x}, -\mathbf{K}\mathbf{x}) - \mathbf{A}_K\mathbf{x} \tag{22}$$

Note for a linear system $\mathbf{\Phi}_K(\mathbf{x}) = \mathbf{0}$. Consider a Lyapunov candidate defined as

$$V(\mathbf{x}) = \mathbf{x}^T\mathbf{P}\mathbf{x} \tag{23}$$

The time derivative of $V(\mathbf{x})$ can be expressed as follows:

$$\frac{\mathrm{d}V(\mathbf{x})}{\mathrm{d}t} = \frac{\mathrm{d}\mathbf{x}^T}{\mathrm{d}t}\mathbf{P}\mathbf{x} + \mathbf{x}^T\mathbf{P}\frac{d\mathbf{x}}{\mathrm{d}t} \tag{24}$$

Substituting from (22) into (24),

$$\frac{\mathrm{d}V(\mathbf{x})}{\mathrm{d}t} = \mathbf{x}^T\left(\mathbf{A}_K^T\mathbf{P} + \mathbf{P}\mathbf{A}_K\right)\mathbf{x} + 2\mathbf{x}^T\mathbf{P}\mathbf{\Phi}_K(\mathbf{x}) \tag{25}$$

Using Eqs. (17) into (25),

$$\frac{\mathrm{d}V(\mathbf{x})}{\mathrm{d}t} = -\mathbf{x}^T\left(\mathbf{Q}^* + \Delta\mathbf{Q}\right)\mathbf{x} + 2\mathbf{x}^T\mathbf{P}\mathbf{\Phi}_K(\mathbf{x}) \tag{26}$$

Rearranging results in the following equation:

$$\frac{\mathrm{d}V(\mathbf{x})}{\mathrm{d}t} = -\mathbf{x}^T\mathbf{Q}^*\mathbf{x} + \left(-\mathbf{x}^T\Delta\mathbf{Q}\mathbf{x} + 2\mathbf{x}^T\mathbf{P}\mathbf{\Phi}_K(\mathbf{x})\right) \tag{27}$$

Rearranging terms from the equation (26),

$$\frac{dV(\mathbf{x})}{dt} = -\mathbf{x}^T\mathbf{Q}^*\mathbf{x} + \left(-\mathbf{x}^T\Delta\mathbf{Q}\mathbf{x} + 2\mathbf{x}^T\mathbf{P}\mathbf{\Phi}_K(\mathbf{x})\right) \tag{28}$$

Consider second term of the expression (28),

$$\mathbf{\Psi}(\mathbf{x}) := \left(\mathbf{x}^T\Delta\mathbf{Q}\mathbf{x} - 2\mathbf{x}^T\mathbf{P}\mathbf{\Phi}_K(\mathbf{x})\right) \tag{29}$$

Using (29) in (28),

$$\frac{\mathrm{d}V(\mathbf{x})}{\mathrm{d}t} = -\mathbf{x}^T\mathbf{Q}^*\mathbf{x} - \mathbf{\Psi}(\mathbf{x}) \tag{30}$$

If $\Omega$ is chosen such that

$$\boldsymbol{\Psi}(\mathbf{x}) = \left(\mathbf{x}^T \Delta\mathbf{Q}\mathbf{x} - 2\mathbf{x}^T \mathbf{P}\boldsymbol{\Phi}_K(\mathbf{x})\right) \geq 0 \tag{31}$$

then

$$\frac{\mathrm{d}V(\mathbf{x})}{\mathrm{d}t} \leq -\mathbf{x}^T \mathbf{Q}^* \mathbf{x} \tag{32}$$

Integrating inequality (32) over the interval, $[t + T_p, \infty)$, it follows that

$$V(\mathbf{x}(t + T_p)) \geq \int_{t+T_p}^{\infty} \mathbf{x}(\tau)^T \mathbf{Q}^* \mathbf{x}(\tau) \mathrm{d}\tau \tag{33}$$

i.e. inequality (19) holds true for any $\mathbf{x}(t + T_p) \in \Omega$.

Note that the two tuning matrices available to the controller designers are the linear gain matrix $\mathbf{K}$ and the additive positive definite matrix $\Delta\mathbf{Q}$, which provide large degrees of freedom for enlarging the terminal region. Since the linear gain matrix $\mathbf{K}$ is chosen arbitrarily, hence the name arbitrary controller-based approach.

## 4 Terminal Region Characterization

Theorem 1 gave conditions for explicit characterization of the terminal region. It is possible to numerically compute the terminal region and subsequently implement the NMPC controller. Steps for characterization of the terminal region using arbitrary controller-based approach are given below:

**S1** Computation of Upper Bound Set:
Compute the largest value of $\gamma$ such that inputs constraints are satisfied in the set $\Omega_\gamma$.

$$\Omega_\gamma \equiv \left\{\mathbf{x} \in \mathbb{R}^{n_x} | \mathbf{x}^T \mathbf{P}\mathbf{x} \leq \gamma, -\mathbf{K}\mathbf{x} \in \mathcal{U}\right\} \tag{34}$$

**S2** Computation of the Terminal Region using inequality-based method:
Compute the largest $\alpha \in (0, \gamma]$ such that

$$\left[\begin{matrix}\min \\ \mathbf{x} \in \Omega\end{matrix} \boldsymbol{\Psi}(\mathbf{x})\right] = 0 \tag{35}$$

The condition given by (35) indicates that minimum of the function $\boldsymbol{\Psi}(\mathbf{x})$ for all states inside the terminal region $\Omega$ is 0 i.e. $\boldsymbol{\Psi}(\mathbf{x})$ is non-negative for all $\mathbf{x} \in \Omega$, which is the necessary condition for nominal asymptotic stability.

Terminal region shape changes according to the computed **P** matrix and its size changes according to the value of $\alpha$. In order to compare the size of the terminal regions, area is computed for state dimension of 2 as

$$A_2 = \frac{\pi\alpha}{\sqrt{\det(\mathbf{P})}} \tag{36}$$

## 5 Demonstration Case Study

Consider a complex second order mathematical example presented in [18]. Equations for the system dynamics are given as follows:

$$\dot{x}_1 = -20x_1 + 10x_2 + 6u + 10x_1^2u \tag{37}$$

$$\dot{x}_2 = 20x_1 - 20x_2 + 6u - 40x_2^2u \tag{38}$$

The terms $x_1^2$ and $x_2^2$ represent the system nonlinearity.

The terminal region is calculated in the neighborhood of the following point:

$$\mathbf{X}_s = \begin{bmatrix} 0 \\ 0 \end{bmatrix} \text{ and } \mathbf{U}_s = \begin{bmatrix} 0 \end{bmatrix} \tag{39}$$

Stage weighting matrices for state and input are chosen as follows:

$$\mathbf{W}_x = \begin{bmatrix} 0.5 & 0 \\ 0 & 0.5 \end{bmatrix} \text{ and } \mathbf{W}_u = \begin{bmatrix} 0.5 \end{bmatrix} \tag{40}$$

The input constraint is stated as follows:

$$\mathcal{U} = \{u \in \mathbb{R} | -0.1 \leq u \leq 0.1\} \tag{41}$$

Linear gain for the approach by Chen and Allgöwer [12] is

$$\mathbf{K}_{\mathrm{CA}} = \begin{bmatrix} 0.4203 \; 0.3538 \end{bmatrix} \tag{42}$$

Tuning parameter is chosen as $\kappa = -0.95 \times \mathrm{Re}[\lambda_{\max}(\mathbf{A} - \mathbf{B}\mathbf{K}_{\mathrm{CA}})] = 0.95 \times 10.5866 = 10.0573$. Terminal region computed using approach given by Chen and Allgöwer [12] is $\Omega_{\mathrm{CA}} = \{\mathbf{x}: \mathbf{x}^T\mathbf{P}_{\mathrm{CA}}\mathbf{x} \leq 0.0039\}$ with an area of 0.1418 where

$$\mathbf{P}_{\mathrm{CA}} = \begin{bmatrix} 0.2281 & 0.3236 \\ 0.3236 & 0.4920 \end{bmatrix} \tag{43}$$

To simplify the numerical computation of the terminal region using the arbitrary controller-based approach, following parameterization is carried out:

$$\widetilde{\mathbf{W}}_x = \rho_x \mathbf{W}_x \text{ and } \widetilde{\mathbf{W}}_u = \rho_u \mathbf{W}_u \tag{44}$$

with $\rho_x > 0$ and $\rho_u > 0$ being the tuning parameters.

Linear gain for the arbitrary controller-based approach is

$$\mathbf{K}_{\mathrm{AC}} = \begin{bmatrix} 0.4203 & 0.3538 \end{bmatrix} \tag{45}$$

Note $\mathbf{K}_{\mathrm{AC}} = \mathbf{K}_{\mathrm{CA}}$ because of the design procedure. Terminal region obtained using the arbitrary controller-based approach with tuning parameters $\rho_x = 0.1$ and $\rho_u = 50$ is $\Omega_{\mathrm{AC}} = \left\{\mathbf{x}: \mathbf{x}^T \mathbf{P}_{\mathrm{AC}} \mathbf{x} \leq 0.0070\right\}$ with an area of 0.3301 where

$$\mathbf{P}_{\mathrm{AC}} = \begin{bmatrix} 0.1817 & 0.1987 \\ 0.1987 & 0.2415 \end{bmatrix} \tag{46}$$

It can be observed that the terminal region obtained using the proposed arbitrary controller-based approach is approximately 2.32 larger by area when compared to the approach from the literature, i.e., approach by by Chen and Allgöwer [12].

Figure 1 presents a graphical comparison of the terminal regions obtained using the literature approach, i.e., approach by by Chen and Allgöwer [12] $\Omega_{\mathrm{CA}}$ (shown by blue color) and proposed arbitrary controller-based approach $\Omega_{\mathrm{AC}}$ (shown by magenta color). It can be observed that the terminal region $\Omega_{\mathrm{CA}} \subset \Omega_{\mathrm{AC}}$ for the current example.

Figure 1 also indicates that there are certain initial conditions which are feasible using the arbitrary controller-based approach and are infeasible using the approach by Chen and Allgöwer [12], for example, points $A$ and $B$ with prediction and control horizon time $T_p$ equal to sampling time $\delta$.

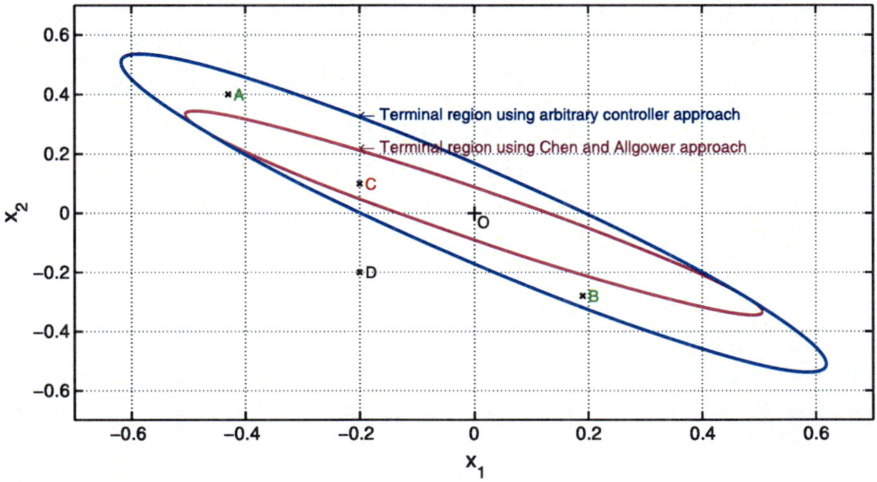


**Fig. 1** Graphical comparison of the terminal regions

**Table 1** Feasibility of the initial conditions using both the approaches

| Point | Chen and Allgöwer [12] | Arbitrary controller based approach |
|---|---|---|
| $A$ | Infeasible | Feasible |
| $B$ | Infeasible | Feasible |
| $C$ | Feasible | Feasible |
| $D$ | Infeasible | Infeasible |

**Table 2** Minimum prediction horizon time required for feasibility of the initial conditions using both the approaches

| Point | Chen and Allgöwer [12] | Arbitrary controller based approach |
|---|---|---|
| $A$ | $T_p > \delta$ | $T_p = \delta$ |
| $B$ | $T_p > \delta$ | $T_p = \delta$ |
| $C$ | $T_p = \delta$ | $T_p = \delta$ |
| $D$ | $T_p > \delta$ | $T_p > \delta$ |

Table 1 presents a summary of the feasibility of the various initial conditions using both the approaches with $T_p = \delta$.

It is well known that smaller the prediction and control horizon time, smaller is the computation time required for MPC optimization routine [22]. Thus, having increased terminal regions is desired for any system. Table 2 presents minimum prediction and control time $T_p$ required for feasibility of the various initial conditions using both the approaches. It may be noticed that for large number of initial conditions such as points $A$ or $B$ i.e., points where $\mathbf{x}(0) \in \Omega_{\text{AC}}$ and $\mathbf{x}(0) \notin \Omega_{\text{CA}}$, prediction horizon time is greatly reduced which results in reduction of the computation time per iteration for the MPC optimization routine. Current example efficiently demonstrates the utility of the proposed approach having larger degrees of freedom for increasing the terminal region.

## 6 Conclusions

Approaches available in the literature for the terminal region characterization for the MPC formulations (continuous time) provide limited degrees of freedom and often result in a conservative terminal regions. An arbitrary controller-based approach is proposed in the current work which exhibits a large degrees of freedom for increasing the terminal region.

Effectiveness of the proposed approach is simulated using a second order complex mathematical system. In the case study, it is observed that terminal region area obtained using the proposed arbitrary controller-based approach is approximately

2.32 times larger by area as compared to the largest terminal region obtained using Chen and Allgöwer's approach from [12]. Future work would involve extending the result for the unequal prediction and control horizon times.

## References


1. Lee JH (2011) Model predictive control: review of the three decades of development. Int J Control Autom Syst 9(3):415–424. https://doi.org/10.1007/s12555-011-0300-6, https://www.springer.com/12555
2. Mayne D, Michalska H (1990) Receding horizon control of nonlinear systems. IEEE Trans Autom Control 35(7):814–824. https://doi.org/10.1109/9.57020
3. Qin SJ, Badgwe-ll TA (2003) A survey of industrial model predictive control technology. Control Eng Pract 11(7):733–764
4. Camacho EF, Bordons CC (2007) Model predictive control. Springer
5. Rawlings J, Mayne D (2009) Model predictive control: theory, computation, and design. Nob Hill Publishing
6. Muske KR, Rawlings JB (1993) Model predictive control with linear models. AIChE J 39(2):262–287. https://doi.org/10.1002/AIC.690390208, https://aiche.onlinelibrary.wiley.com/doi/full/10.1002/aic.690390208
7. Chen WH (2010) Stability analysis of classic finite horizon model predictive control. Int J Control Autom Syst 8(2):187–197. https://doi.org/10.1007/s12555-010-0202-z
8. Mayne DQ (2014) Model predictive control: recent developments and future promise. Automatica 50(12):2967–2986
9. Rawlings JB, Mayne DQ, Diehl M (2017) Model predictive control: theory, computation, and design, 2nd edn. Nob Hill Publishing, LLC
10. Keerthi SS, Gilbert EG (1988) Optimal infinite-horizon feedback laws for a general class of constrained discrete-time systems: stability and moving-horizon approximations. J Optim Theory Appl 57(2):265–293
11. Michalska H, Mayne D (1993) Robust receding horizon control of constrained nonlinear systems. IEEE Trans Autom Control 38(11):1623–1633
12. Chen H, Allgöwer F (1998) A quasi-infinite horizon nonlinear model predictive control scheme with guaranteed stability. Automatica 34(10):1205–1217
13. Johansen TA (2004) Approximate explicit receding horizon control of constrained nonlinear systems. Automatica 40(2):293–300
14. Yu S, Qu T, Xu F, Chen H, Hu Y (2017) Stability of finite horizon model predictive control with incremental input constraints. Automatica 79:265–272. https://doi.org/10.1016/J.AUTOMATICA.2017.01.040
15. Rajhans C, Griffith D, Patwardhan SC, Biegler LT, Pillai H (2018) Two approaches for terminal region characterization in discrete time quasi-infinite horizon NMPC
16. Rajhans C, Griffith DW, Patwardhan SC, Biegler LT, Pillai HK (2019) Terminal region characterization and stability analysis of discrete time quasi-infinite horizon nonlinear model predictive control. J Process Control 83:30–52. https://doi.org/10.1016/j.jprocont.2019.08.002
17. Astrom KJ, Wittenmark B (1997) Computer-controlled systems: theory and design. Prentice Hall
18. Fitri IR, Kim JS (2020) A nonlinear model predictive control with enlarged region of attraction via the union of invariant sets. Mathematics 8(11):2087. https://doi.org/10.3390/MATH8112087, https://www.mdpi.com/2227-7390/8/11/2087
19. Kirk DE (1970) Optimal control theory an introduction. Prentice-Hall Inc., Englewood Cliffs, New Jersey

20. Albertos P, Antonio S (2006) Multivariable control systems: an engineering approach. In: Advanced textbooks in control and signal processing. Springer London. https://books.google.co.in/books?id=mCEGCAAAQBAJ
21. Kailath T, Hall P (1980) Linear systems. Information and system sciences series. Prentice-Hall. https://books.google.co.in/books?id=ggYqAQAAMAAJ
22. Mayne DQ, Rawlings JB, Rao CV, Scokaert POM (2000) Constrained model predictive control: stability and optimality. Automatica 36(6):789–814

# Cascade Controller Design Based on Pole Placement and Model Matching Technique

**Md. Atif Siddiqui, S. H. Laskar, M. N. Anwar, and Abhishek Yadav**

## 1 Introduction

Franks and Worley [1] familiarize the cascade control structure (CCS) in 1956, and since then, it is commonly utilized in the chemical industry. Unlike conventional feed-back control, in the CCS, the corrective action begins before the final output diverges from the target. The CCS has two loops, one is inner or secondary loop, and the other is outer or primary loop. Lee et al. [2] came up with a design approach for controlling stable processes in CCS. In their method, the desired response is first obtained by an ideal controller, and then, the ideal controller is approximated into PID controller. Later on, Lee et al. [3] utilized two controllers and two filters in a CCS to handle the processes having unstable, integrating, and stable dynamics. Both the aforementioned approaches utilize the internal model control (IMC) scheme that is based first-order approximation of the actual process, i.e. order specific.

Complex structures are also used in CCS to control various processes. Kaya [4] suggested the application of the conventional Smith predictor (SP) configuration in the primary loop of CCS, thereby making the CCS a complex structure. Later on, Kaya and Atherton [5] suggested improvement in CCS by the use of an IMC scheme

Md. Atif Siddiqui (✉)
Department of EE, Integral University, Lucknow, U.P, India
e-mail: atifsiddiqui@iul.ac.in

S. H. Laskar
Department of EIE, NIT Silchar, Silchar, Assam, India
e-mail: shlaskar@ei.nits.ac.in

M. N. Anwar
Department of EE, NIT Patna, Patna, Bihar, India
e-mail: nishat@nitp.ac.in

A. Yadav
IEEE Member, Meerut, India
e-mail: shlaskar@ei.nits.ac.in

 
P. H. J. Chong et al. (eds.), *Emerging Electronics and Automation*, Lecture Notes in Electrical Engineering 937, https://doi.org/10.1007/978-981-19-4300-3_5

in the inner loop and SP scheme with PI-PD configuration in the outer loop. Padhan and Majhi [6] also utilized a modified SP cascade structure where the controllers of the both the loops are designed based on the IMC approach and the direct synthesis (DS) approach. A complex delay compensator is used by Çakıroğlu [7] in both the loops (primary and secondary) to achieve advancement in closed-loop performance as compared to [6]. Raja and Ali [8] proposed a customized series CCS having three controllers for the integrating process. Dasari et al. [9] designed the controllers of the complex CCS based on the IMC approach with optimal minimization to achieve notable disturbance rejection. In the process industry, a cascade structure (not a complex one) with less number of filter/controller is usually preferred.

Literature reveals that many researchers have made practical and effective application of the direct synthesis method (DSM) for processes with stable dynamics [10–12]. The DSM is established on the desired closed-loop transfer function (DE-CLTF) to get the required load disturbance or setpoint response. Another popular approach to tune the controllers is the pole placement method, where only poles are placed and zeros are left undetermined [13, 14]. Hence, in this paper, the CCS controllers in form of PID controllers are tuned based on the technique of matching the model in the frequency domain and placing the pole at the anticipated position to achieve improved performance. The secondary controller is designed by performing (i) matching between the DE-CLTF and designed (DG-CLTF) in the frequency domain and (ii) placing the pole at the anticipated position. The same method is also used for calculating the gains of the primary controller by combining the secondary loop and the primary plant. Simulations study indicates the efficacy of the proposed method when compared with the recent literature.

The work is documented as: Sect. 2 defines CLTF for both the loops. The secondary and the primary controller design are summarized in Sects. 3 and 4, respectively. Section 5 narrates guidelines for the time constant followed by Sect. 6, which reports simulation results and Sect. 7 which enlists the conclusion.

## 2 CLTF for Both the Loops

The CCS is given in Fig. 1 where $G_{c1}$, $G_{p1}$, $C_1$, $G_{c2}$, $G_{p2}$, and $C_2$ are the primary controller, outer process, output of outer loop, secondary controller, inner processes, and output of the inner loop, respectively. The output of $G_{c1}$ is $R_2$ which serves as the target value for the inner loop. $R_1$ is the target value for the outer loop. $G_{L1}$ and $G_{L2}$ are disturbance transfer functions. $D_1$ and $D_2$ are the two disturbance points.

The CLTF for the target response of the secondary loop is achieved as

$$\frac{C_2}{R_2} = \frac{G_{c2}G_{p2}}{1 + G_{c2}G_{p2}} \tag{1}$$

The complete CLTF for servo response is obtained as

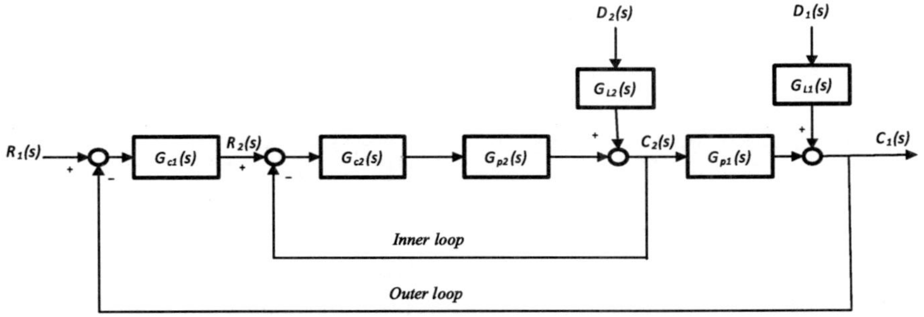


**Fig. 1** CCS block diagram

$$\frac{C_1}{R_1} = \frac{G_{c1}G_{p1}\frac{G_{c2}G_{p2}}{1+G_{c2}G_{p2}}}{1 + G_{c1}G_{p1}\frac{G_{c2}G_{p2}}{1+G_{c2}G_{p2}}} \tag{2}$$

$$\frac{C_1}{R_1} = \frac{G_{c1}G_{p1}\frac{C_2}{R_2}}{1 + G_{c1}G_{p1}\frac{C_2}{R_2}} \tag{3}$$

## 3 Secondary Controller Design Method

$G_{c2}$ is used to suppress effect of disturbances and parametric uncertainty in the secondary system before the controlled variable gets affected.

The $G_{c2}$ is a PID controller here and is given by

$$G_{c2} = K_{p2} + \frac{K_{i2}}{s} + K_{d2}s \tag{4}$$

where $K_{p2}$, $K_{i2,}$ and $K_{d2}$ are the proportional, integral, and derivative gain of the controller, respectively.

The DE-CLTF is selected as

$$\left(\frac{C_2}{R_2}\right)_{\text{Desired}} = \frac{e^{-\theta_2 s}}{\lambda_2 s + 1} \tag{5}$$

Here, the DE-CLTF is in rational polynomial form that contains time delay term $\theta_2$ as it cannot be eliminated by any controller. The DE-CLTF also contains $\lambda_2$ to regulate the speed of the system.

The characteristic equation for the inner loop is

$$1 + G_{c2}G_{p2} = 0 \tag{6}$$

The required transient response may be obtained by placing the closed-loop pole at $s = -1/\lambda_2$. Putting $s = -1/\lambda_2$ in (5) with (4) will yield

$$K_{p2} - K_{i2}\lambda_2 - \frac{K_{d2}}{\lambda_2} = \frac{-1}{G_{p2}(-1/\lambda_2)} = X_1 \tag{7}$$

The desired steady-state performance may be achieved by matching the DE-CLTF and DG-CLTF $\left(\frac{C_2}{R_2}\right)$ at very low-frequency points, say $\omega = 0.001$ rad/sec. The low-frequency point is considered with respect to the bandwidth of the DE-CLTF. For detail description, please refer to [13].

Matching between the DE-CLTF and DG-CLTF is mathematically expressed as

$$\frac{G_{c2}(j\omega)G_{p2}(j\omega)}{1 + G_{p2}(j\omega)G_{c2}(j\omega)} = \left(\frac{C_2}{R_2}(j\omega)\right)_{Desired} \tag{8}$$

Matching done at the lower frequency region say $\omega = 0.001$ rad/sec results in a good approximation of $G_{c2}$ into PID controller. In most methods [2, 3, 9, 17], the Taylor series or Pade approximation is utilized to realize the controller into a PID controller. Thus, the strategy proposed is exempted from order diminution, i.e. higher order into lower order and the rational approximation of dead time.

$$G_{c2}(j\omega) = \frac{\left(\frac{C_2}{R_2}(j\omega)\right)_{\text{Desired}}}{G_{p2}(j\omega)\left(1 - \left(\frac{C_2}{R_2}\right)_{\text{Desired}}\right)} \tag{9}$$

$$K_{p2} + \frac{K_{i2}}{j\omega} + K_{d2}j\omega = \frac{\left(\frac{C_2}{R_2}(j\omega)\right)_{\text{Desired}}}{G_{p2}(j\omega)\left(1 - \left(\frac{C_2}{R_2}\right)_{\text{Desired}}\right)} = A \tag{10}$$

$$j\left(K_{d2}\omega + \frac{-K_{i2}}{\omega}\right) + K_{p2} = \text{imaginary}(A) + \text{real}(A) \tag{11}$$

Separating real and imaginary parts from (11)

$$K_{p2} = \text{real}(A) \tag{12}$$

$$j\left(\frac{-K_{i2}}{\omega} + K_{d2}\omega\right) = \text{imaginary}(A) \tag{13}$$

The Eqs. (7), (12), and (13) may be arranged as

$$\begin{bmatrix} 1 & -\lambda_2 & \frac{-1}{\lambda_2} \\ 1 & 0 & 0 \\ 0 & \frac{-1}{\omega} & \omega \end{bmatrix} \begin{bmatrix} K_{p2} \\ K_{i2} \\ K_{d2} \end{bmatrix} = \begin{bmatrix} X_1 \\ \text{real}(A) \\ \text{imaginary}(A) \end{bmatrix} \tag{14}$$

The parameters of the inner loop PID controller are obtained by solving (14).

## 4 Design of Primary Controller

The inner loop is now combined with the primary plant and then the used for calculating the gains of $G_{c1}$. The CLTF of the outer loop is obtained by (3) where $G_{c1}$ is considered as

$$G_{c1} = K_{p1} + \frac{K_{i1}}{s} + K_{d1}s \tag{15}$$

The characteristic equation is given by

$$1 + G_{c1}G_{p1}\frac{C_2}{R_2} = 0 \tag{16}$$

$$G_{c1}(s) = \frac{-1}{G_{p1}\frac{C_2}{R_2}} \tag{17}$$

Putting $s = -1/\lambda_1$ in (17) yields

$$K_{p1} - K_{i1}\lambda_1 - \frac{K_{d1}}{\lambda_1} = \frac{-1}{G_{p1}\left(\frac{-1}{\lambda_1}\right)\frac{C_2}{R_2}\left(\frac{-1}{\lambda_1}\right)} = X_2 \tag{18}$$

Frequency response matching between $\frac{C_1}{R_1}$ and $\left(\frac{C_1}{R_1}\right)_{\text{Desired}}$ is done. The frequency point chosen is very low, say $\omega = 0.001$ rad/sec to maintain good steady-state performance.

The DE_CLTF for primary loop is considered as

$$\left(\frac{C_1}{R_1}\right)_{\text{Desired}} = \frac{e^{-\theta_1 s}}{\tau_c s + 1},\ \tau_c = \lambda_1 \tag{19}$$

Here, $\theta_1$ and $\lambda_1$ are dead time and tuning parameter, respectively, of DE_CLTF.

$$\left(\frac{C_1}{R_1}\right)_{\text{Desired}} = \frac{\frac{C_2}{R_2}G_{p1}G_{c1}}{1 + \frac{C_2}{R_2}G_{p1}G_{c1}} \tag{20}$$

$$G_{c1}(j\omega) = \frac{\left(\frac{C_1}{R_1}(j\omega)\right)_{\text{Desired}}}{G_{p1}(j\omega)\left(\frac{C_2}{R_2}(j\omega)\right)\left(1 - \left(\frac{C_1}{R_1}(j\omega)\right)_{\text{Desired}}\right)} = B \tag{21}$$

$$K_{p1} + j\left(\frac{-K_{i1}}{\omega} + K_{d1}\omega\right) = \text{real}(B) + \text{imaginary}(B) \tag{22}$$

From (22), separating the real and the imaginary parts

$$K_{p1} = \text{real}(B) \tag{23}$$

$$j\left(\frac{-K_{i1}}{\omega} + K_{d1}\omega\right) = \text{imaginary}(B) \tag{24}$$

Thus, matrix is formed by using the Eqs. (18), (23), and (24)

$$\begin{bmatrix} 1 & -\lambda_1 & \frac{-1}{\lambda_1} \\ 1 & 0 & 0 \\ 0 & \frac{-1}{\omega} & \omega \end{bmatrix} \begin{bmatrix} K_{p1} \\ K_{i1} \\ K_{d1} \end{bmatrix} = \begin{bmatrix} X_2 \\ \text{real}(B) \\ \text{imaginary}(B) \end{bmatrix} \tag{25}$$

The parameters of $G_{c1}$ are thus achieved.

## 5 Guideline for Closed-Loop Time Constant

Closed-loop response speed depends upon $\lambda_1$ and $\lambda_2$. So, $\lambda_1$ and $\lambda_2$ should be carefully adjusted so that the maximum sensitivity comes in the range of 1–2. Maximum sensitivity ($M_s$) is given by

$$M_s = \max_{0<\omega<\infty}\left|\left(1/1 + G_c(j\omega)G_p(j\omega)\right)\right| \tag{26}$$

## 6 Simulation Results

The performance of the proposed scheme is measured by observing integral absolute error (*IAE*) value which is evaluated by

$$\text{IAE} = \int_0^{\infty} |r(t) - c(t)|\text{d}t \tag{27}$$

## 6.1 Example-1

In this example, a first-order plus dead time (FOPDT) process is taken as

$$G_{p1} = \frac{e^{-10s}}{100s+1},\ G_{L1} = \frac{e^{-10s}}{100s+1},\ G_{p2} = \frac{2e^{-2s}}{20s+1},\ G_{L2} = 1 \quad (28)$$

Sadasivarao and Chidambaram [15] have tuned the PID controllers for cascade control systems by applying a genetic algorithm. For comparison, the proposed method is compared with the method of [15] and the method of Krishnaswamy et al. [16]. In Krishnaswamy et al. [16] method, the values of the controller are retrieved by the integral time absolute error method. The gains of the controllers achieved by the proposed scheme [15, 16] are shown in Table 1. Controller designed by the suggested method is found to have maximum sensitivity of 1.66 and 1.64 for inner and outer loop, respectively. The performance of the system is estimated by applying a unit setpoint step change at t = 0 s. The resulting response obtained is shown in Fig. 2. The regulatory performance of the system is evaluated by applying a unit change in the load $D_2$ and $D_1$ at t = 0 s and 150 s, respectively. The resulting results are displayed Fig. 3 which the suggested approach gives superior performance with smaller overshoots and faster disturbance rejection in comparison with [15, 16] under nominal conditions. Assume perturbation of −10% in time constant and + 10% in dead time and gain of both the plants. The response generated under the perturbed condition as shown in Fig. 4 indicates that the proposed scheme and the scheme of Krishnaswamy et al. remain robust whereas the method suggested in [15] becomes unstable. Furthermore, the superiority of the presented approach is also specified by the lesser *IAE* values obtained in comparison with the other two methods as shown in Table 1.

## 6.2 Example-2

Consider a higher plant [17] as

$$G_{p2}(s) = \frac{100}{(s+1)(0.0015s+1)^2},\ G_{p1}(s) = \frac{50}{(0.12s+1)(0.75s+1)} \quad (29)$$
$$G_{L_2} = 1,\ G_{L1} = 1$$

The controller design by Azar and Serrano [17] requires the approximation of the $G_{p1}$ and $G_{p2}$ by using half rule into the second-order and first-order plant model with time delay, respectively. But the presented approach requires no such approximation and is directly applied on the aforementioned plant.

For the purpose of comparison, the setpoint response and the regulatory response are produced by applying a servo change of 1 amplitude at time 0 s and unit load

**Table 1** Comparison of performance

| Method | $G_{c2}$ | | | | $G_{c1}$ | | | | $IAE$ | | |
|---|---|---|---|---|---|---|---|---|---|---|---|
| | $K_{p2}$ | $K_{i2}$ | $K_{d2}$ | $M_s$ | $K_{p1}$ | $K_{i1}$ | $K_{d1}$ | $M_s$ | $D_2$ | $D_1$ | $R_1$ |
| Example-1 | | | | | | | | | | | |
| Proposed | 3.44 | 0.9 | 1.995 | 1.6 | 6.7 | 0.061 | 20.5 | 1.6 | 0.93 | 18.0 | 19.9 |
| | $\lambda_2$=20 | | | | $\lambda_1$=100 | | | | | | |
| Sadasivarao and Chidambaram | 3.99 | 0.8 | 0.136 | 3.1 | 6.97 | 0.058 | 9.25 | 1.6 | 16.9 | 17.1 | 20.1 |
| Krishnaswamy et al. | 2.97 | – | – | 1.6 | 7.3 | 0.036 | – | 1.9 | 4.56 | 35.9 | 40.1 |
| Example-2 | | | | | | | | | | | |
| Proposed | 0.22 | 0.219 | 0.00044 | 1.02 | 0.09 | 0.114 | 0.01 | 1.2 | 0.6 | 0.18 | 0.18 |
| | $\lambda_2$=0.0455 | | | | $\lambda_1$=0.1842 | | | | | | |
| Azar and Serrano | 0.217 | 0.217 | 0.00033 | 1.02 | 0.06 | 0.129 | 0.003 | 8.7 | 7.7 | 0.31 | 0.31 |

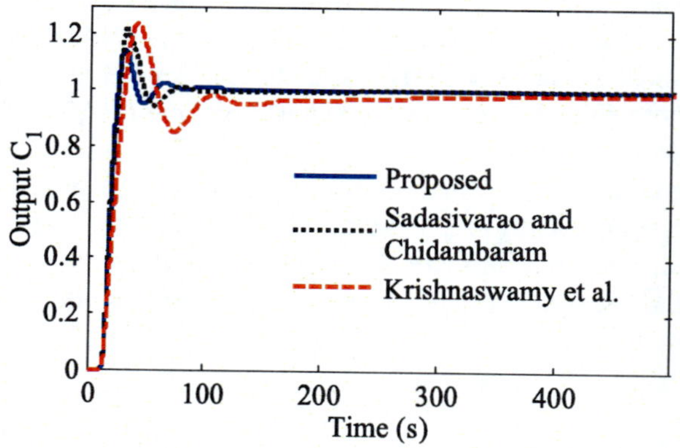


**Fig. 2** Setpoint response for example 1

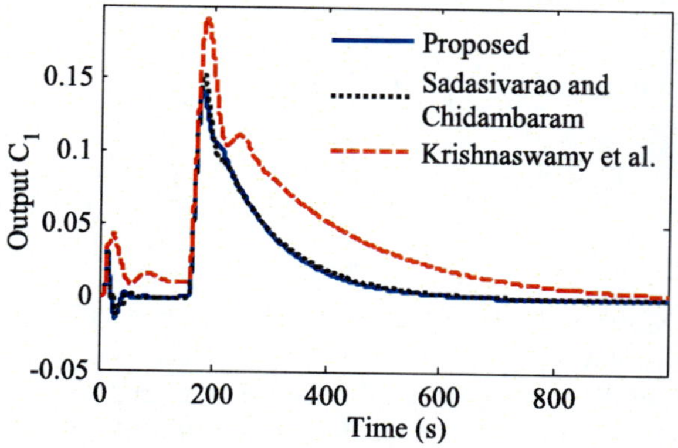


**Fig. 3** Load disturbance response for example 1

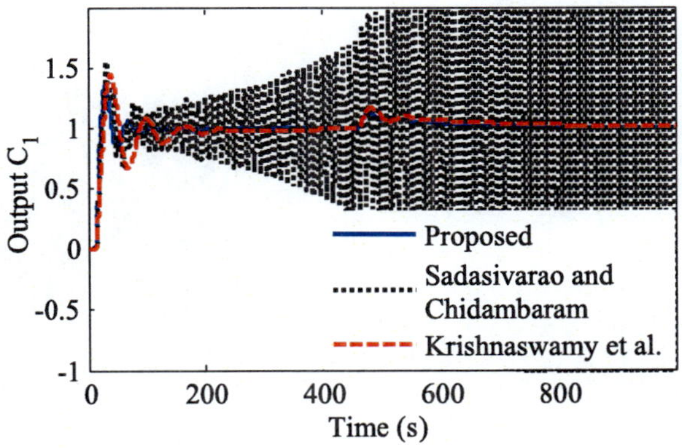


**Fig. 4** Response under perturbed conditions for example 1

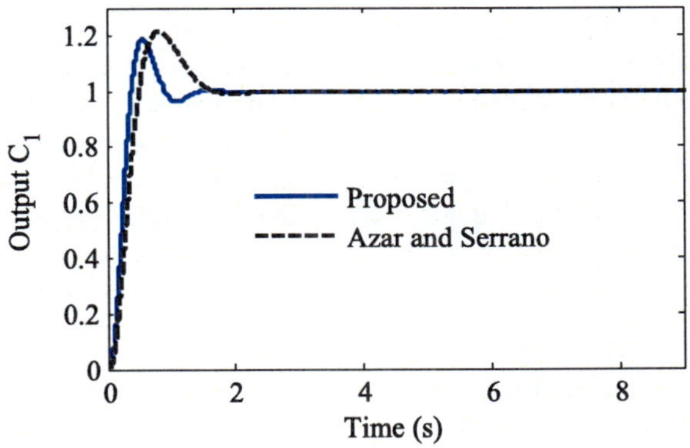


**Fig. 5** Setpoint response for example 2

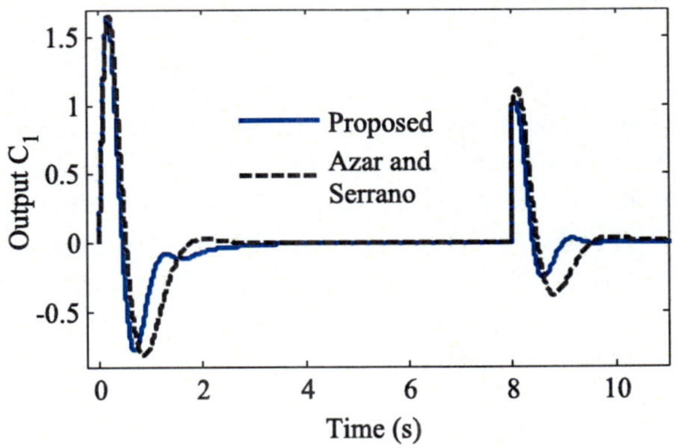


**Fig. 6** Load disturbance response for example 2

disturbances change of amplitude 1 for $D_2$ and $D_1$ at time 0 s and 8 s, respectively. Figures 5 and 6 show the resulting figures. The *IAE* value obtained is shown in Table 1. Results and table indicate that the performance of the suggested approach is better in comparison with that of Azar and Serrano [17]. In the practical world,

**Fig. 7** Noisy response for example 2

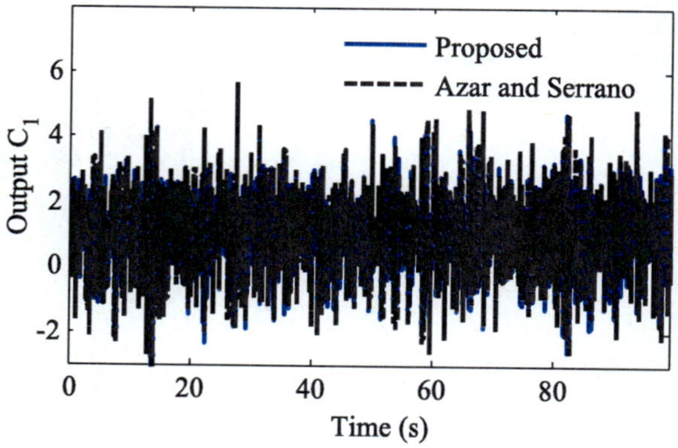


noise emerges from many sources to name a few are the control valves, system itself, other devices utilized in signal measurement, etc. To validate the success of the suggested approach under noisy circumstances, simulation is performed in the noisy environment having seed value of zero and noise power and sample time value of 0.1 on the system output. The response obtained is illustrated in Fig. 7. One can conclude from the figure that the suggested approach is less responsive to the external noise in comparison with the method of Azar and Serrano.

## 7 Conclusion

A simple design method has been proposed for designing the controllers in the CCS. The controllers are designed by placing the pole at the desired location and matching the DE-CLTF with DG-CLTF in frequency domain. With the proposed method, a large range of processes can be controlled in CCS. The method also gives exemption from the approximation of dead time. Thus, the suggested approach is directly applied on the models without approximating them into a lower order. This being the main benefit of the suggested approach. Moreover, the suggested approach succeeds in providing a good setpoint and better load disturbance performance when compared with the existing ones mentioned in the literature.

## References

1. Franks RG, Worley CW (1956) Quantitative analysis of cascade control. Ind Eng Chem 48(6):1074–1079
2. Lee Y, Park S, Lee M, Brosilow C (1998) PID controller tuning for desired closed-loop responses for SI/SO systems. AIChE J 44(1):106–115
3. Lee Y, Oh S, Park S (2002) Enhanced control with a general cascade control structure. Ind Eng Chem Res 41(11):2679–2688

4. Kaya (2001) Improving performance using cascade control and a Smith predictor. ISA Trans 40:223–234
5. Kaya, Atherton DP (2008) Use of smith predictor in the outer loop for cascaded control of unstable and integrating processes. Ind Eng Chem Res 47(6):1981–1987
6. Padhan DG, Majhi S (2013) Enhanced cascade control for a class of integrating processes with time delay. ISA Trans 52(1):45–53
7. Hamamci S, Çakıroğlu O, Güzelkaya M, Eksin İ (2015) Improved cascade controller design methodology based on outer-loop decomposition. Trans Inst Meas Control 37(5):623–635
8. Raja GL, Ali A ()2018 Modified series cascade control strategy for integrating processes. In: 2018 Indian control conference ICC 2018—Proceedings, 252–257
9. Dasari PR, Alladi L, Rao AS, Yoo CK (2016) Enhanced design of cascade control systems for unstable processes with time delay. J Process Control 45:43–54
10. Anwar MN, Siddiqui MA, Laskar SH, Yadav A (2018) PIDA controller design for higher order stable process with inverse response characteristic. In: 2018 international conference on computational and characterization techniques in engineering and sciences, CCTES, pp 236–240
11. Siddiqui MA, Anwar MN, Laskar SH, Mahboob MR (2020) A unified approach to design controller in cascade control structure for unstable, integrating and stable processes. ISA Trans, pp 1–16
12. Siddiqui MA, Anwar MN, Laskar SH (2019) A model-free PI/PID controller based on direct synthesis approach to achieve disturbance rejection. In: IECON 2019—45th annual conference of the IEEE industrial electronics society 1:207–212
13. Siddiqui MA, Anwar MN, Laskar SH (2021) Enhanced control of unstable cascade systems using direct synthesis approach. Chem Eng Sci 232:116322
14. Siddiqui MA, Anwar MN, Laskar SH (2020) Tuning of PIDF controller in parallel control structure for integrating process with time delay and inverse response characteristic. J Control Autom Electr Syst 31:829–841
15. Sadasivarao MV, Chidambaram M (2006) PID Controller tuning of cascade control systems using genetic algorithm. J Indian Inst Sci 86:343–354
16. Krishnaswamy PR, Jha RK, Deshpande PB (1991) When to use cascade control. Ind Eng Chem Res 38(10):38–40
17. Azar T, Serrano FE (2014) Robust IMC–PID tuning for cascade control systems with gain and phase margin specifications. Neural Comput Appl 25(5):983–995

# Design and Performance Analysis of Different Event-Triggering Policies with Sliding Mode Control Strategy

**Asifa Yesmin, Krishanu Nath, and Manas Kumar Bera**

## 1 Introduction

In the past decade, networked-control systems (NCSs) [1] where the control loops are connected by a common digital network are extensively used due to its high reliability, low cost of installation and maintenance. In NCSs, large network transmission load, delayed transmissions, and limited bandwidth are the inherent limitations. To overcome such issues, significant efforts have been devoted on analysis and design of NCSs. A paradigm of control strategy such as event-triggered control (ETC) [2] is one of the prominently suited control technique which is widely used for NCSs to overcome such difficulties. In the event-triggered method, the control signal is updated at discrete instant of time only when some preassigned condition or threshold is violated. This threshold is named as the event condition which is dependent on the system states/outputs. Due to the need-based nature of sampling, the ETC design can reduce the wastage of communication and computational resources.

Based on the design of the threshold condition and its implementation, different event-triggering policies have been introduced in the literature. Each of the triggering polices is associated with different advantages such as higher inter-sampling time, less computational complexity, and minimum hardware requirement for their implementation. A fixed threshold condition-based design of ETC results in static event-triggering mechanism [2, 3] which is the most widely used triggering policy. This triggering mechanism is simple and easy for implementation. To increase the

A. Yesmin (✉) · K. Nath · M. K. Bera
Department of Electronics and Instrumentation Engineering, National Institute of Technology Silchar, Silchar, Assam, India
e-mail: asifa_rs@ei.nits.ac.in

K. Nath
e-mail: krishanu_rs@ei.nits.ac.in

M. K. Bera
e-mail: manas@ei.nits.ac.in



P. H. J. Chong et al. (eds.), *Emerging Electronics and Automation*, Lecture Notes in Electrical Engineering 937, https://doi.org/10.1007/978-981-19-4300-3_6

inter-sampling time, a time-varying triggering law is introduced in literature [4]. In the time-varying triggering mechanism, the event condition is designed with a time-varying function which usually consists of a constant scalar with a decaying function. It will coincide with static triggering scheme when the decaying function is absent in time-varying triggering mechanism. Other than the time-varying triggering policy, a concept of dynamic threshold condition has been introduced which aims to enlarge the inter-sampling time known as dynamic triggering policy [5]. In the dynamic triggering policy, the event condition involves the constant design parameter along with the dynamic variable which is usually described by a first order differential equation. The solution of the differential equation is used to enable the triggering mechanism. It has been observed that this scheme can achieve better result with respect to static triggering in terms of reduced number of event generation. To implement the abovementioned triggering policies, a dedicated hardware is required for the continuous monitoring of the states/outputs. To relax the requirement of this dedicated hardware, a new triggering mechanism is proposed, called self-triggering policy [6, 7]. In self-triggering policy, the triggering instants are generated based on the last sampled information of the states/output. Most of the early design of ETC does not consider the effect of disturbance on the system. Also, the system stability has been derived based on input-to-state-stability.

To deal with the uncertainties and disturbances, the event-triggered mechanisms were designed with robust control techniques and one of which is ET-SMC. SMC can ensure the robust performance of the system with matched perturbations [8]. The philosophy of SMC is to design a control law such that the system trajectories evolves on a hyper-plane, called the sliding manifold in the state space which ensures stability and robustness. The detailed design procedures for SMC can be found in the book [9]. In ET-SMC, the system doesn't slide exactly on the hyper-plane but remains in its vicinity. This notion of sliding mode named as practical sliding motion. Many works can be found on event-triggered design and implementation of SMC [10–16]. The initial designs of ET-SMC [14] used static triggering policy, where the authors have showed that the event-triggered design of SMC ensured a practical sliding motion and the system trajectories will converge inside a band. The band size will be decided by the event design parameter only. This result is different from the discrete-time SMC, where the band of quasi sliding motion is dependent on the bound of the disturbance and sampling time. To achieve the robust stability with reduced number of control updates, ET-SMC was designed with time-varying triggering policy in [12]. The design has achieved larger inter-sampling time with the cost of greater discontinuous gain. The design of ET-SMC with dynamic event-triggering policy was introduced in [10]. The design has ensured the greater inter-event time with partially relaxing the dependency on the switching gain of the event design parameter. In [15], the self-triggered-based SMC has been designed and analyzed. It was shown that the system can achieve reduced number of control updates compared to the time-triggered implementation. The disadvantage of the self-triggering-based SMC is that it over-approximates the disturbance to its maximum value and computes the next triggering instance which causes lower inter-sampling time compared to the event-triggered designs.

In this work, the sole objective is to introduce and discuss the various triggering policies for ET-SMC. We also include a comparative analysis of different triggering policies by simulation of numerical example.

## 2 System Description and Problem Formulation

Consider a perturbed linear time-invariant system dynamics given by the following

$$\dot{z}(t) = Az(t) + B(u(t) + f(t)) \tag{1}$$

where $z(t) \in \mathbb{R}^n$ is the states of the system and $u(t) \in \mathbb{R}$ is the control input respectively. $A \in \mathbb{R}^{n\times n}$ and $B \in \mathbb{R}^n$ are known constant matrices. $f(t)$ is the matched perturbation which is bounded. The following assumptions are considered which are valid through out the work.

**Assumption 1** *The matrix pair* $(A, B)$ *is controllable.*

**Assumption 2** *The matched perturbation is bounded and is given such that* $\sup_{t\geq 0} |f(t)| \leq f_0$.

The system (1) is connected to the controller using a shared communication channel as shown in Fig. 1. The event generator monitors the system states and generates the triggering instances. As the triggering occurs, the state information is sampled as $z(t_i)$ which is sent to the controller, and then, the controller computes the control law and gets updated at the actuator end using zero-order hold (ZOH). To design an event-based sliding mode control strategy, we use the emulation-based approach. Thus, it is required to design a continuous time controller followed by a suitable triggering law such that the stability of the event sampled system can be ensured.

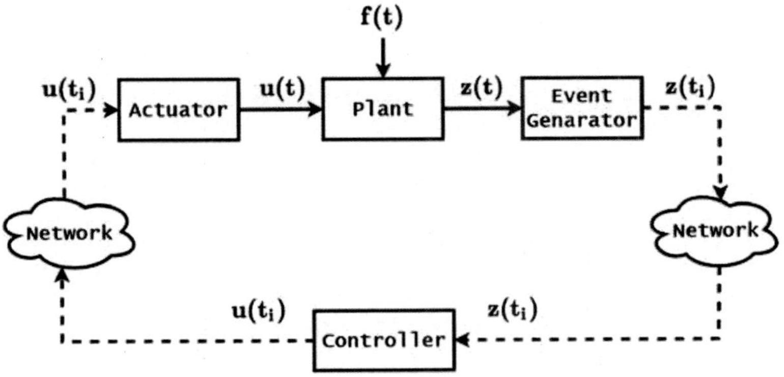


**Fig. 1** Event-based SMC system

With the assumption 1, the regular form of the system (1) can be written as

$$\dot{z}_1(t) = A_{11}z_1(t) + A_{12}z_2(t) \tag{2a}$$
$$\dot{z}_2(t) = A_{21}z_1(t) + A_{22}z_2(t) + B_2(u(t) + f(t, z)) \tag{2b}$$

where $z_1(t) \in \mathbb{R}^{n-1}$, $z_2(t) \in \mathbb{R}$, and $B_2 \neq 0$, respectively. It is observed that from the assumption 1, the pair $(A_{11}, A_{12})$ is controllable.

Let $C_1^\top \in \mathbb{R}^{n-1}$ be chosen such that the matrix $(A_{11} - A_{12}C_1)$ is Hurwitz. Now, the sliding surface is designed as $s(t) = Cz(t) = [C_1\ 1]z(t)$ where $C^\top = [C_1\ 1]^h T \in \mathbb{R}^n$. Then, the below-mentioned set

$$S := \{z(t) \in \mathbb{R}^n : s(z) = Cz(t) = 0\} \tag{3}$$

defines a stable sliding manifold for the system (2). Thus, the states asymptotically reach to the sliding manifold toward the origin, and there exists a $\tau \geq 0$ such that $z(t) \in S$ for all $t \geq \tau$. This motion is called as *sliding mode* in the system. The control law $u(t)$ which forces the trajectories toward the sliding manifold in some definite time is given by

$$u(t) = -(CB)^{-1}(CAz(t) + Q\text{sign}(s(t))) \tag{4}$$

where $Q > f_0\|CB\|$. The control law (4) is discontinuous in nature, and the solutions of the closed-loop system with this control signal can be understood in the sense of Filippov [17]. During sliding motion $s(z) = 0$, thus $s(z) = C_1z_1 + z_2 = 0$. From this, it can be written further as $z_2 = -C_1z_1$. From equation (2a), substituting $z_2 = -C_1z_1$, the dynamics of the reduced-order system can be shown as

$$\dot{z}_1(t) = (A_{11} - A_{12}C_1)\, z_1(t) \tag{5}$$

By the appropriate choice of $C_1$, the asymptotic stability of the reduced-order system $A_{cs} = (A_{11} - A_{12}C_1)$ can be ensured. Now, in the event-triggering mechanism, the states are sampled only at some discrete instant of time when the preassigned threshold is violated. At that instant of time, the control signal will be updated. The control law (4) can be written with the event-triggered sampled states as

$$u(t) = -(CB)^{-1}\,(CAz(t_i) + Q\text{sign}(s(t_i))) \tag{6}$$

for all $t \in [t_i, t_{i+1})$ and $i \in \mathbb{Z}_{\geq 0}$. Here $\Delta T_i := t_{i+1} - t_i$ is defined as the inter-event execution time. The control signal is updated at the triggering instant $t_i$, and it remains same till the next instant $t_{i+1}$ using a ZOH, i.e., $u(t) = u(t_i)$ for $\forall\, t \in [t_i, t_{i+1})$. The error $e(t) = z(t_i) - z(t)$. Thus, the error is the difference between the present and the previous measured state of the system, which grows for all $t \in [t_i, t_{i+1})$ due to the event-triggered implementation of the controller. This error is used to design the different triggering policies.

In the next section, we introduce the design of different triggering policies for the implementation of the control law (6).

## 3 Design of Event-Triggering Policies

The different event-triggering policies have different event design rules based on which the triggering instants are generated and control signal is updated. The different triggering mechanisms are as follows.

### *3.1 Static Event-Triggering Policy*

The first form of triggering policy is the static triggering mechanism where the error $\mathsf{e}(t)$ is monitored until it reaches a threshold which is constant. The triggering policy is given by

$$t_{i+1} = \inf\{t \in [t_i, +\infty): \|\mathsf{e}(t)\| \geq \sigma\alpha\} \tag{7}$$

where $\sigma \in (0, 1)$ and $\alpha > 0$ are design constants. This type of triggering usually results in a greater number of control updates during the transient phase of the system response and settle around a steady band as the system reaches steady state.

### *3.2 Time-Varying Triggering Policy*

The time-varying triggering introduces a time-dependent function. Usually, the design function has a decaying nature, to achieve larger inter-sampling time in the transient phase of the response. The triggering rule is given by

$$t_{i+1} = \inf\{t \in [t_i, +\infty): \|\mathsf{e}(t)\| \geq \gamma(t)\} \tag{8}$$

where $\gamma(t)$ is a time-varying function. One of the widely used function is given by

$$\gamma(t) = \rho \mathrm{e}^{-at} + \varsigma \tag{9}$$

where $\rho > \varsigma > 0$ and $a > 0$ are the design parameters. If $\rho$ is chosen to be zero, this converges to the static triggering rule. Further, it may be observed that at steady state, the effect of the decaying term vanishes, and thus, the time-varying triggering and static triggering will have same behavior. This triggering mechanism can achieve improved performance if the event design parameters are appropriately chosen and also, can guarantee the minimum inter-event execution time even if the constant design parameters become zero.

### 3.3 Dynamic Event-Triggering Policy

The dynamic event-triggering rule is based on introduction of a scalar dynamics that acts as a filter for the event condition. The output of the filter is used in the triggering rule to achieve a larger inter-event time. This triggering mechanism can achieve sparser-triggering instants which may result in reduced control computation. The sampling mechanism is governed by

$$t_{i+1} = \inf\{t \in [t_i, +\infty) \colon \|\mathsf{e}(t)\| \geq \sigma\alpha + \varpi(t)\} \tag{10}$$

where $\varpi(t)$ is an internal dynamic variable. The dynamics of the variable can be chosen as

$$\dot{\varpi}(t) = -\chi(\varpi(t)) + \phi\,(\alpha, |s(t)|, \|\mathsf{e}(t)\|)\,, \qquad \varpi(0) = \varpi_0 \tag{11}$$

where $\chi(\cdot)$ is a continuous class $\mathcal{K}_\infty$ function, $\phi\,(\cdot)$ is a design function, $\alpha > 0$ and $\varpi_0 > 0$ are the design parameter. The output $\varpi(t)$ for the above dynamics can be perceived as the nonlinear-filtered output of $\phi\,(\alpha, |\mathsf{s}(t)|, \|\mathsf{e}\|)$. The dynamics can be chosen in many ways to achieve stability and sparser-triggering instants.

### 3.4 Self-triggering Policy

The above three triggering polices are dependent on the signal $\mathsf{e}(t)$, and the triggering is achieved by continuous monitoring of $\mathsf{e}(t)$ which requires a dedicated hardware. To alleviate the problem of continuous monitoring of $\mathsf{e}(t)$, the self-triggering mechanism is designed by choosing the next sampling as the calculated minimum inter-event execution time. The next triggering instant generation is evaluated based on information at $t = t_i$. Thus, the next sampling instance $t_{i+1}$ for self-triggering scheme is given by

$$t_{i+1} = \inf\{t \in [t_i, +\infty) \colon \|\mathsf{e}(t)\| \geq t_i + \Theta(t, t_i)\} \tag{12}$$

where $\Theta(t, t_i)$ is the lower bound of minimum inter-event time which is dependent on $\mathsf{f}_0$. In practical scenario, the upper bound of the disturbance $\mathsf{f}_0$ is unknown. In such case, the over estimation of $\mathsf{f}_0$ may lead to smaller inter-event execution time.

The abovementioned four triggering polices have been considered here for the design of ET-SMC. Next, we introduce the notions of stability for the closed-loop system with the plant (1), sliding variable (3), control law (6) for the different triggering polices.

## 4 Stability Analysis

This section presents an outline of the stability analysis. The stability analysis of ET-SMC is carried out by choosing a Lyapunov candidate as

$$V(s) = \frac{1}{2}s^{\top}(t)s(t) \quad (13)$$

and evaluating its behavior in the interval $t \in [t_i, t_{i+1})$. The system trajectories are then shown to be convergent by establishing the reachability condition (i.e., $s(t)\dot{s}(t) < -\eta|s|$) by ensuring a sufficient switching gain which compensates the disturbance and the sampling error $\|e(t)\|$. Once this reachability is met, it can be assured that $s(t)$ hits the manifold $\mathbf{S}$ in finite time. As the trajectories cross the manifold, it will grow, and the growth is restricted as the next triggering occurs, and the trajectory falls back toward the manifold. Thus, the sliding variable growth in the interval $t \in [t_i, t_{i+1})$ can be calculated by finding $|s(t) - s(t_i)|$. Thus, the system trajectories of $s(t)$ are ultimately bounded in the vicinity of the manifold $\mathbf{S}$. A general form of the ultimate set of the sliding variable is given by

$$\Sigma = \left\{z \in \mathbb{R}^n : |s(t)| \leq \Delta\right\}. \quad (14)$$

where $\Delta$ is a positive value dependent on the event design parameters. In literature, $\Delta$ is often referred to as practical sliding mode band. This form of stability analysis is valid for static triggering policy, time-varying triggering policy, and self-triggering policy. In case of a dynamic triggering policy, a Lyapunov candidate is chosen in the form

$$W(s, \varpi) = V(s) + \varpi. \quad (15)$$

The stability analysis is similar to that of the other cases where the reachability condition has to be satisfied for finite time reaching of the sliding variable which means $\dot{W}(s, \varpi) < -\eta|s(t)|$. The detailed proof for each of the cases is given in [10, 13].

Once the stability of $s(t)$ is guaranteed, we need to establish the stability of the reduced-order system which is governed by

$$\dot{z}_1(t) = (A_{11} - A_{12}C_1)\, z_1(t) + A_{12}s(t) \quad (16)$$

The stability of the above dynamics can be shown by treating $s(t)$ as a non-vanishing bounded perturbation (see [18]). To avoid the Zeno phenomenon of the system, $\Delta T_i$ must be greater than zero. The rate of maximum growth of $\|e(t)\|$ is calculated to establish that fact and the differential inequality can be solved using comparison lemma (see [18]). The minimum inter-event time is calculated as $\Theta(t, t_i)$ which is used to define the triggering law for the self-triggered policy.

To have a quantitative comparison of the ET-SMC designed with different triggering policies, a numerical example is chosen to simulate it.

## 5 Numerical Examples

This section includes the simulation results of the event-triggered SMC with different triggering policies for a second-order perturbed system. We consider here a double integrator system which is analogous to many practical systems in the domain of control engineering. The system is described by

$$\dot{z} = \begin{bmatrix} 0 & 1 \\ 0 & 0 \end{bmatrix} z + \begin{bmatrix} 0 \\ 1 \end{bmatrix} (u(t) + f(t)) \tag{17}$$

where $z = [z_1 \; z_2]^\top$ is the state vector with the initial condition is considered as $z(0) = [1 \; -1]^\top$. The perturbation is considered as $f(t) = 0.2\sin(10t) + 0.1\text{sign}(z_1 z_2)$. We start with the design of a stable sliding surface given by $s = z_1 + z_2$. The control law (6) with the design parameters $C = [1 \; 1]^\top$ and the switching gain $Q = 1.2$ is applied on the perturbed double integrator system for all four triggering policies. We choose $\alpha = 0.1$ and $\sigma = 0.99$ for the static triggering case. The time-varying triggering condition is designed using the function $\gamma(t) = 0.9e^{-0.8t} + 0.1$ such that at steady state, the value converges to 0.1 ensuring the same steady-state band of the sliding variable as in the static case. In the dynamic event-triggered case, we continue with the design with same value of $\alpha$, i.e., 0.1, and the dynamics of the dynamic variable is given by

$$\dot{\varpi}(t) = -2\varpi(t) - \varpi(t)|s(t)| + (\alpha - \|C\|\|A\|\|e(t)\|)|s(t)|$$

with $\varpi(0) = 100$ as in [10]. To implement the self-triggering scheme, we use the value of $\Theta(t, t_k) = \frac{1}{\|A\|} ln\left[1 + \frac{\sigma\alpha}{\|C\|(\zeta\|z(t_i)\| + \beta)}\right]$ from [15] to generate the triggering sequence with $\sigma$=0.99 and $\alpha = 0.1$, where $\beta = \|B(CB)^{-1}Q\| + \|B\| f_0$ and $\zeta = \|A - B(CB)^{-1}CA\|$. The simulated results for the four type of triggering schemes, i.e., static triggering, time-varying triggering, dynamic triggering, and self-triggering are shown in Figs. 2, 3, 4 and 5, respectively.

In case of static triggering policy, the system-state trajectories are plotted in Fig. 2a. It may be noticed that both the states $z_1$ and $z_2$ converge and achieve practical stability. The sliding variable is plotted in Fig. 2b where it converges to the practical sliding manifold in definite time and stay thereafter. In Fig. 2c, the evolution $\|e(t)\|$ is presented along with its envelope $\sigma\alpha$. As $\|e(t)\|$ grows and hits the envelope, a triggering occurs, and the state information is sent to the controller, and the control signal is updated. Simultaneously, the sampled value of states is also updated which causes $e(t) = 0$. Now, the system evolves in an open-loop manner, and $e(t)$ grows

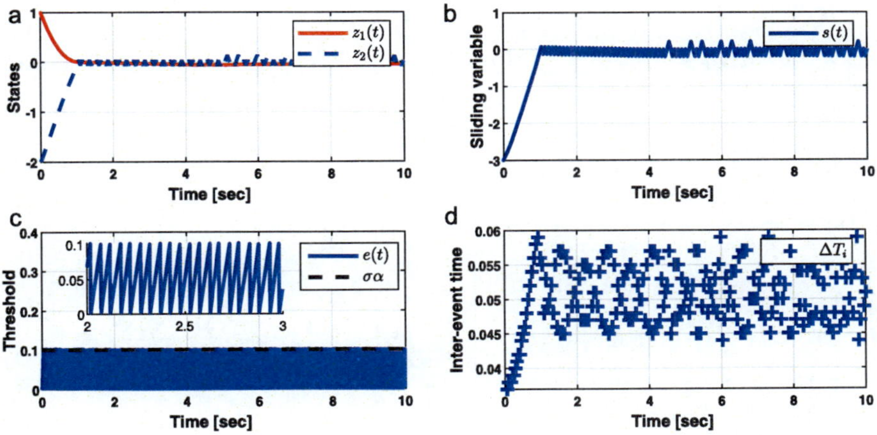


**Fig. 2** Evolution of **a** system states, **b** sliding variable, **c** threshold, **d** inter-event execution time for event-triggered SMC with static triggering condition

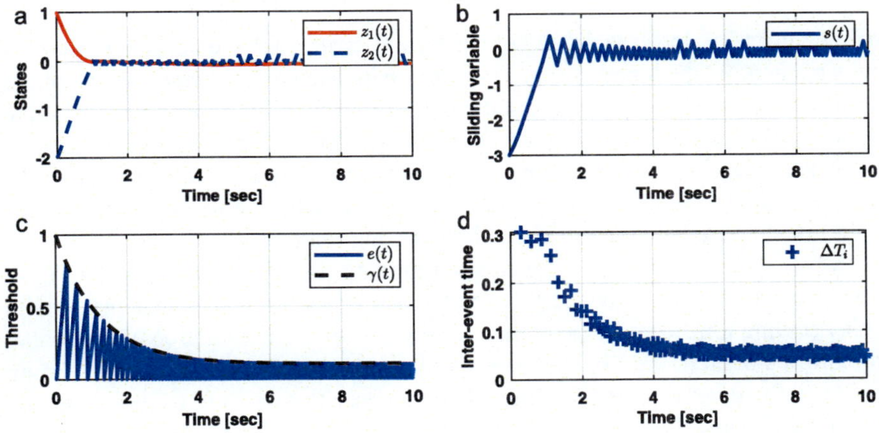


**Fig. 3** Evolution of **a** system states, **b** sliding variable, **c** threshold, **d** inter-event execution time for event-triggered SMC with time-varying triggering condition

and subsequently reaches the threshold value resulting in next sampling instance. The inter-event execution time is plotted in Fig. 2d. The inter-event time is observed to be in a band with all values greater than zero ensuing Zeno-free behavior of the closed-loop system.

The response of the system (17) with the same control law and a time-varying (an exponentially decaying) threshold condition (9) is shown in Fig. 3. The state trajectories and sliding variable are plotted in Fig. 3a, b where the states and the sliding variable are convergent. As the band of the sliding variable is governed by the event design function $\gamma(t)$, it can be observed to have a larger band size during the initial phase of response. The plot of $\|\mathrm{e}(t)\|$ and $\gamma(t)$ is given in Fig. 3c. Similar

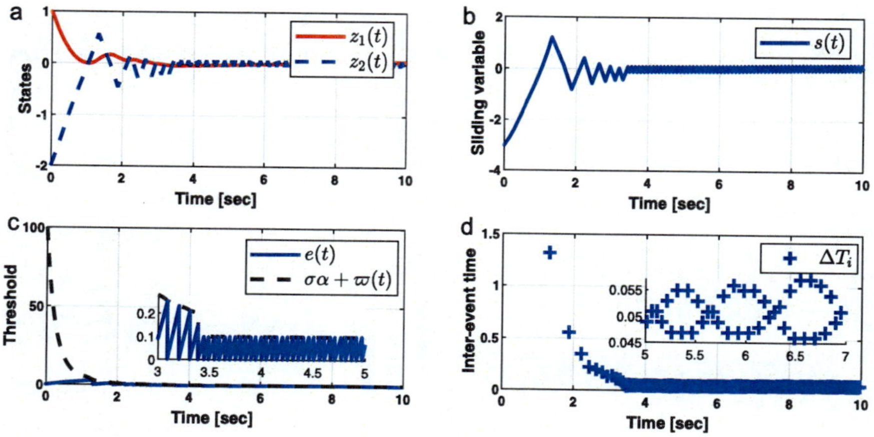


**Fig. 4** Evolution of **a** system states, **b** sliding variable, **c** threshold, **d** inter-event execution time for event-triggered SMC with dynamic triggering condition

to the static triggering case, $\|\mathbf{e}(t)\|$ grows from 0 to $\gamma(t)$, and the sampling instances are generated. As the function $\gamma(t)$ converges to its steady-state value, the triggering occurs in a similar fashion to the static case. From the Fig. 3d, it is observed that the inter-event execution time starts with a larger value at around 0.3 s compared to the static case where it was approximately 0.35 s. This saves the burden on the communication networks during initial phase only. Finally, the steady-state behavior of $\Delta T_i$ is same for both the static and time-varying cases.

The simulated results with dynamics triggering policy are presented in Fig. 4. From Fig. 4a, it can be noticed that the system states are convergent but require higher settling time to converge to their ultimate bound. The evolution of the sliding variable is plotted in Fig. 4b, where the system trajectories achieve practical sliding motion but with greater reaching time than the other two cases of triggering policy which are discussed above. The dynamic variable is plotted in Fig. 4c, where the black-dashed line is of the envelope $\sigma\alpha + \varpi$. As $\sigma\alpha \approx 0.1$ during the initial phase, the triggering is mainly governed by $\varpi$ which decays to zero as time progresses. Finally, the envelope converges to $\sigma\alpha$ which means at steady state, the sampling occurs as like in the static triggering case. The plot of inter-event execution time with dynamic triggering policy is given in Fig. 4d. Here in the initial phase, the inter-event time is greater than the case of static triggering and the time-varying triggering case. This reduction of control updates is dependent on the initial condition of $\varpi(t)$. The decay of the $\varpi(t)$ cannot be controlled by the designer because it is coupled with the evolution of $s(t)$. On the other hand, in the design of the time-varying triggering policy, the designer has flexibility on the choice of the function $\gamma(t)$ where the decay rate can be fixed arbitrarily.

The simulated results for self-triggered SMC are presented in Fig. 5 where the response of the system is similar to that of the static triggering case. The implementation of this scheme does not require the continuous information of the states and is

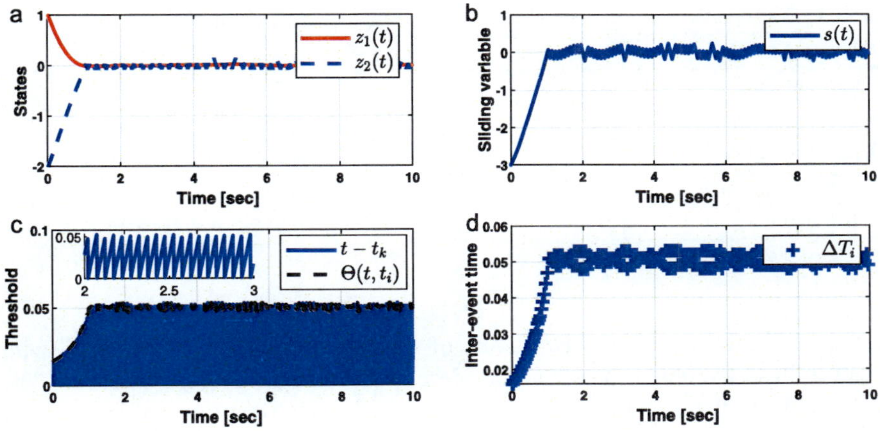


**Fig. 5** Evolution of **a** system states, **b** sliding variable, **c** threshold, **d** inter-event execution time for event-triggered SMC with self-triggering condition

dependent on the last sampled value. One shortcoming of the strategy is that if the bound of the disturbance is exactly not known, then the inter-sampling time may be unnecessary small, leading to over-utilization of the network resources.

To have a quantitative comparative analysis of the system performance with different triggering policies, we introduce some performance indices which are $N_{\text{up}}$ as the total number of control updates; $N_{\text{use}}$ as the total network usage; $t_{\text{avg}}$ as the average time between two consecutive sampling and $S_{\text{IA}}$ as the integral of the absolute sliding variable. In this comparison, we have added the case of time-triggered controller implementation with the sampling time of 0.01 s. The performance indices are shown in Table 1. It can be observed that ET-SMC with all event-triggering policies require reduced number of control updates compared to time-based triggering. Among the event-triggered policies, dynamic triggering has the least number of control updates followed by time-varying triggering, which leads to greater average value of inter-sampling time and reduced usage of the communication network. The effect on the performance of the closed-loop system can be understood using $S_{\text{IA}}$, which is larger for the dynamic triggering case and smallest for time-triggered case. It is clearly visible that the performance of the system deteriorates with the cost of increased inter-sampling time. This trade-off needs to be taken care based on the application. If precise tracking/stabilization is required, the magnitude of the threshold condition must be smaller and vice versa.

## 6 Conclusions

In this paper, the design of ET-SMC strategy with different triggering policies has been analyzed for a perturbed LTI system. The design methods include different event conditions, namely static triggering, time-varying triggering, dynamic trig-

Table 1 Effect on the performance indices with different triggering policies

| Triggering policy | $N_{up}$ | $N_{use}$ (%) | $t_{avg}$ (s) | $S_{IA}$ |
|---|---|---|---|---|
| Time-triggering | 1000 | 100 | 0.010 | 1.69 |
| Static triggering | 199 | 1.99 | 0.053 | 2.48 |
| Time-varying triggering | 141 | 1.41 | 0.071 | 2.71 |
| Dynamic triggering | 140 | 1.40 | 0.072 | 2.91 |
| Self-triggering | 217 | 2.17 | 0.046 | 2.33 |

gering, and self-triggering. The outline of the systems stability is presented with a generalized Lyapunov function considering all the triggering mechanisms. The numerical simulation results are analyzed for different triggering mechanisms. A comparative table has been presented to show the effect of different triggering policies on the system performance, and it has been observed that with the appropriate choice of event design parameters, among all the triggering policies, the dynamic ET-SMC outperforms with respect to control update and network usage.

## References


1. Zhang XM, Han QL, Yu X (2016) Survey on recent advances in networked control systems. IEEE Trans Ind Inform 12(5):1740–1752
2. Tabuada P (2007) Event-triggered real-time scheduling of stabilizing control tasks. IEEE Trans Autom Control 52(9):1680–1685
3. Heemels WP, Johansson KH, Tabuada P (2012) An introduction to event-triggered and self-triggered control. In: 2012 IEEE 51st IEEE conference on decision and control (CDC). IEEE, pp 3270–3285
4. Zhang J, Feng G (2014) Event-driven observer-based output feedback control for linear systems. Automatica 50(7):1852–1859
5. Girard A (2015) Dynamic triggering mechanisms for event-triggered control. IEEE Trans Autom Control 60(7):1992–1997
6. Almeida J, Silvestre C, Pascoal AM (2014) Self-triggered output feedback control of linear plants in the presence of unknown disturbances. IEEE Trans Autom Control 59(11):3040–3045
7. Gommans T, Antunes D, Donkers T, Tabuada P, Heemels M (2014) Self-triggered linear quadratic control. Automatica 50(4):1279–1287
8. Gao W, Hung JC (1993) Variable structure control of nonlinear systems: a new approach. IEEE Trans Ind Electron 40(1):45–55
9. Edwards C, Spurgeon S (1998) Sliding mode control: theory and applications. CRC Press
10. Yesmin A, Behera AK, Bera MK, Bandyopadhyay B (2021) Dynamic event-triggering based design of sliding mode control. Int J Robust Nonlin Control 31(12):5910–5925
11. Yesmin A, Bera MK (2020) Design of event-based sliding mode controller with logarithmic quantized state measurement and delayed control update. ISA Trans
12. Yesmin A, Bera MK (2019) Design of event-triggered sliding mode controller based on reaching law with time varying event generation approach. Euro J Control 48:30–41
13. Bandyopadhyay B, Behera AK (2018) Event-triggered sliding mode control: a new approach to control system design, vol 139. Springer
14. Behera AK, Bandyopadhyay B (2017) Robust sliding mode control: an event-triggering approach. IEEE Trans Circuits Syst II Express Briefs 64(2):146–150

15. Behera AK, Bandyopadhyay B (2015) Self-triggering-based sliding-mode control for linear systems. IET Control Theory Appl 9(17):2541–2547
16. Behera AK, Bandyopadhyay B, Cucuzzella M, Ferrara A, Yu X (2021) A survey on event-triggered sliding mode control. IEEE J Emerg Sel Topics Ind Electron
17. Filippov AF (1988) Differential equations with discontinuous right-hand sides. Kluwer, Dordrecht, The Netherlands
18. Khalil HK (2001) Nonlinear systems. Prentice Hall

# Control and Robotics Track—2

# Effect of Sampling and Simulation of Wireless Networked Control System Using TrueTime and MATLAB: A Comparative Study

Vijay R. Dahake, Mukesh D. Patil, and Vishwesh A. Vyawahare

## 1 Introduction

### *1.1 What Is WNCS?*

Recently, there has been tremendous progress in wireless networking, sensing control and computing which are revolutionizing how control systems interact with information and physical process such as cyber physical system (CPS), tactile Internet and Internet of things [1]. Wireless or wired networked control systems (NCS) are completely distributed real-time feedback control system in which sensors to controller (S–C) and controller to actuator (C–A) communication, i.e., exchange of feedback signal, control signals take place among system components like sensor, actuator in the form of information packets through wireless network which always have limited bandwidth. Sensors continuously read the output of process and due to limited bandwidth of wireless network, sensors do not send output as it is, to the controller through wireless network. Instead, sensor's output is continuously sampled at fixed rate such that this rate is sufficient to recover the system output successfully, and then, it is sent to the controller (S–C) through wireless network. The controller then compares this output with reference input, accordingly generates the control signal and sends this signal to the actuator. The actuator accordingly reacts with and generates actuating signal to adjust input signal to the plant so that it will produce desired output. Wireless NCS is a very vast area. Research in WNCS may be divided into two main types: (a) control of network and (b) control through or over network [2].

(a) Control of network

Under this class, the researcher can work on the issues of communication network such as routing control, congestion control and communication protocol.

V. R. Dahake (✉) · M. D. Patil · V. A. Vyawahare
Ramrao Adik Institute of Technology, Navi Mumbai, India 400706
e-mail: vijay.dahake@rait.ac.in



P. H. J. Chong et al. (eds.), *Emerging Electronics and Automation*, Lecture Notes in Electrical Engineering 937, https://doi.org/10.1007/978-981-19-4300-3_7

(b) Control through or over network

Under this class, the researcher can pay attention on the design and control of systems that uses wireless network as a medium for transmission of data.

Control over network guarantees the quality of service (QoS) and the quality of control (QoC) both. QoS concerned with transmission rates and error rates. QoC concerned with the stability of the system subjected to different conditions. Simultaneous achievements of both QoS and QoC are a major objective of research in NCS.

Research on networked control system allows us to work on different areas like (a) methods for design of robust controllers for NCS, (b) modeling of time delay in the NCS, (c) modeling of packet loss in networked control systems, (d) analysis of stability in networked control systems, (e) synthesis of network types and communication protocols. Wireless communication network shows different performance limitations [3] and effects on the linear or non-linear NCS which are given as—(i) Time delays and jitter induced due to network (ii)Disordering of the sent and received communication packets (iii) Limited bandwidth of wireless network (iv) Types of protocol used during communication, e.g. Ethernet (v) synchronization and timing characteristics of networked devices e.g. sensors and actuator (vi) Controller's limited computational power (vii) Reliability, consistency and availability of wireless network.

## *1.2 Related Work*

Many authors contribute to research in WNCS by using TrueTime toolbox for practical implementation. They extended and appreciated the work of Cervin et al. [4]. TrueTime is a powerful popular tool used to do analysis of real-time wireless NCS, developed by Dr. Anton Cervin team at Lund University in year 2002. It is continuously upgraded with addition of new features since year 2002.

In paper [5], author Magdi S. Mahmoud et al. use TrueTime toolbox for experimental investigation of distributed NCS. They established experimental setup for NCS on pilot-scale plants. They investigate tracking errors under different wired and wireless network. They performed simulation to investigate stability of DC servo motor in networked scenario against different loss probabilities.

In paper [6], author Russul N. Abdul-Hussain et al. proposed particle swarm optimization (PSO)-based PID and FOPID controllers to control plants in WNCS using TrueTime simulation. The objective of the proposed controllers was to minimize the time delays and to keep the system stable when the network handles a large number of nodes in the system with large variable length packet sizes.

In paper [7], author Nyan Phayo Aung et al. used PD controller for simulation of WNCS using TrueTime. In this paper, simulation is carried out on AC servo motor control system using 802.11b (WLAN) as communication network medium. They concluded that overshoot in output response increases if we increase sampling time at sensor node.

In paper [8], author Martin Andersson et al. used different network simulators like TOSSIM, NAB, RTSIM and TrueTime for wireless embedded system. They presented two simulation case studies: a simple communication scenario and a mobile robot soccer game using TrueTime simulator.

### *1.3 Objective and Motivation*

Many researchers have been working on the topic of wireless NCS and constantly contributing to this area. Two decades ago, research on WNCS was limited, but due to the remarkable progress in wireless network, different protocols structure and control system, this topic today has huge potential, and an opportunities becomes available to work in different areas like control, communication and computer system. Thus, this is the main motivation for this research study. Extension of this area is Cyber-physical system (CPS). Research in CPS becomes easy if we worked first in WNCS, because WNCS is background for CPS.

Quality of service (QoS) is concerned with transmission rates and error rates and quality of control (QoC) is concerned with the stability of the system under different conditions. Simultaneous guarantees and achievement of both QoS and QoC in any application is the main objective of research in WNCS.

The main objective of this paper is to explain how communication happens between sensor to controller and controller to actuator by using wireless network and to show, how simultaneously, with less delay and low packet loss, (QoS) can be achieved by maintaining stability (QoC) of the given system while doing simulation using TrueTime. There are many software which can be used for simulation of WNCS, e.g., PiccSimulator, RTNS, LabNS2.

### *1.4 Organization of the Paper*

This paper is organized as follows. Modeling of wireless networked control system using state space analysis by considering delay and packet loss at controller and actuator is explained in Sect. 2. Importance of sampling and why sampling rate at sensor and in an initialization file of the Kernel block, configured as discrete PID controller (Init file), must be same is explained in Sect. 4. Simulation using MATLAB and TrueTime toolbox for different transfer function like second order, non-minimum phase and unstable system and response is given in Sect. 5. Finally, result analysis and conclusion are given in Sect. 6 followed by the references.

## 2 Modeling of Wireless NCS

Modeling and analysis of linear time invariant plant for wired networked control system are an easy task since negligible delay or no delay is involved for communication between sensor to controller (S–C) and controller to actuator (C–A). Modeling of such system can be easily done using following state space analysis equation given by Eqs. (1) and (2),

$$\frac{dx(t)}{dt} = Ax(t) + Bu(t) \tag{1}$$

$$y(t) = Cx(t) \tag{2}$$

where $x$ is state vector, $y$ is output vector, $u$ is input or control vector, $A$ = system matrix, $B$ = input matrix, $C$ = output matrix.

In modeling and analysis of linear time invariant (LTI) plant system in wireless networked control system, delay is involved for communication between sensor to controller $\tau_{sc}(t)$ and indicator of packet loss $\gamma(t))$ at controller must be considered. Also delay involved for communication between controller to actuator $\tau_{ca}(t)$ and indicator of packet loss $\alpha(t)$ at actuator must be considered. Due to this, complexity in modeling analysis of linear time invariant plant in WNCS increases to a great extent. Modeling of LTI plant in wireless NCS can be done with following modified state space equation [9] given by Eqs. (3) and (4),

$$\frac{dx(t)}{dt} = Ax(t) + Bu^a(t) \tag{3}$$

where $u^a(t) = \alpha(t)u^c[t - \tau_{ca}(t)]$ OR

$$\frac{dx(t)}{dt} = Ax(t) + B\alpha(t)u^c[t - \tau_{ca}(t)] \tag{4}$$

$$y(t) = C\gamma(t)x[t - \tau_{sc}(t)] \tag{5}$$

where

$$\gamma(t) = \begin{cases} I^{n\times n} & \text{if controller received } x(t) \text{ from the plant at time } t \\ 0^{n\times n} & \text{Otherwise} \end{cases}$$

$$\alpha(t) = \begin{cases} I^{m\times m} & \text{if actuator received controlled command from controller at time } t \\ 0^{m\times m} & \text{Otherwise} \end{cases}$$

with $x(t) \in \mathbb{R}^{n\times n}$ represents system state, $u^c(t)) \in \mathbb{R}^{m\times m}$ represents control inputs computed at controller, $u^a(t) \in \mathbb{R}^{m\times m}$ represents control inputs received at actuator and $y(t) \in \mathbb{R}^{n\times n}$ represents output of the plant. Note that $A \in \mathbb{R}^{n\times n}$ and $B \in \mathbb{R}^{n\times m}$

denote system matrices. Here, we assume that sum of both network induced delay be bounded, i.e., $\tau_{sc}(t) + \tau_{ca}(t) < bT_s$ [9] where $b$ denotes delay bound and $T_s$ is sampling interval.

## 3 TrueTime: Simulator for NCS

Networked control system is an overlapping subject of research study between two areas: information systems and feedback control system as indicated in Fig. 1. Different simulators used for independent area and overlapping area are also shown in Fig. 1.

TrueTime is a powerful popular tool used to do analysis of real-time wireless NCS, developed by Dr. Anton Cervin and his team at Lund University in year 2002. It is continuously upgraded with addition of new features since year 2002. TrueTime toolbox is a MATLAB/Simulink-based library of simulation blocks that extends usability of MATLAB/Simulink to simulate networked process control, co-simulation of controller task execution in real-time kernels, network transmissions and continuous plant dynamics. The library of TrueTime contains different blocks like TrueTime kernel block, TrueTime send block, TrueTime receive block, TrueTime network block and TrueTime wireless network block as shown in Fig. 2.

Responsibility of TrueTime kernel block is to look after for I/O and network data acquisition, i.e., data processing and calculations. TrueTime kernel block, which is the brain of every simulink system, can be realized by using a control algorithm/logic. Several periodic and non-periodic independent tasks can be executed in the kernel, which can co-operate on the same goal. TrueTime network blocks include wired networks as CSMA/CD (Ethernet), CSMA/AMP (CAN), Round Robin, FDMA, TDMA, Switched Ethernet, FlexRay, PROFINET, NCM. TrueTime wireless network blocks include wireless network as IEEE 802.11b/g (WLAN), IEEE 802.15b/g

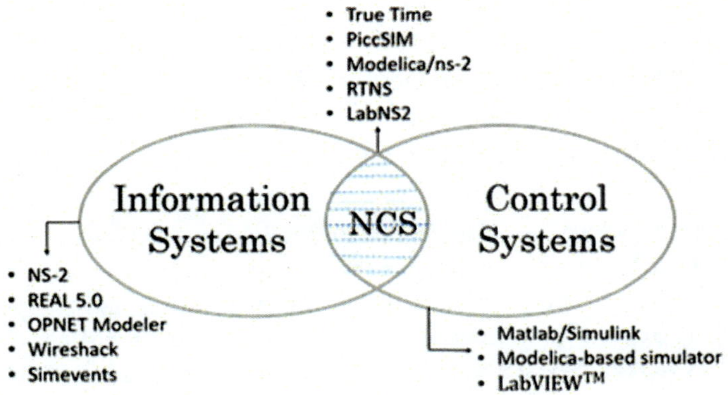


**Fig. 1** Simulators for NCS [3]

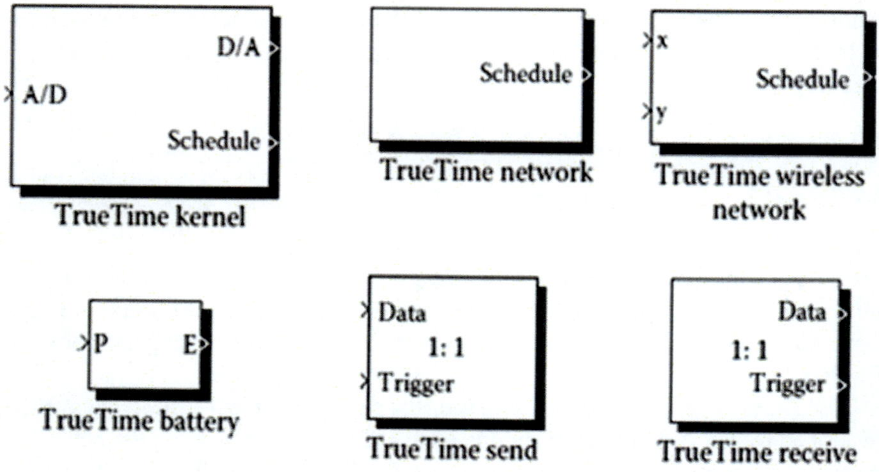


**Fig. 2** Toolbox of TrueTime [4]

(WLAN). TrueTime software consists of a kernel block and a network block, both variable-step S-functions written in C++.

Different blocks in TrueTime can be connected with ordinary continuous Simulink blocks in MATLAB to form a real-time control system as shown in Fig. 8. The TrueTime kernel block executes user defined tasks such as interrupts handlers, representing I/O tasks, control algorithms and communication tasks. The TrueTime kernel block simulates a computer with an event-driven real-time kernel, A/D and D/A converters, a network interface and external interrupt channels. Generally, scheduling policy of the TrueTime block kernel is arbitrary and decided by the user. The dialog box of kernel block in TrueTime is shown in Fig. 3, and dialog box of wireless network block in TrueTime is shown in Fig. 4. Note that there is also a dialog box of wired network block in TrueTime not shown here.

## 4 Effect of Sampling in Wireless NCS

### *4.1 Types of Sampling Used in NCS*

In WNCS, sensor output signals are continuous time signals which are required to be sampled due to bandwidth constraint of the wireless network. Choice of sampling shall be decided by network traffic, influence of disturbances and computational load [1, 10]. There are two methods of sampling in WNCS to sample the continuous time signals: (a) time triggered sampling and (b) event triggered sampling. In time triggered sampling, the next sampling instant occurs after the elapse of fixed time interval, regardless of plant state. Time triggered sampling also called as ‘Riemann

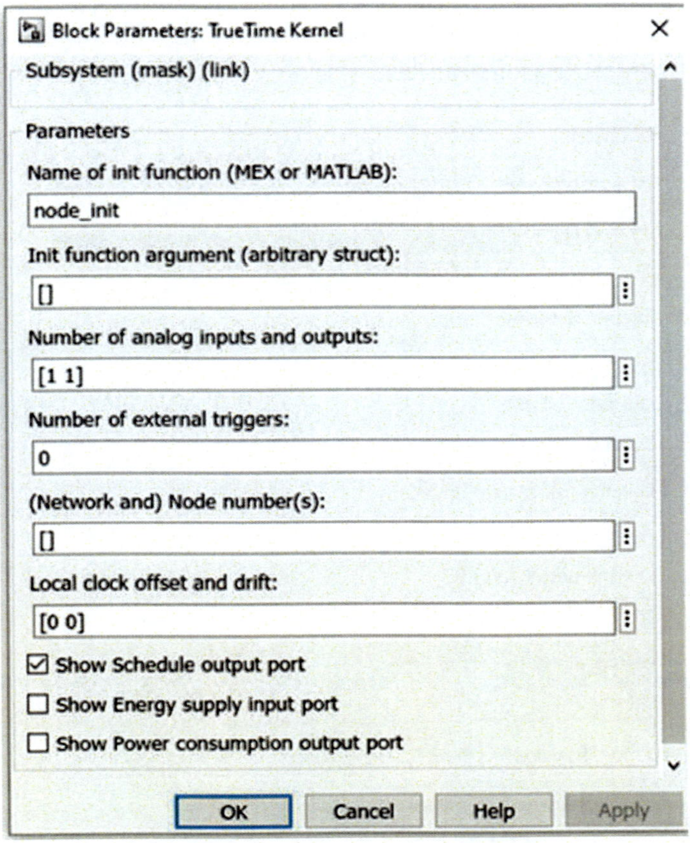


**Fig. 3** Dialog box of kernel block in TrueTime

sampling' again may be classified into two categories: hard sampling period and soft sampling period, by considering relationship between sampling period and message delay [11]. In the event triggered sampling, the execution of controlled task is determined by the occurrence of an event rather than the elapse of a fixed time period. Events are triggered only when stability or pre-known control performance is about to be lost [12, 13]. Event triggered sampling can reduce the traffic load of the network with minor control performance degradation [14, 15].

From research experience, many people state that for simulations in WNCS selection of sampling period is $\omega.h = 0.1–0.6$ is adequate where $\omega$ is desired natural frequency of the closed loop system, and h is sampling period. In digital control system based on wired NCS, smaller value of sampling period is preferred [16]. But in WNCS small value of sampling period increases the network traffic, which may increase the delay and loss probability of information degrading the performance of WNCS. It is possible to decreased in sampling period reduces message delay and message

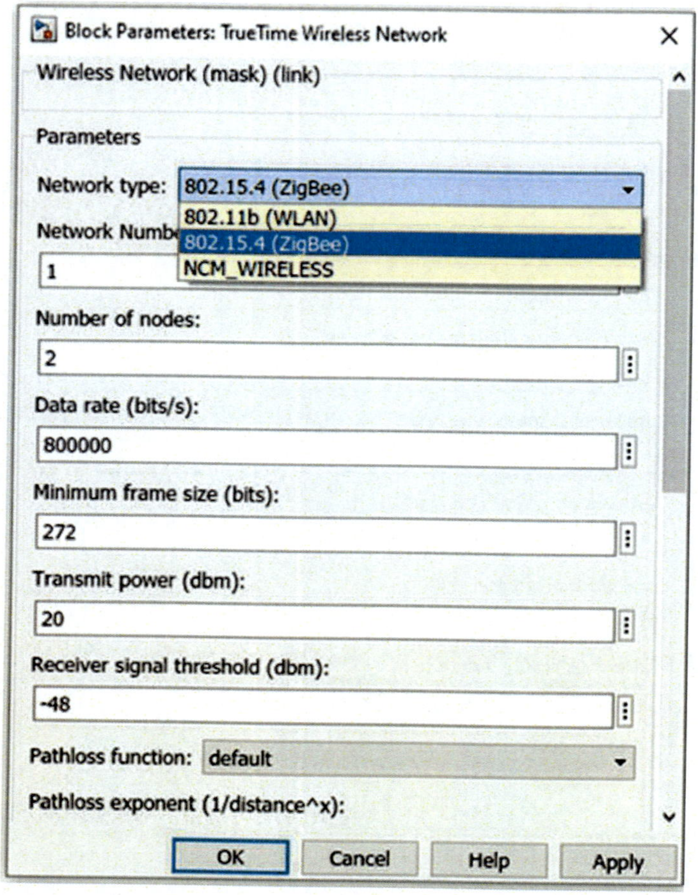


**Fig. 4** Dialog box of wireless network block in TrueTime

dropout which improves the performance of the control system, but for this, we have to pay in terms of higher energy consumption in the communication system [17].

## *4.2 Importance of Sampling Used in NCS*

It is observed in wireless networked control system that, when communication between sensor to controller (S–C) and controller to actuator (C–A) takes place through wireless network, then the sampling process and sampling rate $h$ play a very important role. To get expected output from plant or process, the sampling rate ($h$) used at sensor node and used at PID controller must be the same. This can be supported by doing practical simulation using TrueTime. Responses for three different

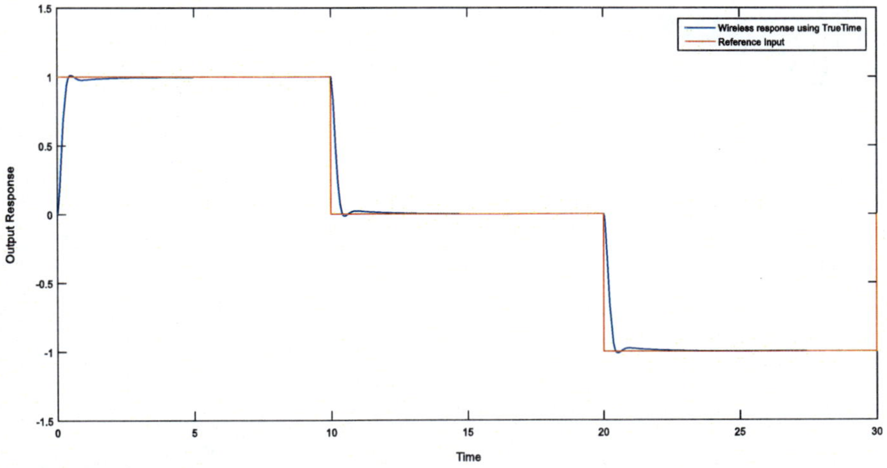


**Fig. 5** Simulated response using TrueTime for plant1 with $h_c = 10$ ms and $h_s = 10$ ms

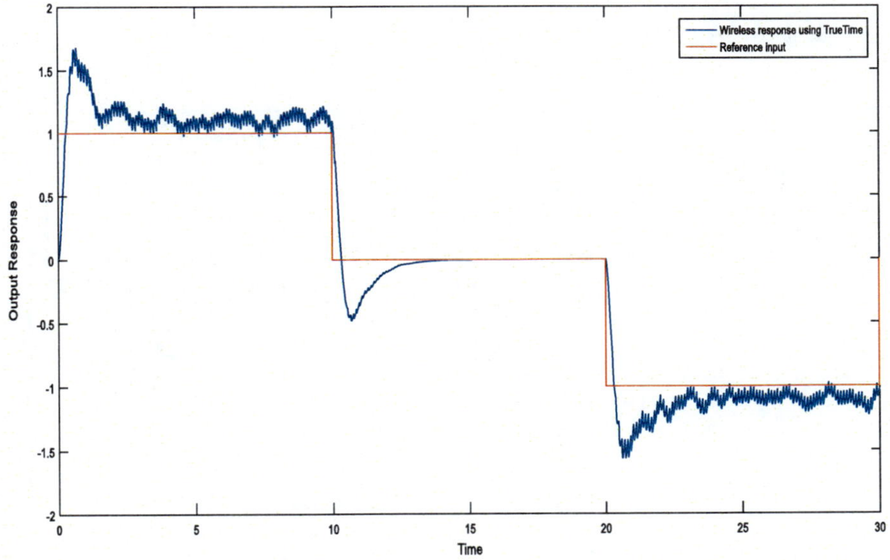


**Fig. 6** Simulated response using TrueTime for plant 1 with $h_c = 10$ ms and $h_s = 11$ ms

cases are shown in Figs. 5, 6 and 7. As shown in Fig. 5, response is as expected if sampling rate at sensor and in initialization file of Kernel block configured as discrete PID controller (Init file) is the same i.e. $h_C = 10$ ms and $h_S = 10$ ms. As shown in Figs. 6 and 7, the sampling rate at sensor and in initialization file of Kernel block is not the same and hence output responses are degraded due to packet loss that happens due to different sampling rate (Fig. 8).

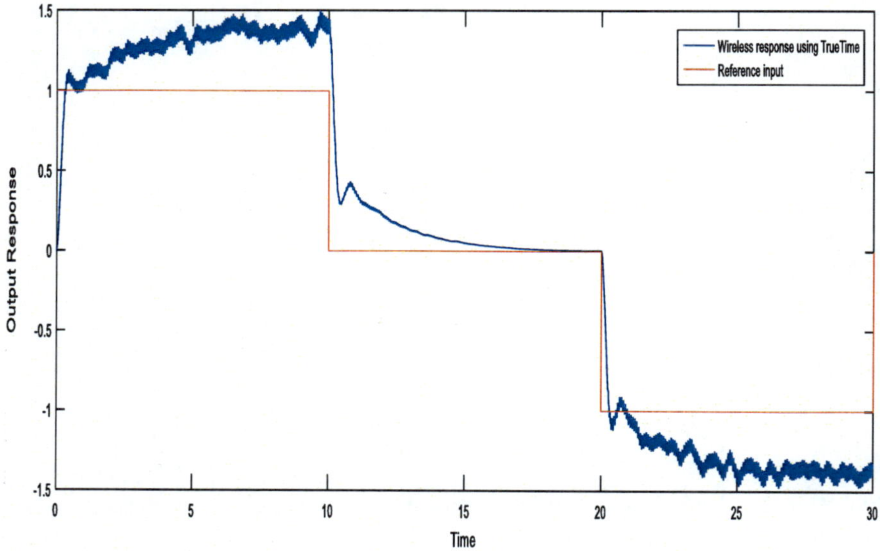


**Fig. 7** Simulated response using TrueTime for plant 1 with $h_c = 10$ ms and $h_s = 14$ ms

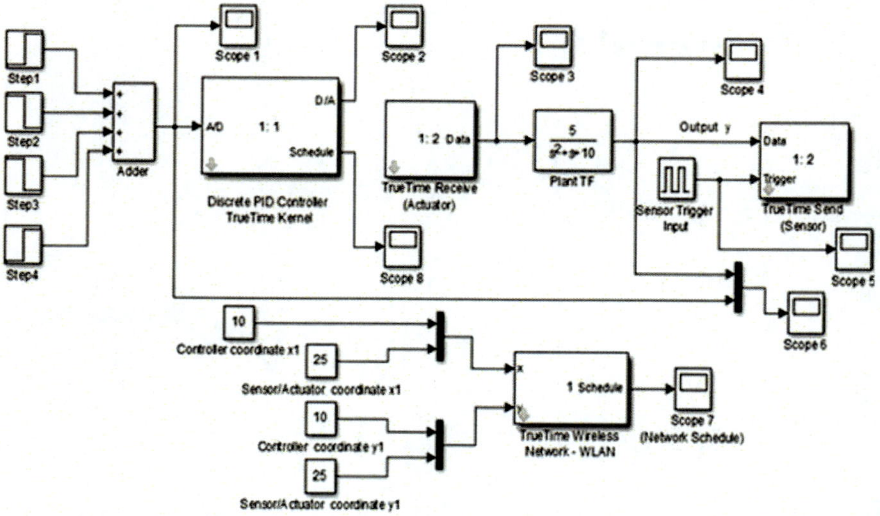


**Fig. 8** Simulink diagram using TrueTime

While simulation using TrueTime, it is observed that for successful recovery of process output, optimum sampling rate $h = 10$ ms must be used at sensor node and in $m$-file of Kernel while realizing Kernel block as discrete PID controller. Suppose sampling rate at sensor node is $h_S = 12$ ms and sampling rate at PID controller is 10 ms then after every 12 ms, sensor samples output of process, does the necessary conversion and sends the information and feedback signal to PID controller through wireless network. Now since sampling rate used at controller is $h_C = 10$ ms then after every 10 ms, controller checks for necessary information signal from sensor and thus there is a mismatch in checking the process of controller which results in data loss at the controller i.e., packet dropout. Due to this, there is no expected output response from the plant. Hence the same sampling rate must be used at sensor node and PID controller.

$$\text{Plant 1}: G_1(S) = \frac{5}{s^2 + s + 10} \quad \text{(second order system)}$$

## 5 Simulation Using MATLAB and TrueTime

In this section, we will investigate simulation and closed loop response using MATLAB simulink [18] and using TrueTime for five different transfer functions. For both cases, closed loop responses are given in Figs. 9, 10, 11, 12 and 13. While doing simulation using MATLAB, discrete PID controller [19] in ideal form with sampling 0.01 sec is selected, discrete PID controller is tuned, PID gain parameters $K$, $K_i$, $K_d$ [18, 19] are updated and noted, and then, simulink is run for the required time. Closed loop output responses using MATLAB simulink are given in Figs. 9, 10, 11, 12 and 13 for five different transfer functions like second order, non-minimum phase, unstable system.

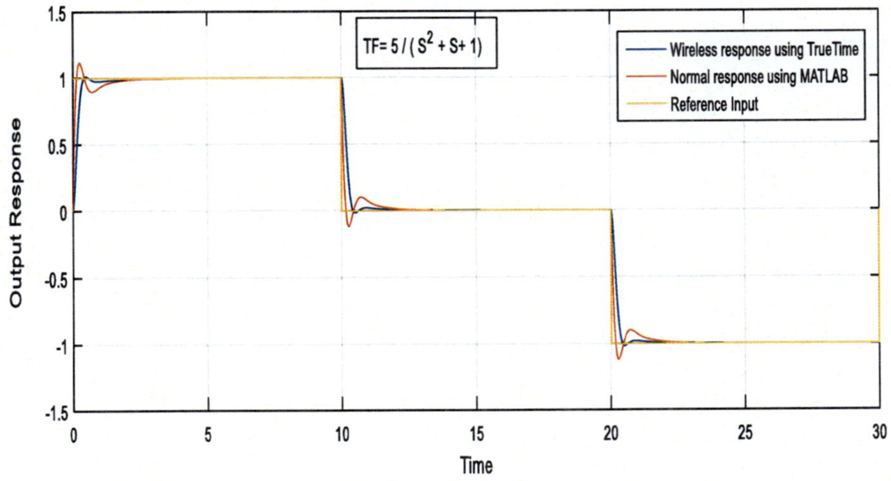


**Fig. 9** Simulated response using TrueTime and MATLAB for plant 1

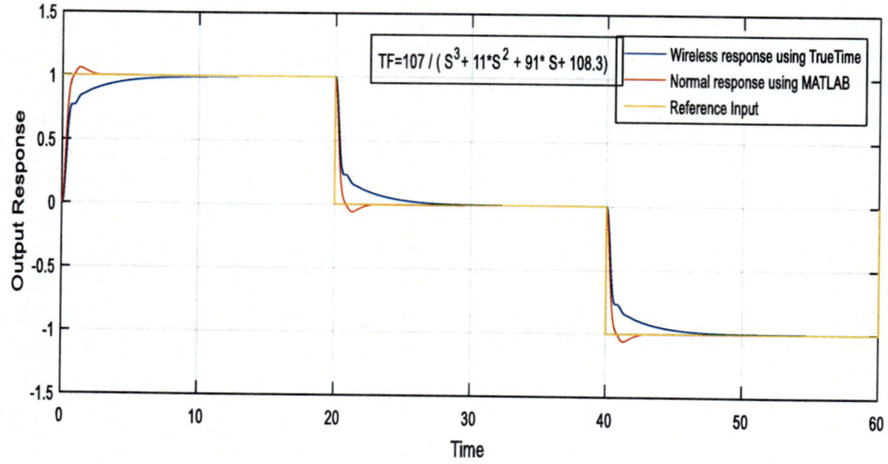


**Fig. 10** Simulated response using TrueTime and MATLAB for plant 2

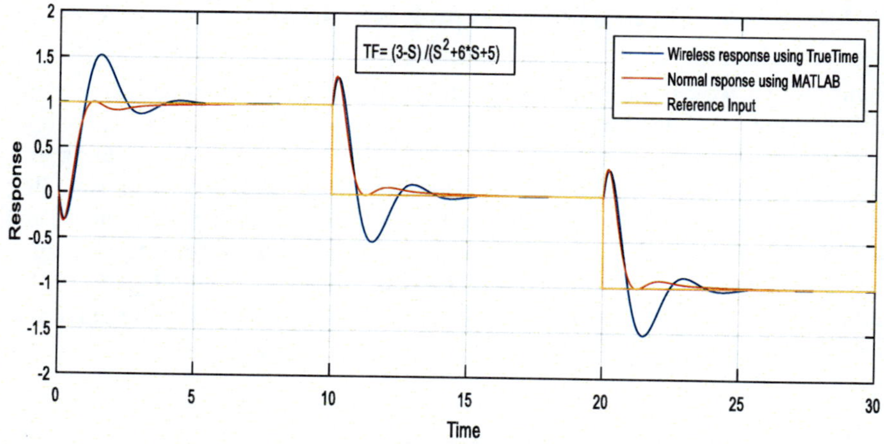


**Fig. 11** Simulated response using TrueTime and MATLAB for plant 3

For the same transfer function, simulated response using TrueTime is also given in Figs. 9, 10, 11, 12 and 13. For wireless simulation using TrueTime, it must be noted that the same PID gain parameters $K$, $K_i$, $K_d$ are used for coding and development of *m*-file to realize Kernel block as discrete PID controller. Sampling rate at sensor is selected as 0.01 s. Responses for both cases using same PID gain parameters $K$, $K_i$, $K_d$ cannot overlap. This is due to sensor to controller and controller to actuator delay and packet dropout due to use of wireless network for communication. But it is must be noted that the nature of response using TrueTime is comparable to that of response using MATLAB though they are not same. Both responses may have different overshoots, rise time, setting time, etc. It is possible to overlap both graphs if we use different PID gains parameters $K$, $K_i$, $K_d$ for simulation using TrueTime which is not shown in this paper.

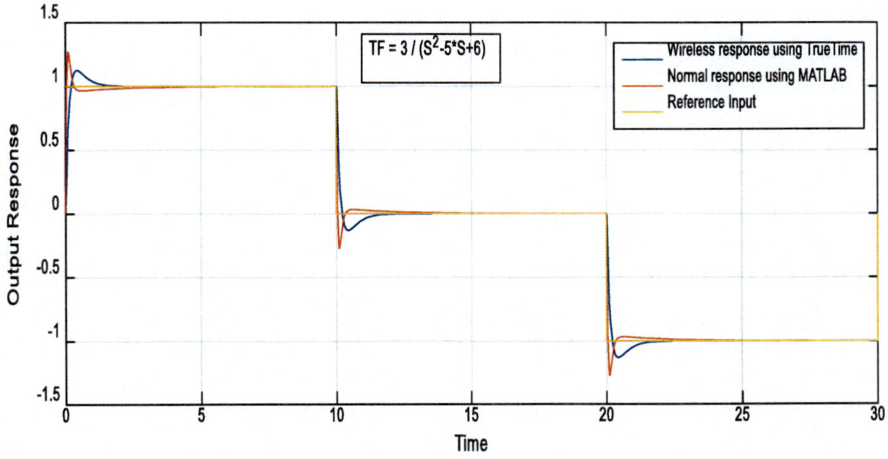


**Fig. 12** Simulated response using TrueTime and MATLAB for plant 4

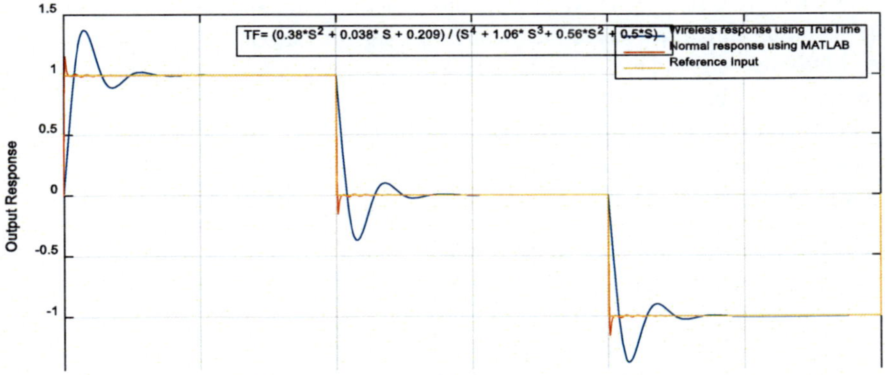


**Fig. 13** Simulated response using TrueTime and MATLAB for plant 5

Simulink diagram using TrueTime is shown in Fig. 8, and parameters used for wireless network while doing simulation using TrueTime are given as follows—Type of Wireless Network: 802.11b (WLAN), Data Rate: 800,000 Bits/s, Minimum Frame Size: 272 Bits, Transmit Power: 25 dBm, Receiver Signal Threshold: − 48 dBm, Path Loss Exponent: 3.5, ACK Timeout: 0.00004 s, Error Coding Threshold: 0.03.

**Development and working of** Fig. 8

It must be noted that simulink diagram given in Fig. 8 is a wireless networked feedback control system where in feedback path wireless network is present. Kernel block in TrueTime is programmed as discrete PID controller by writing initialization code (init file) and function code using TrueTime, because kernel block worked as computer and is responsible for all functions like I/O operation, scheduling, A/D and D/A conversion, set up timers, change task attributes, etc. Note that one can write

C++ code also. No other block in TrueTime requires coding. TrueTime receive block functions as actuator and TrueTime send block functions as sensor, both blocks are connected to Plant TF block. Written number 1:1, 1:2 and 1:2 written, respectively, on a kernel block, TrueTime receive and TrueTime send block indicates that the kernel block (PID controller) acts as node 1 and TrueTime receive and TrueTime send both block act as node 2 and for communication among each other they use wireless channel 1 which is 802.11b (WLAN). So first 1 denotes wireless channel, and second 1 or 2 denotes node number. Note that sampling rate at sensor node and in init file of discrete PID controller must be equal and its preferred value is 0.01 s.

Output signal of plant TF block is continuously sampled by sensor block (TrueTime send), and sampled values are sent to the kernel block (discrete PID controller) via wireless network 802.11b (WLAN). It must be noted that this is sensor to controller communication (S–C), delay and packet loss may take place at the controller. Depending upon received values from sensor, discrete PID controller reacts generate signal for actuator and send this signal to actuator again via wireless network 802.11b (WLAN). It must be noted that this is controller to actuator communication (C–A), delay packet loss may also take place at the actuator. Actuator accordingly generates the actuating signal to control the plant to get desired output from the plant or process. Different signals like process output, the network schedule signal at different blocks can be observed using scope.

Plant 1: $G_1(S) = \frac{5}{s^2+s+10}$ (second order system)
Plant 2: $G_2(S) = \frac{107}{s^3+11s^2+91s+108.3}$
Plant 3: $G_3(S) = \frac{3-s}{s^2+6s+5}$ (non-minimum phase system)
Plant 4: $G_4(S) = \frac{3}{s^2-5s+6}$ (unstable system).
Plant 5: $G_1(S) = \frac{0.38*s^2+0.038*s+0.209}{s^4+1.06*s^3+0.56*s^2+0.5*s}$.

For plant 1, $K = 11.2893$, $K_i = 1.3046$, $K_d = 0.1778$ (Fig. 9).
For plant 2, $K = 2.0268$, $K_i = 1.9291$, $K_d = 0.07680$ (Fig. 10).
For plant 3, $K = 2.6286$, $K_i = 0.6767$, $_d = 0.0623$ (Fig. 11).
For plant 4, $K = 70.403$, $K_i = 0.5222$, $K_d = 0.1512$ (Fig. 12).
For plant 5, $K = 17.3826$, $K_i = 4.6692$, $K_d = 7.7383$ (Fig. 13).

## 6 Conclusion

Both closed loop responses (using TrueTime MATLAB) for the same PID gain parameters $K$, $K_i$, $K_d$, cannot overlap due to sensor to controller controller to actuator delay and packet dropout due to use of wireless network for communication. But it must be noted that the nature of output response using TrueTime is comparable with the responses using MATLAB. Both responses may have different overshoots, rise time, setting time, etc. It is possible to overlap both responses if we use different PID gains parameters $K$, $K_i$, $K_d$ in init file of Kernel block using TrueTime which is not shown in this paper. It can also be noted that for this investigation, same PID gain parameters $K$, $K_i$, $K_d$ obtained using MATLAB simulink for given plant TF were used to develop init file for kernel block in TrueTime.

It is observed that in wireless networked control system, when communication between sensor to controller (S–C) and controller to actuator (C–A) takes place through wireless network, e.g., 802.11b WLAN, ZigBee, then the sampling process and the sampling rate ($h$) play a very important role. To get the expected output from plant or process, the sampling rate used at sensor node and at PID controller must be the same.

## References

1. Park P, Coleri S, Fischione EC, Lu C, Johansson KH (2017). Wireless network design for control system: a survey. https://doi.org/10.1109/COMST.2017.2780114
2. Gupta RA, Chow M-Y (2010) Networked control system: overview and research trends. IEEE Trans Ind Electron 57(7)
3. Mkondweni NS (2013) Design and implementation of linear robust networked control system. Ph.D. thesis submitted to Cape Peninsula University of Technology, Cape Town, South Africa, March 2013
4. Cervin A, Henriksson D, Ohlin M (2016) TrueTime 2.0 reference manual. Department of Automatic Control, Lund University, Sweden
5. Mahmoud MS, Sabih M (2013) Experimental investigation for distributed networked control system. IEEE Syst J. https://doi.org/10.1109/jsyst.2012.2228122
6. Abdul-Hussain RN, Awad OA (2019) Studying the effect of sampling time and network load on wireless networked control systems. J Al-Qadisiyah Comput Sci Math 11(3):31–42
7. Aung NP, Naing ZM, Wai MMM, Tun HM (2016) Simulation of wireless networked control system using TrueTime and MATLAB. Int J Sci Technol Res 5(06)
8. Anderssonr M, Henriksson D, Cervin A, Arzen K-E (2005) Simulation of wireless networked control systems. In: Proceedings of the 44th IEEE conference on decision and control, and the European control conference 2005 Seville, Spain, 12–15 Dec 2005
9. Sarangpani J, Xu H (2016) Optimal networked control system with MATLAB. Taylor & Francis
10. Wittenmark B, Astrom KJ, Arzen K-E (2002) Computer control: an overview. Technical report, IFAC Professional Brief
11. Arzen K (1999) A simple event-based PID controller. In: IFAC world congress
12. Wang X, Lemmon MD (2009) Self-triggered feedback control systems with finite-gain L2 stability. IEEE Trans Autom Control 54(3):452–467
13. Tabuada P (2007) Event-triggered real-time scheduling of stabilizing control tasks. IEEE Trans Autom Control 52(9):1680–1685
14. Araujo J, Mazo M, Anta A, Tabuada P, Johansson KH (2014) System architectures, protocols and algorithms for aperiodic wireless control systems. IEEE Trans Ind Inform 10(1):175–184
15. Lunze J, Lehmann D (2010) A state-feedback approach to event-based control. Automatica 46(1):211–215
16. Franklin G, Powell J, Workman M (1997) Digital control of dynamic systems. Addison Wesley Longman
17. Park P, Marco PD, Fischione C, Johansson KH (2013) Modelling and optimization of the IEEE 802.15.4 protocol for reliable and timely communications. IEEE Trans Parallel Distrib Syst 24(3):550–564
18. A reference manual of MATLAB R 2016. MathWorks, USA
19. Katsuhiko O (2013) Modern control engineering, 5th edn. Prentice Hall

# Comparative Analysis of Different Methods of Two Degree of Freedom Controller Designs on Varied Systems

**Debasish Sur**

## 1 Introduction

A good system will have a good ability to track set point as well as eliminate the disturbance input in it. Moreover, it should also have small settling time and less peak overshoot about its set point. Disturbance can be very effectively eliminated by increasing the integral constant, but it leads to high-peak overshoot. So we need to have a compromise between the set point tracking and disturbance rejection in case of one degree of freedom (1 DOF) controller. Thereby the concept of two degree of freedom (2 DOF) controller [1] comes into purpose. If a controller is able to control two closed loop transfer functions simultaneously and independently, then, it can be termed as a 2 DOF controller. So 2 DOF controllers can provide good set point tracking response as well as effective disturbance rejection [2]. Usually, the disturbances are low-frequency load disturbance signals. General PI-PD controllers do not have the ability of simultaneous set point tracking and disturbance rejection [3]. 2 DOF configurations are able to provide proper set point tracking as well as rejection of external noise disturbances. In this paper, we have considered various methods of design of 2 DOF controller designs and have carried out analysis on stable, unstable, and on processes with time delays. Here, dynamic behavior of a system is represented by linear time invariant (LTI) system.

D. Sur (✉)
Department of Electrical Engineering, Academy of Technology, Adisaptagram, West Bengal, India
e-mail: debasishsur88@gmail.com



P. H. J. Chong et al. (eds.), *Emerging Electronics and Automation*, Lecture Notes in Electrical Engineering 937, https://doi.org/10.1007/978-981-19-4300-3_8

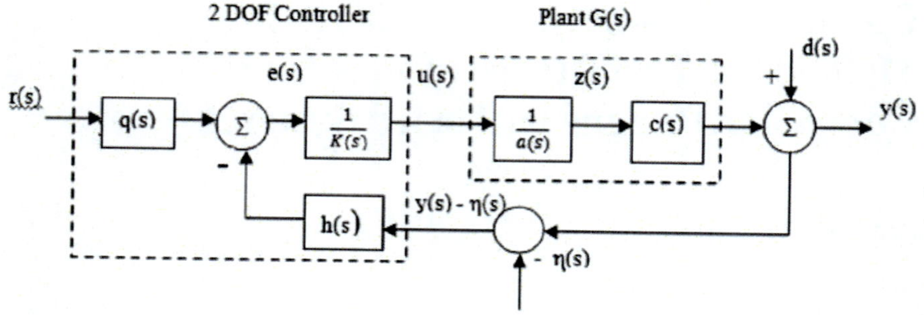


**Fig. 1** Control system [1]

## 2 Two Degree of Freedom (2 DOF) Control

Designing of control system is a multi-objective problem, so two degree of freedom (2 DOF) control systems are advantageous compared to one degree of freedom (1DOF) control systems.

**Assumptions**: (i) nth order LTI system represents the plant defined by strictly proper, rational transfer function $G(s) = \frac{y(s)}{u(s)} = \frac{c(s)}{a(s)}$ where degree $[c(s)] < n =$ degree$[a(s)]$.

(ii) Noise input enters the sensor and produces noisy output $[y(t) - \eta(t)]$, and controller output is produced by $r(t)$ and $[y(t) - \eta(t)]$. Dynamic behavior of the system with reference to Fig. 1 can be written as $u(s) = \frac{[q(s)-h(s)]}{k(s)}\begin{bmatrix} r(s) \\ y(s) - \eta(s) \end{bmatrix}$ and degree[q(s)–h(s)] ≤ degree$[k(s)]$ to avoid high-frequency noise amplifications. Here, $k(s)$ is a polynomial which defines order of the controller, and we need to determine compensator polynomials to produce desired closed loop performance [1].

2 DOF controller structure can be of different forms. Feedforward configuration and two parameter configuration have been considered for analysis. Multiple integration magnitude optimum method of tuning [4] is used for feedforward configuration and linear algebraic method (LAM) [5], and 2 DOF PI-PD compensation [1] techniques are used. We have made a comparative analysis among the three methods of tuning of 2 DOF controller.

## 3 Different 2 DOF Controller Design Methods

### *3.1 Multiple Integration Magnitude Optimum Method [4]*

Stable and fast closed loop response can be obtained using this method. It is a non-parametric approach in time domain using multiple integrations of process input

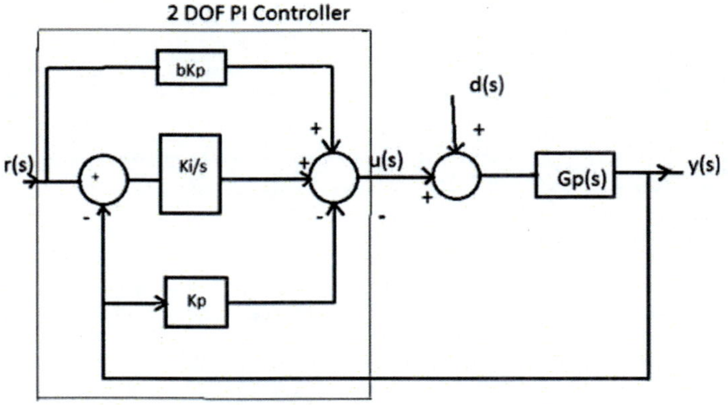


**Fig. 2** Two DOF controller in feed-forward configuration [4]

and output signals, and so it is termed as multiple integration magnitude optimum method. Through this method, a comparatively fast and non-oscillatory closed loop response can be achieved. The controller structure is shown below: (Fig. 2).

The closed loop transfer function from reference $R(s)$ to plant output $Y(s)$ is $G_{\mathrm{CL}}(s) = \frac{Y(s)}{R(s)} = \frac{G_P(s)\left\{bK_P + \frac{K_i}{s}\right\}}{1+G_P(s)\left\{K_P+\frac{K_i}{s}\right\}}$. The controller is designed in so that

$$G_{CL}(0) = 1 \tag{1}$$

and

$$\lim_{\omega\to 0}\left[\frac{d^{2k}|G_{CL}(j\omega)|^2}{d\omega^{2k}}\right] = 0;\ k = 1, 2, .., k_{\max} \tag{2}$$

Suppose the rational plant transfer function be $G_P(s) = \frac{K_C}{s}\frac{1+q_1s+q_2s^2+\ldots+q_ms^m}{1+p_1s+p_2s^2+\ldots+p_ns^n}e^{-sT_{\mathrm{del}}}$.

Solving for k = 1 and k = 2, the PI parameters obtained are $K_P = \frac{-P_1+\sqrt{P_1^2+\xi}}{\xi}$ $K_i = 0.5P_0(1-q^2)K_P^2$ $\xi = (1-q^2)(P_0P_2 - P_1^2)$.

where $P_0 = K_C$; $P_1 = K_C(p_1 - q_1 + T_C)$ $P_2 = K_C(q_2 - p_2 - T_Cq_1 + \frac{T_C^2}{2})$ and $P_0$, $P_1$, and $P_2$ are known as the characteristic area.

## 3.2 *Linear Algebraic Method (LAM) for 2 DOF Controller Design [5]*

If $G(s) = \frac{\mathrm{Num(s)}}{\mathrm{Den(s)}}$ where Num(s) and Den(s) are co-prime set point response (Fig. 3)

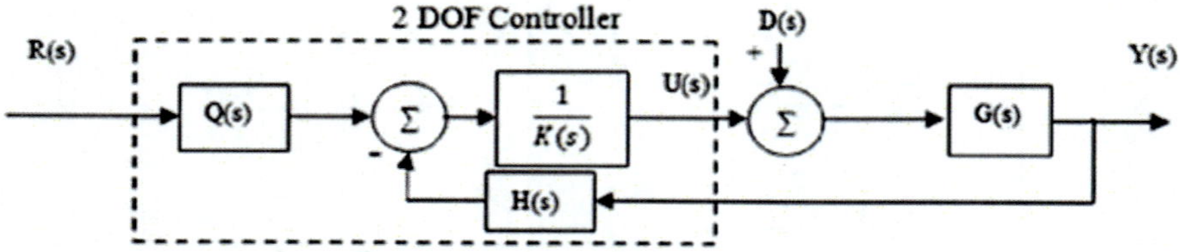


**Fig. 3** Two parameter configuration of 2 DOF controller

$$\frac{Y(s)}{R(s)} = \frac{Num(s)Q(s)}{K(s)Den(s) + H(s)Num(s)} \tag{3}$$

where $K(s) = \left(K_0 + K_1 s + K_2 s^2 + \ldots\right)$ and $H(s) = \left(H_0 + H_1 s + H_2 s^2 + \ldots\right)$. Let desired set point response be $G_0(s) = \frac{Num_0(s)}{Den_0(s)}$. Differentiators will not be required if the system is proper, and if the overall closed loop transfer function is proper, then high-frequency noise amplification can be avoided.

$$\frac{G_0(s)}{Num(s)} = \frac{Num_0(s)}{Den_0(s)Num(s)} = \frac{Num_p(s)}{D_{enp}(s)} \tag{4}$$

If $Num_0(s)$ and $Den_0(s)$ are co-prime, all common factors between $Num_0(s)$ and Num(s) get canceled. If degree of $Den_p(s) = p < 2n-1$, then an arbitrary polynomial $\overline{Den_p}(s)$ of degree (2n-1-p) is introduced. If degree $Den_p(s) = p = 2n\text{-}1$, set $\overline{Den_p}(s) = 1$. Therefore,

$$G_0(s) = Num(s) * \frac{\left[Num_p(s)\overline{Den_p}(s)\right]}{\left[D_{enp}(s)\overline{Den_p}(s)\right]} \tag{5}$$

Solving them the expressions formed are.

$Q(s) = Num_p(s)\overline{Den_p}(s)$ and $K(s)Den(s) + H(s)Num(s) = Den_p(s)\overline{Den_p}(s)$. Therefore, the controller parameters Q(s), H(s), and K(s) can be obtained. If H(s) be the transfer function from the disturbance d to the output *y*. So $A(s) = \frac{Num(s)K(s)}{K(s)Den(s)+H(s)Num(s)}$. Step disturbance rejection is achieved if and only if $A(0) = 0$. As $Num(0) \neq 0$, to get $A(0) = 0$, the necessary condition is $K(0) = 0$, i.e., $K_0 = 0$, and it is obtained by increasing the degree of the compensator by one.

## 3.3 *PI-PD Compensation for 2 DOF Controller Design [1]*

From Fig. 4, $U(s) = \frac{K_P s + K_I}{s}[R(s) - Y(s)] - \frac{K_D s}{\frac{K_D}{N}s + 1}Y(s)$ where $\frac{K_D}{N}$ is the low-pass filter time constant.

So

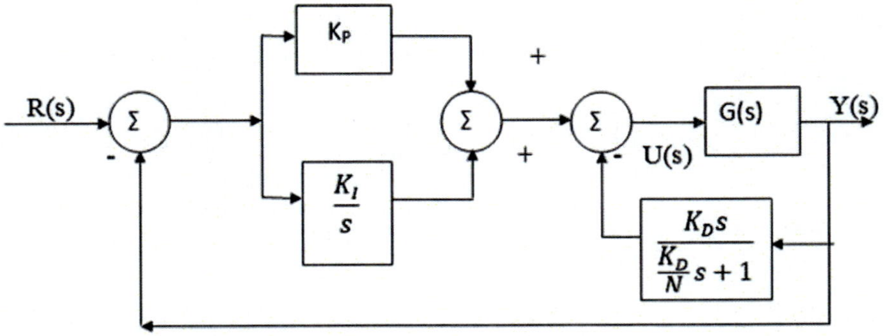


**Fig. 4** Practical PI-PD controller [1]

$$s\left(s+\frac{N}{K_D}\right)U(s) = (K_P s + K_I)\left(s+\frac{N}{K_D}\right)[R(s) - Y(s)] - Ns^2Y(s) \quad (6)$$

From Fig. 4,

$$K(s)U(s) = Q(s)R(s) - H(s)Y(s) \quad (7)$$

So the value of the parameters is $K(s) = \left(s + \frac{N}{K_D}\right)$;

$$H(s) = (N + K_P)s^2 + \left(K_I + \frac{NK_P}{K_D}\right)s + \frac{NK_I}{K_D} \approx Ns^2 + \frac{NK_P}{K_D}s + \frac{NK_I}{K_D} \quad (8)$$

$$Q(s) = K_P s^2 + \left(K_I + \frac{NK_P}{K_D}\right)s + \frac{NK_I}{K_D} \quad (9)$$

So the parameters $K(s)$ and $H(s)$ are determined by Eq. (8), but as $Q(s)$ does not affect the loop performance of the system, so $Q(s)$ can be any arbitrary polynomial so that $Q(0) = q_0 = H(0) = \frac{NK_I}{K_D}$ which ensures a robust steady-state error to zero for step changes in reference input.

## 4 Experimental Results

**System 1**: Applying multiple integration magnitude optimum method on a delayed process $G_P(s) = \frac{e^{-0.5s}}{s(1+s)(1+2s)}$.

The characteristic areas calculated are $P_0 = 1$, $P_1 = 3.5$, and $P_2 = 8.625$. For $q = 0.5$, the parameters calculated are (Tables 1 and 2).

**Table 1** Calculated values for multiple integration magnitude optimum method

| $q$ | $K_P$ | $K_I$ |
|---|---|---|
| 0.5 | 0.152 | 0.0086 |

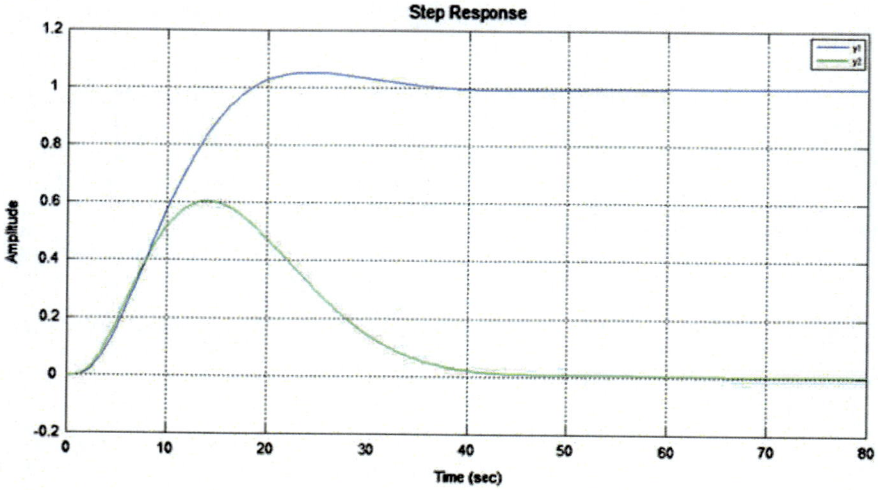


**Fig. 5** Plot of Y(s) w.r.t R(s) and Y(s) w.r.t D(s)

**System 2**: Applying multiple integration magnitude optimum method on a fourth-order integrating process $G_P(s) = \frac{1}{s(s+1)^4}$ The characteristic areas calculated are $P_0 = 1$, $P_1 = 4$, and $P_2 = 10$. For different values of $q$, the parameters as calculated are. (Fig. 5).

**Table 2** Calculated values for multiple integration magnitude optimum method

| $q$ | $K_P$ | $K_I$ |
|---|---|---|
| 0.2 | 0.138 | 0.00914 |
| 0.5 | 0.135 | 0.00683 |
| 0.7 | 0.1316 | 0.0044 |

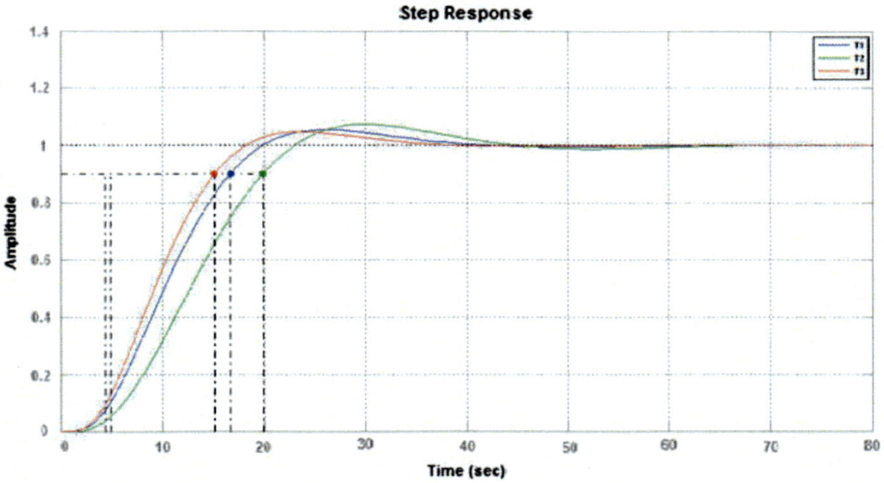


**Fig. 6** Plots of $Y(s)$ with respect to $R(s)$ for increasing values of '$q$'

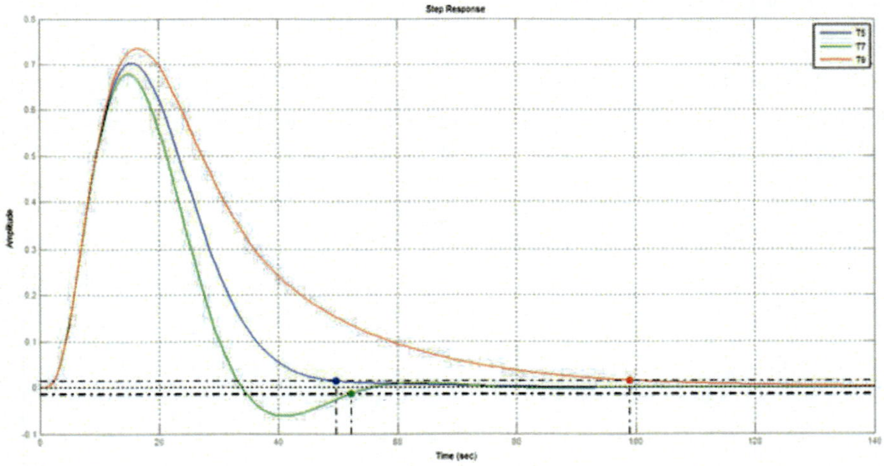


**Fig. 7** Plots of $Y(s)$ with respect to $D(s)$ for increasing values of '$q$'

**System 3**: Considering a plant $G_p(s) = \frac{2}{s(s^2+0.25s+6.25)}$ using LAM and let the desired set point response be $G_0(s) = \frac{20}{(s+10)(s^2+2s+2)}$. Using LAM, the parametric values are (Figs. 6 and 7).

$Q(s) = 10s^3 + 600s^2 + 12000\ s + 80{,}000$; $K(s) = s^3 + 71.8s^2 + 1918.7\ s$ and.

$H(s) = 11{,}406.1s^3 + 55{,}806.8s^2 + 100000\ s + 80{,}000$.

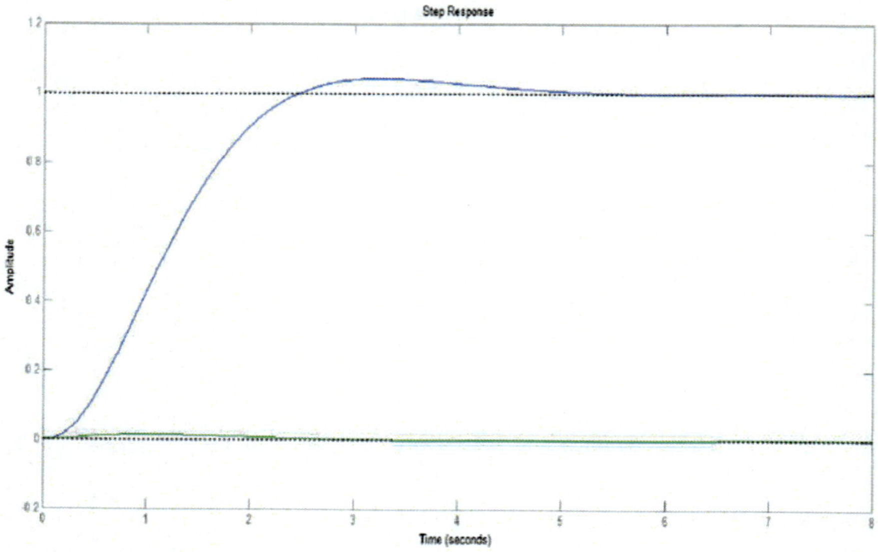


**Fig. 8** Plot of Y(s) w.r.t R(s) and Y(s) w.r.t D(s)

**System 4**: Considering a plant $G_P(s) = \frac{e^{-0.25s}}{s+1}$ using LAM and let the desired response be $G_0(s) = \frac{e^{-0.25s}}{0.85s+1}$. Linear algebraic method gives the controller parameters and responses as: Q(s) = 1; K(s) = 0.9 s and H(s) = s + 1.

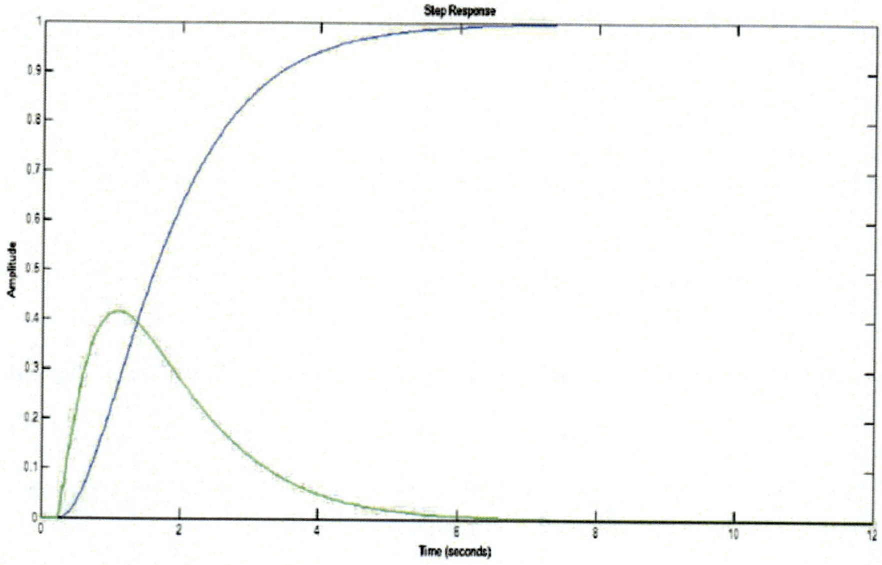


**Fig. 9** Plot of Y(s) w.r.t R(s) and Y(s) w.r.t D(s)

**System 5**: Considering a plant $G(s) = \frac{5}{s^2+6s+1}$ using PI-PD compensation method for 2 DOF controller $K_P = 2.4$, $K_I = 2$, $K_D = 0.4$. So $K(s) = s\left(s + \frac{N}{K_D}\right) = s^2 + 25\,s$ and $H(s) = Ns^2 + \frac{NK_P}{K_D}s + \frac{NK_I}{K_D} = 10s^2 + 60\,s + 50$ and $q_0 = \frac{NK_I}{K_D} = 50$ where $N = 10$ (Figs. 8, 9, 10 and 11).

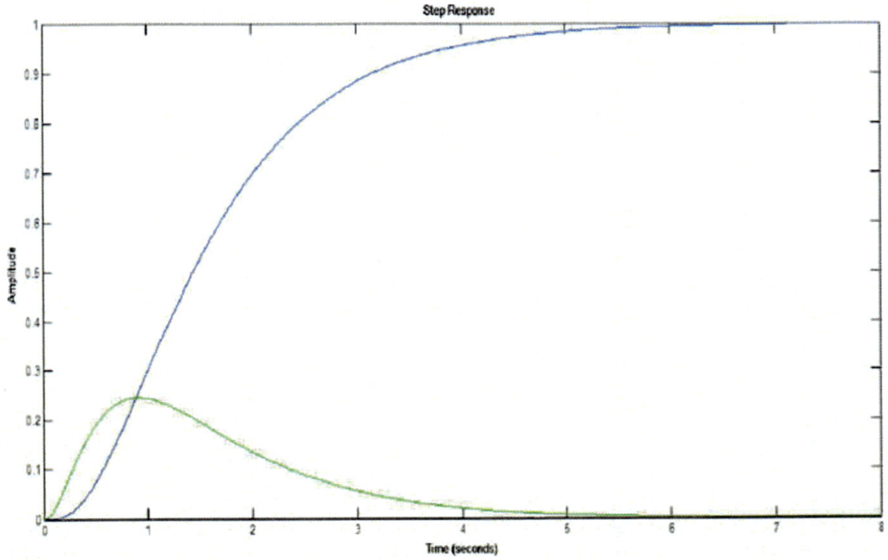


**Fig. 10** Plot of Y(s) w.r.t R(s) and Y(s) w.r.t D(s)

**System 6**: Considering a plant $G(s) = \frac{e^{-s}}{s+1}$ by PI-PD compensation method for 2 DOF controller, $K_P = 1.11$, $K_I = 0.7655$, $K_D = 0.35187$, and $N = 10$.

So $K(s) = s\left(s + \frac{N}{K_D}\right) = s^2 + 28.419s$; $H(s) = Ns^2 + \frac{NK_P}{K_D}s + \frac{NK_I}{K_D} = 10s^2 + 31.5457\,s + 21.755$ and $q_0 \frac{NK_I}{K_D} = 21.755$.

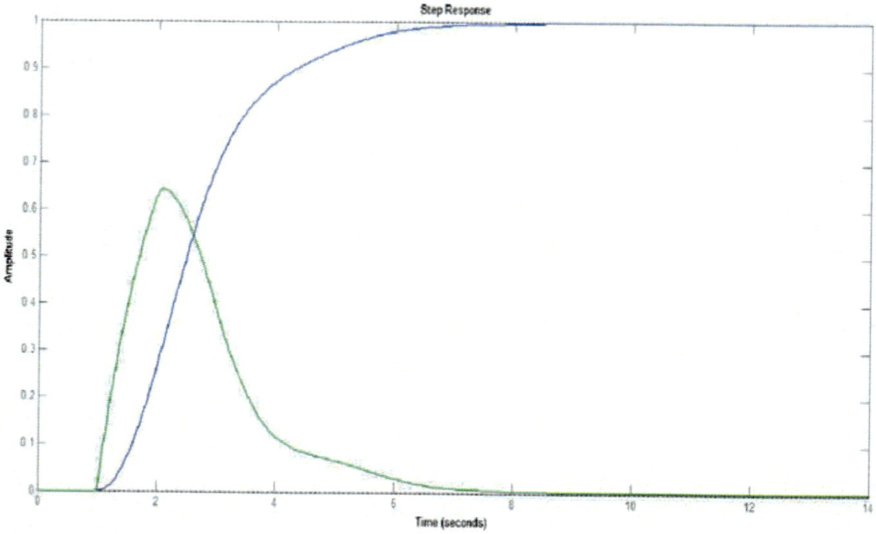


**Fig. 11** Plot of Y(s) w.r.t R(s) and Y(s) w.r.t D(s)

## 5 Conclusion

Accurate set point tracking can be obtained with a 2 DOF controller along with good disturbance rejection. Multiple integration magnitude optimum method, LAM, and PI-PD compensation method are used here for tuning. Multiple integration magnitude optimum method gives satisfactory performance, but overshoot for disturbance response is large. In PI-PD compensation technique, the general PI-PD controller parameters provide the values of the parameters by comparison with the 2 DOF structure. Here, pole placement technique is adopted [2].The LAM [5] gives stable responses even for unstable plants. The best part of this method is its simple method of determination of the parameters. Therefore, it can be concluded that the 2 DOF controllers give much better response both for set point and disturbance rejection when tuned using the LAM.

## References


1. Wolovich WA (2011) Automatic control systems, basic analysis and design. Oxford University Press
2. Chen D, Seborg DE (2002) PI/PID controller design based on direct synthesis and disturbance rejection. Industr Eng Chem Res 41(19):4807–4822
3. Lee Y, Park S, Lee M, Brosilow C (1998) PID controller tuning for desired closed loop responses for SI/SO systems. AIChE J 44(1)
4. Vrancic D, Strmcnik S (2011) Design of 2-DOF PI controller for integrating processes. In: proceedings of 2011, 8th Asian Control Conference (ASCC) Kaohsiung, Taiwan, 15–18

5. Chen CT (1987) Introduction to linear algebraic method for control system design. IEEE Control Syst Mag
6. Dukkipati RV (2006) Analysis and design of control systems using MATLAB. New Age Inter Publishers
7. Araki M, Taguchi H (2003) Two degree of freedom PID controllers. Inter J Control, Autom Syst Vol 1
8. Vrancic D, Strmcnik S, Juricic DA (2001) Magnitude optimum multiple integration tuning method for filtered PID controller. Tech Communique Automatica 37:1473–1479
9. Chen CT, Seo B (1990) Applications of linear algebraic method for control system design, IEEE Control Syst Mag
10. Vilanova R, Alfaro VM, Arrieta O (2011) Analytical robust tuning approach for two degree-of-freedom PI/PID controllers. Eng Lett 19:3
11. Shibasaki H, Endo J, Hikichi Y, Tanaka R, Kawaguchi K, Ishida YA (2012) Modified internal model for an unstable plant with an integrator in continuous-time system. Inter J of Inf Electron Eng 3(4)

# Robust Stabilization of an Inverted Pendulum Using ANFIS Controllers

**Ruchika Lamba, Sunil Kumar Singla, and Swati Sondhi**

## 1 Introduction

The dynamics of an inverted pendulum (IP) is found to be nonlinear. So, testing the performance of these systems using different control schemes is challenging [1]. The IP control is related to the control of physical systems like self-balancing robots and segways. Inverted pendulums can be found as linear, spherical, vertical, and rotary [2–4] etc.

The IP control is of three types, *i.e.,* the swing up control [5, 6], the stabilization control [7, 8], and the tracking control [9]. Due to ease of implementation and well-known simple structure, PID controllers are most prevalent [10]. Wang discussed the control of inverted pendulums using PID controllers [9]. A double-loop PID controller scheme has been proposed by Nasir [11] and Prasad [12] for an inverted pendulum system. Ghosh et al. used the control theory based on pole placement for controlling this system [13]. Sliding mode control strategy for such a system has been stated by Wai and Chang [14], Tsai et al. [15] and Wang [16], respectively. Linear quadratic regulator [17], Lyapunov approach [18], and full-state feedback [19] method have also been implemented for regulating the system.

This paper proposes ANFIS controllers which are robust against changes in varying pendulum mass and frictional forces. The performance of the proposed ANFIS controller has been compared with PID controller described by Wang [9].

R. Lamba (✉) · S. Kumar Singla · S. Sondhi
Department of Electrical and Instrumentation Engineering, Thapar Institute of Engineering and Technology, Patiala, Punjab 147004, India
e-mail: ruchika.mehta@thapar.edu

S. Kumar Singla
e-mail: sssingla@thapar.edu

S. Sondhi
e-mail: swati.sondhi@thapar.edu



P. H. J. Chong et al. (eds.), *Emerging Electronics and Automation*, Lecture Notes in Electrical Engineering 937, https://doi.org/10.1007/978-981-19-4300-3_9

The organization of the paper is as follows:

The dynamics of linear IP system has been explained in Sect. 2. The design procedure for the proposed controller has been discussed in Sects. 3 and 4 discusses the simulation results.

## 2 Dynamic Model of an IP System

The representation of an IP system is given in Fig. 1.

Here, $l$ represents the length of the pendulum from center mass, $M$ the cart mass, and m the pendulum mass. Further, $F$ represents force in $x$ direction. Angle $\theta$ signifies pendulum angle from the vertical axis.

The Lagrange's equations are given as [9]:

$$(M + m)\ddot{x} + \mathrm{ml}\cos\theta\ddot{\theta} - \mathrm{ml}\sin\theta\dot{\theta}^2 = F \tag{1}$$

$$\cos\theta\ddot{x} + l\ddot{\theta} - g\sin\theta = 0 \tag{2}$$

Assuming $x_1 = x$ and $x_3 = \theta$, the state equations are as follows:

$$\dot{x}_1 = x_2 \tag{3}$$

$$\dot{x}_2 = \frac{-\mathrm{mg}\cos x_3 \sin x_3 + \mathrm{ml}\sin x_3 x_4^2 + F}{M + m\sin^2 x_3} \tag{4}$$

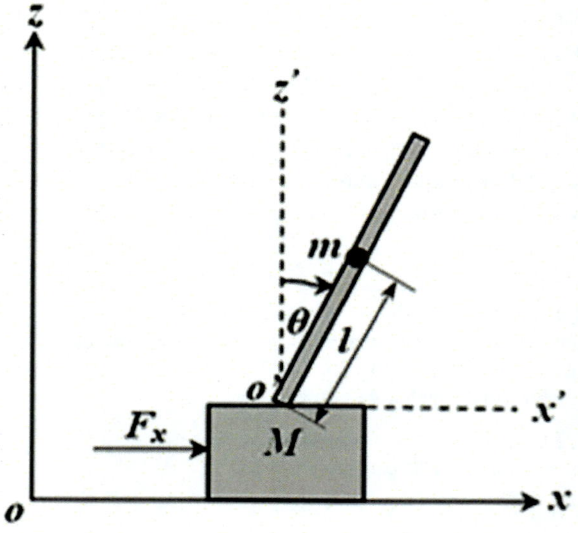


**Fig. 1** Inverted pendulum [9]

$$\dot{x}_2 = \frac{-\text{mg} \cos x_3 \sin x_3 + \text{ml} \sin x_3 x_4^2 + F}{M + m \sin^2 x_3} \tag{5}$$

$$\dot{x}_4 = \frac{-\text{ml} \cos x_3 \sin x_3 x_4^2 - \cos x_3 F + (M + m) g \sin x_3}{\text{Ml} + \text{ml} \sin^2 x_3} \tag{6}$$

The above state equations are used in building the Simulink representation of the system.

## 3 ANFIS Controller Design

The blend of neural networks and fuzzy logic leads to ANFIS and was introduced in 1993 by Jang. The ANFIS controller design is based on inferencing using a suited membership function to match the input–output dataset with minimal error.

The method of designing an ANFIS controller is described as follows.

### 3.1 Dataset Generation

To generate the dataset, conventional PID controllers proposed in the literature [9] have been used. Two different datasets are acquired for designing ANFIS controllers as one controller is designed for position control of cart, and second controller will keep the pendulum in the vertical upright position. Also, the dataset is obtained by changing the value of pendulum mass and retuning the PID controller to achieve the optimal performance.

The dataset is created using each control input $\boldsymbol{u}$, the operating range of $\boldsymbol{e}$ and its variation $\mathbf{d}\, \boldsymbol{e}/\, \mathbf{d}\, \boldsymbol{t}$.

$$\text{Robust dataset}_i = \begin{bmatrix} e^T & \dot{e}^T & u_k^{0.1} \\ e^T & \dot{e}^T & u_k^{0.2} \\ \vdots & \vdots & \vdots \\ e^T & \dot{e}^T & u_k^{0.6} \end{bmatrix} \forall k = 1, 2, \ldots, n \tag{7}$$

Here, $U_k$ is the control vector for $k$th motor and $n$ the number of control inputs.

### 3.2 Defining Parameters of ANFIS

The type and number of membership functions, epochs, error tolerance, and the training algorithm used are mentioned in this step. For each input variables, the number of membership functions has been chosen as three and is *Gaussian shaped* membership function. The training algorithm is *hybrid* with no tolerance in error. The number of epochs has been chosen as 12.

### 3.3 ANFIS Training

The training of the network is done using "*anfis" command.* A "*.fis*" file is created after ANFIS has been trained. This "*.fis*" file is used for evaluating the system response.

## 4 Results and Discussion

Two ANFIS controllers have been designed; one controller is efficiently maintaining the pendulum in upright position, and at the same time, the other controller displaces the cart by 1 m distance linearly. Table 1 represents the IP system parameters.

The parametric variations are achieved by varying the pendulum mass from 0.1 to 0.5 kg. Figure 2 shows the Simulink model representing the control structure for the IP system utilizing two ANFIS controllers. The actuating signal will be difference in output of the two controllers as the pendulum and cart move in opposite directions.

The performance of the designed ANFIS controller for the position and angle control of the IP system has been compared with PID controllers suggested by Wang.

**Table 1** IP system parameters

| $M$ (in kg) | $m$ (kg) | $l$ $(m)$ | $g$ (m/s$^2$) |
|---|---|---|---|
| 1 | 0.1 | 0.3 | 9.8 |

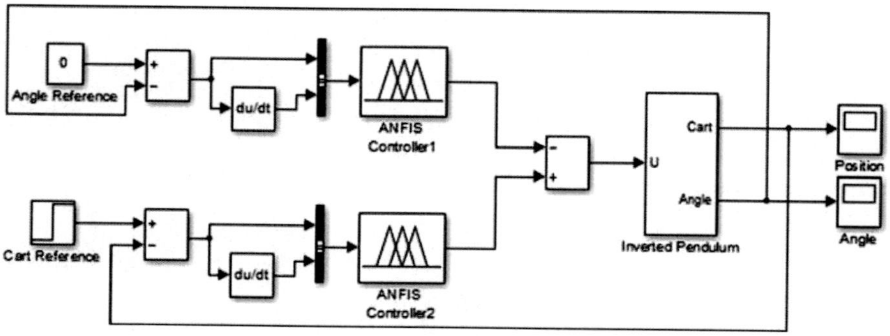


**Fig. 2** Control structure of the IP system with ANFIS controllers

Fig. 3a–b and Fig. 4a–b show the response of the controller when pendulum mass is 0.1 kg and 0.5 kg, respectively.

Figure 5 presents simulation results in the presence of friction.

The quantitative parameters of the IP system using the designed controller and the PID controller existing in literature have been presented in Tables 2 and 3. A significant reduction has been observed in most of the performance parameters, respectively.

It can be concluded that controller performance and robustness is much better for ANFIS controller than conventional PID controller.

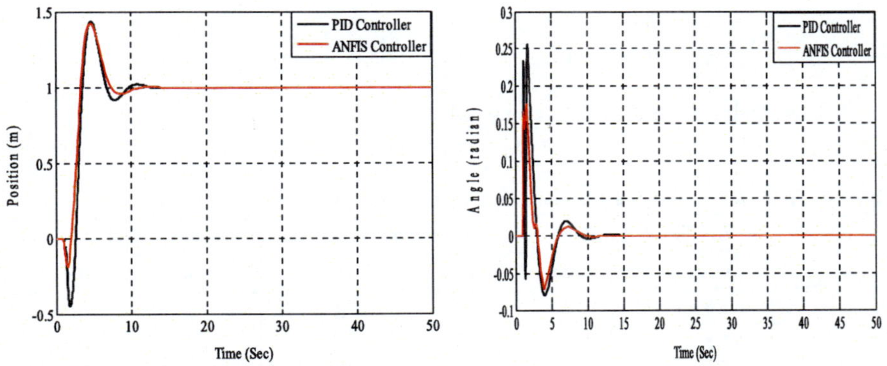


(a) position control (b) angle control

**Fig. 3** Stabilization of the IP system when pendulum mass m = 0.1 kg

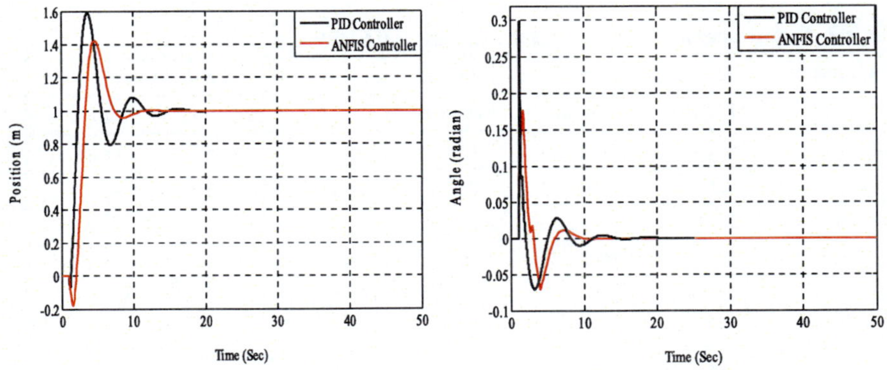


(a) position control (b) angle control

**Fig. 4** Stabilization of the IP system when pendulum mass is 0.5 kg

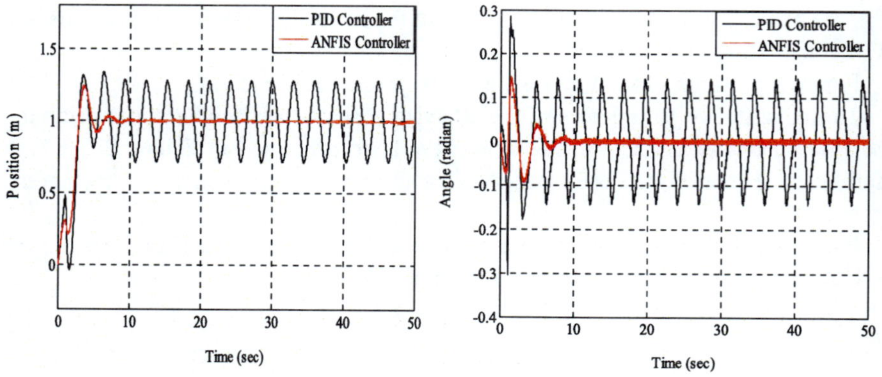


(a) position control (b) angle control

**Fig. 5** Stabilization of the IP system in presence of friction

**Table 2** Performance analysis when pendulum mass is 0.1 kg

| Parameters | ANFIS controller | | PID controller | | % Improvement | |
|---|---|---|---|---|---|---|
| | Position | Angle | Position | Angle | Position | Angle |
| $M_p$ | 0.42 | 0.17 | 0.42 | 0.256 | 0 | 50.58 |
| $M_u$ | 0.18 | 0.067 | 0.41 | 0.076 | 127.7 | 13.43 |
| $t_r$ (Sec) | 3.9 | – | 3.9 | – | 0 | – |
| $t_s$ (Sec) | 8.9 | 9.1 | 9.2 | 9.1 | -3.37 | 0 |
| $e_{ss}$ | 0 | 0 | 0 | 0 | 0 | 0 |

**Table 3** Performance analysis when pendulum mass is 0.5 kg

| Parameters | ANFIS controller | | PID controller | | % Improvement | |
|---|---|---|---|---|---|---|
| | Position | Angle | Position | Angle | Position | Angle |
| $M_p$ | 0.42 | 0.175 | 0.60 | 0.3 | 42.85 | 71.42 |
| $M_u$ | 0.18 | 0.067 | 0.04 | 0.068 | 77.77 | −1.49 |
| $t_r$ (Sec) | 3.6 | – | 2.2 | – | 38.88 | – |
| $t_s$ (Sec) | 9 | 9.1 | 11.6 | 10.0 | – 28.88 | −9.89 |
| $e_{ss}$ | 0 | 0 | 0 | 0 | 0 | 0 |

## 5 Conclusion

In this paper, ANFIS controllers have been designed for stabilizing an IP. The dataset required is obtained from PID controllers proposed by Wang. The controller performance has been related with PID controller for varying pendulum masses. The simulation results show that the ANFIS controller has outclassed the PID controller in attaining improved performance and robustness to discrepancies in pendulum mass.

Further, it has been observed that upon considering the frictional forces the conventional controller fails to respond, while the proposed controller performs well. The proposed controllers can be instigated on other types of inverted pendulums which can be a future scope for further analysis.

## References

1. Astrom KJ, Furuta K (2000) Swinging up a pendulum by energy control. Automatica 36(2):287–295
2. Kumar V, Jerome J (2013) Robust LQR controller design for stabilizing and trajectory tracking of inverted pendulum. Procedia Eng 64:169–178
3. Liu GY, Nesic D, Mareels I (2008) Non-linear stable inversion-based output tracking control for a spherical inverted pendulum. Int J Control 81(1):116–133
4. Maravall D (2004) Control and stabilization of the inverted pendulum via vertical forces. In: Robotic welding, intelligence and automation. Lecture notes control information sciences, Berlin: Springer-Verlag 299:190–211
5. Mason P, Broucke M, Piccoli B (2008) Time optimal swing-up of the planar pendulum. IEEE Trans Autom Control 53(8):1876–1886
6. Muskinja N, Tovornik B (2006) Swinging up and stabilization of a real inverted pendulum. IEEE Trans Industr Electron 53(2):631–639
7. Bloch AM, Leonard NE, Marsden JE (2000) Controlled lagrangians and the stabilization of mechanical Systems I: the first matching theorem. IEEE Trans Autom Control 45(12):2253–2270
8. Chang LH, Lee AC (2007) Design of nonlinear controller for bi-axial inverted pendulum system. IET Control Theory Appl 1(4):979–986
9. Wang JJ (2011) Simulation studies of inverted pendulum based on PID controllers. Simul Model Pract Theory 19(1):440–449
10. Astrom KJ, Hagglund T (1995) PID controllers: theory, design and tuning. In: Instrument society of america, 2nd edition, USA
11. Nasir A (2007) Modeling and controller design for an inverted pendulum system. Faculty of Electrical Engineering, Universiti Teknologi Malaysia, Master of Technology Degree
12. Prasad LB, Tyagi B, Gupta HO (2011) Optimal control of nonlinear inverted pendulum dynamical system with disturbance input using PID controller & LQR. In: IEEE international conference on control system, computing and engineering 540–545
13. Ghosh A, Krishnan TR, Subudhi B (2012) Robust PID compensation of an inverted cart-pendulum system: an experimental study. IET Control Theory Appl 6(8):1145–1152
14. Wai RJ, Chang LJ (2006) Adaptive stabilizing and tracking control for a nonlinear inverted pendulum system via sliding-mode technique. IEEE Transaction on Industrial Electronics 53(2):674–692
15. Tsai CC, Yu CC, Chang CS (2011) Aggregated hierarchical sliding-mode control for spherical inverted pendulum. In: Proceedings of 2011 8th Asian control conference (ASCC), pp 914–919
16. Wang JJ (2012) Stabilization and tracking control of X-Z inverted pendulum with sliding-mode control. ISA Trans 51(6):763–770
17. Tang TF, Chong SH, Pang KK (2020) Stabilization of a rotary inverted pendulum system with double-PID and LQR control: experimental verification. Int J Autom Control 14(1):18–33
18. Ibanez CA, Gutierrez O, Sossa-Azuela H (2006) Lyapunov approach for the stabilization of the inverted spherical pendulum. In: 45th IEEE conference on decision & control, pp 6133–6137
19. Angeli D (2001) Almost global stabilization of the inverted pendulum via continuous state feedback. Automatica 37(7):1103–1108

# Sensors Energy and Power Electronics

# Design of a PV-Battery-Supercapacitor Water Pumping System Based on the Varying Operating Factor of the PV Module

Madhumita Das and Ratan Mandal

## 1 Introduction

Solar water pumping applications are one of the most valuable solar applications for off-grid agricultural applications [1, 2]. The design of the solar PV system water pumping system is important to ensure the reliability of supplying the required amount of water [3]. The proper selection and sizing of the appropriate components of a PV system are necessary to increase the system's reliability to deliver loads irrespective of the disturbances in the source of the system, efficient usage, and cost-effectiveness [4]. A PV system's various design and sizing methodologies have been studied based on technical, economic, and environmental parameters [5]. Numerous design components related to PV systems are PV module or array, load demand, DC/DC converter, maximum power point tracker, inverter, battery storage unit, etc. [6]. The primary design of the system takes account of the radiation and the ambient temperature of the site. The design of a PV system is primarily based on the meteorological parameters of the site. The system design's accuracy depends on the solar radiation model [7, 8] and the temperature on the PV panel model of the system. Solar radiation, environmental, and climate data greatly influence solar PV systems [9]. So, a precise solar radiation and climate data model are necessary for the proper and accurate design of SPVWPS.

Energy storage in a PV system improves the energy quality of the system. An optimal control, power, and energy management of PV systems with energy storage devices are reported [10, 11]. Battery and supercapacitors are also utilized in power flow control and energy feeding in the grid-tied system [12, 13]. In a stand-alone

M. Das (✉)
Applied Electronics and Instrumentation Engineering, Haldia Institute of Technology, Purba Medinipur, Haldia, WB 721657, India
e-mail: Madhumita666@gmail.com; Madhumita666@yahoo.co.in

R. Mandal
School of Energy Studies, Jadavpur University, Kolkata, WB 700032, India



P. H. J. Chong et al. (eds.), *Emerging Electronics and Automation*, Lecture Notes in Electrical Engineering 937, https://doi.org/10.1007/978-981-19-4300-3_10

system, SC and battery are used as a buffer against rapid fluctuations in solar radiations [14]. SC and battery are also used in solar water pumping systems [15, 16].

The size of the PV module is a vital factor for charging the battery, supercapacitor, and delivering the load in a PV-battery-supercapacitor hybrid system. The design considerations for the PV-battery-supercapacitor hybrid system in the existing works of literature do not consider the impact of the radiation and temperature, sun hours, etc., on the sizing of the PV module required to deliver to the battery, supercapacitor, and pump. The operating factor of the PV module determines the sizing of the PV module required for a system installed for a specific site.

| Symbol | Nomenclature | Symbol | Nomenclature |
|---|---|---|---|
| SPVWPS | Solar photovoltaic water pumping system | HE | Hydraulic energy (Wh/day) |
| $O_{PV}$ | Operating factor of the PV panel | $P_m$ | Pump capacity (in watts) |
| LLP | Loss of load probability | $\eta_p$ | Pump efficiency (%) |
| $H_t$ | Solar radiation incident on a tilted surface (kWh/m$^2$/day) | $\rho_1$ | Density of water (1000 kg/m$^3$) |
| $\delta$ | Declination angle(°) | $g$ | Acceleration due to gravity (9.81 m/s$^2$) |
| $n$ | Julian days | $W_{req}$ | Total water required (m$^3$) |
| $\phi$ | Latitude (°) | $H$ | Dynamic head (m) |
| $\omega_s$ | Sunshine hour angle (°) | DoD | Depth of discharge (%) |
| $S_0$ | Solar constant (1.367 kW/m$^2$) | $\beta$ | Tilt angle of the module (°) |
| $H_0$ | Solar radiation on a horizontal surface in extra-terrestrial region (kWh/m$^2$/day) | $C_{bat}$ | Battery capacity (Ah) |
| $K_t$ | Monthly average daily clearness index | $A_d$ | Days of autonomy |
| $H_g$ | Monthly average daily solar radiation on a horizontal surface (kWh/m$^2$/day) | $V_{bat}$ | Battery voltage (volts) |
| $H_d$ | Diffused radiation (kWh/m$^2$/day) | $\eta_{bat}$ | Battery efficiency (%) |
| $\Upsilon$ | Azimuth angle (°) | $E_{sc}$ | Energy stored in supercapacitors (Wh) |
| STC | Standard test conditions | TCF | Temperature correction factor |
| $T_{Cref}$ | Reference cell temperature (°C) | SC | Supercapacitor |
| $H_b$ | Beam radiation or direct radiation (kWh/m$^2$/day) | $\eta_{cc}$ | Charge controller efficiency (%) |
| $R_b$ | Beam radiation conversion factor | $E_{preq}$ | Peak energy required |
| $\rho$ | Ground albedo | $V_p$ | Voltage of the pump (Volts) |
| PR | Performance ratio | $I_p$ | Peak current of the pump (Amps) |

(continued)

(continued)

| Symbol | Nomenclature | Symbol | Nomenclature |
|---|---|---|---|
| FS | Full sun (1000 W/m$^2$) | PV | Photovoltaic |
| $sun_h$ara> | Sun hours (hours) | $C_{SC}$ | Capacitance of the supercapacitor (F) |
| $Y_f$ | Yield factor | $V_{max}$ | Maximum voltage of supercapacitor (V) |
| $Y_r$ | Reference yield factor | $V_{min}$ | Minimum voltage of supercapacitor (V) |
| $PV_{pump}$ | PV panel size required by the pump (in Watts) | $T_{sch}$ | Time of charging of the supercapacitor (seconds) |
| mf | Mismatch factor | $\eta_{sc}$ | Supercapacitor efficiency (%) |
| $PV_{bat}$ | PV panel size required by the battery (in Watts) | $PV_{int}$ | Initial PV panel size (in Watts) |
| $PV_{sc}$ | PV panel size required by the supercapacitor (in Watts) | $PV_T$ | Total PV panel size (in Watts) |
| $E_{sc}$ | Energy required by the load (Wh) | DE | Deficit energy (Wh) |
| $E_{load}$ | Energy required by the load (Wh) | $E_{gen}$ | Energy generated (Wh) |

In the existing works of literature, a few work have reported the design of a stand-alone PV system with a varying operating factor of the module ($O_{PV}$) which is a factor of radiation and temperature of the PV module, sun hours, etc., [17]. A PV module does not give the actual output at its rated value. It operates at a lower value which is 60–90% lower than its rated value [18]. It varies due to various factors associated with the PV module like radiation, the temperature of the panel, dust on the PV module, and the mismatch. So, proper corrections in the value of $O_{PV}$ are needed to get the optimum design of the PV system and hence to get optimum output from the system. In this paper, the optimal PV module size for an SPVWPS with battery and supercapacitor energy storage has been computed using the varying operating factor of the module ($O_{PV}$) considering the variations of the radiation and the temperature throughout the year in MATLAB software. The optimum size of the PV module is to be ensured to cater to the energy demand of the supercapacitor, battery, and pump. The addition of a supercapacitor to the system may prove beneficial that may take care of the transients of the solar radiation or low-solar radiation, thus reducing the peak load stress on the battery and enhancing its lifetime.

Section 2 discusses the objective and methodology of the work. Section 3 reports the modeling of the design parameters and the components of the PV-battery-supercapacitor water pumping system. Section 4 comprises the results and discussion with conclusions in Sect. 5.

## 2 Objective and Methodology of the Work

A SPVWPS using battery-supercapacitor HESS is shown in Fig. 1, and the methodology of the system is shown in Fig. 2. In this system, the solar PV array is connected to the DC bus bar, which is connected to the pump. The energy storage devices, battery, and supercapacitors are connected to the bus bar through a charge controller and a DC/DC converter. In this work, a sizing methodology is proposed for a battery-supercapacitor-based SPVWPS which aims to find the optimum PV module size considering the operating factor of the PV module ($O_{PV}$) to guarantee water output from the pump and to fulfill the charging characteristics of the battery and supercapacitor throughout the year. After determining the optimized PV module, the system's loss of load probability (LLP) is calculated considering the designed module size throughout the year. The design of the system parameters and components is simulated in a MATLAB environment. The objective of this work is summarized as follows: the primary system design parameter modeling; $H_t$, TCF, and $O_{PV}$ modeling, the preliminary sizing of the system components: battery, supercapacitor pump, and PV module and the comparison of the PV module size with a constant $O_{PV}$ and yearly average $O_{PV}$ of the panel considering LLP of the system.

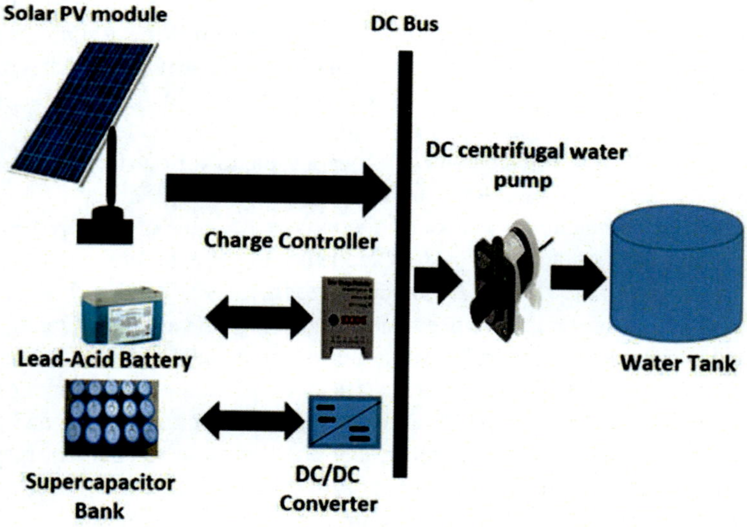


**Fig. 1** Schematic of a PV-battery-supercapacitor water pumping system

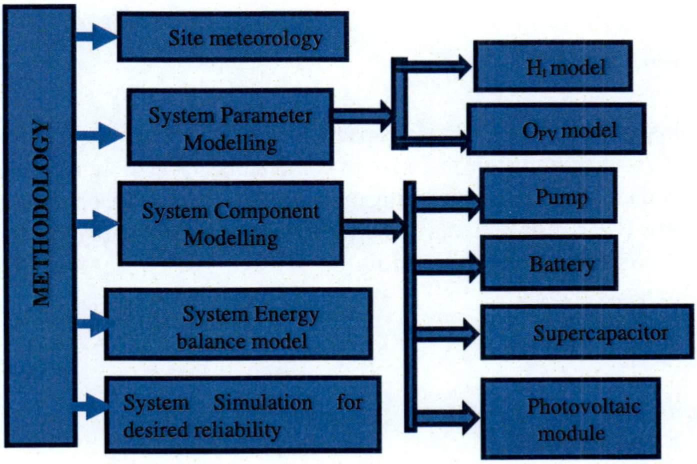


**Fig. 2** Methodology of the work

## 3 Radiation and the Ambient Temperature of the Site

Haldia is situated in the eastern part of the West Bengal state of India. Haldia receives higher radiation from February to May and the lowest radiation from June to September. Its ambient temperature revolves around 20–30 °C. The typical global solar radiation for Haldia and the ambient temperature of Haldia are shown in Fig. 3.

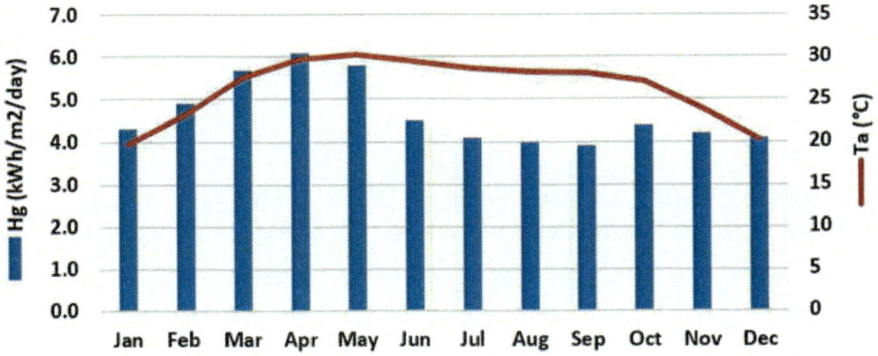


**Fig. 3** Meteorological data of Haldia

# 4 Modeling of the PV-Battery-Supercapacitor Water Pumping System

The modeling of the PV-battery-supercapacitor water pumping system consists of the modeling of design parameters and system components. The design parameters that are considered are tilted radiation falling on the panel ($H_t$) and the operating factor of the PV module ($O_{PV}$). The system components such as pump, battery, supercapacitor, and the PV module are also modeled in this section. The typical flowchart for the design of the system is shown in Fig. 4.

## 4.1 *Modeling of Parameters*

The radiation falling on the PV surface and the operating factor of the PV module are considered two primary design parameters for the system which are discussed in this section.

**Tilted radiation ($H_t$) incident on the solar panel**. The latitude and longitude of Haldia are 22.06°N and 88.08°E. The annual ambient temperature of Haldia averages around 26.4 °C. The monthly average daily solar radiation incident on a tilted surface ($H_t$) of the solar module at a tilted angle equal to the latitude ($\phi$) is calculated from Eq. (1)–(9). Considering the Julian days ($n$) throughout the year, the declination angle ($\delta$) is given by Eq. (1).

$$\delta = 23.5(\sin(360(284 + n)365)) \tag{1}$$

The sunshine hour angle ($\omega_s$) is given by Eq. (2),

$$\omega_s = \cos^{-1}(-\tan\phi\tan\delta) \tag{2}$$

Considering the solar constant $S_0$ as 1.367 kW/m$^2$, the solar radiation on a horizontal surface in the extra-terrestrial region ($H_0$) is given by Eq. (3),

$$H_0 = \frac{24}{\pi} S_0 \left(1 + 0.033\cos\left(\frac{360n}{365}\right)\right)(\omega_s \sin\phi\sin\delta + \cos\phi\cos\delta\sin\omega_s) \tag{3}$$

The monthly average daily clearness index, $K_t$, is given by Eq. (4),

$$K_t = \frac{H_g}{H_0} \tag{4}$$

where $H_g$ is the monthly average daily solar radiation on a horizontal surface. The diffused radiation $H_d$ in India is more [19, 20], which can be determined from Eqs. (5) and (6).

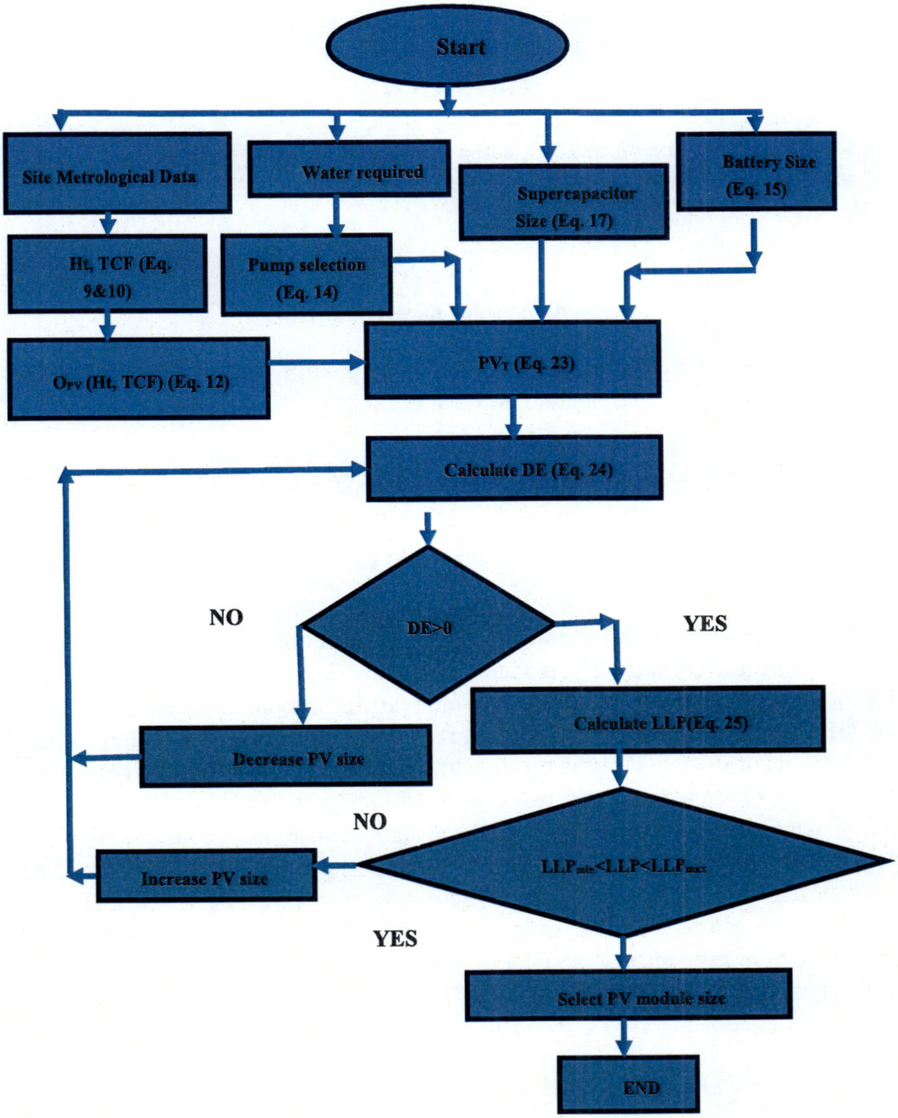


**Fig. 4** Flowchart of the design of the PV-battery-supercapacitor water pumping system

For ($\omega_s \leq 81.4°$),

$$H_d = H_g\big(1.391 - 3.560K_t + 4.189K_t^2 - 2.137K_t^3\big) \tag{5}$$

For ($\omega_s$>81.4°),

$$H_d = H_g(1.311 - 3.022K_t + 3.427K_t^2 - 1.821K_t^3) \tag{6}$$

The beam radiation or direct radiation $H_b$ can be written as in Eq. (7)

$$H_{\mathrm{b}} = H_{\mathrm{g}} - H_{\mathrm{d}} \tag{7}$$

For the tilt angle of the module ($\beta$) same as the latitude ($\phi$) of the site and faced at the south ($\gamma = 0$), the conversion factor ($R_b$) for the beam radiation is calculated as in Eq. (8),

$$R_{\mathrm{b}} = \frac{\cos\delta\sin\omega_{\mathrm{s}}}{\cos\phi\cos\delta\sin\omega_{\mathrm{s}} + \omega_{\mathrm{s}}\sin\phi\sin\delta} \tag{8}$$

As given by Liu and Jordon, in 1962 [21], incorporating the isotropic model, the tilted radiation on the solar module is given by Eq. (9)

$$H_{\mathrm{t}} = H_{\mathrm{b}}R_{\mathrm{b}} + \rho H_{\mathrm{g}}\left(1 - \frac{\cos\beta}{2}\right) + H_{\mathrm{d}}\left(1 + \frac{\cos\beta}{2}\right) \tag{9}$$

The tilt angle for this experiment is kept fixed throughout the year which is same as the latitude of the site (22°).

**Operating factor of the solar panel ( $O_{\mathbf{PV}}$):** The operating factor is a design factor for a PV system. The solar PV module does not operate at its rated power output due to radiation, the temperature of a PV module, dust accumulated on the module, and mismatch factors. The operating factor of the module is used to determine the actual output of the module which varies from 0.6 to 0.9. It indicates that at normal prevalent conditions of a site, the output power from the solar module varies from 60 to 90%. However, for the simplicity of the design estimations, a constant operating actor of the module of 75% is considered [18, 22]. In the modeling of the operating factor of the PV module, it is assumed that the solar module is daily cleaned, thus avoiding dust accumulation on the solar module, and a charge controller is installed in the system to get a mismatch factor of unity.

The temperature correction factor (TCF) is given by as in Eq. (10).

$$\mathrm{TCF} = \alpha(T_{\mathrm{C}} - T_{\mathrm{Cref}}) \tag{10}$$

And, the performance ratio which is defined as the effect of overall losses on the PV module's nominal power output under ideal operating conditions varies from 80 to 90% [23–25] is given as in Eq. (11), where $Y_{\mathrm{f}}$ is the yield factor, and $Y_{\mathrm{r}}$ is the reference yield factor.

$$\mathrm{PR} = \frac{Y_{\mathrm{f}}}{Y_{\mathrm{r}}} \tag{11}$$

The performance ratio value is considered as 80%.

Considering the operating factor of the PV module as a factor of the tilted radiation on the module, the performance ratio (PR), and the temperature correction factor

(TCF) of the module as given in Eq. (12).

$$O_{PV} = \frac{H_t \times \mathrm{PR} \times \mathrm{TCF}}{\mathrm{FS} * \mathrm{sun_h}} \tag{12}$$

where FS is full sun (1000 W/m$^2$), $\mathrm{sun}_h$ is the site sun hours available.

So, it can be observed from Eq. (12) that the operating factor of the PV module ($O_{PV}$) is a varying factor throughout the year. Thus, it depends on the site radiation, temperature, and the sun hours.

## 4.2 *Modeling of System Components*

The main system components of the PV-battery-supercapacitor water pumping system consist of the submersible water pump, battery, supercapacitor, and PV module. The modeling of these system components is done in this section.

**Pump**: The required hydraulic energy (HE) considering the total water required ($W_{req}$) in m$^3$/day, water pumping to a dynamic height '$H$' in $m$, where $\rho 1$ is the density of water (1000 kg/m$^3$), as Eq. (13) and Eq. (14).

$$\mathrm{HE} = \frac{W_{req} \times H \times \rho 1 \times g}{3600} \tag{13}$$

$$P_W = \frac{\mathrm{HE}}{\mathrm{sun_h} \times \eta_P \times Opv} \tag{14}$$

**Battery**: Let the pump capacity be $P$ Watts, and the battery be $V_{bat}$ volts, battery depth of discharge (DoD), battery efficiency as $\eta_{Bat}$ and charge controller efficiency as $\eta_{cc}$, the battery capacity for the system is given by Eq. (15)

$$C_{Bat} = \frac{P_m \times V_{bat} \times DoD}{sun_h \times \eta_{Bat} \times \eta_{cc}} \tag{15}$$

In this system, lead acid battery is chosen which has a DoD of 50%.

**Supercapacitor**: The supercapacitor capacity depends on the peak power required by the pump and the radiation variation throughout the day. The peak energy required by the system is during the high transients of the solar irradiation on a partly cloudy day or a cloudy day. So for this system, let the peak energy required be $E_{preq}$ which is calculated as in Eq. (16).

$$E_{preq} = V \times I_p \times t_p \tag{16}$$

where $V$ is the voltage of the pump (in Volts), $I_p$ is the peak current (in Amps) and $t_p$ (in seconds) is the duration of the peak energy required. Considering the required peak energy as $E_{preq}$, , the supercapacitor bank capacity ($C_{SC}$) in farads is determined using Eq. (17), where $V_{max}$ and $V_{min}$ are the supercapacitor maximum and minimum voltage, respectively, is given in Eq. (17).

$$C_{SC} = \frac{2 \times E_{preq}}{V_{max}^2 - V_{min}^2} \tag{17}$$

The energy stored in the supercapacitor in joules is given by Eq. (18),

$$E_{SC} = \frac{1}{2} C_{SC} (V_{max}^2 - V_{min}^2) \tag{18}$$

**PV module**: The PV module sizing depends on the energy demand to charge the battery and supercapacitor and to deliver power to the pump during sunshine hours. Considering the standard test conditions (STC), the required PV wattage deliver to a pump with the hydraulic energy (HE), with pump efficiency ($\eta_P$), with a mismatch factor ($m_f$) of unity, the PV panel sizing for pump load ($PV_{Pump}$) is given as given in Eq. (19),

$$PV_{Pump} = \frac{HE}{sun_h \times \eta_P \times m_f} \tag{19}$$

PV module sizing for charging the battery ($PV_{Bat}$) is given as in Eq. (20),

$$PV_{Bat} = \frac{C_{Bat} \times V_{bat} \times DoD}{sun_h \times \eta_{Bat} \times \eta_{cc}} \tag{20}$$

PV module sizing for charging the supercapacitor ($PV_{SC}$), considering the efficiency of the supercapacitor bank as $\eta_{SC}$, $T_{sch}$ as the time of charging of the supercapacitor bank by the solar panel, is given by Eq. (21),

$$PV_{SC} = \frac{E_{SC}}{T_{sch} \times \eta_{SC}} \tag{21}$$

The total initial PV module size ($PV_{int}$) is given in Eq. (22) as

$$PV_{int} = PV_{Pump} + PV_{Bat} + PV_{SC} \tag{22}$$

The overall PV module size ($PV_{int}$) is calculated considering the $O_{PV}$ of the module is given in Eq. (23).

$$PV_T = \frac{PV_{Pump} + PV_{Bat} + PV_{SC}}{O_{PV}} \tag{23}$$

### 4.3 Loss of Load Probability (LLP)

The loss of load probability (LLP) of a system is defined as the difference of the energy required by the load to the energy generated by the source to the energy required by the load over a period of time as given in Eqs. (24)–(25), where DE is the deficit energy, $E_{\text{load}}$ is the energy required by the load, and $E_{\text{gen}}$ is the energy generated by the PV module.

$$\text{DE} = (E_{\text{load}} - E_{\text{gen}}) \tag{24}$$

$$\text{LLP} = \frac{\sum \text{DE}}{\sum E_{\text{load}}} \tag{25}$$

If the LLP is zero, the energy generated by the generator is just sufficient to fulfill the load demand. If it is negative, the PV power generated is more to the energy required by the load. If the LLP is positive, the PV power generated is less and insufficient to fulfill the energy demand of the load.

## 5 Results and Analysis

The proposed design algorithm of the PV-battery-supercapacitor water pumping system with the operating factor of the PV module is simulated in a MATLAB environment. The simulations of the parameters and the system components are presented in this section.

### 5.1 Simulation of Parameters

The tilted radiation on the PV module and the module's operating factor ($O_{\text{PV}}$) for the different months and the climatic conditions of Haldia is simulated in the MATLAB environment presented here.

**Tilted radiation on the PV module ($H_t$), the temperature of the module ($T_c$), and module efficiency ($\eta_{pv}$)**

The monthly average daily tilted radiation on the module for the different months for the site is simulated using Eq. (9), and the module efficiency is simulated using Eq. (11). The simulated tilted radiation falling on the module and the global solar radiation on a flat surface for the Julian days of the year are shown in Fig. 5. It can be observed from the figure that the tilted radiation on the solar module is maximum in panel with respect to the $H_g$ (kWh/m$^2$/day) of the site for

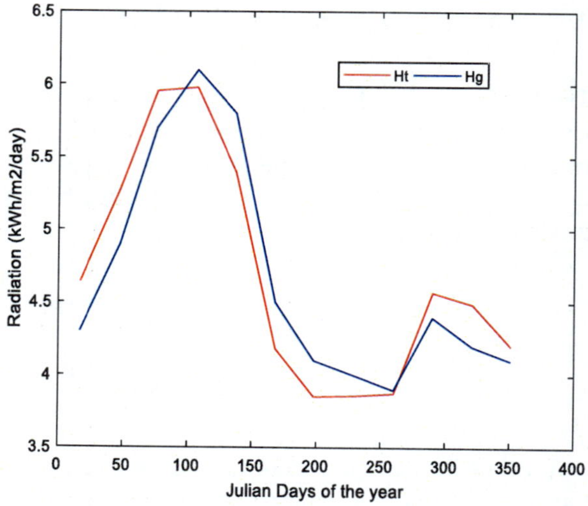


**Fig. 5** Simulated $H_t$ (kWh/m$^2$/day) in comparison with $H_g$ (kWh/m$^2$/day)

April and the minimum for June to August. $H_t$ is received more than $H_g$ for January to April and less from mid of April to August for a tilted angle of 22° with south-facing of the panel. The module efficiency and temperature for a polycrystalline material at Haldia are shown in Fig. 6. The temperature on the module is highest around 62 °C and lowest for December and January around 40 °C. The temperature of the panel is around 52 °C from July to October. Due to the variations of the PV module temperature, the efficiency of the PV module also varies. The efficiency of the PV module varies from 12.4 to 13.9%. The panel's efficiency is found to be minimum in April as the module temperature is highest (60 °C). It is maximum in January and December.

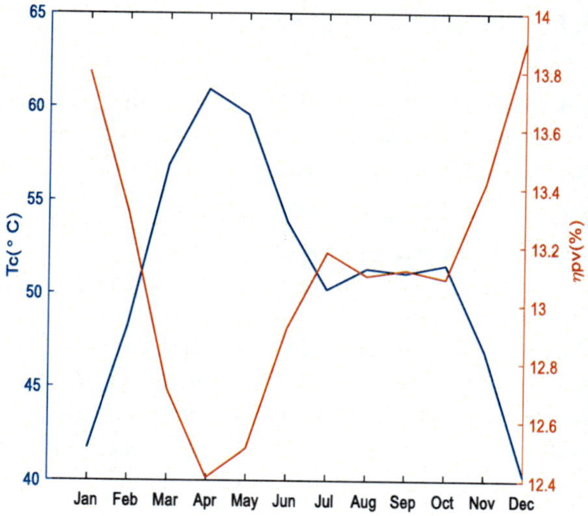


**Fig. 6** Simulated temperature on the PV panel and efficiency of the panel

**Operating factor of the PV module ($O_{PV}$)**

The operating factor of the PV panel depends on the radiation and the temperature of the PV module considering the use of a solar charge controller and regular cleaning of the PV panel. The operating factor thus obtained from Eq. (10) is shown in Fig. 7 for different months. It varies from a maximum of 0.81 in March to a minimum of 0.54 from July–September. Figure 8 shows the monthly PV size for the corresponding $O_{PV}$. In April, the PV panel size is estimated to be the lowest at 31 Wp corresponding to an $O_{PV}$ of 0.81. The PV panel size is obtained highest for July–September for $O_{PV}$ 0.54. However, a particular sizing of PV panel is necessary to deliver to the loads throughout the year. For the various PV panel sizes for different months, the LLP of the system is also displayed in Fig. 9. The desired LLP range for the design of the system is set to 0–0.03.

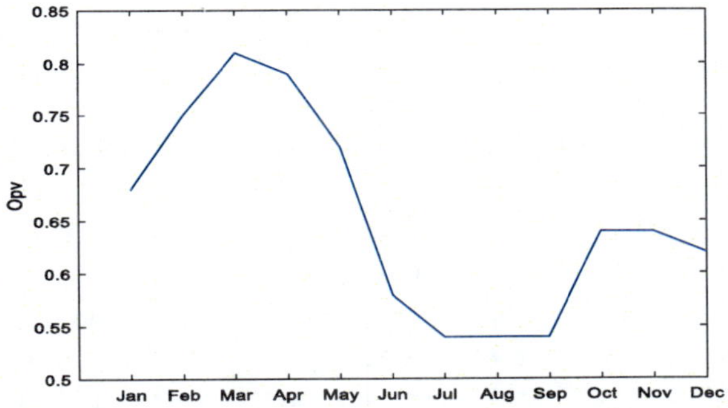


**Fig. 7** Operating factor of the PV panel ($O_{PV}$) for each month

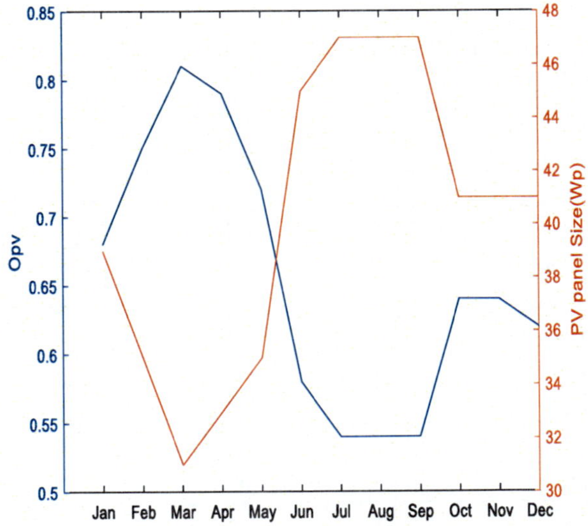


**Fig. 8** Operating factor of the PV panel ($O_{PV}$) and the PV panel size for each month

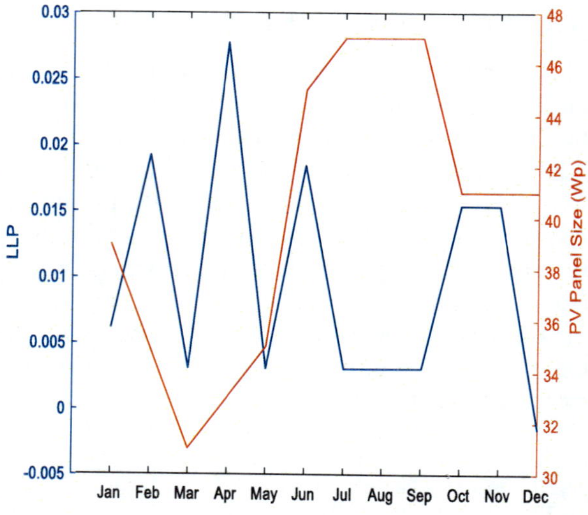


**Fig. 9** Simulated PV panel size and corresponding LLP of the system for each month

## 5.2 *Simulation of the Components of the System*

The PV-battery-supercapacitor system design is simulated in a MATLAB environment. The various inputs and the system component design output from the simulation are given in Table 1. From this simulation results, a battery of 10 Ah, 12 V, a supercapacitor module of 210 F, 12 V, and an 8 W, 12 V DC submersible pump are chosen. The PV size selection is based on the LLP of the system. It is studied in consecutive sections.

**Table 1** Various inputs were considered for the design of the PV-battery-supercapacitor water pumping system

| Inputs | Value | Units | Outputs | Value | Units |
|---|---|---|---|---|---|
| $W_{req}$ | 3 | $m^3$ | HE | 17.15 | Wh/day |
| $t_p$ | 960 | s | $C_{bat}$ | 10.66 | Ah |
| $V_{max}$ | 13 | V | $E_{preq}$ | 3.84 | Wh/day |
| $V_{min}$ | 6 | V | $C_{sc}$ | 207.87 | F |
| $V_{pump}$ | 12 | V | $PV_{pump}$ | 4.9 | Wp |
| $P_{pump}$ | 14.4 | W | $PV_{bat}$ | 15.8 | Wp |
| $\eta_P, \eta_{CC}, \eta_{BAT}$ | 70,90,90 | % | PVSC | 11.36 | Wp |
| $m_f$ | 1 | – | $PV_{int}$ | 32 | Wp |
| $A_d$ | 1 | day | Pw | 8 | W |
| $O_h$ | 4 | h | | | |
| DoD | 50 | % | | | |
| $sun_h$ | 5 | h | | | |

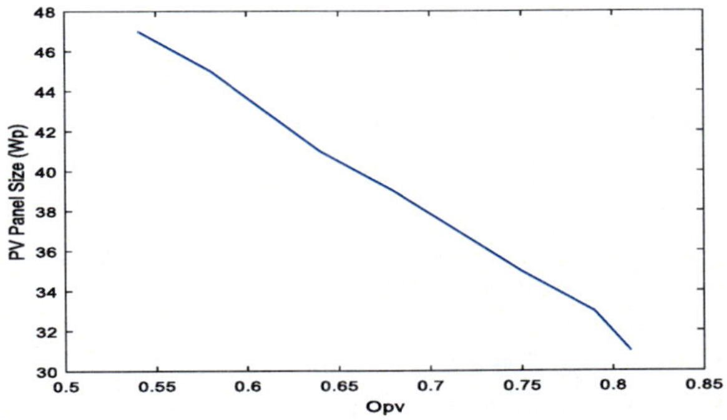


**Fig. 10** Simulated operating factor of the PV panel ($O_{PV}$) for each month

## 5.3 *Effect of $O_{PV}$ on PV Size and LLP of the System*

The effect of the $O_{PV}$ on the PV size is shown in Fig. 10. In this study, the $O_{PV}$ is varied from 0.54 to 0.81. As $O_{PV}$ increases, the PV module size is decreased. For a variation of $O_{PV}$ from 0.54 to 0.81, the PV size decreases from 47 to 31 Wp. However, for the proper designing of the system under the climatic condition of Haldia, optimal $O_{PV}$ has to be selected.

## 5.4 *System Design with Fixed $O_{pv}$ and Yearly Average $O_{pv}$: A Comparison*

A comparative analysis is done for the design of the PV-battery-supercapacitor WPS for various $O_{PV}$ values consideration. The three considerations are based on fixed $O_{PV}$ (0.75), yearly average $O_{PV}$ (0.65), and monthly average $O_{PV}$. The analysis for these $O_{PV}$ considerations is shown in Table 2. From Table 2, it is observed that the yearly average LLP for the PV size 35 Wp is 0.019, whereas PV size obtained *Opv* of 0.65 is 39 Wp with an annual average LLP of 0. The system design considerations using the monthly average of $O_{PV}$ give a PV size of 40.16 Wp with negative annual that PV module size for the system's load for $O_{PV}$ 0.75 is undersized for the climatic conditions of Haldia. A 40 Wp panel will be optimal for designing the above specified PV-battery-supercapacitor water pumping system with an LLP value of zero. However, more than 40 Wp, PV modules may be oversizing for this system.

**Table 2** PV-battery-supercapacitor water pumping system design with various $O_{PV}$ considerations

| $O_{PV}$ considerations | $O_{PV}$ | PV size (Wp) | LLPy |
|---|---|---|---|
| Fixed | 0.75 | 35 | 0.019231 |
| Annual average | 0.65 | 39 | 0 |
| Monthly average | 0.54–0.81 | 40.16 | −0.009 |

## 6 Conclusion

Proper PV module sizing is necessary for a PV/battery/SC system to charge the energy storage devices when sunshine is available. A system with constant $O_{PV}$ to calculate the module size may lead to under or oversizing the PV module. In this work, PV/battery/SC-based water pumping system is designed considering the variable operating factor of the PV module throughout the year. The optimal module size is determined to charge the storage devices and also to pump water during the daytime based on the LLP of the system. The $H_t$ and the $O_{PV}$ throughout the year are simulated in MATLAB. The effect of the varying $O_{PV}$ on the LLP of the system and PV module size is determined. A battery of 10 Ah, 12 V, a supercapacitor module of 210 F, 12 V, and an 8 W, 12 V DC submersible pump is chosen for a $3m^3$ of water pumping and hydraulic energy of 17.15 Wh/day for the climatic conditions of Haldia. The PV module size of 35 Wp is obtained with an annually average LLP of 0.01, 39 Wp for annually average LLP of zero, and 40.16 Wp for annually average LLP of -0.009. 40 Wp PV module size is found optimal for the PV-battery-SC water pumping system.

## References


1. Santra P (2021) Performance evaluation of solar PV pumping system for providing irrigation through micro-irrigation techniques using surface water resources in hot arid region of India. Agric Water Manag 245:106554
2. Rathore PKS, Das SS, Chauhan DS (2018) Perspectives of solar photovoltaic water pumping for irrigation in India. Energ Strat Rev 22:385–395
3. Benghanem M, Daffallah KO, Joraid AA, Alamri SN, Jaber A (2013) Performances of solar water pumping system using helical pump for a deep well: a case study for Madinah, Saudi Arabia. Energy Conv Manage 65:50–56
4. Rawat R, Kaushik SC, Lamba R (2016) A review on modeling, design methodology and size optimization of photovoltaic based water pumping, standalone and grid connected system. Renew Sustain Energy Rev 57:1506–1519
5. Khatib T, Ibrahim IA, Mohamed A (2016) A review on sizing methodologies of photovoltaic array and storage battery in a standalone photovoltaic system. Energy Convers Manage 120:430–448
6. Rana J, Kamruzzaman M, Oliver MH, Akhi K (2021) Financial and factors demand analysis of solar powered irrigation system in Boro rice production: a case study in Meherpur district of Bangladesh. Renew Energy 167:433–439
7. Labed S, Lorenzo E (2004) The impact of solar radiation variability and data discrepancies on the design of PV systems. Renew Energy 29(7):1007–1022
8. Yesilata B, Firatoglu ZA (2008) Effect of solar radiation correlations on system sizing: PV pumping case. Renew Energy 33(1):155–161
9. Abdulkarim HT, Sansom CL, Patchigolla K, King P (2020) Statistical and economic analysis of solar radiation and climatic data for the development of solar PV system in Nigeria. Energy Rep 6:309–316
10. Guentri H, Allaoui T, Mekki M, Denai M (2021) Power management and control of aphotovoltaic system with hybrid battery-supercapacitor energy storage basedonHeuristicsmethods. J Energy Storage 39:102578

11. Cabrane Z, Kim J, Yoo K, Ouassaid M (2021) HESS-based photovoltaic/batteries/supercapacitors: energy management strategy and DC bus voltage stabilization. Sol Energy 216:551–563
12. Masaki MS, Zhang L, Xia X (2018) Hierarchical power flow control of a grid-tied photovoltaic plant using a battery-supercapacitor energy storage system. Energy Procedia 145:32–37
13. Jaszczur M, Hassan Q (2020) An optimisation and sizing of photovoltaic system with supercapacitor for improving self-consumption. Appl Energy 279:115776
14. Cabrane Z, Ouassaid M, Maaroufi M (2016) Analysis and evaluation of battery-supercapacitor hybrid energy storage system for photovoltaic installation. Int J Hydrogen Energy 41(45):20897–20907
15. Das M, Mandal R (2018) A comparative performance analysis of direct, with battery, supercapacitor, and battery-supercapacitor enabled photovoltaic water pumping systems using centrifugal pump. Sol Energy 171:302–309
16. Das M, Mukherjee D, Chaudhuri SRB (2017) An approach to study the performance of photovoltaic water pumping using supercapacitor. Mater Today Proc 4(9):10400–10406
17. Lopez Ramirez I (2017) Operating correction factor of PV system: effects of temperature, angle of incidence and invertor in PV system performance
18. Solanki CS (2008) Renewable energy technologies: a practical guide for beginners. PHI Learning Pvt. Ltd.
19. Modi V, Sukhatme SP (1979) Estimation of daily total and diffuse insolation in India from weather data. Sol Energy 22(5):407–411
20. Garg HP, Garg SN (1985) Correlation of monthly-average daily global, diffuse and beam radiation with bright sunshine hours. Energy Conv Manage 25(4):409–417
21. Liu B, Jordan R (1961) Daily insolation on surfaces tilted towards equator. ASHRAE J (United States) 10
22. Solanki CS (2015) Solar photovoltaics: fundamentals, technologies and applications. Phi learning Pvt. Ltd.
23. Kumar BS, Sudhakar K (2015) Performance evaluation of 10 MW grid connected solar photovoltaic power plant in India. Energy Rep 1:184–192
24. Singh VP, Kumar D, Ravindra B, Performance assessment of 5 MW grid connected photovoltaic plant in Western Region of India
25. Van Sark WGJHM, Reich NH, Müller B, Armbruster A, Kiefer K, Reise C (2012) Review of PV performance ratio development. In: World renewable energy congress. Denver CO, USA, pp 4795–4800

# Portable Stress Measurement and Analysis System (PSMAS): The Correlation of Body and Mind Analysis Using GSR Sensor

**Shakila Azim, I. D. Soubache, B. Annapurna, S. V. G. V. A. Prasad, Capt. K. Sujatha, and Raushan Kumar Singh**

## 1 Introduction

Certain factors have a significant impact on a human's mental process and behavioural style. Some of the primary positive elements of mental health include happiness, confidence, self-esteem, interpersonal relationships, and comfort [1]. Our perceptions of society, and even of ourselves, are primarily influenced by these elements. The foregoing criteria have a big impact on our behaviour and psychological presentations. Several negative mental parameters influence our beliefs and social behaviours [2]. Social withdrawals, severe paranoia, dramatic mood swings, and other negative aspects are some of the most common mental characteristics. There is a deciding factor that governs both positive and negative mental characteristics, and that component is stress. Stress is a factor that, when induced in abundance, reduces the stillness of mental state, it is crucial to keep it in a manageable range.

---

S. Azim
Department of Psychology, MDDM College, Muzaffarpur, Bihar, India

I. D. Soubache (✉)
Department of BME, RGCET, Pondicherry, India
e-mail: idsoubache@gmail.com

B. Annapurna
Department of CSE, Aditya College of Engineering, Surampalem, AP, India

S. V. G. V. A. Prasad
Department of Physics, Ideal College of Arts and Sciences, Kakinada, AP, India

Capt. K. Sujatha
Department of Mathematics, St. Joseph's College for Women (A), Visakhapatnam, AP, India

R. K. Singh (✉)
Spectrum Solutions, Pondicherry, India
e-mail: spectrumsolutions.md@gmail.com



P. H. J. Chong et al. (eds.), *Emerging Electronics and Automation*, Lecture Notes in Electrical Engineering 937, https://doi.org/10.1007/978-981-19-4300-3_11

## 1.1 Major Cause Effects of Stress

Stress is an unavoidable element of human life for a variety of reasons. It is fairly frequent in today's environment, but when stress begins to negatively impact our health, it is critical to find a solution; otherwise, significant anger difficulties, anxiety, depression, and even suicidal tendencies can result. Increased occurrences of road rage and suicide may always be traced back to stress if properly investigated [3] (Fig. 1).

Family and work-related concerns are the main sources of stress. The severity of stress is determined by the size of the problem. It might be as minor as relocating to a new home to as major as the death of a loved one or the loss of a career. Divorce, financial commitments, chronic illness, traumatic events, natural disasters, violence, discrimination, and a variety of other factors all contribute to stress. Whatever the cause of stress, it undoubtedly causes a slew of social, emotional, econoic, and professional issues, necessitating the development of a sophisticated and precise system for measuring and analyzing stress levels.

Biomedical sensors are way smart now, and they are also capable of detecting minor fluctuations in body parameters. Complex diseases like foot ulcers can also be treated with the help of electronic types of equipment. These electronic sensors play an important role to correct our body illness if used practically [4].

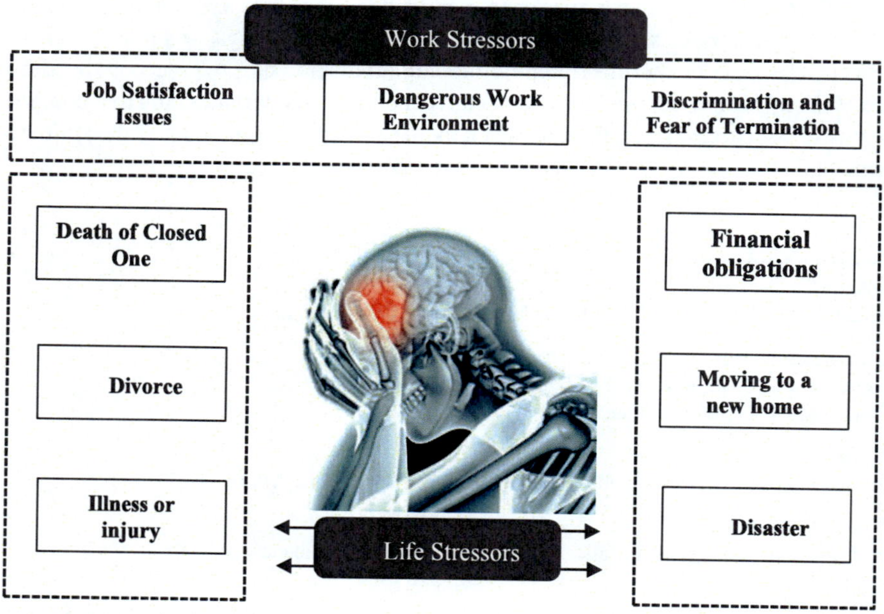


**Fig. 1** Major stressors and their integration in people's life

## 2 Related Works

Stress is very common today due to several reasons associated with over-ambition, life dissatisfaction, family and financial imbalance, etc. Especially, professional issues bring a lot of health concerns with it. People with high-tension jobs are always at a high risk of getting overstressed. The management of stress should start long back before it starts to cause health-related problems. At the current time, sensor technology is smart enough to detect the level of stress based on physical symptoms. The detailed analysis of stress data collected from GSR sensor is used in the paper. Stress identification, reductionist approach, and major issues with the practicality of using GSR are well evaluated. Experiments are conducted to study the real GSR data collected from the field. The identification of stress in all three categories of acute, episodic acute, and chronic is well explained so that it could be identified during practice. The GSR graph can provide a clear picture of tracking the human mind for its different phases in terms of stress like normal, aroused, stressed, and relaxing. Sensor data in several conditions like during exercise and ending exercise on different users for different time lengths are recorded and observed. The data analysis and observation are found to be highly ambiguous from the study, whereas stress patterns can be more reliable with more controlled settings [5].

Cognitive stress at an interpersonal level is an important issue that needs urgent evaluation to reach social acceptance. The major device used for the analysis is galvanic skin response (GSR), electroencephalogram, and photoplethysmogram (PPG). Each sensor is applied to evaluate the low- and high-cognitive stress. The major findings and targeted sectors are to identify the stress parameters through major sensors like EEG, GSR, and PPG signals that correlate with stress. The cross verification is performed for multiple sensors to validate the accuracy of the readings and outputs. The main aim of the work was to discriminate between low and high stress using multiple devices. During the study, GSR and EEG were performed decently. This gave us an idea that different devices are present to analyze the stress parameter but GSR serves the task accurately [6].

With the advancement of technology, biomedical and physiological sensors became smarter than ever. The advancement also brought many realistic digital games and this caused to have a debate of their actual nature and its impact on the human brain. This stress needed to be analyzed to identify, at which point of the game how much stress is experienced by the user. Generally, analyzing the GSR data is very complex as it gives the long-term measure of the user. The perceived notion of the user is analyzed in this paper to match up with the GSR data in this paper. The individual approach provides accurate, precise, and relevant stress analysis data in the study [7].

## 3 System Design and Methodology

Galvanic skin response (GSR) is also known as a skin conductance measurement device. It is an indirect measure of emotional arousal because the state of emotion is directly linked with sweat glands. A stressed state of mind severely stimulates the sweat glands. The sweating skin has lower resistance and higher conductance, and dry skin has higher resistance and lower conductance. The minor fluctuation in skin conductance and resistance is minutely observed by the galvanic skin response (GSR) sensor [8]. This makes the whole system extremely accurate to track the emotional parameter. The sensors were apt to be used for stress analysis in our proposed system. We look forward to exploring the practical usage and stress measurements for several stressed and comparatively less stressed subjects from different age groups [9] (Fig. 2).

The above graph clearly states the variation in the skin conductance graph with a deep breath. During a deep breath, our brain gets enough oxygen and diverts our concentration towards other necessary biological and mental refreshing settings. Deep breathing is already proved to affect the stress is utilized in our concept [11].

### 3.1 *Working Principle*

Skin conductance input is taken generally from hand and foot areas (Fig. 3). To collect GSR data Ag/AgCl (silver chloride), contact points are connected with skin. The ionic activities are accurately transferred through these electrodes. The signal received from the skin is transmitted through the lead wire to the GSR device. The data rate is usually with the sampling of 1–10 Hz, and the unit is micro-siemens (μS) [9]. The skin conductance data are either analogue values in terms of numbers

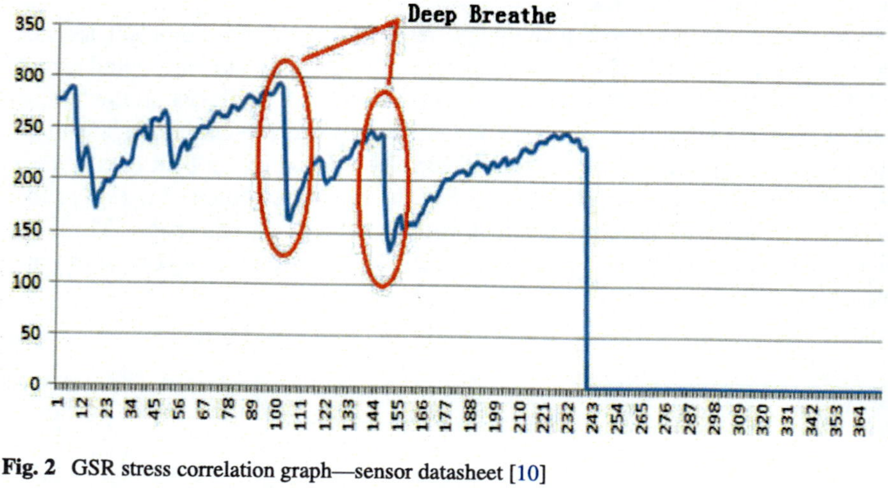


**Fig. 2** GSR stress correlation graph—sensor datasheet [10]

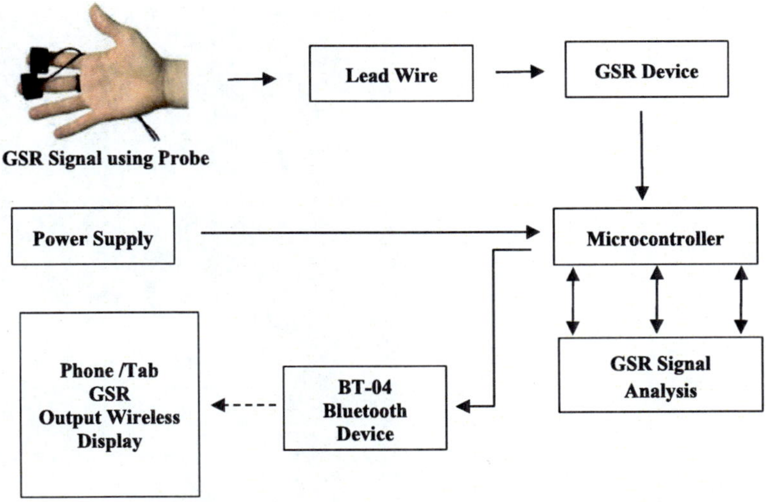


**Fig. 3** Overall system block diagram

or voltage. We utilize numbers because of higher analysis points. The graphical representation of numerical data looks much easy to handle due to its high-variation profile.

## 4 Development and Testing

Development of the module started with finalizing the power required to efficiently run the overall systems till assembling all the components. During testing phase, the detailed analysis of different instances of stress was analyzed.

### *4.1 Development*

We linked the GSR sensor with the Arduino system to analyze its data in graphical and numerical format. We utilized an Android app to see the reading in numerical format with audio instruction on our phones. We used TP4056 circuitry and a rechargeable 18,650—2600 mAh battery to power the system. This makes the whole module a portable unit to have a handheld status. The wireless display enables it to provide better visibility of the overall process (Figs. 4, 5, and 6).

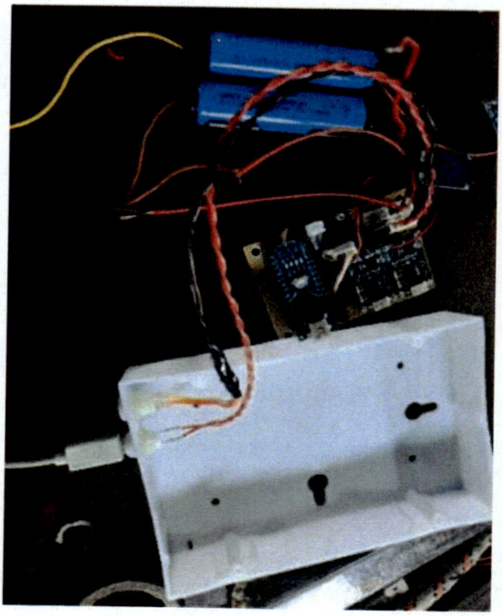

Fig. 4 Initial developments

Fig. 5 End product

## 4.2 Testing

There have been several studies on how the human body reacts with stress and its measurement with GSR sensors [12, 13]. Our analysis worked on how different the GSR behaves with different emotions, with relaxation, sleep, music, and other parameters. The test is subjected to multiple individuals in multiple instances and time

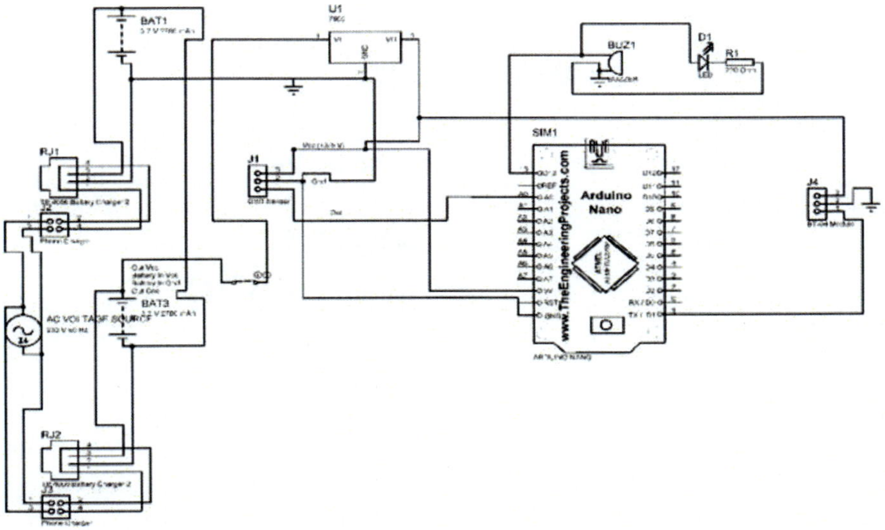


**Fig. 6** Overall circuit diagram

durations. Result obtained is altered in higher and lower values of GSR, but increase and decrease in skin conductance are identical in similar situations during the testing phase, we tried to analyze the GSR value in different instances of time, mental condition, with stress controllable environment, and with deep breathing/meditation and music effect. We have analyzed the system for a controlled environment. Controlled environment parameters are like regulating stress by cycling, exercise, listening to music, deep breath, etc. (Figs. 7 and 8).

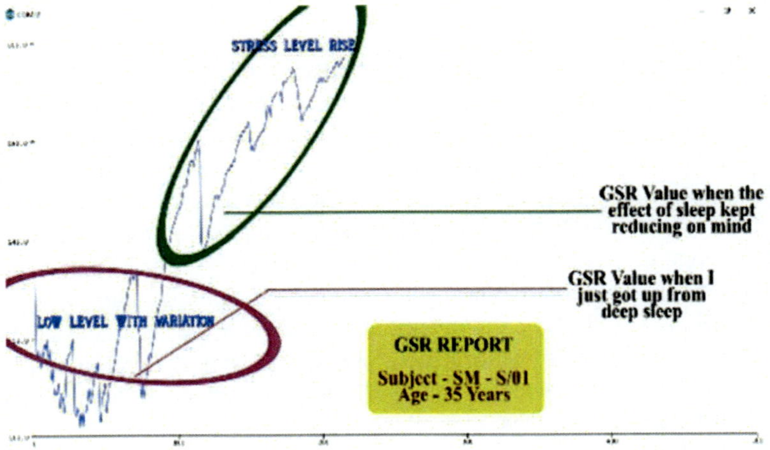


**Fig. 7** Stress versus sleep analysis

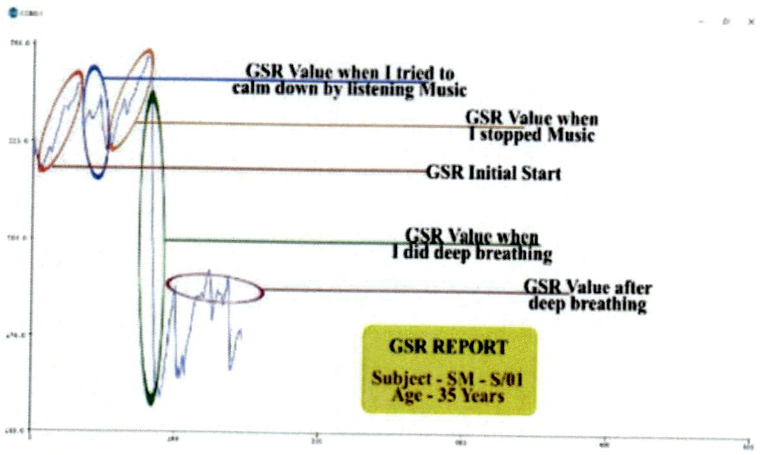


**Fig. 8** Stress versus deep breathing and music

- When the GSR is measured initially, it is almost constant with minor variations.
- There is a stress level drop recorded when the subject listened to the music and rose almost to the point when music was turned off.
- Deep breathing showed a huge drop in stress level that too instantaneously. The most impressive this is, it remained on the minimum level for a very long time.
- The stress level was found to be minimal when the subject just got up from sleep.
- As the person wakes up and mind comes out of sleep subject's mind, the stress level keeps on increasing till it reached the persons average stress level.

## 5 Conclusion

Stress is the sole constant component that persists across life lengths and oscillates with positive and negative mental features at random. Several scales have been used in various studies to determine the amount of stress using questionnaires. Scale-based systems are excellent for long-term analysis, but our technique works for both immediate stress analysis and long-term study. The system is used in real-time for data collection and analysis. The result suggests that the GSR sensor is accurate enough to be used in day to day life for stress analysis. We developed a handheld portable device called portable stress measurement and analysis system (PSMAS), which is efficient enough to measure and plot the accurate stress level of a person. It is powered by rechargeable batteries to keep it working around the clock. It has a Bluetooth device that links with Android applications to provide voice announcements as well as a visual display. Soon, the device will become popular like thermometer and SpO2 device because of the increasing stress level amongst the people. The system is tested for real-time usage and stress analysis, which gives an acceptable output. As a future enhancement, we plan to add IoT services to the system to make it a device with worldwide coverage.

## References


1. Ram Chandra K (2013) Meaning reorganization view vis-a-vis hidden reality view-revisiting the allotropes of psychodynamics of insight. Int J Hum Resour Manag Res 3(4):69–74. Available: http://www.tjprc.org/view_archives.php?year=2013&id=34&jtype=2&page=3
2. Ram Chandra K (2008) Hetero-balancing approach to curriculum planning using systemic-functional analysis, voices of the world. In: International system linguistic forum, Sydney, pp 78–82. ISBN-978-0-9805447-1-8
3. Watson S (2021) Common causes of stress & their effect on your health. WebMD. Accessed 22 Aug 2021. https://www.webmd.com/balance/guide/causes-of-stress
4. Soubache ID, Thirumurugan T, Annapurna B, Singh RK (2020) Diabetic foot ulcer treatment device using peltier and embedded electronics. J Phys Conf Ser 1712(1). https://doi.org/10.1088/1742-6596/1712/1/012041.
5. Bakker J, Pechenizkiy M, Sidorova N (2011) What's your current stress level? Detection of stress patterns from GSR sensor data. In: Proceedings of the IEEE international conference on data mining, ICDM, no 1, pp 573–580. https://doi.org/10.1109/ICDMW.2011.178
6. Das D, Bhattacharjee T, Datta S, Choudhury AD, Das P, Pal A (2018) Classification and quantitative estimation of cognitive stress from in-game keystroke analysis using EEG and GSR. In: 2017 IEEE life science conference LSC 2017, pp 286–291. https://doi.org/10.1109/LSC.2017.8268199
7. Satti FA, Hussain M, Hussain J, Kim TS, Lee S, Chung TC (2021) User stress modeling through galvanic skin response. In: Proceedings of the 2021 15th international conference ubiquitous information management communication, IMCOM 2021
8. Zhai J, Barreto AB, Chin C, Li C (2005) Realization of stress detection using psychophysiological signals for improvement of human-computer interactions. Undefined 415–420. https://doi.org/10.1109/SECON.2005.1423280
9. NMSBA (2021) What is GSR (galvanic skin response) and how does it work?—NMSBA. Accessed 22 Aug 2021, https://www.nmsba.com
10. "Grove - GSR Sensor". Accessed 19 Nov 2021, https://seeeddoc.github.io/Grove-GSR_Sensor/
11. Perciavalle V et al (2017) The role of deep breathing on stress. Neurol Sci 38(3):451–458. https://doi.org/10.1007/s10072-016-2790-8
12. Villarejo MV, Zapirain BG, Zorrilla AM (2012) A stress sensor based on Galvanic skin response (GSR) controlled by ZigBee. Sensors (Basel) 12(5):6075. https://doi.org/10.3390/S120506075
13. Hosseini SA, Khalilzadeh MA (2010) Emotional stress recognition system using EEG and psychophysiological signals: using new labelling process of EEG signals in emotional stress state. Int Conf Biomed Eng Comp Sci ICBECS. https://doi.org/10.1109/ICBECS.2010.5462520

# Performance Comparison of P&O and Fuzzy Logic-Based MPPT Control Technique for Stand-Alone Photovoltaic System

**Nibedita Swain**

## 1 Introduction

Nowadays, developing countries require the energy from fossil fuels like diesel, coal, petrol, and gas. Due to increasing demand of energy, many problems occur like change in climatic conditions and energy crisis. To make a clean, sustainable, and affordable environment, carbon dioxide emission has to be reduced, and greenhouse gas production has to be increased. Renewable energy is a good choice for this environment. Renewable energy includes sunlight from sun, wind, rain, tides, waves, and geothermal heat. This paper focuses the energy coming from the sun. Solar energy is abundantly available and can be utilized properly in various fields. It can be used for stand-alone system or can be a grid connected system [1]. Thus, it utilizes the solar energy to produce power in rural areas where the accessibility of grids is low. Solar energy uses the energy coming from the sun and converts it in to electrical energy depending upon the level of irradiance. This is possible with the help of PV module that consists of number of solar cells [2].

The 100% solar energy cannot be converted into electrical energy, and only 35–45% of energy is converted into electrical energy. So to store the rest of the solar energy, it requires an efficient battery charging system with lesser charging time. For enhancing the efficacy of the solar panel, MPPT algorithms should be employed [3]. The commonly used algorithms are hill-climbing process, fractional open circuit voltage method, P&O method, incremental conductance process, neural network control, fuzzy logic-based control, etc. This paper explains two MPPT methods named as P&O and fuzzy logic-based MPPT (FMPPT).

N. Swain (✉)
Department of Electrical Engineering, Silicon Institute of Technology, Bhubaneswar, Odisha, India
e-mail: nswain@silicon.ac.in



P. H. J. Chong et al. (eds.), *Emerging Electronics and Automation*, Lecture Notes in Electrical Engineering 937, https://doi.org/10.1007/978-981-19-4300-3_12

Table 1 Electrical characteristic of PV module

| | |
|---|---|
| Maximum power ($P_{max}$) | 130 W |
| Voltage at $P_{max}$ ($V_{max}$) | 17.6 V |
| Current at $P_{max}$ ($I_{max}$) | 7.39 A |
| Short circuited current ($I_{sc}$) | 8.02 A |
| Open circuited voltage ($V_{oc}$) | 21.9 V |
| Series connected solar cells ($N_s$) | 7 |
| Parallel connected solar array ($N_p$) | 1 |

Table 2 Circuit parameters of boost converter

| | |
|---|---|
| Inductance ($L$) | 350 micro-henry |
| Capacitance ($C$) | 560 micro-farad |
| Resistance ($R$) | 15 Ω |
| Operating frequency ($f_s$) | 5 kHz |
| Duty cycle ($d$) | 0.5 |
| Output voltage ($V_o$) | 42 V |

## 2 Design of PV Fed Boost Converter

The boost converter is fed with a photovoltaic module with electrical specification given in Table 1. The total module uses two different MPPT algorithms to track maximum power and also to track output voltage for different load conditions [4]. This paper discusses the regulation of output voltage by electronic controller such as P&O method and fuzzy MPPT method. The parameter specification for the converter is given in Table 2.The step-up converter consists of two energy storing elements along with one switch as in [5].

## 3 Maximum Power Point Tracking (MPPT) Technique

Maximum power point tracking technique is used to boost the efficiency of the solar panel by using maximum power transfer theorem [6]. MPPT controllers force the solar panel to operate at the most efficient voltage, thereby delivering maximum power to the load. The input side is connected to a boost converter excited by PV source in order to improve the output voltage. The MPPT techniques are based on different variables to track MPP and some based on various techniques to track MPP.

(a) ***Perturb and Observe (P&O) Method***

P&O is the simplest method used for tracking the MPP. This algorithm utilizes the PV voltage and current to track MPP. The PV array voltage is sensed by a voltage sensor and changes the MPP due to perturbation and ends up calculating

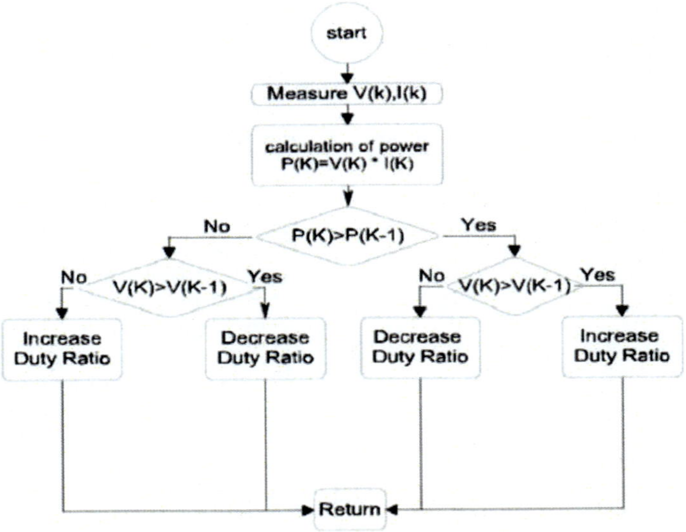


**Fig. 1** Flow diagram of P&O algorithm

the wrong MPP. In this practice, if the voltage is varying unidirectionally and power is increasing at a time, then the design keeps on changing in the same path [7], and the perturbation path is opposite if the current power is less than the earlier one. The oscillation around MPP occurs when the module power reaches at MPP. Figure 1 shows the flow diagram of P&O design.

(b) ***Fuzzy Logic Control Method***

Fuzzy logic-based MPPT (FMPPT) is an intelligent technique due to capable of handling nonlinearity associated with the system. The FMPPT is mostly preferred for tracking the MPP in the PV system. The algorithm computes the variable step size by calculating the slope of the PV graph of photovoltaic module [8]. After that, it controls the duty ratio perfectly by tracking the voltage output.

The fuzzy controller contains three functional components such as fuzzification, rule base, and defuzzification. Here, the two input variables to the FLC are error (*e*) and the change in error (*ce*). The FLC output is the duty ratio. The main function of this controller block is to track the voltage output and maximize the power output [9]. The slope of *P*–*V* curve and changes in slope will be acting as two inputs to the FMPPT. Figure 2 depicts the flow diagram of FMPPT design.

In fuzzification block, input and output variables are converted into fuzzy variables by assigning the values of membership function. Different fuzzy levels are used for each linguistic variable such as negative big (NB), negative small (NS), zero (ZE), positive small (PS) and positive big (PB). Here, triangular membership functions are used. The fuzzy rule uses a combination of if-then rules. In this work, there

Fig. 2 Flowchart of fuzzy logic-based MPPT

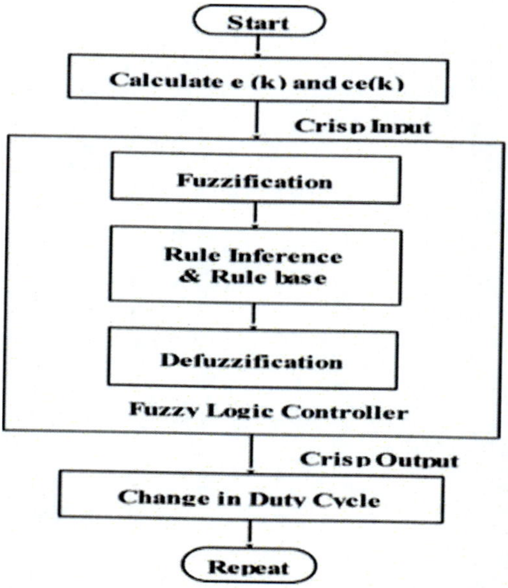


are 25 fuzzy rules. Mamdani FIS is used in this work with max-min defuzzification approach. The rule base is formed from this PV curve illustrates in Fig. 3.

The PV curve shows three different regions, namely region1, region 2, and region 3. Region1 is having negative slope; hence, duty ratio has to be increased to record the MPP, because the variation in error determines the value of duty ratio has to be enhanced. If error is negative and change in error is positive that indicates the operating point shifts toward MPP from the right side. Hence, the response is zero in order to avoid the system from oscillations [10]. In similar ways, the rules are mentioned for other two different regions. The rule base for fuzzy MPPT is shown in Table 3.The error is the slope of the PV curve. The operating point corresponds to a point

Fig. 3 *P–V* curve of PV module

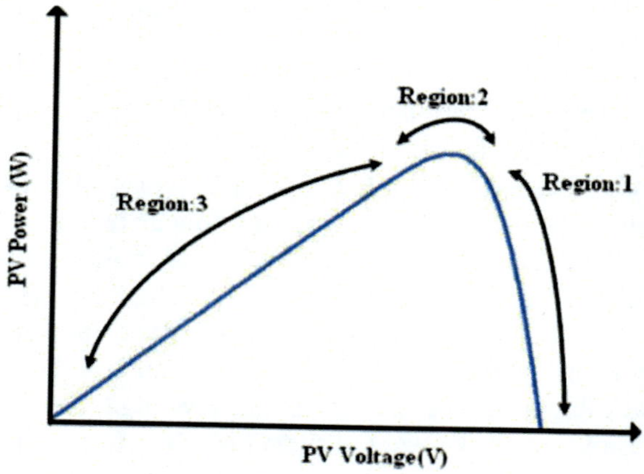

**Table 3** Fuzzy rule base for MPPT

| Error/change in error | NB | NS | ZE | PS | PB |
|---|---|---|---|---|---|
| NB | PB | PB | PB | PB | PB |
| NS | PB | PS | PS | ZE | ZE |
| ZE | PS | ZE | ZE | ZE | NS |
| PS | ZE | ZE | NS | NS | NB |
| PB | PB | ZE | NS | NB | NB |

on the power–voltage (*P–V*) curve, and for generating the maximum power output at a particular irradiance and temperature, the operating point should correspond to the maximum of the *P–V* curve.

## 4 Simulink Model and Result Analysis

The boost converter produces high-output voltages from a low-input voltage. The continuous conduction mode (CCM) of the step-up converter is considered in this work. The detailed circuit structure is depicted in Fig. 4.

The transfer function of step-up regulator is given by $V_o/V_{in} = 1/(1 - d)$ representing $d$ as duty cycle of switch S.

The PWM technique is realized within the MPPT to produce gate pulses and operate the converter switch with proper duty cycle [11]. Figure 5 shows Simulink model of PV fed step-up converter with P&O MPPT.

Figures 6 and 7 illustrate the voltage output of step-up regulator using P&O algorithm for 10% fluctuation in the load. It is viewed that all the output voltages are almost constant by varying the load conditions.

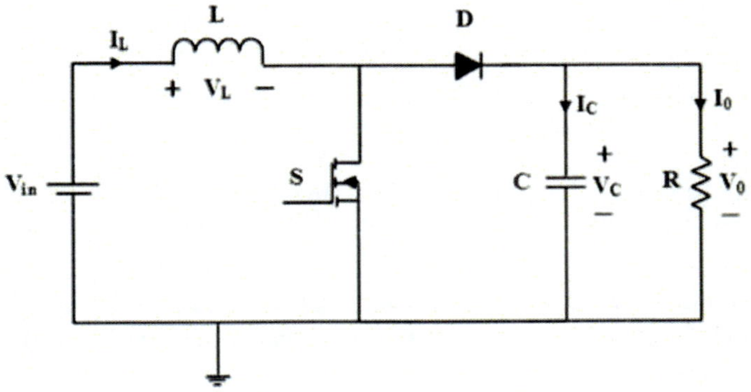


**Fig. 4** Schematic configuration of non-isolated step-up converter

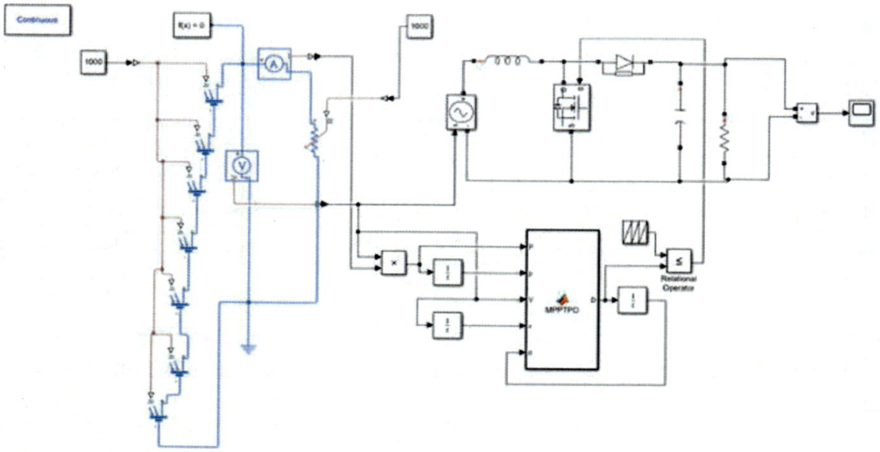

**Fig. 5** Simulink model of step-up converter with P&O MPPT

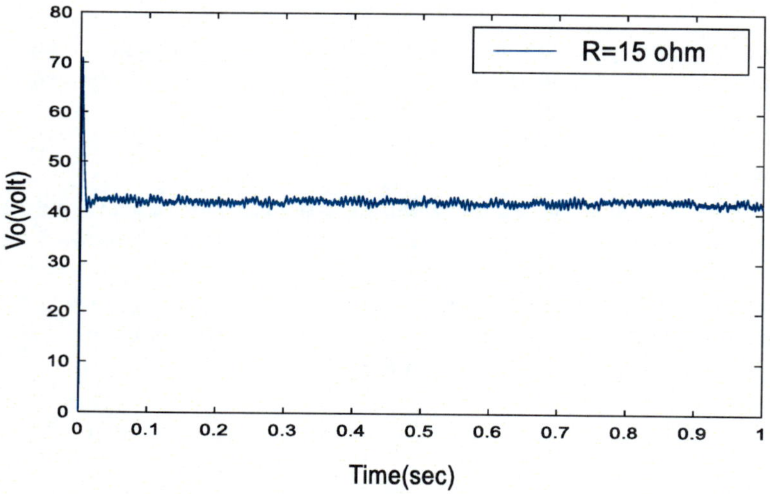


**Fig. 6** Voltage output with P&O MPPT ($R = 15\ \Omega$)

From Figs. 6, 7, and 8, it is noted that the voltage output reaches to 42 V in a very short duration of time with an overshoot of 66.6%. The output voltage contains a lot of disturbance.

The Simulink model of PV system with fuzzy-based MPPT is shown in Fig. 9. In this method, the step size is irregular in nature, and the changes are depending on the control rules [12].

The output voltage waveform of the PV system with FMPPT for various loads is depicted in Figs. 10, 11, and 12, respectively. There is more fluctuations in the

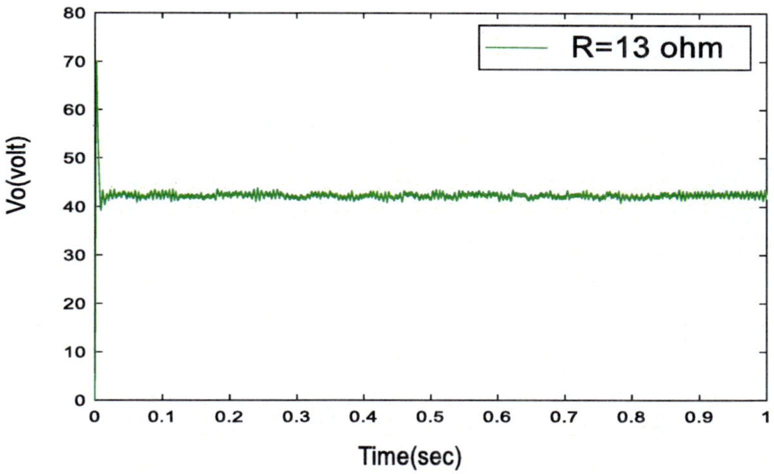


**Fig. 7** Voltage output with P&O MPPT ($R = 13\ \Omega$)

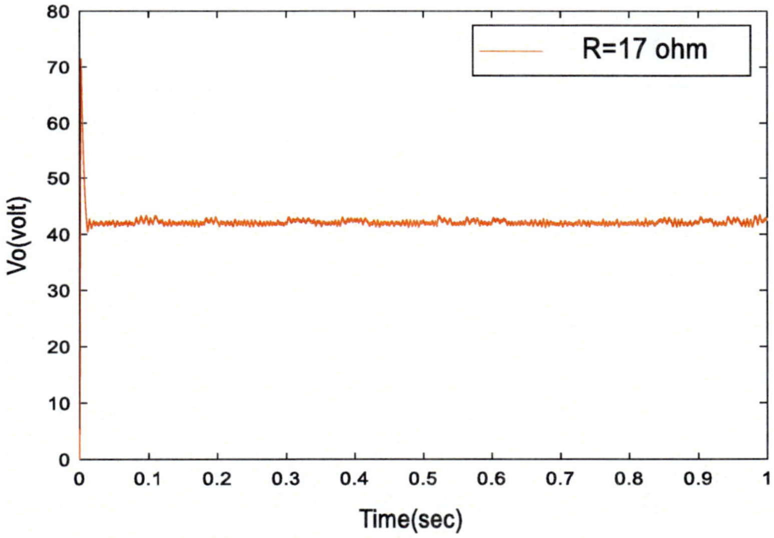


**Fig. 8** Voltage output with P&O MPPT ($R = 17\ \Omega$)

system responses when the load is decreased. The average output voltage is of about 42 V.

For a load resistance of 15 Ω, there will be no overshoot, less steady-state error, and less settling time. When the load is reduced from 15 to 13 Ω, large fluctuations are occurring in the output but it has no overshoot.

When the load is increased from 15 to 17 Ω, fluctuations in the output voltage will be reduced, and it tracks the output voltage very fast.

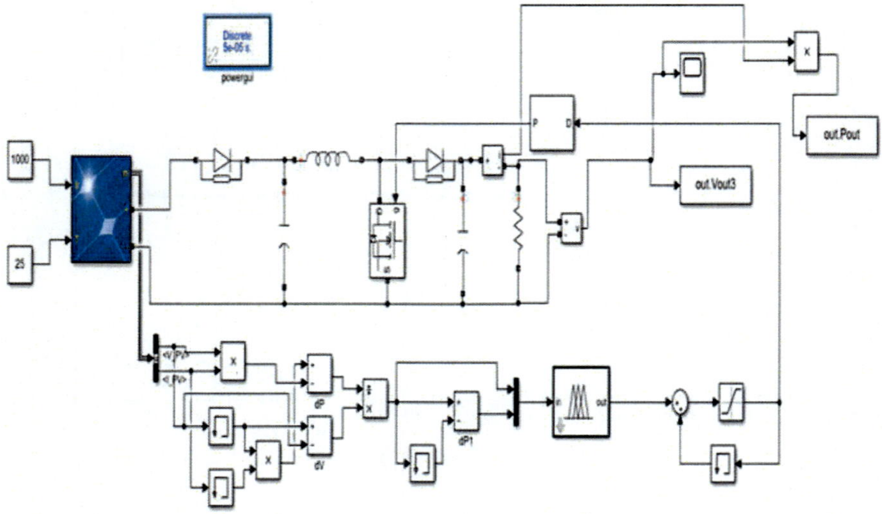


Fig. 9 Simulink diagram of step-up converter using FMPPT

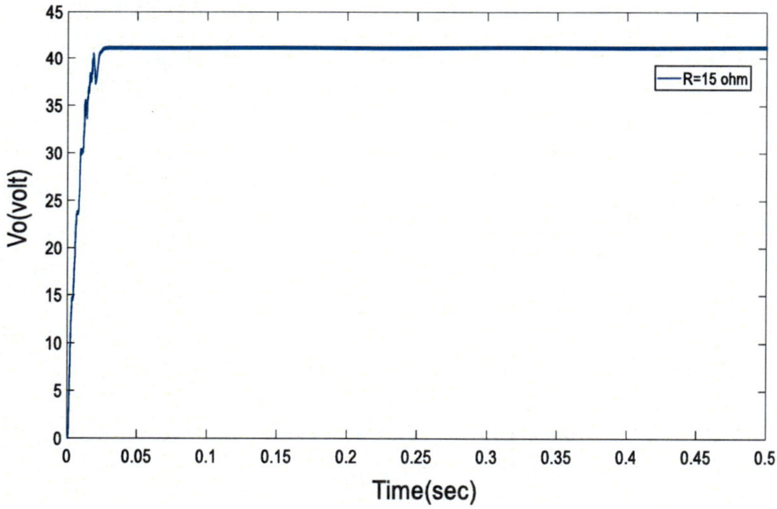


Fig. 10 Output voltage of boost converter with FMPPT with $R = 15\ \Omega$

It can be observed from Figs. 5 and 10 that application of FLC highly boosted the output by allowing a fixed magnitude. It is not oscillatory for reference load of 15 Ω. This implied that the described rule base is yielding a better control to record the voltage output and MPP.

The power output of the converter using two different MPPT techniques is shown in Figs. 13 and 14, respectively.

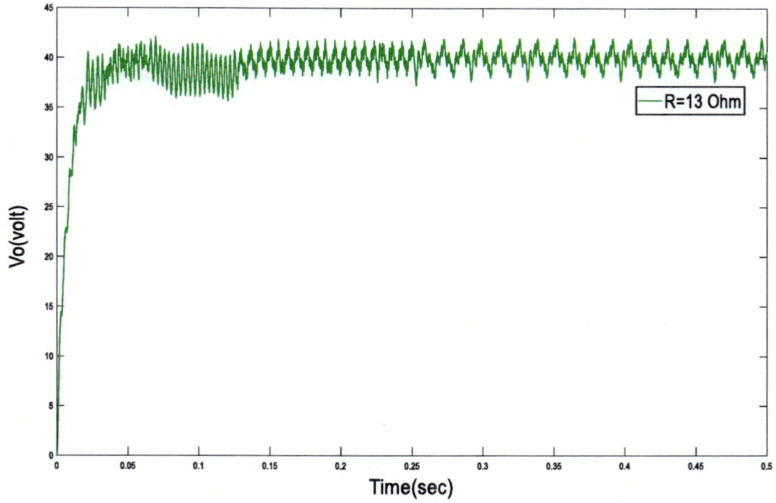


**Fig. 11** Output voltage of boost converter with FMPPT with $R = 13\ \Omega$

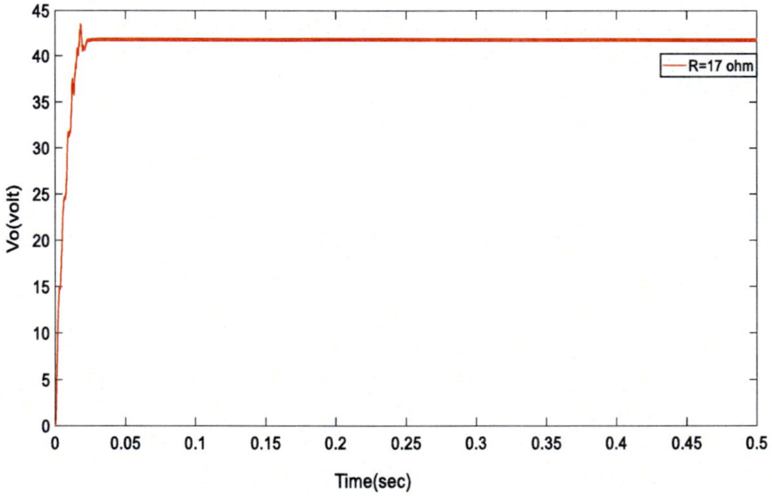


**Fig. 12** Output voltage of boost converter with FMPPT with $R = 17\ \Omega$

By using FMPPT controller, the output power is100 watt, and it settles at 0.03 s. The power output is 77% of the maximum power. Figure 14 shows the output power graph using P&O MPPT.

The power output is 130 W, but it is an oscillatory one. Hence fixed power, we are getting from the converter using FMPPT.

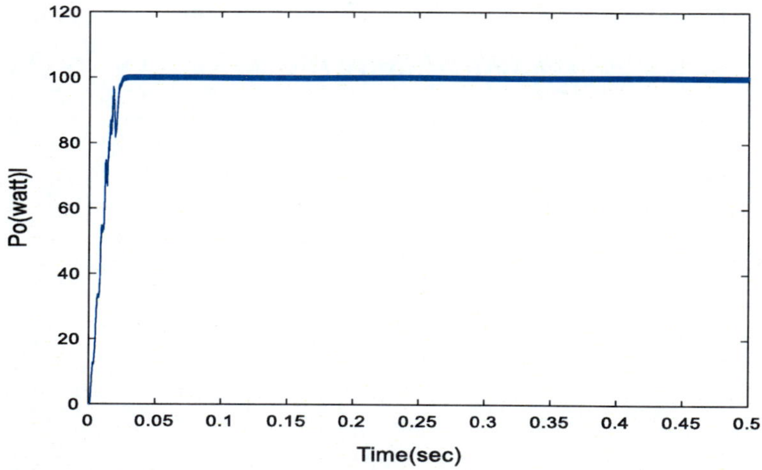


**Fig. 13** Power output of step-up converter with FMPPT

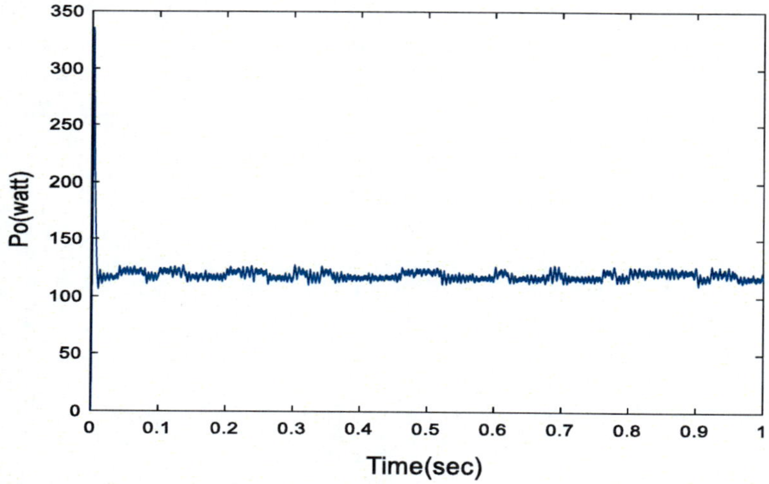


**Fig. 14** Output power of boost converter with P&O MPPT

# 5 Performance Identification Between P&O and Fuzzy MPPT

Table 4 compares the perturbation in the performance using P&O and FMPPT techniques [13–15]. The table shown below describes that the response settles very rapidly in open loop step-up converter with an overshoot of 42.8% and a steady-state error of 6.66%. By using P&O MPPT, output voltage settles at 0.015 s and having a steady-state error of 0.005 and achieves power output of 115 W which is nearest to maximum

**Table 4** Performance comparison

| | Input voltage 21 V, load resistance = 15 Ω | | | | | |
|---|---|---|---|---|---|---|
| | $t_s$ (s) | $M_p$ | SS error | Voltage output (V) | Power output (W) | Stability |
| Using P&O MPPT | 0.015 | 66.6% | 0.005 | 42 | 115 | Unstable |
| Using FMPPT | 0.04 | 0 | 0 | 42 | 100 | Stable |

power of the PV module. By using FMPPT controller, output voltage settles at 0.04 s and having a steady-state error of 0 with zero overshoot and output power of 100 W. FMPPT tracks the output voltage with less steady-state error, and the power output is close to maximum power. So it is a good choice in comparison with P&O.

## 6 Conclusion

The simulation results show that the output voltage at different load conditions and also the power characteristic at reference load resistance. The output power fluctuation is more using P&O method, but it matches with the maximum power. By the use of FMPPT, the fluctuations are reduced. But the output power is fixed at 100 W. The performances of the MPPT techniques were measured and matched by specifying the time response specifications. From the simulation waveforms, it can be concluded that the FMPPT provides better voltage regulation and a stable output at increasing load but in P&O method, step size is fixed, and it tracks the maximum power.

## References

1. Datta M, Tomonobu S, Atsushi Y, Toshihisa F, Kim KCH (2009) A coordinated control method for leveling PV output power fluctuations of PV diesel hybrid systems connected to isolated power utility. IEEE Trans Energy Conv 24:153–162
2. Roman E, Alonso AR, Pedro ESI, Damian G (2006) Intelligent PV module for grid-connected PV systems. IEEE Trans Ind Electron 53:1066–1073
3. Ganesh D, Moorthi S, Sudheer H (2012) A voltage controller in photo-voltaic system with battery storage for stand-alone applications. Int J Power Electr Drive Syst 2:9–18
4. Hauke B (2014) Basic calculation of a boost converter's power stage. Texas Instruments Application Report, Dallas, TX, Texas Instruments, pp 1–8
5. Vaucourt C (2004) Choosing inductors and capacitors for DC/DC converters. Texas Instruments Application Report, Dallas, TX: Texas Instruments
6. Salas V, Olias E, Barrado A, Lazaro A (2006) Review of the maximum power point tracking algorithms for stand-alone photovoltaic systems. Elsevier Sol Energy Mater Sol Cells 90:1555–1578
7. Subudhi B, Pradhan R (2013) A comparative study of maximum power point tracking techniques for photovoltaic system. IEEE Trans Sustain Energy 4:89–98
8. Bendiba B, Krimb F, Belmilia H, Almia MF, Bouloumaa S (2014) Advanced fuzzy MPPT controller for a stand-alone PV system. Elsevier Energy Proc 50:383–392

9. Sadeghzadeh SM, Ghassami ASAA (2013) A high performance maximum power point tracker for PV systems. Elsevier Electric Power Energy Syst 53:237–243
10. Abubakkar Siddik A, Shangeetha M (2012) Implementation of Fuzzy logic controller in photovoltaic power generation using boost converter and boost inverter. Int J Power Electron Drive Syst 2:249–256
11. Kolsi S, Samet H, Ben Amar M (2014) Design analysis of DC–DC converters connected to a photovoltaic generator and controlled by MPPT for optimal energy transfer throughout a clear day. J Power Energy Eng 2:27–34
12. Kumari JS, Babu CS (2012) Mathematical modeling and simulation of photovoltaic cell using Matlab-Simulink environment. Int J Electr Comput Eng 2:26–34
13. Tsai HL, Tu CS, Su YJ (2008) Development of generalized photovoltaic model using Matlab/Simulink. In: Proceedings of the world congress on engineering and computer science 2008, San Francisco, CA, pp 468–518
14. Alajmi BN, Ahmed KH, Finney SJ (2011) Fuzzy-logic control approach of a modified hill climbing method for maximum power point in Micro-grid standalone photovoltaic system. IEEE Trans Power Electron 26:1022–1030
15. Rezk H, Eltamaly AM (2015) A comprehensive comparison of different MPPT techniques for photovoltaic systems. Elsevier Sol Energy 112:1–11

# Optimization of Hybrid Solar, Wind, and Diesel Energy System from Cost Analysis of Micro-Grid Using Homer Software

P. Dinesh and Yashwant Sawle

## 1 Introduction

Electricity has been a problem in many distant regions in recent years. Demand for energy transmission from diesel fuel and fossil fuel is dominated in remote areas by concerns about air pollution, water pollution, deforestation, and global environmental deterioration. Conventional energy, on the other hand, can be a good substitute for fossil fuels and is the best option from a social and economic standpoint. "Wind and Photovoltaic are being used to replace fossil fuels, which are abundant in nature" [1]. Micro-grids for rural areas are used to fulfill required load. Non-conventional sources are used to power these micro-networks [2]. Because certain distant locations lack electricity due to transmission line problems, the off-grid method is mostly employed for isolated areas [3]. "Solar energy is obtained from the sun through the use of a photovoltaic system, and it can also be used to power a wind turbine" [4, 5]. Hybrid systems that are connected to the grid are more likely to provide continuous power. The efficiency and dependability of a hybrid renewable energy system (HRES) are largely dependent on the components and their precise modeling, as well as the use of proper optimization methods [6]. Renewable energy supplies are limitless and will never run out. As a result, we must make the most of such tremendous amounts of energy [1]. There are two optimization techniques available in this software: search space and homer optimization. The COE can be decreased by employing a load-dependent optimization approach [7, 8]. The author developed a off-grid system to electrify remote areas located in, India [9].

A feasibility analysis for different hybrid renewable cases through genetic algorithm (GA) and particle swarm optimization using different DG strategies. [10].

P. Dinesh · Y. Sawle (✉)
Department of Electrical Engineering, Vellore Institute of Technology, Vellore, India
e-mail: yashsawle@gmail.com

Y. Sawle
Electrical Engineering Department, Madhav Institute of Technology & Science, Gwalior, India



P. H. J. Chong et al. (eds.), *Emerging Electronics and Automation*, Lecture Notes in Electrical Engineering 937, https://doi.org/10.1007/978-981-19-4300-3_13

**Table 1** Ramanathapuram city climate details

| District | Ramanathapuram |
|---|---|
| State | Tamil Nadu |
| Country | India |
| Weather | 31 °C |
| Latitude | 9°21.8 |
| Longitude | 78°50.4E |
| Wind speed | 6.49 |

The author suggested a off-grid system is the best economical scenario among the designed six various cases for the location of Barwani in India [11]. The author discussed various issues related to the design of hybrid systems and reviewed various software and artificial techniques to perform optimization [12]. The authors explored about smart micro-grid's design, integration and optimization, as well as numerous constraints for its components and objectives [13]. The author performed a various system design of off-grid systems for remote areas located in the Tamil Nadu district, India [14]. Using the HOMER platform, the author performed optimization and sensitive analysis for four different integrated renewable sources for the location of Ukai, Gujarat, Tamil Nadu [15, 16].

The author reviewed various control strategies and the effects of techno-economic analysis of hybrid renewable systems and performed a case study for a remote location using HOMER and SO technology [17]. By utilizing different renewable sources, the author performed a novel methodology based on multi-objective functions using different artificial intelligence algorithms [18]. The major goal of this study is to use the HOMER software to create a solar, wind turbine, and diesel generator combination. To meet the load needs of a specific area, several systems have used a diesel generator and a battery bank as a backup (Table 1 and Fig. 1).

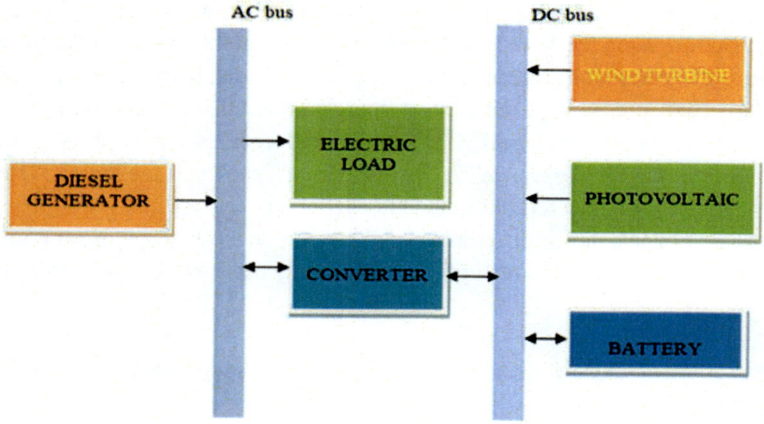


**Fig. 1** Off-grid hybrid system

## 2 District of Analysis and Load Description

The chosen territory is the district of Ramanathapuram in the Indian state of Tamil Nadu, with its location on the map at latitudes 9°21.8 N and longitude 78°50.4E, where the usefulness of green energy sources is investigated. Each of these would be utilized in a stand-alone capacity. A measurement that is standard of typical energy use is 11.25kWh/day; average load system is 0.47 kW, the production of peak demand is 2.32 kW. The load factor is 0.2. Figure 2a and b displays the yearly demand information based on the load value measured over 24 h.

### 2.1 *Potential of the Sun*

From Fig. 3 and Table 2, the sun-based information, which includes the clearness index and sun radiation, varies from 0.454 to 0.603 and 4.120 kWh/m$^2$/day to 6.180 kWh/m$^2$/day, respectively. The November month has the lowest sun radiation

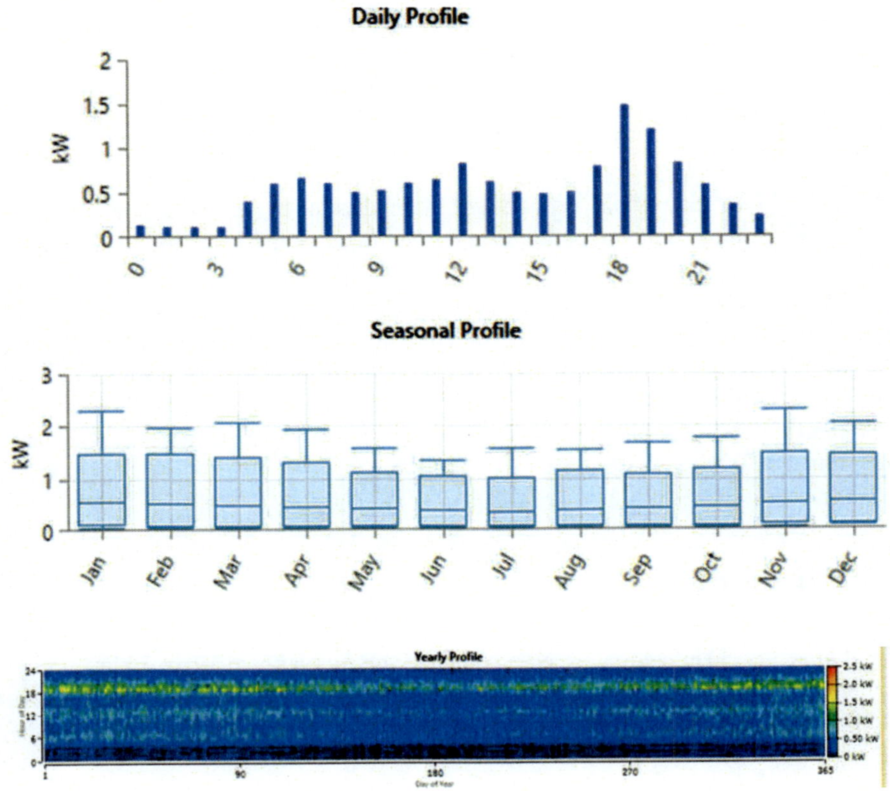


**Fig. 2** **a** Annual load profile. **b** Hourly and monthly demand profile

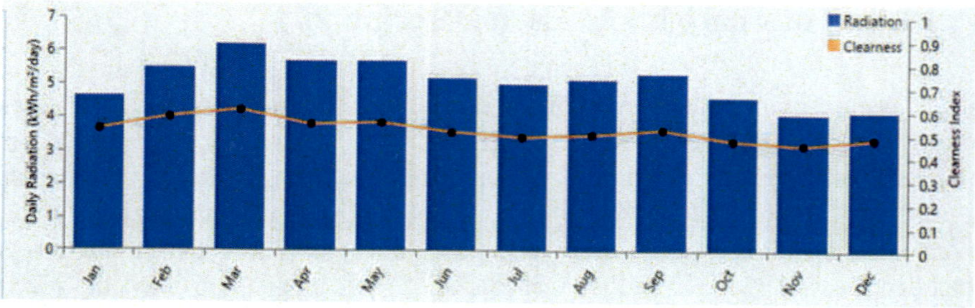


**Fig. 3** Monthly averages for global solar PV data for the location

**Table 2** Month-wise solar data

| S. No. | Month | Index clearness | Everyday solar isolation kWh/m$^2$/day |
|---|---|---|---|
| 1 | January | 0.520 | 4.660 |
| 2 | February | 0.571 | 5.500 |
| 3 | March | 0.603 | 6.180 |
| 4 | April | 0.542 | 5.690 |
| 5 | May | 0.548 | 5.690 |
| 6 | June | 0.507 | 5.180 |
| 7 | July | 0.486 | 4.980 |
| 8 | August | 0.496 | 5.150 |

of 4.620 kWh/m$^2$/day, while the month of March has the highest solar irradiation of 6.180 kWh/m$^2$/day. A month of low solar concentrated radiation is enough to employ sun-based energy for power generation.

The solar panel converts irradiance of sunlight into sun-powered energy, to meet load. It only works when the sun is available, and the excess energy generated by the solar photovoltaic has been utilized to charge the backup batteries, which feed to load. Especially at night where there is not available of solar energy. The photovoltaic type used in this study is a generic flat PV plate, which has a ecomomic details are mentioned in Table 1.

## 2.2 *Wind Turbine*

Wind speed is calculated using NASA data given by National Renewable Energy Laboratory (NERL). An anemometer at a height of 50 m recorded 8.360 m/s as the monthly average wind speed for the specified site (Latitude 9.25S, Longitude 78.75E) (Fig. 4).

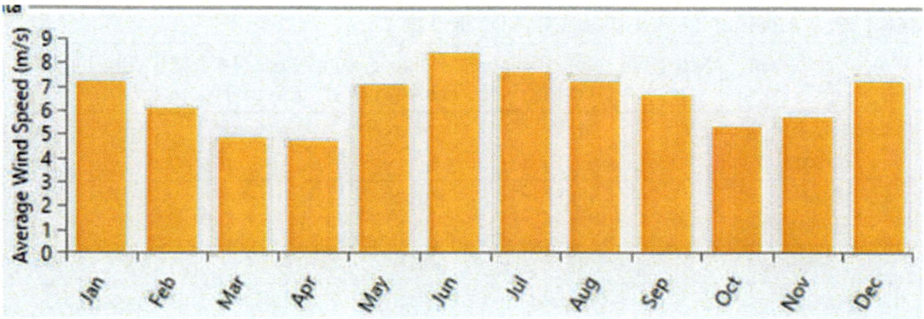


**Fig. 4** Shows monthly average wind speed

# 3 System Description

HOMER was used to assess and quantify the costs of numerous hybrid power system models, both on and off the grid, in this research. In order to analyze the optimization outcomes for various combinations, the HOMER simulation software requires specific input data, as stated in the following section [19]. We investigated PV, WT, and DG resources in study. The NASA resources Web site was utilized to get monthly average solar and wind statistics for the specified location based on latitude and longitude [20, 21] (Fig. 5).

**Fig. 5** Design of a hybrid power system model

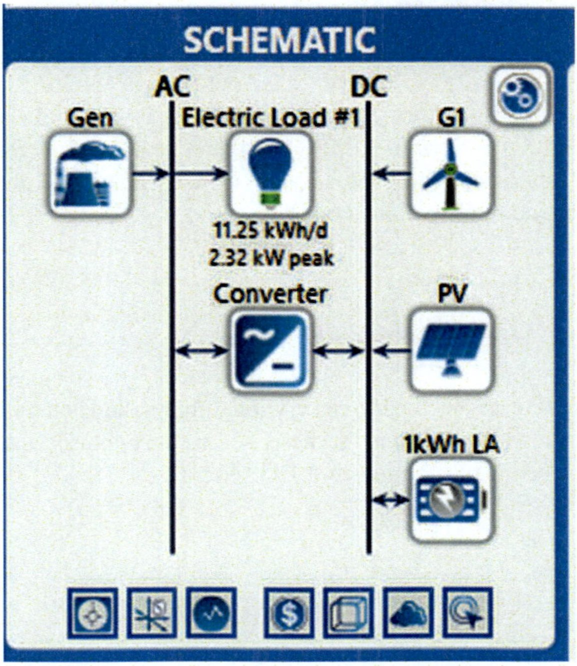

**Table 3** Cost details of the system compounds [26, 27]

| S. No. | Equipment | Size (kW) | Initial capital cost ($) | Replacement cost ($) | Cost of O&M ($ year) | Time elapsed |
|---|---|---|---|---|---|---|
| 1 | Wind turbine | 1 | 300 | 300 | 20 | 20 |
| 2 | Solar photovoltaic | 1 | 630 | 630 | 10 | 20 |
| 3 | Converter | 1 | 300 | 300 | 03 | 15 |
| 4 | Battery | 1 | 162.66 | 250 | 10 | 3550 (h) |
| 5 | Generator | 1 | 300 | 300 | 05 | 15,000 (h) |

## 3.1 Wind Turbine

Wind power system form wind turbines primary input is air flow. The two major characteristics that a wind turbine is works with the air speed and air direction. Due to the size of the turbine blades, sufficient speed and directed air flow are necessary. A number of essential characteristics affect the overall wind turbine system [22]. The speed of the wind does not remain consistent throughout the year. The table illustrates the replacement expenses as well as the operating and maintenance costs.

## 3.2 Photovoltaic

The intensity of light, or solar power concentration, is crucial for producing the highest amount of electricity [23]. This is accomplished using solar panels linked in serial and parallel. MPPT [24] can be used to increase the PV module's output. Despite its rising popularity, it is mostly employed in rural regions. Table 3 shows the economic parameter of PV.

## 3.3 Diesel Generator (DG)

In the case that the PV array, wind turbines, and battery bank are insufficient to fulfill the load requirement, the diesel generator supplies backup power [25]. Table 3 shows the economic parameter of DG.

### 3.4 *Battery*

When both the sun and the electric grid are unavailable, the battery bank storage capacity is chosen to meet the energy load demands.

### 3.5 *Converter*

To calculate the inverter's capacity, the size of the converters is employed as a decision variable. The quantity of AC power obtained by the system after converting DC electricity is referred to as the inverter's capacity. The user can additionally choose the rectifier capacity or the full DC power that can be altered from ac power. Unlike surge abilities, these capacities are nonstop and can last for a longer length of time.

## 4 Economic Evaluation

Financial factors such as COE, system NPC, system capital expense, and salvage cost might be employed, to examine the hybrid system. The best feasible outcome based on these economic variables simulates HOMER.

### 4.1 *Net Present Cost (NPC)*

The NPC is a significant expenditure factor in software used [28]. The NPC is computed by subtracting value of totally expenditures absorbed by the program from the value of all revenues earned over the life of the endeavor. The NPC includes the initial capital cost, replacement, operations and maintenance expenditures, fuel, and other factors. Equation (1) expresses the net current cost numerically as [29].

$$\text{Net present cost} = \frac{C_{\text{annual.Tot}}}{\text{CRF(I, L project)}} \tag{1}$$

where

"$C_{\text{annual.Tot}}$ is absolute annual cost
CRF is a capital recovery factor
i is the interest rate
L project is designed project life time."

### 4.2 COE (Cost of Energy)

The COE is the defined as the quantity of useable energy engendered by the designed system. According to HOMER platform, COE means ratio of the system's total yearly electricity expenditure to the total usable power generated [30]. Equation (2) represents the cost of energy [31].

$$\text{Cost of energy (COE)} = \frac{C_{\text{annual.Tot}}}{\text{Eprimary.AC} + \text{Eprimary.DC}} \tag{2}$$

where

$C_{\text{annual.Tot}}$ is total annual cost
EprimaryAC is primary AC load served
Eprimary DC is primary DC load served.

### 4.3 Renewable Fraction (RF)

The RF refers to the percentage of energy given to the load that comes from non conventional sources. The RF is calculated by HOMER using the following equation [32]:

$$f_{\text{ren}} = 1 - \frac{E_{\text{nonren}} + H_{\text{nonren}}}{E_{\text{served}} + H_{\text{served}}} \tag{3}$$

where:

$E_{\text{nonren}}$ = nonrenewable electrical production [kWh/year]
$E_{\text{grid,sales}}$ = energy sold to the grid [kWh/year] (included in $E_{\text{served}}$)
$H_{\text{nonren}}$ = nonrenewable thermal production [kWh/year].

## 5 Findings of the Homer Simulation

When HOMER simulates and optimizes, it discovers that the need profile must be met. The optimal estimation for over a framework is 4.04 kW solar photovoltaic panel, no value of generators, 18KWh lead-acid battery, and 2.25 kW converters. The initial cost is $7950; the annual running cost is $467.50; the cost of power is $0.264, and there is a renewable component with a 100 percent optimum arrangement. In addition, the hybrid system generates 6065 kWh per year, with the PV array and diesel generator providing 27.6% and 72.4% of total energy, respectively (Fig. 6).

The AC vital load is taken care of with the system's power. The AC load is used for an average of 4104 KWh per year. The average annual power generated by PV system is 6065 kWh, whereas the average annual electricity generated by a

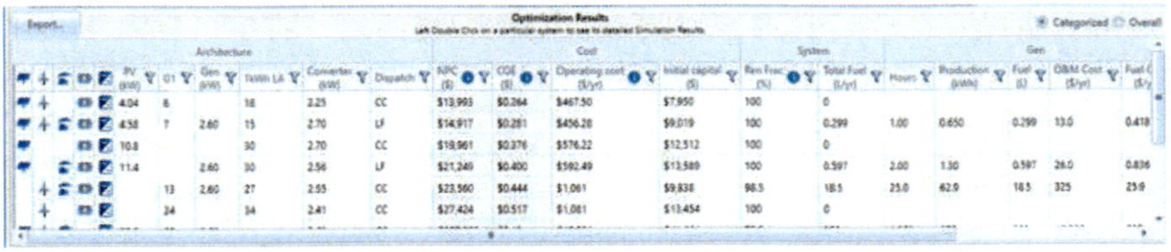


**Fig. 6** Optimization result from homer

DG is 15,915 kWh. The additional power available with 0.0404% unmet loads is 17,450 KWh per year. Figure 7: Monthly electric production energy system power generation. Figure 8 illustrates that the planned integrated renewable scheme protects money during the project's life span, with a operating cost of $118.134/year.

A DG against $467.50/year for the most cost-effective solution Fig. 9 depicts the ideal configuration's total net present expenses, which are expected to be $13,993 against $1.53 M for the basic case with a diesel generator. In the ideal crossover situation, by taking into account NPC, the cost of these systems is reduced by around half.

**Fig. 7** Power generation of monthly electric production energy system

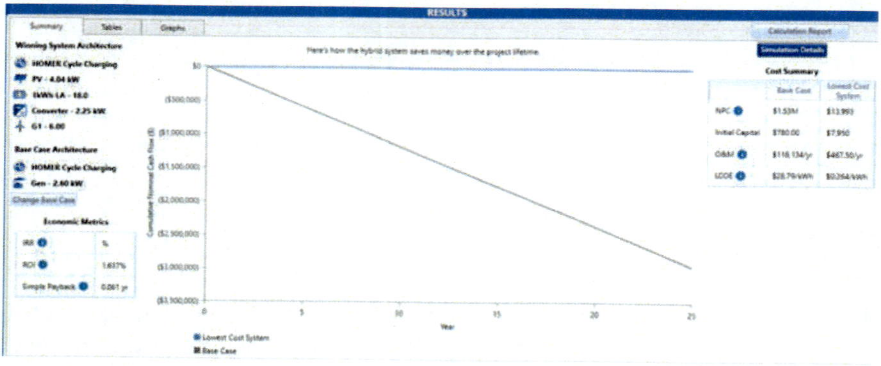


**Fig. 8** Result of hybrid system saves money over the project life time

**You may choose a different base case using the Compare Economics button on the Results Summary Table.**

| | Architecture | | | | | | | | | | Cost | |
|---|---|---|---|---|---|---|---|---|---|---|---|---|
| | PV icon | Wind icon | Gen icon | Battery icon | Converter icon | PV (kW) | G1 | Gen (kW) | 1kWh LA | Converter (kW) | NPC ($) | Initial capital ($) |
| Base system | | | ✓ | | | | | 2.60 | | | $1.53M | $780.00 |
| Proposed system | ✓ | ✓ | ✓ | ✓ | ✓ | 4.58 | 7 | 2.60 | 15 | 2.70 | $14,917 | $9,019 |

**Fig. 9** Net present cost result

## 6 Conclusion

In this work, we utilized HOMER to evaluate the outcomes of modeling and a hybrid system optimization that generates electrical power using solar photovoltaic's, a diesel engine generator, and a battery. With NPC $13,993, cost of energy 0.264 $/kWh and a cost of work of $467.50/year, the solution to this analysis is a 4.04 kW PV, a Nil value of capacity diesel generator, 18 kWh lead-acid battery, and a 2.25 kW converter which is less expensive than the considered base scenario with COE $28.79. As a consequence of examining the load demands of these off-grid and the adequate renewable sources available in their unique locations, the hybrid PV/DG/SB framework has resulted in a economically more attractive structure.

**Acknowledgements** "I want to thank the Vellore Institute of Technology (VIT) and Prof. Yashwant Sawle for their kind support."

## References


1. Wibowo IA, Sebayang D (2015) Optimization of solar-wind-diesel hybrid power system design using HOMER. Int J Innov Mech Eng Adv Mater 1:27–31
2. Ghenai C, Salameh T, Merabet A, Hamid AK (2017) Modeling and optimization of hybrid solar-diesel-battery power system. In: 7th IEEE international conference on modeling, simulation, and applied optimization (ICMSAO), 2017, pp 1–5
3. Nazir R, Laksono HD, Waldi EP, Ekaputra E, Coveria P (2014) Renewable energy sources optimization: a micro-grid model design. Energy Procedia 52:316–327
4. Heydari A, Askarzadeh A (2016) Optimization of a biomass-based photovoltaic power plant for an off grid application subject to loss of power supply probability concept. Appl Energy 165:601–611
5. Tanezaki S, Matsushima T, Muroyama S (2003) Stand-alone hybrid power supply system composed of wind turbines and photovoltaic modules for powering radio relay stations. In: IEEE annual conference on magnetics, 2003, pp 457–462
6. Kazem HA, Al-Badi HAS, Al Busaii AS, Chaichan MT (2017) Optimum design and evaluation of hybrid solar/wind/diesel power system for Masirah Island. Environ Develop Sustain 19(5):1761–1778
7. Sen R, Bhattacharyya SC (2014) Off-grid electricity generation with renewable energy technologies in India: an application of HOMER. Renew Energy 62:388–398
8. Gan LK, Shek JKH, Mueller MA (2015) Hybrid wind–photovoltaic–diesel–battery system sizing tool development using empirical approach, life-cycle cost and performance analysis: a case study in Scotland. Energy Convers Manage 106:479–494
9. Sawle Y, Gupta SC (2014) Optimal sizing of photovoltaic/wind hybrid energy system for rural electrification. In: 2014 6th IEEE power india international conference (PIICON), Delhi, India, December 5–7, 2014 (IEEE), 1–4. https://doi.org/10.1109/34084.poweri. 2014.7117758
10. Sawle Y, Gupta SC, Bohre AK (2018) Techno-economic scrutiny of HRES through GA and PSO technique. Inderscience Int J Renew Energ Technol 9(1/2):84–107. https://doi.org/10.1504/ijret.2018.090106
11. Sawle Y, Thirunavukkarasu M (2020) Techno-economic comparative assessment of an off-grid hybrid renewable energy system for electrification of remote area. Anal Appl Renew Energy Syst 2020:199. https://doi.org/10.1016/B978-0-12-824555-2.00027-7
12. Sawle Y, Gupta SC, Bohre AK|Wei Meng (Reviewing Editor) (2016) PV-wind hybrid system : a review with case study. Cogent Eng 3:1. https://doi.org/10.1080/23311916.2016.1189305
13. Thirunavukkarasu M, Sawle Y (2021) Smart microgrid integration and optimization. In: Baseem (ed) Active electrical distribution network: a smart approach. Wiley Publishing Group, pp 203–234
14. Thirunavukkarasu M, Sawle Y (2021) A comparative study of the optimal sizing and management of off-grid solar/wind/diesel and battery energy systems for remote areas. Front Energy Res 9:752043. https://doi.org/10.3389/fenrg.2021.752043
15. Sawle Y, Jain S, Babu S, Nair AR, Khan B (2021) Prefeasibility economic and sensitivity assessment of hybrid renewable energy system. IEEE Access 9:28260–28271. https://doi.org/10.1109/ACCESS.2021.3058517
16. Jain S, Sawle Y (2021) Optimization and comparative economic analysis of standalone and grid-connected hybrid renewable energy system for remote location. Front Energy Res 9:724162. https://doi.org/10.3389/fenrg.2021.724162
17. Sawle Y, Gupta SC, Bohre AK (2018) Review of hybrid renewable energy systems with comparative analysis of off-grid hybrid system. Renew Sustain Energy Rev 81 (Part 2) 2217–2235. https://doi.org/10.1016/j.rser.2017.06.033
18. Sawle Y, Gupta SC, Bohre AK (2018) A novel methodology for scrutiny of autonomous hybrid renewable energy system. Int J Energy Res 42(570):586. https://doi.org/10.1002/er.3841
19. Kellogg WD, Nehrir MH, Gerez V, Venkataramanan GV (1998) Generation unit sizing and cost analysis for stand-alone wind, photovoltaic and hybridwind/pvsystems. IEEE Trans Energy Convers 13:70–75

20. Nehrir MH, Wang C, Strunz K, Aki H, Ramakumar R, Bing J, Miao Z, Salameh Z (2011) A review of hybrid renewable/alternative energy systems for electric power generation: configuration, control, and applications. IEEE Trans Sustain Energy 2(4):392–403
21. Maherchandani JK, Agarwal C, Sahi M (2012) Economic feasibility of hybrid biomass/PV/wind system for remote village using HOMER. Int J Adv Res Electric Electron Instr Eng 1(2):49–53
22. Duong MQ, Grimaccia F, Leva S, Mussetta M, Sava G, Costinas S (2014) Performance analysis of grid-connected wind turbines, Scientific Bulletin-Politehnica University of Bucharest Series C. Electr Eng Comp Sci 76(4):169–180
23. Badwawi RA, Abusara M, Mallick T (2016) A review of hybrid solar PV and wind energy system. Smart Sci 3(3):127–138
24. Lazaroiu GC, Longo M, Roscia M, Pagano M (2015) Comparative analysis of fixed and sun tracking low power PV systems considering energy consumption. Energy Conv Manage 92:143–148
25. Rajoriya A, Fernandez E (2013) Hybrid energy system size optimization and sensitivity evaluation for sustainable supply in a remote region in India. Int J Sustain Energ 32(1):27–41
26. Thirunavukkarasu M, Sawle Y (2020) Design, analysis and optimal sizing of standalone PV/diesel/battery hybrid energy system using HOMER. IOP Conf Ser Mater Sci Eng 937:012034. https://doi.org/10.1088/1757-899X/937/1/012034
27. Mala R, Saini RP (2020) Dispatch strategies based performance analysis of a hybrid renewable energy system for a remote rural area in India. J Clean Prod 259:12069
28. Givler T, Lilienthal P (2005) Using HOMER software, NREL's micropower optimization model, to explore the role of gensets in small solar power systems. Case Study: Sri Lanka, National Renewable Energy Lab., Golden, CO (US)
29. Sawle Y, Gupta SC, Bohre AK (2018) A novel methodology for scrutiny of off-grid hybrid renewable system, John Wiley & Sons. Int J Energy Res 42:570–586
30. Das BK, Hoque N, Mandal S, Pal TK, Raihan MA (2017) A techno-economic feasibility of a stand-alone hybrid power generation for remote area application in Bangladesh. Energy 134:775–788
31. Sawle Y (2022) Scrutiny of PV biomass stand-alone hybrid system for rice mill electrification. In: Deregulated electricity market. Apple Academic Press, pp 135–152
32. https://www.homerenergy.pro/doc/latest

# Dynamic Economic Dispatch with Electric Vehicle

**Sulabh Sachan and Sanchari Deb**

## Nomenclature

| **Notation** | **Discription** |
|---|---|
| $w_j$ | Waiting time at $j$th CS |
| $R_j^t$ | Reputation of $j$th CS at time t |
| $\text{obj}_i^{\text{EV}}$ | Objective function of $i$th EV |
| $\text{obj}_j^{\text{CS}}$ | Objective function of $j$th CS |
| $\pi_j(.)$ | Payoff function of $j$th CS |
| $i$ | $i$th Number of EV |
| $j$ | $j$th number of CS |
| Ø | Øth route for CS |
| $t$ | $t$th time instant |

## 1 Introduction

The evolution of electric vehicle is a big challenge to the power system operation and control. Electric vehicles are fully driven by electricity. As it is powered through the electricity so that, it cannot produce harmful gases such as CO, $CO_2$, and $SO_2$. So it reduces the emission of carbon in the environment, therefore it gives no sound that it is sound free, and it also improves energy efficiency as compared to the conventional vehicles.

S. Sachan (✉)
GLA University, Mathura, India
e-mail: sulabh.sachan@gla.ac.in

S. Deb
Power and Control Lab, University of Warwick, Coventry, UK



P. H. J. Chong et al. (eds.), *Emerging Electronics and Automation*, Lecture Notes in Electrical Engineering 937, https://doi.org/10.1007/978-981-19-4300-3_14

As we know that Government imposed a new policy to enhance the development of electric vehicles. Studies specify that with the development of electric vehicles, the share market in U.S.A. will increased to 40% in 2021, 52% in 2032, and 63% in 2051 [1–3]. Electric vehicles can be used as an energy storage device [4–6]. According to vehicle to grid technology, electric vehicles discharge to the power grid at the maximum load and charge at minimum load from the grid [7, 8]. DSM can be defined as activities that are planned to affect the use of electricity at the customer side. DSM requires active residential customer engagement and a novel communications framework for information exchange and control, which are also considered to be some of the main components in smart grids [9]. Controlling EV charging load, so-called smart charging, is one such DSM strategy that can also be employed at the residential customer to increase the load matching [10–13]. In [11], a distributed smart charging scheme has been developed for EVs with an objective to minimize the net-load variance for residential buildings with a PV system and EV charging demand. This method proved, through simulations, that the self-consumption could be increased and peak loads could be reduced. However, the benefit of this method has not been tested on an actual LV distribution system.

In this paper, we have made a dispatch model by considering the both economic as well as environmental factors, and the proposed model is solved using GAMS software algorithms. In this paper, we mostly focus on static dispatch, although this review depicts that dynamically economic dispatch replica is more reliable for the dispatching of power system problems, by considering the electric vehicle. In this paper, a dynamic economic dispatch is evolved by considering the electric vehicles. Beside this power balance limitation, the demand of user travel, the battery charging on board, etc., are involved in the limitation. The cost of total fuel and the emission of total pollutant are the two optimized objectives of this paper.

## 2 Economic Dispatch

This section focusses on modelling all the parameters affecting the price game for EVs and CSs. The selected schematic model for price game is presented in Fig. 1. In this, each CS optimizes charging price in order to maximize profit. Economic dispatch is the minimum-term commitment of optimal output of various electrical, to accomplish the system load, at the cost which is minimum, subject to transmission and operation limitations. The problem of economic dispatch is solved by computer software GAMS which is specialized, and this software can satisfy the operation and system limitation of the resources which are available and it corresponds to the capabilities of transmission.

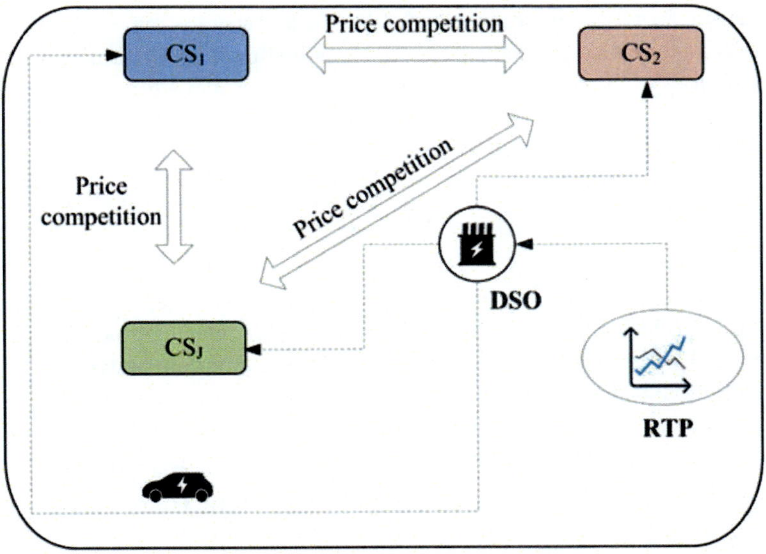


**Fig. 1** System model replicates the operation of CSs

## *2.1 Classification of Economic Dynamic Dispatch*

Classification of economic dynamic dispatch is depicted below:

1. Load-based dynamic economic dispatch (LDED)
2. Price-based dynamic economic dispatch (PDED).

In LDED, objective is to minimize the load. The load of the $q$th thermal unit is defined below:

$$\mathrm{EM}\sum_{\mathrm{g,t}} d_{\mathrm{g}} P_{\mathrm{g,t}}^{2} + e_{\mathrm{g}} P_{\mathrm{g,t}} + f_{\mathrm{g}}$$

The operation limits of the $q$th unit are as follows:

$$P_{\mathrm{g_{min}}} \le P_{\mathrm{g,t}} \le P_{\mathrm{g_{max}}}$$

The main aim of the price-based dynamic economic dispatch (PDED) is to maximize the total benefits of the generating units of the generating company by other products and the energy. In this, there is no hourly supply-demand balanced complexity. Besides this, the other technical complexity should be satisfied.

The EV's selection of CS accounts for a variety of factors such as the offered price by CSs, the distance from CSs, the expected waiting time required for charging EV, and the reputation of CSs that is newly described in this work. In the realistic scenario, the distance between $j$th CS and $i$th EV ($d_{i,j}$) is a very important parameter for the selection of CS as it can increase the driving time and required charging. Therefore, each EV tries to reduce the distance in the objective function. Similarly, the offered charging price by the $j$th CS ($p_j$) is also important in the competitive game. Each EV tries to reduce the charging cost and distance. Considerations of these two parameters can be written for $i$th EV as following:

$$\text{obj}_i = \underset{j \in J}{\text{argmin}}(\varphi_1 \cdot p_j + \varphi_2 \cdot d_{i,j}) \tag{1}$$

The formulation of waiting time is done with the consideration of EVs and CS behaviour on $t$th time interval. Mainly, there are few parameters that directly affect the waiting time such as EVs arrival at CS, EV average required power $P_i$, number of charger $N_j^s$, and charging rate $\text{CR}_j$. We assume that $d_{i,j} \times \lambda^t$ EVs are approaching the $j$th CS to charge their EVs at time $t$. The average waiting time $w_{i,j}$ will increase with increase in the required power $\left(d_{i,j} \times \lambda^t \times P_i\right)$. Whereas, number of charger $\left(N_j^s\right)$ and charging rate $(\text{CR}_j)$ will reduce the average waiting time. This phenomena can be formulated as the following:

$$w_{i,j} = \frac{d_{i,j} \times \lambda^t \times P_i}{N_j^s \times \text{CR}_j} \tag{2}$$

Note that we use $\lambda^t$ to denote the EV density (EVs/km) on road because it changes with respect to time.

## 3 Modelling of EV

There is a problem in controlling a single electrical vehicle so, that the two-layer dispatch strategy model is employed, that is, the top layer is the power grid dispatch centre, middle layer is the electrical vehicle mechanism, and the bottom is the user unit (Fig. 2).

The total power of electric vehicle which is rated in a dispatch interval can be depicted below:

$$P_{\text{rated}} = \sum P_{\text{rated}}(i)$$

where $P_{\text{rated}}$ ($i$) is the rated power of $i$th electrical vehicle and prated is the total rated power

In each dispatch interval, the charging and discharging of electrical vehicle not happened at the same time, the conditions can be expressed as follows:

Where $x < 0$ means charging condition of an electrical vehicle. Where $x > 0$ means the discharging state.

At $x = 0$, electrical vehicle does not engage into a power system dispatch.

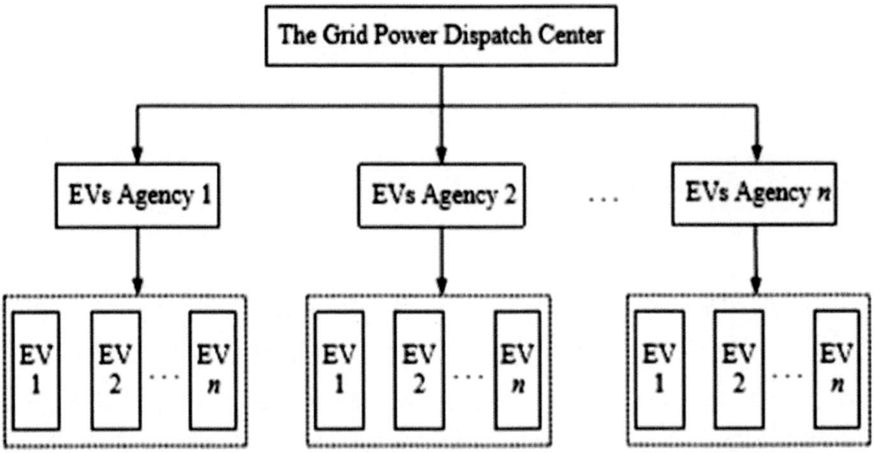


**Fig. 2** Modelling of EV

## 4 Modelling of Dynamic Environmental Economic Dispatch Difficulty

Since electrical vehicles engaged in a power system network dispatch, so the goal become dynamic

$$\text{Minimum } F_c = \sum_{n=1}\sum_{j=1}\left(a_j + b_j d_{j,n} + c_j d^2\right)$$

where $a_j$, $b_{j,}$ and $c_j$ are coefficients of costs of $j$th power generator, $K$ is the total period of dispatching, $d_{j,n}$ is the output power of the $j$th unit at $n$th time. $F_c$ is the total system fuel cost, and $P$ is the total thermal generator units.

The whole emission of pollution goal can be expressed as

$$\text{Minimum } E_m = \sum_{n=1}^{K}\sum_{j=1}^{P}\left[(a_j \beta_j d_{j,n} + \gamma_j d_{j,n}^2 + \zeta_j \exp(\phi_j, d_{j,n}))\right]$$

where $\alpha_j$, $\beta_j$, $\gamma_{j,}$ and $\zeta_j$ and $\phi_j$ are the coefficients of emissions of the $J$th power generator. $E_m$ is the entire emission of the pollution of the system.

## 4.1 Limitations

Because of the existence of electrical vehicle charging and discharging power which varies from the conventional environment economic dispatch. The system limitations are not disrupting but it has problematic attributes.

The entire power output of thermal generators, the power output of electric vehicles charging and discharging power must meet with the entire demand load.

$$D\left[\sum_{n=1}^{p}\left(D_{j,n} + D_{\text{dis},n} + D_{w} \geq D_{d,n} + D_{n} + D_{\text{ch},n}\right)\right] \geq \zeta_{1}$$

$D_{\text{ch},n}$ is the charging power of electrical vehicles in a $n$th time interval, $D_{\text{dis},n}$ is the discharging power of electrical vehicles in a $n$th time interval, $D_{\text{w}}$ is the active power, $D_{\text{d},n}$ is the power demanded in $n$th time interval, $\zeta_1$ is the level of confidence that shows the power system load demand meets at what level.

The electrical vehicles battery which is on board the power which is remaining $Q_t$ at time interval t is expressed as

$$Q_{t} = \left(Q_{t-1} + \gamma_{c} D_{\text{ch},tt} - \left((1/\gamma_{\text{d}})\, D_{\text{dis},t*t}\right) - Q_{\text{Trip},t}\right.$$

where $\gamma_{\text{c}}$ and $\gamma_{\text{d}}$ are the charging and discharging efficiencies coefficients, respectively, t is the interval of dispatch, $Q_{\text{trip},t}$ is the consumption of the electrical vehicles in the mechanism of running power, which can be given as

$$Q_{\text{Trip},t} = Q * L$$

where $Q$ is the total power consumption (which is approximately average) of unit distance, and $L$ is the distance up to which electrical vehicles drive.

For safety purpose and for life time services of the battery, the power which is remaining $Q_t$ is limited by maximum and minimum which is represented as

$$Q_{\min} \leq Q_{t} \leq Q_{\max}$$

The discharging and charging power of the electrical vehicle are approximately less as compared to the rated power.

$$D_{\text{ch},t} \leq \text{DN, ch}$$

$D$ discharging, $t \leq$ DN, discharging.

Where DN, Ch, and DN, discharging is the charging and discharging power, respectively, which is rated in nature of electrical vehicles, respectively.

The generator output power limitation is expressed as (Fig. 3):

$$D_{\min j,t} \leq D_{j,t} \leq D_{\max j,t}$$

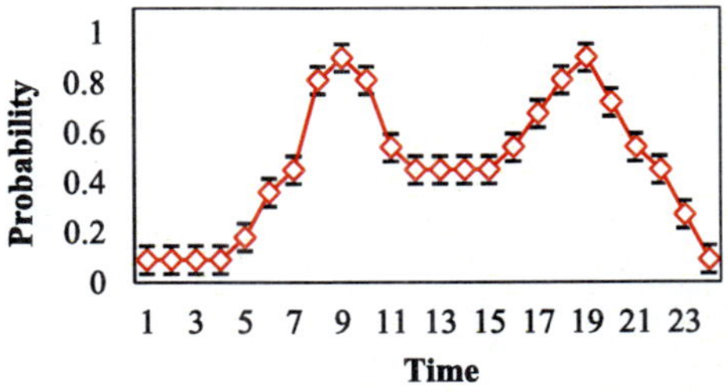


**Fig. 3** Probability of availability of EV on road

## 4.2 Problem Formulation

Depending on the goals and limitations, the dynamic environmental economic dispatch issue with electrical vehicles can be mathematically prepared as

$$\text{Minimum [FC, EM]}$$
$$\text{Concern to } h_j\,(D_{\text{G}}) \leq 0 \,\text{Where}\, j = 1, 2,\ 3\,\text{J}$$
$$\text{Gi}(D_{\text{G}}) = 0 \quad \text{Where}\, i = 1, 2,\ 3\,\text{I}$$
$$D_{\text{G min}} \leq D_{\text{G}} \leq D_{\text{G max}}$$

where $D_{\text{G}}$ is the decision dimensional variables which have to be optimized, and this include rated output power of generator (thermal generators) units. Where $h_{\text{J (DG)}}$ and $G_{\text{I (DG)}}$ are inequality and equality limitations, respectively. It represents the pare to optimized front for a distinctive many objective issue.

# 5 Results and Discussion

As we know that electrical vehicles have many environmental benefits as compared to the vehicles which operated on petroleum and diesel. Also, electrical vehicles have low maintenance and operating costs. Therefore, we can save money by using the electrical vehicles. The air pollution due to electrical vehicles is approximately zero. It also minimizes the dependency on petroleum as well as on the diesel. Electrical vehicles have one more benefit that, due to electrical vehicles, there is no emission of greenhouse gases. The efficiency of electrical vehicles is also more, it has great impact on the health of the public of any country, and due to this, there will be minimum damage of ecology. We can charge the electrical vehicles through renewable energy for example: solar energy, wind energy, etc., due to that emissions are decreasing more rapidly. We can charge the electrical vehicles at rooms, at work places, or even on the road. So, there is no requirement to go to the stations. The moving parts incorporated in an electrical vehicles are approximately less as compared to the vehicles which operated on diesel and petroleum. By the use of electrical vehicles, the emission of air pollution content is minimized, and also the time and places of

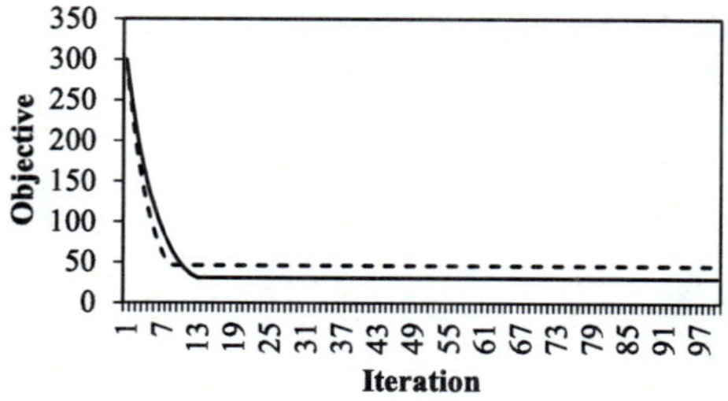


**Fig. 4** Learning curve comparison

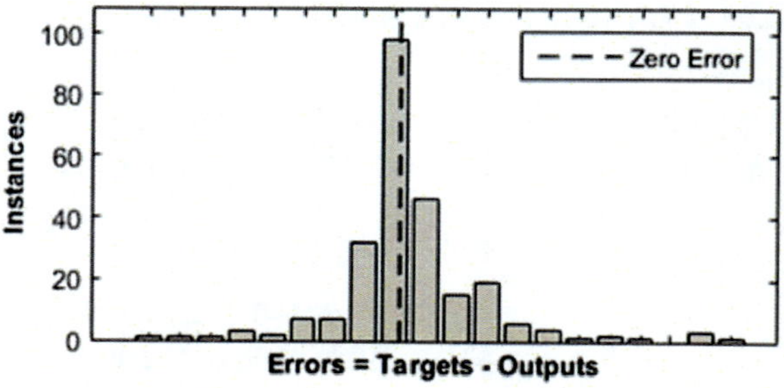


**Fig. 5** Discharging of EV

the emission are gets changed. The emission of pollution through the vehicles which operated on diesel and petroleum is during maximum-driving hours which is in the amidst of the city. The emission from an electrical vehicles during fuelling is on off-maximum running hours at the power plants (Figs. 4 and 5).

## 6 Conclusion

In this paper, numerous-objective dynamic environmental economic dispatch replica is composed to study the electrical vehicles. Whilst dealing with the unreliability of power, in this paper, distribution function strategy is used. For the discharging and charging command, we use the two-layer strategy for dispatch. For the solving of high level of dynamic environmental economic dispatch issue, with numerous limitations, this paper has better double-step limitation rectification technique. Here, we use 2 generator unit system. The result is obtained through the general algebraic modelling system (GAMS) software. From the conclusion of the above case study, this work acquired the impacts of electrical vehicle discharging and charging for the associated power grids. Although the cost of the battery and degradation of the battery are the essential factors in the total price of an electric vehicle. Our upcoming future duties will examine different policies of discharging and charging of an electric vehicle, cost of battery, and cost for the management services are more reliable with the real results. Another essential path for the future research is to investigate the limitations handing method and to upgrade the other numerous-objective optimized algorithms.

## References


1. Deb S, Tammi K, Gao XZ, Kalita K, Mahanta P (2020) A hybrid multi-objective chicken swarm optimization and teaching learning based algorithm for charging station placement problem. IEEE Access 8:92573–92590. https://doi.org/10.1109/ACCESS.2020.2994298
2. Singh M, Kumar P, Kar I (2012) Implementation of vehicle to grid infrastructure using fuzzy logic controller. IEEE Trans Smart Grid 3(1):565–577
3. Sachan S, Amini MH (2020) Optimal allocation of EV charging spots along with capacitors in smart distribution network for congestion management. Int Trans Electr Energy Syst 30:e12507. https://doi.org/10.1002/2050-7038.12507
4. Gupta V, Konda SR, Kumar R, Panigrahi BK (2017) Multi-aggregator collaborative electric vehicle charge scheduling under variable energy purchase and EV cancelation events. IEEE Trans Industr Inf 14(7):2894–2902
5. Sachan S, Deb S, Singh SN (2020) Different charging infrastructures along with smart charging strategies for electric vehicles. Sustain Cities Soc 102–238
6. Sachan S (2018) Stochastic charging of electric vehicles in smart power distribution grids. Sustain Cities Soc 40:91–100
7. Li S, Bao K, Fu X, Zheng H (2013) Energy management and control of electric vehicle charging stations. Electr Power Compon Syst. https://doi.org/10.1080/15325008.2013.837120
8. Sachan S, Deb S, Singh PP, Alam MS, Shariff SM (2021) A comprehensive review of standards and best practices for utility grid integration with electric vehicle charging stations. Wiley Interdiscipl Rev Energy Environ e424. https://doi.org/10.1002/wene.424
9. Sedighizadeh M, Mohammad Pour A, Alavi SM (2018) A daytime optimal stochastic energy management for EV commercial parking lots by using approximate dynamic programming and hybrid big bang big crunch algorithm. Sustain Cities Soc. https://doi.org/10.1016/j.scs.2018.12.016
10. Kuran MŞ, Viana AC, Iannone L, Kofman D, Mermoud G, Vasseur JP (2015) A smart parking lot management system for scheduling the recharging of electric vehicles. IEEE Trans Smart Grid 6(6):2942–2953
11. Andrenacci N, Ragona R, Valenti G (2016) A demand-side approach to the optimal deployment of electric vehicle charging stations in metropolitan areas. Appl Energy 182:39–46
12. Sachan S, Deb S, Singh SN, Sharma DD (2021) Planning and operation of EV charging stations by chicken swarm optimization driven heuristics. Energy Conv Econ 1–9. https://doi.org/10.1049/enc2.12030
13. Yuan W, Huang J, Zhang YJ (2017) Competitive charging station pricing for plug-in electric vehicles. IEEE Trans Smart Grid 8(2):627–639

# A Review and Analysis of Electrical Equivalent Circuit Models of Vanadium Redox Flow Battery

**Jusmita Das and Rajdeep Dasgupta**

## 1 Introduction

The electricity demand throughout the world is increasing rapidly due to population growth and industrialization [11]. Renewable energy sources such as solar and wind, unlike conventional power plants, are intermittent, with electricity production depending on the environment and time. With the rapid rise in renewable energy production, largely due to lower solar photo voltaic (PV) costs, utilities are looking for strategies to address grid instability and poor reliability that come with integrating the sustainable energy sources [7, 16]. Large-scale battery storage systems play a significant role in renewable power generation, continuous power supply, and smart grid application. These battery storage technologies are potential candidates capable of large-scale energy storage and play an essential role in electricity production and consumer needs for immediate electricity distribution [6]. Compared to the other different types of battery storage technologies such as lead-acid,nickel-cadmium (Ni-Cd), lithium-ion (Li-ion), and sodium-sulfate (Na-S), the redox flow batteries [18] offer various advantages. Redox flow batteries (RFBs) are attractive because of their flexible power and energy storage deriving from their extraordinary architecture [1]. The power output of RFBs can be scaled up by adding more cells to stack or expanding the cell areas. Similarly, energy can be scaled up by increasing the tank size or by adding more electrolyte solutions [20].

J. Das (✉) · R. Dasgupta
Department of Electronics and Instrumentation Engineering, National Institute of Technology, Silchar, Assam, India
e-mail: jusmita_rs@ei.nits.ac.in

R. Dasgupta
e-mail: rajdeep@ei.nits.ac.in



P. H. J. Chong et al. (eds.), *Emerging Electronics and Automation*, Lecture Notes in Electrical Engineering 937, https://doi.org/10.1007/978-981-19-4300-3_15

The **Redox Storage System Development Project** of NASA first identified the potential of RFB as an energy storage system. Due to their extensive study on Fe–Cr RFBs, the consumer-level production of the test system started in Japan during the late 80s [14]. Though these systems were easily scalable and had uniform state of charge for each cells, the cells had some shortcomings like low energy density, cell degradation, and membrane aging. To improve upon these shortcomings, various researchers developed RFBs with different metal ion combinations, like

(a) Fe–Cr RFB
(b) Metal-free quinone–Br RFB
(c) All-polymer RFB
(d) Zn–bromide RFB
(e) Zinc–cerium RFB
(f) Magnesium–vanadium RFB
(g) Vanadium–cerium RFB
(h) Vanadium–polyhalite RFB, etc.

where they used two separate active materials in their respective half-cells [9]. For that reason, these redox flow batteries experienced cross-contamination problems. So, to overcome the cross-contamination problem, Maria Skyllas-Kazacos and her research team developed the vanadium redox flow battery (VRFB) for the first time in the 1980s [13].

Among the various kinds of RFBs, VRFB was the most researched and successful technology for energy storage applications [17]. They have used the same active material in both the half-cells ($V^{2+}$,$V^{3+}$,$V^{4+}$,$V^{5+}$) [21].

Also, VRFB has some excellent features, like

(a) long cycle life (>20,000), which is almost like a solar photovoltaic (PV) power plant [8]
(b) easy scalability
(c) high efficiency
(d) excellent electrochemical reversibility
(e) non-corrosive
(f) negligible degradation of electrolytes
(g) safe operation and environment friendly [12].

Over the decades, various researchers have made efforts to improve the efficiency of VRFB; some of these are the battery stack design, modifications of electrode, electrolyte, and membrane. For understanding the practical VRFB system performance and its operation, an electrical equivalent circuit model is essential. Nevertheless, very few pieces of literature are available which have worked with electrical equivalent circuit model and their parameter estimation techniques. Therefore, this paper intends to provide a systematic review and analysis of some selected electrical equivalent circuit models of VRFB [15].

## 2 System Description and Operational Principle of VRFB

For energy storage systems, VRFB is one of the most promising technology. As illustrated in Fig. 1, it mainly consists of two electrolyte reservoirs (the electrolytes are circulated via pumps), a negative and positive electrode, an ion-exchange membrane (IEM), and bipolar plates. Redox couples in the electrolyte are used to store and release energy, while the charging–discharging process, respectively. The two electrolytes are stored separately in reservoirs called an anolyte and catholyte outside of the electrochemical cell. The IEM is used to transport ions from one side to the other and does not allow unwanted micro-molecules to pass through it.

An external electrical power source is connected to supply the energy to charge the VRFB; similarly, for discharging, an external load is connected. Pumps flow the electrolytes in both the half-cells, and the two half-cells are partitioned by an ion-selective membrane. The electrolytes are pumped into the battery stack continuously; in the battery stack, the redox reactions occur at the inert electrodes. While charging the battery, the electrolytes are chemically oxidized and reduced within the half-cells to convert electrical energy to chemical energy, and vice versa happens while discharging [13].

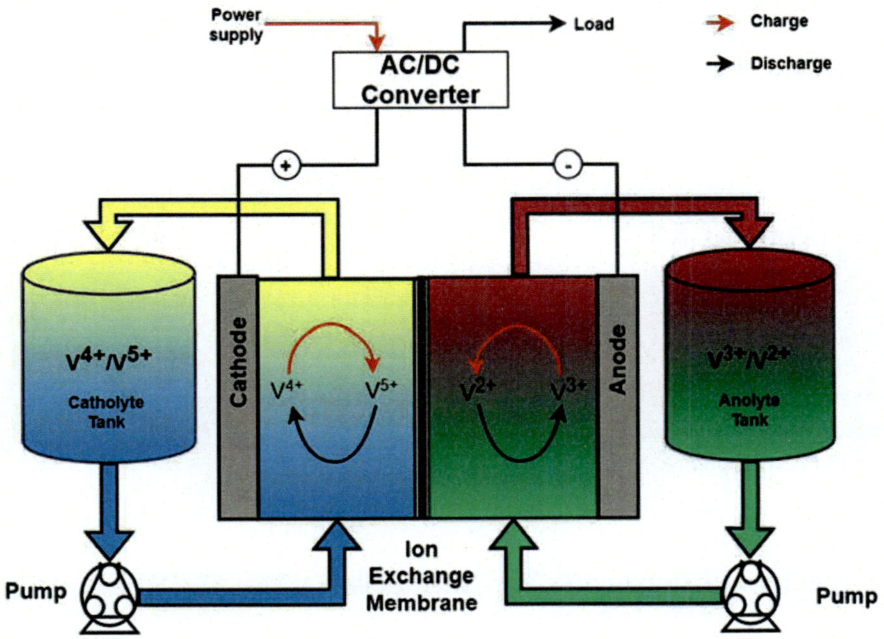


**Fig. 1** Schematic representation of VRFB

Therefore, the chemical reaction occurs are as follows:
At positive electrode,

$$VO^{2+} + H_2O \longleftrightarrow VO_2^+ + 2H^+ + e^- \tag{1}$$

At negative electrode,

$$V^{3+} + e^- \longleftrightarrow V^{2+} \tag{2}$$

The overall cell equation is,

$$VO_2^+ + V^{2+} + 2H^+ \longleftrightarrow VO^{2+} + V^{3+} + H_2O \tag{3}$$

Cell potential,

$$E^0(\text{Cell}) = E_{\text{Cathode}} - E_{\text{Anode}} \tag{4}$$

where $E^0(\text{Cell})$ denotes the standard potential of the overall cell reaction, $E_{\text{Cathode}}$ and $E_{\text{Anode}}$ are the standard potential of the cathode and anode respectively [5]. The Nernst equation is used to find out the potential of the two reduction reactions which can be shown as follows:
In positive half-cell

$$E^+ = E^{0+} - \frac{RT}{F}\ln\left[\frac{[VO^{2+}]}{[VO_2^+][H^+]^2}\right] \tag{5}$$

In negative half-cell

$$E^- = E^{0-} - \frac{RT}{F}\ln\left[\frac{[V^{2+}]}{[V3^+]}\right] \tag{6}$$

where value of $R$ is the universal gas constant, $F$ is Faraday's constant, $T$ is temperature, $[V]$ are the concentration of different vanadium species, $[H^+]$ is the concentration of hydrogen ion.

## 3 Electrical Equivalent Circuit Models (ECMs) of VRFB

This section gives a detailed description about the recent ECMs of VRFB.

1. ECM 1:

   - **Model Description**: In the paper [4], an equivalent circuit network as shown in Fig. 2 has been proposed which consists of a voltage source representing

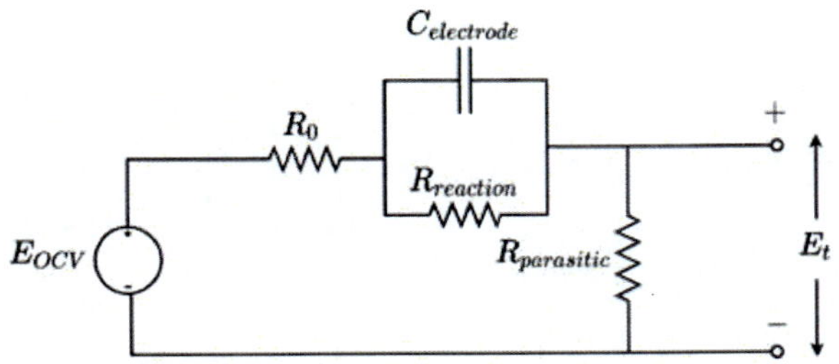


**Fig. 2** ECM 1 of VRFB in [4]

the OCV of the VRFB. The resistor $R_0$ signifies the internal resistance which is in cascade with a RC branch which symbolizes the electrode capacitance and the reaction resistance. A resistor ($R_{parasitic}$) is connected at the output terminal which stands for the pump power loss which occurs during charging and discharging.

- **Estimated Parameters**: Stack potential is obtained from Nernst potential equation, and state of charge (SOC) of the vanadium redox flow battery is estimated by dynamically varying the time step.
- **Limitations**: The model doesn't consider self-discharge and cross-contamination effect. Also, the flow rate during the running condition is considered to be static which is likely to make the model inaccurate during operations with dynamic flow rate.
- **Experimental Validation**: The authors have not validated the proposed model for a practical system.

2. ECM 2

- **Model Description**: An ECM as shown in Fig. 3 has been proposed in [10], which consists of an open-circuit voltage (OCV), an internal ohmic resistance $R_0$, and two parallel RC network ($R_{ap}$, $C_{ap}$, $R_{cp}$, $C_{cp}$) which depicted the dynamic behavior and the polarization characteristics of the VRFB.

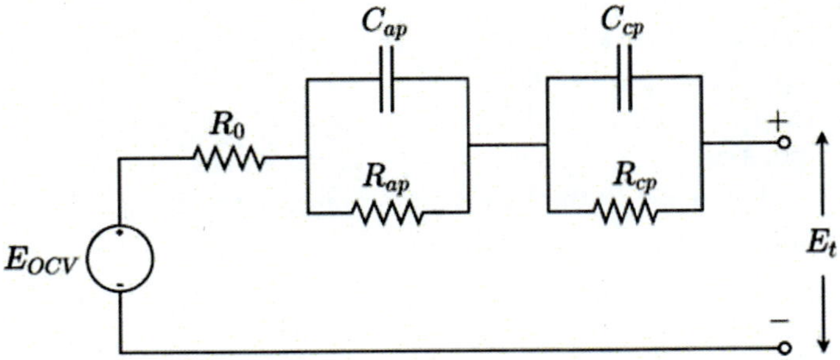


**Fig. 3** ECM 2 of VRFB in [10]

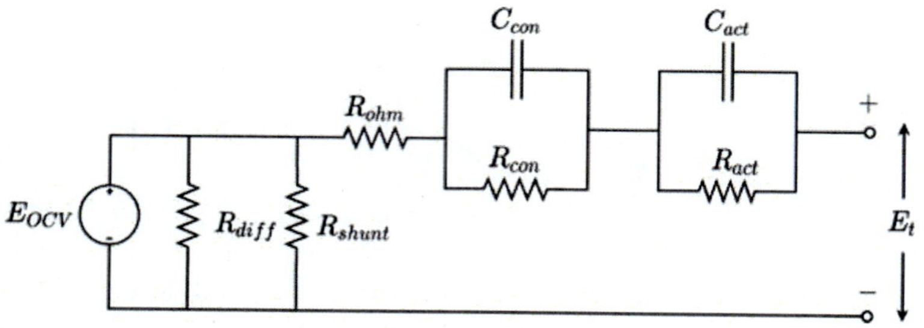


**Fig. 4** ECM 3 of VRFB in [19]

- **Estimated Parameters**: The open-circuit cell potential and the internal resistance are calculated by direct measurement. And the parameters of the RC branches are estimated by a recursive algorithm of extended kalman filter (EKF).
- **Limitations**: The model doesn't contemplate the effect of practical parameters like temperature, electrolyte concentration, and the effect of dynamic flow rate is also not considered while charging and discharging of the VRFB.
- **Experimental Validation**: The authors have shown the experimental validation of their proposed VRFB system with the charge–discharge characteristics.

3. ECM 3

- **Model Description**: A modified version of the previous ECM as shown in Fig. 4 has been proposed in [19]. This model consists of same architecture as in the previous case but has considered adding two shunt resistors across the open-circuit voltage source, which signify diffusion of vanadium ions across the ion-exchange membrane. The total resistance offered by $R_{\text{diff}}$ and $R_{\text{shunt}}$ is the equivalent self-discharge resistance. Moreover, the proposed model considered thermal effect. The electrochemical behaviors of the VRFB during the charging–discharging are expressed by the equivalent electrical network, where as the heat transfer process is represented by the thermal sub-circuit.
- **Estimated Parameters**: The model parameters are identified by particle swarm optimization (PSO) technique.
- **Limitations**: The model has not been considered the dynamic flow rate during the charge–discharge operation.
- **Experimental Validation**: The proposed equivalent circuit model is validated with a practical VRFB system; hence, the results describe that the proposed model is able to predict the VRFB performance considering the dynamic temperature conditions.

4. ECM 4

- **Model Description**: An electrical equivalent circuit model has proposed as shown in Fig. 5 [3], which consists of a voltage source representing the OCV

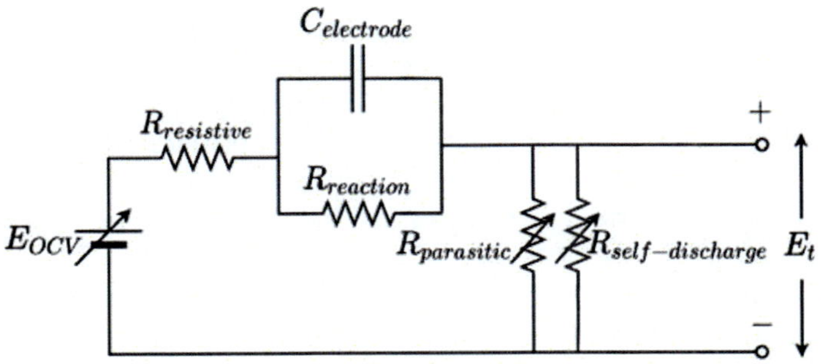


**Fig. 5** ECM 4 of VRFB in [3]

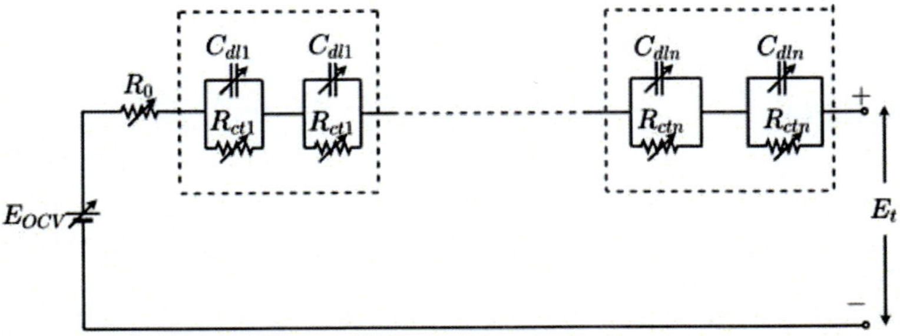


**Fig. 6** ECM 5 of VRFB in [2]

of the VRFB, an active electrolyte resistance $R_0$ in series with the first-order RC network, in which $R_{\text{reaction}}$ represents the electrolyte resistance and $C_{\text{electrode}}$ represents the double-layer capacitance, and two parallel resistance which represented the parasitic loss due to pump power and self-discharge.

- **Estimated Parameters**: Open-circuit voltage, terminal voltage is estimated. Optimization of different flow rates while charge–discharge process is done. Along with that, different state of charge is calculated with respect to different flow rates. Estimation of the self-discharge and pump power loss is done.
- **Limitations**: This equivalent circuit model does not consider the temperature effect, and it has been assumed that the generated heat due to the electrolytic flow is uniformly distributed through the reservoir to cell stacks.
- **Experimental Validation**: The proposed model is validated successfully with a practical VRFB system.

5. ECM 5

- **Model Description**: The model as shown in Fig. 6 [2] is developed by considering the dynamic internal circuit parameters. The model consists of an open-circuit voltage source, nth order of parallel RC networks series with an internal resistance, where $R_0$ represents the internal resistance, $R_{ct}$ and $C_{dl}$ represents charge transfer resistance and double-layer capacitance, respectively.

- **Estimated Parameters**: In this work, charge–discharge characteristics have been used to extract the dynamic internal circuit parameters of the proposed model. Stack voltage is determined by the Nernst equation.
- **Limitations**: This literature expresses that the temperature effect in the VRFB system is negligible. Their research considered that the heat generation due to the continuous flow of electrolytes into the stack to tanks is uniformly distributed.
- **Experimental Validation**: The proposed model is validated with a practical system, which exhibits robustness for large-scale applications by further extracting the internal parameters.

## 4 Discussion

The accurateness of the presented models in this paper varies according to the different applications of the vanadium redox flow battery. The presented models could further be modified, considering a few practical parameters, like temperature effect, self-discharge effect, and crossover effect of different vanadium species. The models might act more precisely with the practical system if the effects of the side reactions were also considered. Moreover, the thermal effect influences the open-circuit voltage and also the battery capacity. Battery capacity increases when the temperature increases. So, battery models can improve the battery's performance if the temperature effect is considered. The electrolyte flow rate is one of the most critical aspects of VRFB operation. Considering the dynamic flow rate while charge–discharge of the battery might improve its performance, so the better extraction of the parameters of the electrical equivalent circuit model is possible. Its impact on the overall efficiency of the VRFB system has been modeled. The simulation result shows how the electrolyte flow rate changes when the SOC changes. This aids in the development of a tracking algorithm for ensuring maximum overall VRFB system efficiency by maintaining dynamic ideal flow rates throughout charging and discharging. From the literature survey, it has been seen that the pump power loss is also an essential factor for the battery; the efficiency of the battery varies with the pump power loss. Numerous significant parameters, comprising electrolyte concentration, SOC, output voltage, current, and internal loss, can be simulated using the equivalent circuit models under various operating situations and applied to simulations of short- and long-term operations for creating a scalable VRB system. The analysis of above-mentioned electrical equivalent circuit models of VRFB (ECM 1, ECM 2, ECM 3, ECM 4, ECM 5) is shown in Table 1.

**Table 1** Analysis of electrical equivalent circuit models of VRFB

| Models studied | Self-discharge consideration | Dynamic flow rate consideration | Thermal effect consideration | Crossover of vanadium species | Experimental Validation |
|---|---|---|---|---|---|
| ECM 1 | No | No | No | No | No |
| ECM 2 | No | No | No | No | Yes |
| ECM 3 | Yes | No | Yes | Yes | Yes |
| ECM 4 | Yes | No | No | Yes | Yes |
| ECM 5 | Yes | Yes | No | Yes | Yes |

## 5 Conclusion

In this paper, the recently developed equivalent circuit models of the vanadium redox flow battery have been presented. These are the standard equivalent circuit models, which include the first-order RC equivalent circuit, second-order RC circuit, and nth-order RC circuit. The number of RC branches may differ according to the consideration of the dynamic behaviors of practical parameters. A brief idea of the parameter identification methods is also presented in this paper. The charging and discharging characteristics of VRFB are used as input to the proposed model to extract the dynamic internal electrical circuit parameters.

## References


1. Alotto P, Guarnieri M, Moro F (2014) Redox flow batteries for the storage of renewable energy: a review. Renew Sustain Energy Rev 29:325–335
2. Bhattacharjee A, Roy A, Banerjee N, Patra S, Saha H (2018) Precision dynamic equivalent circuit model of a vanadium redox flow battery and determination of circuit parameters for its optimal performance in renewable energy applications. J Power Sources 396:506–518
3. Bhattacharjee A, Saha H (2017) Design and experimental validation of a generalised electrical equivalent model of vanadium redox flow battery for interfacing with renewable energy sources. J Energy Storage 13:220–232
4. Challapuram YR, Quintero GM, Bayne SB, Subburaj AS, Harral MA (2019) Electrical equivalent model of vanadium redox flow battery. In: 2019 IEEE green technologies conference (GreenTech). IEEE, pp 1–4
5. Corcuera S, Skyllas-Kazacos M (2012) State-of-charge monitoring and electrolyte rebalancing methods for the vanadium redox flow battery. Eur Chem Bulletin 1(12):511–519
6. Divya K, Østergaard J (2009) Battery energy storage technology for power systems-an overview. Electric Power Syst Res 79(4):511–520
7. Dunn B, Kamath H, Tarascon JM (2011) Electrical energy storage for the grid: a battery of choices. Science 334(6058):928–935
8. Esan OC, Shi X, Pan Z, Huo X, An L, Zhao T (2020) Flow batteries: modeling and simulation of flow batteries. Adv Energy Mat 10(31):2070133
9. Menictas C, Skyllas-Kazacos M (2011) Performance of vanadium-oxygen redox fuel cell. J Appl Electrochem 41(10):1223

10. Mohamed M, Ahmad H, Seman MA, Razali S, Najib M (2013) Electrical circuit model of a vanadium redox flow battery using extended Kalman filter. J Power Sources 239:284–293
11. Qazi A, Hussain F, Rahim NA, Hardaker G, Alghazzawi D, Shaban K, Haruna K (2019) Towards sustainable energy: a systematic review of renewable energy sources, technologies, and public opinions. IEEE Access 7:63837–63851
12. Ravikumar M, Rathod S, Jaiswal N, Patil S, Shukla A (2017) The renaissance in redox flow batteries. J Solid State Electrochem 21(9):2467–2488
13. Rychcik M, Skyllas-Kazacos M (1988) Characteristics of a new all-vanadium redox flow battery. J Power Sources 22(1):59–67
14. Skyllas-Kazacos M, Chakrabarti M, Hajimolana S, Mjalli F, Saleem M (2011) Progress in flow battery research and development. J Electrochem Society 158(8):R55
15. Skyllas-Kazacos M, Kazacos G, Poon G, Verseema H (2010) Recent advances with unsw vanadium-based redox flow batteries. Int J Energy Res 34(2):182–189
16. Suberu MY, Mustafa MW, Bashir N (2014) Energy storage systems for renewable energy power sector integration and mitigation of intermittency. Renew Sustain Energy Rev 35:499–514
17. Ulaganathan M, Aravindan V, Yan Q, Madhavi S, Skyllas-Kazacos M, Lim TM (2016) Recent advancements in all-vanadium redox flow batteries. Adv Mat Interf 3(1):1500309
18. Weber AZ, Mench MM, Meyers JP, Ross PN, Gostick JT, Liu Q (2011) Redox flow batteries: a review. J Appl Electrochem 41(10):1137–1164
19. Xiong B, Yang Y, Tang J, Li Y, Wei Z, Su Y, Zhang Q (2019) An enhanced equivalent circuit model of vanadium redox flow battery energy storage systems considering thermal effects. IEEE Access 7:162297–162308
20. Zhang J, Jiang G, Xu P, Kashkooli AG, Mousavi M, Yu A, Chen Z (2018) An all-aqueous redox flow battery with unprecedented energy density. Energy Environ Sci 11(8):2010–2015
21. Zhang Y, Zhao J, Wang P, Skyllas-Kazacos M, Xiong B, Badrinarayanan R (2015) A comprehensive equivalent circuit model of all-vanadium redox flow battery for power system analysis. J Power Sources 290:14–24

# AI and Soft Computing

# Power Transformer Protection Based on Fuzzy Logic System

**Vijay Kumar Sahu and Yogesh Pahariya**

## 1 Introduction

In electrical power system, power transformers are expensive as well as one of the most critical equipment. The failure of such an equipment will have a severe negative impact on the power supply. This may even lead to high maintenance costs and even massive power blackouts. So, to maintain, the reliability of power transformers should be considered as priority in the electrical power system. Different suitable methods for the protection and to detect fault must be ensured for stable as well as reliable energy delivery [1].

The percentage differential logic is the common protection technique used for transformer protection as shown in Fig. 1. This protection technique can distinguish between normal operating condition, an external fault, and internal fault. But, simply to detect a differential current will not be sufficient to differentiate between internal faults from other similar situations which can produce such currents. Few of the situations may appear when transformer energization (inrush currents) take place, over excitation, among others, these can result in wrong tripping. For modern protection of the power transformer, fast and correct distinction of internal faults from the other similar situations mentioned is a great challenge. Different algorithms developed recently to differentiate between internal fault current and other similar situations which can produce similar current should be understood, so effective maintenance of transformer will be possible [2].

Different methods have been developed to improve the differential protection [3–14], used for power transformer. Some have implemented the same [4–6], by using of wavelet transform and other various transform [8, 9]. Few have used hybrid systems and restraint method [2, 3]. For correct operation of differential protection, distinction between the magnetizing inrush current and the internal faults in transformer is

V. K. Sahu (✉) · Y. Pahariya
Electrical & Electronics Department, SOET, Sandip University, Nashik, India
e-mail: vijay25051991@gmail.com



P. H. J. Chong et al. (eds.), *Emerging Electronics and Automation*, Lecture Notes in Electrical Engineering 937, https://doi.org/10.1007/978-981-19-4300-3_16

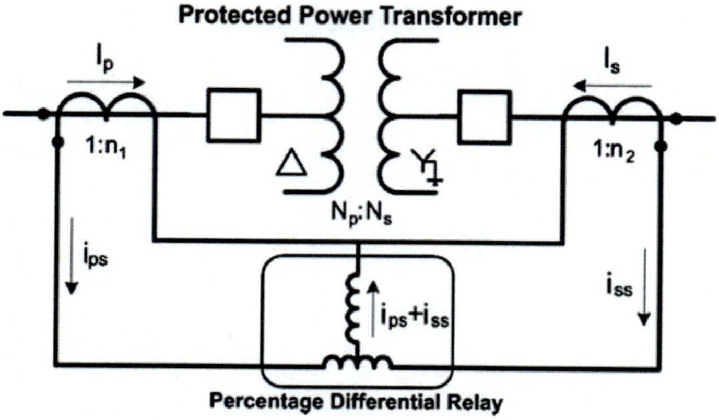


**Fig. 1** Differential relay connection diagram

necessary [14–18], as the reliability enhances are discussed. Few more for fault detection methods were proposed [19–22]. Fault identification algorithms are also available [23–25]. For the diagnosis of the faults methods were presented [26, 25, 27]. A review of protection methods of the transformer and the CT saturation detection methods are also available [28, 29], respectively.

## 2 Method Proposed

### *2.1 Flowchart of Relay*

The implementation of proposed algorithm was done in MATLAB Simulink, and it is illustrated in Fig. 2. The current signals values are acquired from transformer by the use of current measurement blocks. After collecting the current data, the differential currents are calculated by processing using Clarke's transformation. These differential currents obtained are sent to the fuzzy system. The threshold value of 0.5 is set for the fuzzy system. If fuzzy output is higher than threshold value, i.e., 0.5, the relay will send signal to trip the circuit breaker.

Each block is described in the following section.

### *2.2 Data Acquisition*

For the proposed method, data acquisition is done for currents from the secondary of the CTs connected on primary side and even secondary side of power transformer,

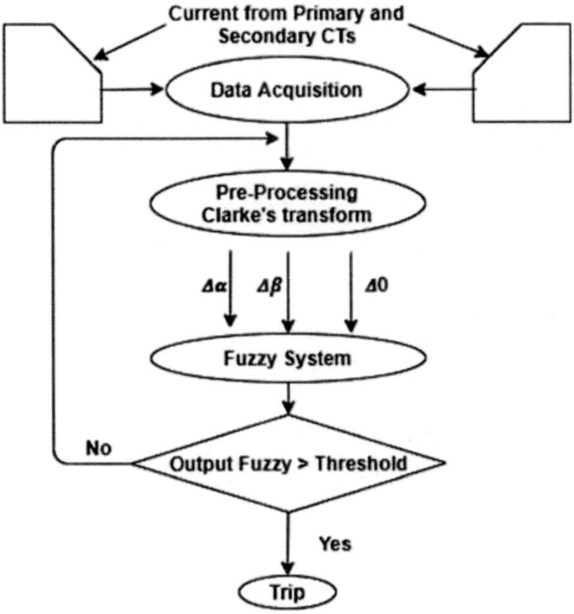


**Fig. 2** Basic relay algorithm

which is required for differential protection. Current data are obtained from current measurement block.

## 2.3 *Preprocessing: Clarke's Transform*

As the data have been acquired, a preprocessing will be executed, to obtain the desired signals for the fuzzy logic system. The signal will be acquired by applying Clarke's transformation to the 3-phase currents in secondary winding current of the CTs in both transformer ends, i.e., primary and secondary. The equations are represented as

$$\begin{bmatrix} i_\alpha \\ i_\beta \\ i_0 \end{bmatrix} = \begin{bmatrix} \frac{2}{3} & \frac{-1}{3} & \frac{-1}{3} \\ 0 & \frac{1}{\sqrt{3}} & \frac{-1}{\sqrt{3}} \\ \frac{1}{3} & \frac{1}{3} & \frac{1}{3} \end{bmatrix} \begin{bmatrix} i_a \; i_b \; i_c \end{bmatrix} \tag{1}$$

where $I_a$, $I_b$, and$I_c$ represent three sinusoidal balanced currents:

$$I_a = I_m \sin(\omega t) \tag{2}$$

$$I_b = I_m \sin\left(\omega t - \frac{2\pi}{3}\right) \tag{3}$$

$$I_c = I_m \sin\left(\omega t + \frac{2\pi}{3}\right) \quad (4)$$

The phase currents are flowing in their, respectively, windings of a three-phase winding, as the Fig. 3 shows.

The rotating space vector $I_S$ is represented by its components, $i_\alpha$ and $i_\beta$, in a fixed reference frame whose $\alpha$ axis is aligned with phase A axis. The amplitude of $I_S$ being proportional to the rotating mmf produced by the three-phase currents. It is computed as follows:

$$I_s = i_\alpha + j.i_\beta = \frac{2}{3}\left(i_a + i_b.e^{\frac{j2\pi}{3}} + i_c.e^{\frac{-j2\pi}{3}}\right) \quad (5)$$

Clarke's transformation can be used to both phasor as well as instantaneous values. The main concept is using Clarke's transformation is to discriminate normal operation, internal faults, and energization. The differential $\alpha$-$\beta$-0 components of the current are used.

$$\Delta\alpha_{ph} = \left|\sum_{n=0}^{N}[I_\alpha(n) + i_\alpha(n)]\right| \quad (6)$$

$$\Delta\beta_{ph} = \left|\sum_{n=0}^{N}[I_\beta(n) + i_\beta(n)]\right| \quad (7)$$

$$\Delta 0_{ph} = \left|\sum_{n=0}^{N}[I_0(n) + i_0(n)]\right| \quad (8)$$

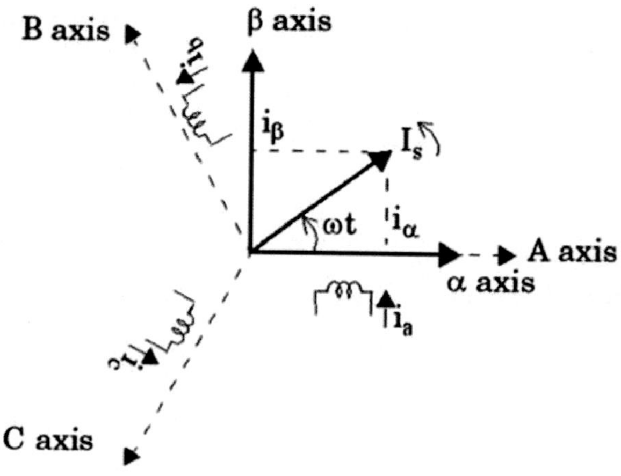


**Fig. 3** abc to αβ0 axis representation

where $I_\alpha(n)$, $I_\beta(n)$, $I_0(n)$, $i_\alpha(n)$, $i_\beta(n)$, and $i_0(n)$ are $\alpha$-$\beta$-0 current components of the primary current and secondary current acquired by a power transformer and N is the number of samples used to get the values.

The values computed of the differential $\alpha$-$\beta$-0 components of the phase currents approaches zero during a normal operation and for different specific situations fluctuation are observed in differential current values. Hence, various phenomena of power transformer could be distinguished. With input of the differential $\alpha$-$\beta$-0 components of the current, fuzzy system is used, so the fault condition can be determined accurately compared to the conventional methods, which has predefined rules to discriminate between steady state and fault conditions.

## *2.4 Design of Fuzzy System*

To point out the fault condition by considering all data and accurately, the fuzzy logic system is used. Steps used for the fuzzy logic are as follows:

**Fuzzification**: In proposed method, three fuzzy inputs for the fuzzy system: (1) $\Delta\alpha$, (2) $\Delta\beta$, and (3) $\Delta 0$. These are obtained from Eqs. (6)–(8). Figure 4a–c shows the inputs membership functions. For fuzzification of an input variable defined in Eq. (6), the range is between $-15$ and 220, and the membership value ranges from 0 to 1. The remaining two variables are from Eqs. (7) and (8), having range of $-120$ to 75 and $-5$ to 12, respectively. The range of fuzzy inputs was amplified for better analysis. Figure 4d shows the output variable.

**Inference method**: Here, proposed method will use 12 rules to distinguish between internal faults condition and steady-state condition. Mamdani method was chosen [30], to perform the mathematical operation. Table 1 gives the rules used in proposed relay, and Table 2 gives the specification of the proposed relay.

**Defuzzification**: Here, the method requires a crisp value for the control purposes. The technique used is centroid according to [31]

$$\text{Output} = \frac{\sum_{j=0}^{N} y_j \mu_F(y_j)}{\sum_{j=0}^{N} \mu_F(y_j)} \tag{9}$$

where $y_j$ is the value of every point on a domain of final output fuzzy set.

$\mu_F(y_j)$ is membership value at every point.

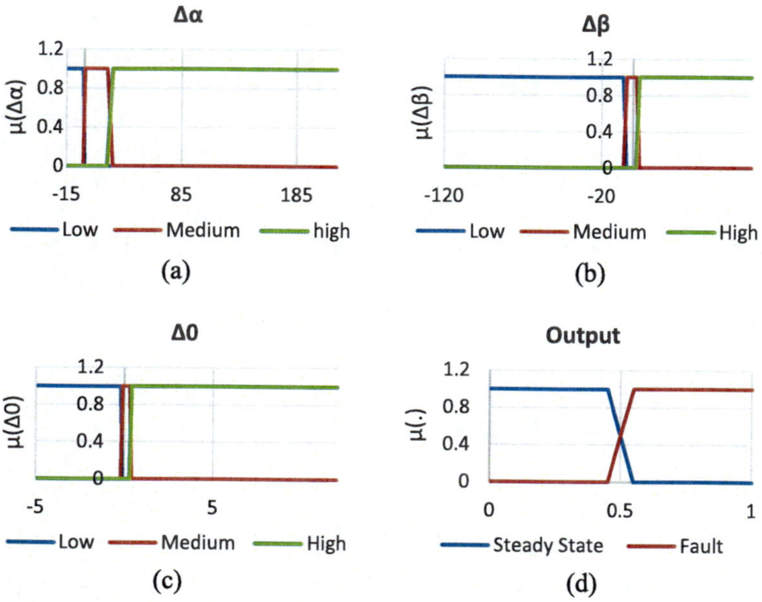


**Fig. 4** Membership functions of fuzzy. **a** $\Delta\alpha$, **b** $\Delta\beta$, **c** $\Delta 0$, **d** output fuzzy

**Table 1** Summary of fuzzy rules

| Rules | Inputs | | | Output |
|---|---|---|---|---|
| | $\Delta\alpha$ | $\Delta\beta$ | $\Delta 0$ | |
| 1 | Low | Low | Low | Fault |
| 2 | Low | Low | Medium | Fault |
| 3 | Low | Low | High | Fault |
| 4 | Low | Medium | Low | Fault |
| 5 | Low | Medium | Medium | Fault |
| 6 | Medium | Low | Low | Fault |
| 7 | Medium | Medium | Medium | Steady state |
| 8 | Medium | High | High | Fault |
| 9 | High | Low | Low | Fault |
| 10 | High | Low | High | Fault |
| 11 | High | Medium | Medium | Fault |
| 12 | High | Medium | High | Fault |

**Table 2** Specification of proposed relay

| Measurement taken | Primary and secondary currents of transformer |
|---|---|
| Output of the relay | Trip signal |
| Threshold of operation | Fuzzy output >0.5 |

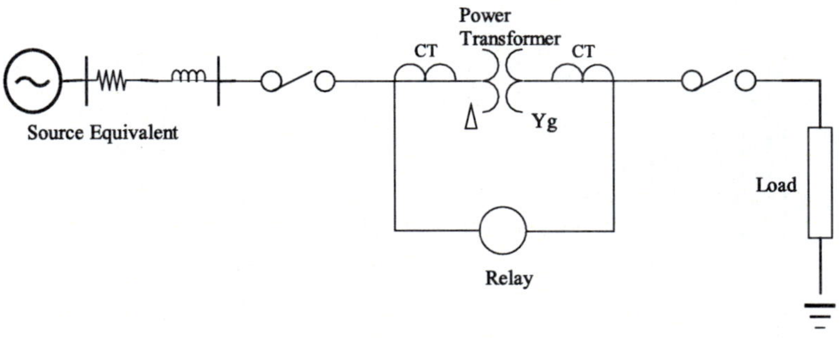


**Fig. 5** Line diagram of power system considered

## 3 Simulated Power System

The MATLAB software was used to simulate the electrical system. Figure 5 shows the simulated power system to generate data for the fuzzy system developed.

For a power transformer, a complete differential protection scheme is presented. The electrical system comprises of a 138 kV and 90 MVA generator, three-phase power transformer of 138:13.8 kV and 25 MVA, and a load of 10 MVA having power factor of 0.92 lagging. The primary winding has delta connection, and secondary winding has star connection of the power transformer. In accordance with the winding, CTs connected for protection are star in primary side and delta in secondary side.

## 4 Results

In this section, different results are presented for few conditions that can occur in the power transformer during its operation.

Figure 6a shows the primary current when unloaded transformer (secondary winding open) is being energized at zero crossing. For the same, no trip signal was obtained, Fig. 6b from the proposed method.

When LG fault occurs in primary of the power transformer, so the primary as well as the secondary current waveforms are shown in Fig. 7. The fault is created at 0.02 s, and the fault is detected as early as 0.0277 s, Fig. 7c.

Figure 8 shows the primary as well as secondary currents, respectively, when LLL fault is present on secondary side of power transformer. The fault was again developed at 0.02 s, and the fault is detected at 0.02785 s, Fig. 8a.

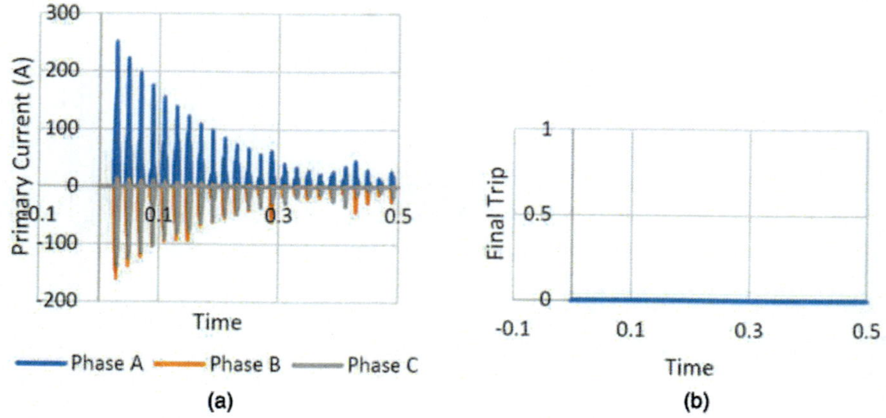


**Fig. 6** Energization at zero crossing of unloaded transformer **a** primary current. **b** final trip

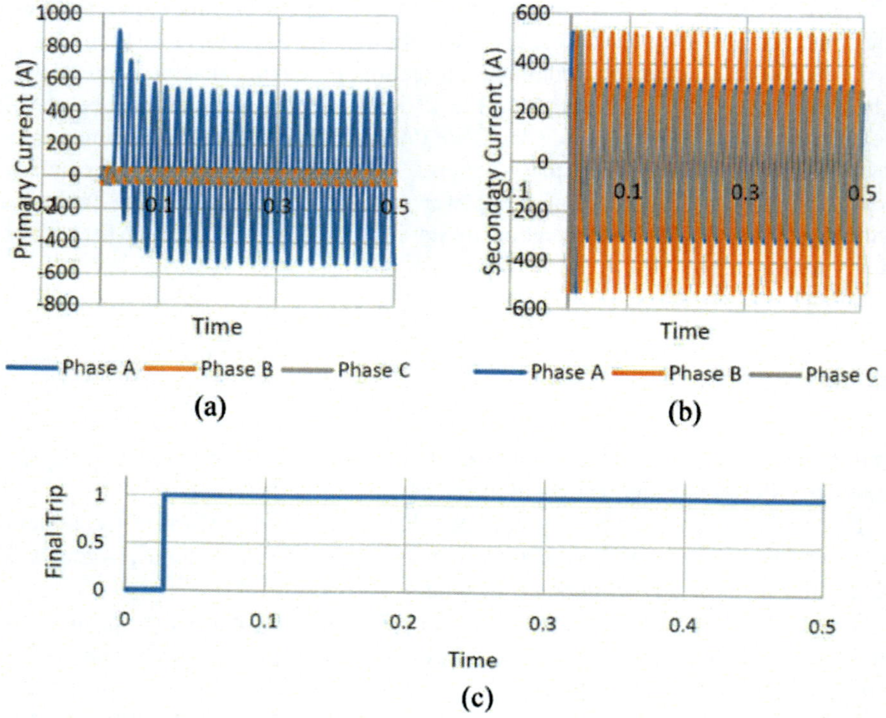


**Fig. 7** LG fault in phase A on primary of transformer **a** primary current, **b** secondary current, **c** final trip

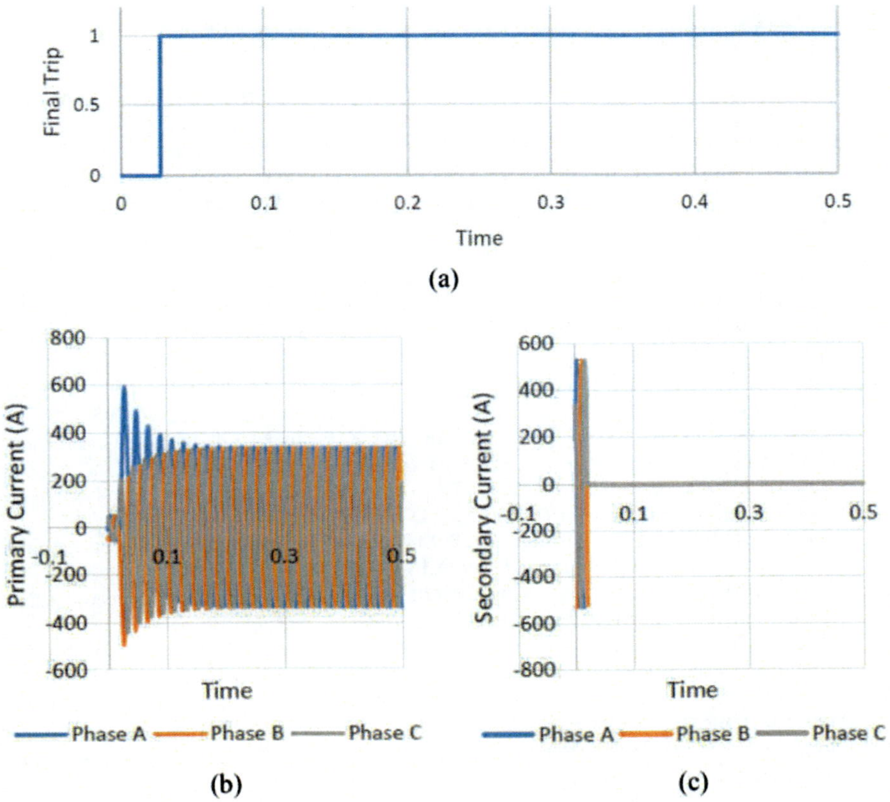


**Fig. 8** LLL fault on secondary of transformer **a** final trip, **b** primary current, **c** secondary current

## 5 Conclusion

The paper is presenting a different methods used for power transformer protection. The same is being done by a-b-c to $\alpha$-$\beta$-0 transformation and fuzzy logic. The $\alpha$-$\beta$-0 components were helpful for discrimination between different faults from other operating conditions like energization and overexcitation. During this operating conditions, the relay was able to avoid the trip and was efficiently able to detect the various faults within the protected region. By observing the results for different faults, it can also be concluded that most of the faults were detected within half cycle of fault development, which is considerably fast.

The algorithm is also advantageous as harmonic components are not being used as the basis of relay decision as in most of the algorithm for energization detection, and it is a simple method which can be easily understood.

## References


1. Lin M-J, Chen LB, Yu C-T (2020) A Methodology for diagnosing faults in oil-immersed power transformers based on minimizing the maintenance cost. IEEE Access, 8:209570–209578
2. Barbosa D, Coury DV, Oleskovicz M (2012) New approach for power transformer protection based on intelligent hybrid systems. IET Gener Transm Distrib 6(10):1003–1018
3. Phadke AG, Thorp JS (1983) A new computer-based flux-restrained current-differential relay for power transformer protection. IEEE Trans Power App Syst PAS-102 (11):3624–3629
4. Medeiros RP, Costa FP (2018) A wavelet-based transformer differential protection: internal fault detection during inrush conditions. IEEE Trans Power Delivery 33(6):2965–2977
5. Medeiros RP, Costa FB (2018) A wavelet-based transformer differential protection with differential current transformer saturation and cross-country fault detection. IEEE Trans Power Delivery 33(6):789–799
6. Medeiros RP, Costa FB, Silva KM (2016) Power transformer differential protection using the boundary discrete wavelet transform. IEEE Trans Power Delivery 31(5):2083–2095
7. Murugan SK, Simon SP, Nayak PSR, Sundareswaran K, Padhy NP (2016) Power transformer protection using chirplet transform. IET Gen Trans Distrib 10(10):2520–2530
8. Ashrafian A, Vahidi B, Mirsalim M (2014) Time–time-transform application to fault diagnosis of power transformers. IET Gen Transmission Distrib 8(6):1156–1167
9. Hayder T, Schaerli U, Feser K, Schiel L (2008) Universal adaptive differential protection for regulating transformers. IEEE Trans Power Delivery 23(2):568–575
10. Vázquez E, Mijares II, Chacón OL, Conde A (2008) Transformer differential protection using principal component analysis. IEEE Trans Power Delivery 23(1):67–72
11. Tripathy M, Maheshwari RP, Verma HK (2010) Power transformer differential protection based on optimal probabilistic neural network. IEEE Trans Power Delivery 25(1)
12. Zhalefar F, Sanaye-Pasand M (2010) A new fuzzy-logic-based extended blocking scheme for differential protection of power transformers. Electric Power Comp Syst 38:675–694
13. Manoj Tripathy RP, Maheshwari, Verma HK (2008) Neuro-fuzzy technique for power transformer protection. Electric Power Comp Syst 36:299–316
14. Etumi AAA, Anayi FJ (2021) Current signal processing-based methods to discriminate internal faults from magnetizing inrush current. Electric Eng 103:743–751
15. Samet H, Ghanbari T, Ahmadi M (2015) An auto-correlation function based technique for discrimination of internal fault and magnetizing inrush current in power transformers. Electric Power Comp Syst 43(4):399–411
16. Fani B, Hamedani Golshan ME, Saghaian-Nejad M (2011) Transformer differential protection using geometrical structure analysis of waveforms. Electric Power Comp Syst 39:204–224
17. Hooshyar A, Afsharnia S, Sanaye-Pasand M, Ebrahimi BM (2010) A new algorithm to identify magnetizing inrush conditions based on instantaneous frequency of differential power signal. IEEE Trans Power Del 25:2223–2233
18. Hooshyar A, Afsharnia S, Sanaye-Pasand M, Ebrahimi BM (2010) A new algorithm to identify magnetizing inrush conditions based on instantaneous frequency of differential power signal. IEEE Trans Power Delivery 25(4):2223–2233
19. Ghoneim SSM, Mahmoud K, Lehtonen M, Darwish MMF (2021) Enhancing diagnostic accuracy of transformer faults using teaching-learning-based optimization. IEEE Access 9:30817–30832
20. Kari T, Gao W, Zhao D, Abiderexiti K, Mo W, Wang Y, Luan L (2018) Hybrid feature selection approach for power transformer fault diagnosis based on support vector machine and genetic algorithm. IET Gen Transm Distrib 12(21):5672–5680
21. dos Santos GM, de Aquino RRB, Lira MMS (2018) Thermography and artificial intelligence in transformer fault detection. Electric Eng 100:1317–1325
22. Moravej Z, Abdoos AA (2012) An improved fault detection scheme for power transformer protection. Electric Power Comp Syst 40:1183–1207
23. Bigdeli M, Siano P, Alhelou HH (2021) Intelligent classifiers in distinguishing transformer faults using frequency response analysis. IEEE Access 9:13981–13991

24. Patel D, Chothani N, Mistry K (2018) Discrimination of inrush, internal, and external fault in power transformer using phasor angle comparison and biased differential principle. Electric Power Comp Syst 46(7):788–801
25. Kari T, Gao W (2017) Power transformer fault diagnosis using FCM and improved PCA. J Eng 2017(14):2605–2608
26. Zhou L, Hu T (2020) Multifactorial condition assessment for power transformers. IET Gen Transm Distrib 14(9):1607–1615
27. Kim Y, Park T, Kim S, Kwak N, Kweon D (2019) Artificial intelligent fault diagnostic method for power transformers using a new classification system of faults. J Electric Eng Technol 14:825–831
28. Sahu VK, Pahariya Y (2021) A review of various protection schemes of power transformers. Turkish J Comput Math Educ 12(10):7533–7541
29. Dashti H, Pasand MS, Davarpanah M (2009) Fast and reliable CT saturation detection using a combined method. IEEE Trans Power Delivery 24(3):1037–1044
30. Zilouchian A, Jamshidi M (eds) (2000) Intelligent control systems using soft computing methodologies. CRC, Boca Raton, FL
31. Kovacic Z, Bogdan S (2005) Fuzzy controller design. CRC, Boca Raton, FL

# Human Activity Recognition: A Step Toward Ambient Assisted Living

**Rohan Mandal, Uday Maji, and Saurabh Pal**

## 1 Introduction

In this modern digital era, self-dependency is one of the foremost stipulations of mankind. The aged and differently abled persons are also not out of this ensemble. The field of technology which helps to meet their craving is named ambient assisted living [1]. In the field of ambient assisted living, the primary requirement is human activity recognition. Excluding AAL, the real-time monitoring of human activity finds major application in the area of personal fitness, gaming, and entertainment. Human activity recognition can be done using two types of sensors; the first one is high-end audio visual sensor nodes with camera [2, 9], microphone [3], etc., and the second type is wearable sensors like gyroscope, accelerometer [4–8], thermal sensors, radio frequency identifier [10, 11], etc. Among these, the gyroscope and accelerometer are aptly felicitous for their small size, low-power consumption, fingertip availability with smart phone and smart watches, and they do not constrain user mobility while protecting their privacy. For these reason, these sensors found huge application in the field of AAL, while image or video processing-based activity recognition is useful for surveillance and security purpose. As this work deals with detection of daily life activities for assisted living, tri-axial accelerometer is used as sensor.

From the detail study of the previous proposed works for activity detection, the pr methods can broadly be categorized as the generative model and the discriminative approach. Majority of the state-of-the-art works comes under generative model category. Application of Gaussian mixture modeling (GMM) and Gaussian mixture regression (GMR) on accelerometer data [5], hidden Markov models (HMM)

R. Mandal (✉) · U. Maji
Department of Applied Electronics and Instrumentation Engineering, Haldia Institute of Technology, Haldia, West Bengal, India
e-mail: rohan_instru02@yahoo.com

S. Pal
Department of Applied Physics, University of Calcutta, Kolkata, West Bengal, India



P. H. J. Chong et al. (eds.), *Emerging Electronics and Automation*, Lecture Notes in Electrical Engineering 937, https://doi.org/10.1007/978-981-19-4300-3_17

on RFID tag generated data in smart home environment [11], dictionary learning approach on accelerometer data [7] are some of them. In the discriminative approach for recognition of activities of daily life, we find application of support vector machine [3, 9, 12], KNN [13], decision tree [14], ANN [8], and many more classifiers. In a HMM-based HAR system, [15] classification accuracy of 69.18% is achieved, while in a principal component analysis (PCA) of video signal being accompanied with SVM classifier, 95% accuracy is reported [9]. In many HAR systems, PCA [9, 16, 17] makes a salient appearance in dimensionality reduction of large number of feature set extracted from sensor data. Walse et al. [8] shown that use of PCA reduces the dimension of feature set from 561 to 70 causing a visible reduction of modeling time while compromising very less with the accuracy of the system. PCA-based decision tree classifier is proven to be useful in classification of cancer data [18] and other medical data [19]. Being inspired by the efficient application of PCA in dimensionality reduction without losing information, it is used in this proposed work to reduce the three dimensional data of tri-axial accelerometer into one dimensional data keeping major information content within them. In this aspect, it is novel approach to wield PCA on raw sensor data unlike on feature set as seen in most state-of-the-art works. The high-classification accuracy in various discriminative approaches lead us toward the use of a very popular bagging technique named random forest classifier. Application of random forest algorithm on combined feature set of data generated by accelerometer and gyroscope is reported [20, 23] to be capable of activity recognition with accuracy of 90–98% depending on selection of activity type. In a random forest-based sports activity, recognition system around 99% detection accuracy was achieved using sensors worn on arm, forearm, belt, and dumbbell [21]. A comparative study of different classifier algorithm [22] on inertia sensors data, worn on wrist, chest, and ankle shows random forest which gives sufficiently high accuracy with very less training time.

In rest of this paper, Sect. 2 describes the data collection and pre-processing using PCA, Sect. 3 contains feature extraction, and Sect. 4 emphasis on classification of activities of daily life using random forest classifier. Result and comparative study with other related works are depicted in Sect. 5.

## 2 Data Collection and Pre-processing

The data for this noble activity detection algorithm are acquired from a popular benchmark dataset named human motion primitive (HMP) dataset collected from archive.ics.uci.edu [24]. In this work, five common daily life activities are considered for classification. These are (i) descend stairs, (ii) climb stairs, (iii) comb hair, (iv) brush teeth, and (v) drink glass.

Around 25 dataset of each activities are taken. These dataset contains readings of tri-axial accelerometer worn on right wrist of the volunteer, with a sampling frequency of 32 Hz. The three columns of the dataset contain coded values of accelerometer reading along X, Y, and Z direction. These data are normalized between the ranges

−1.5 to +1.5 g and filtered using a median filter of size 3 to reduce noise. The sensitivity of this data is 6 bits per axis. The sample plots of data of each selected activities are shown in Fig. 1.

Observation of these plots adjudicates that though the signals of last three activities are quite different from each other, the nature of signals of first two activities is very much similar. Also the data of the three axis for a particular activity are very much

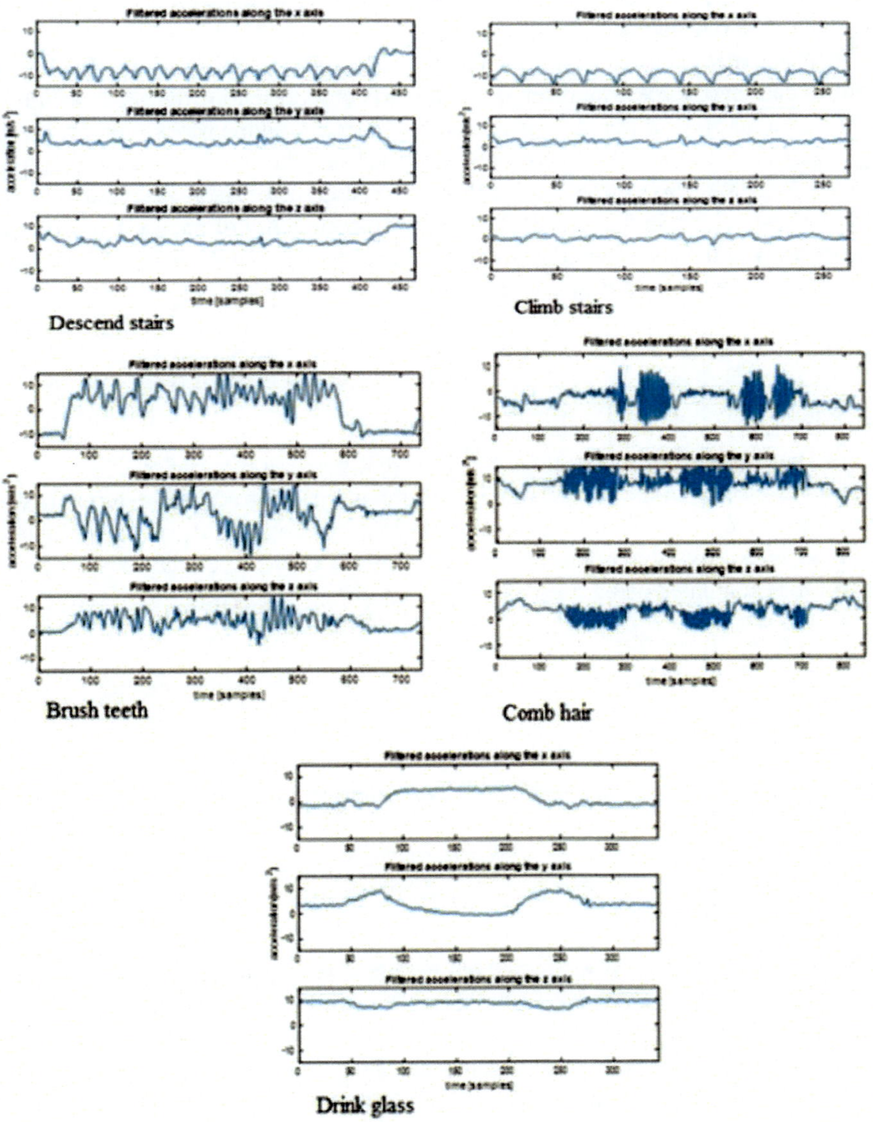


**Fig. 1** Sample plot of tri-axial accelerometer data for selected activities

random in nature, and it is difficult to establish any mathematical relationship in between them. Therefore in most of the related work, features are extracted from all the three columns of sensor data, and a large dimensional feature set is developed which increases complexity and execution time of the classifier algorithm.

This work uses a different approach by reducing the dimensionality of the raw data while protecting the major information content within them. Here, principal component analysis (PCA) comes out to be the best possible tool to extract a single dimensional data from the three dimensional data without reducing the information content in the raw dataset.

PCA is a method of representing a given dataset A of a certain dimension into a derived dataset of same dimension where the axis of representing the vector samples of the data matrix is different. For example, if A is an $n$ dimensional matrix, given by $A = \{A1, A2, A3 \ldots An\}$, PCA converts it into another n dimensional matrix $B = \{B1, B2, B3 \ldots Bn\}$ where most of the information of dataset $A$ is conserved into the first few columns of matrix $B$. Thus, the dimensionality of matrix $A$ is reduced significantly with a small tradeoff information content in initial data.

$A$ can be converted to $B$ by a matrix computation

$$A = C \cdot B \tag{1}$$

where $C$ is an $n \times n$ coefficient matrix.

Its values are determined in such a way that maximum variance of new data is ensured in first column of matrix $B$, and this variance gradually reduces as we proceed toward 1 to $n$. Here, information content in a data is basically represented by its variance. But direct determination of $C$ is not possible, therefore covariance matrix $K$ is computed using following equation.

$$K_{n \times n} = \frac{1}{n-1} \sum_{j=1}^{n} (A_j - A')^T \cdot (A_j - A') \tag{2}$$

where $A'$ is the mean of the considered data samples and is given by

$$A' = \frac{1}{n} \sum_{j=1}^{n} A_j \tag{3}$$

Next, the eigenvalues of $K_{n \times n}$ for initial data $A$ are calculated, and let they are denoted by $\lambda 1, \lambda 2, \ldots \lambda n$. Here, $\lambda 1 \geq \lambda 2 \geq \lambda 3 \ldots \lambda n - 1 \geq \lambda n$. Then, eigenvectors corresponding to n eigenvalues are computed among them m eigenvectors $S1, S2 \ldots Sm$ associated with m highest eigenvalues define the representation of data $A$ into m dimensional axis known as principal axis. In this work, the raw dataset is 3 dimensional, and PCA decomposes each dataset into three different uncorrelated array of data, which are actually the eigenvectors SCORE1, SCORE2, SCORE3.

The plot of these for a 'comb hair' dataset is shown in Fig. 2. Among these only the array of data with highest eigenvalue, i.e., the first one is taken for further analysis which is also known as principal component of data. Plot of this for the same dataset is shown in Fig. 3. This single dimensional principal component data with highest variance are taken for further analysis.

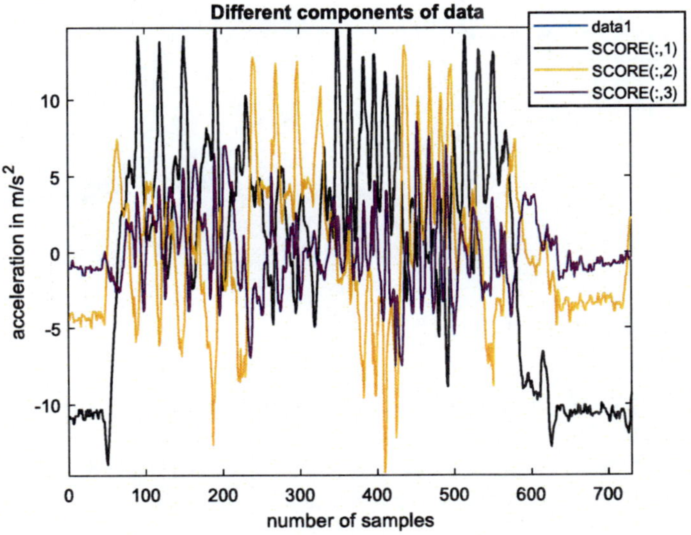


**Fig. 2** Principal component plot of acceleration data

**Fig. 3** Plot of the first dimension of PCA

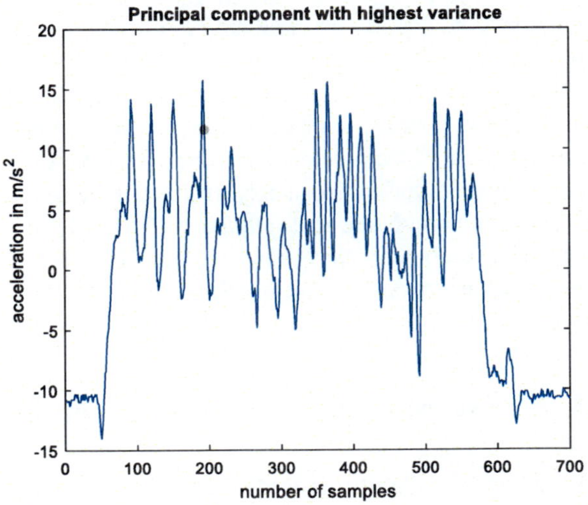

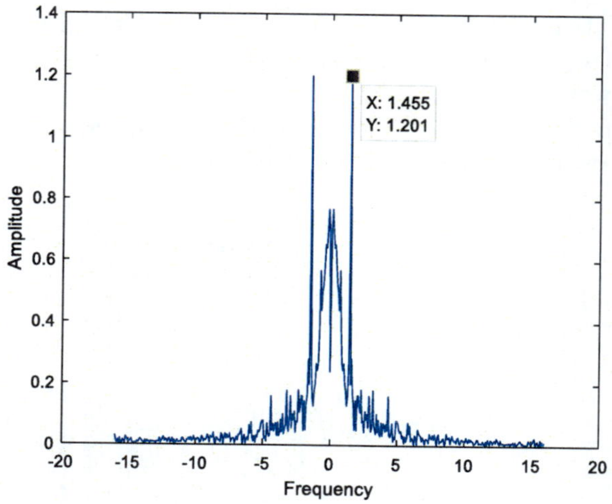


**Fig. 4** Frequency spectrum of a descend stair signal after PCA

## 3 Feature Extraction

Several time domain and frequency domain features of this principal component of accelerometer data are calculated. Mean, median, variance, standard deviation, kurtosis, and RMS value are the time domain features estimated here, among them the variance and standard deviation are found to come up as very much significant in the classification algorithm.

The significant frequencies of this principal component of data are determined using fast Fourier transform technique.

The frequency spectrum got from this is shown in Fig. 4. The peak amplitude in this spectrum and its corresponding frequency is recorded as two significant feature for classification. A complex feature is derived by taking the ratio of this recorded amplitude and frequency denoted as ratio of amplitude to frequency (RAF).

The most significant features which are used in classification process are, respectively, variance (var), standard deviation (std), peak amplitude of frequency spectrum (amp), corresponding frequency of this peak amplitude (principal frequency), and ratio of this amplitude and frequency (RAF). The feature table with three sample dataset of each activity is shown in Table 1. In actual practice, 25 such datasets of each activity are analyzed to form a feature set of 125 ($25 \times 5$) rows and 5 columns.

## 4 Classification Based on Random Forest

Random forest classifier is used to classify the five different types of activity. Random forest is a popular bagging technique. Here, if a $n \times m$ dimensional feature set is used to train a classifier, then it randomly selects some ensembles of feature samples with dimension $p \times q$ where $p < n$ and $q < m$. Presence of feature samples in these

**Table 1** Features table

| | Features | | | | |
|---|---|---|---|---|---|
| Activity | Var (PRC) | Std (PRC) | Amp | Principal frequency | RAF |
| Descend stairs 1 | 7.228648 | 2.68861 | 1.7965 | 0.0483 | 37.1668 |
| Descend stairs 2 | 7.222931 | 2.68755 | 1.0513 | 1.7445 | 0.60264 |
| Descend stairs 3 | 8.428257 | 2.90314 | 1.2700 | 1.9439 | 0.65335 |
| Climb stairs 1 | 3.676245 | 1.91735 | 0.7965 | 1.3941 | 0.57136 |
| Climb stairs 2 | 3.517628 | 1.87553 | 0.9123 | 1.5269 | 0.59748 |
| Climb stairs 3 | 4.354583 | 2.08676 | 1.1940 | 1.5924 | 0.74986 |
| Comb hair 1 | 60.45058 | 7.77499 | 3.9826 | 0.0639 | 62.3113 |
| Comb hair 2 | 52.37828 | 7.23728 | 2.8604 | 0.0704 | 40.5833 |
| Comb hair 3 | 23.79908 | 4.87843 | 1.7204 | 0.0544 | 31.5783 |
| Brush teeth 1 | 20.14694 | 4.48853 | 0.6069 | 4.1231 | 0.14720 |
| Brush teeth 2 | 21.04398 | 4.58737 | 0.7256 | 4.539 | 0.15972 |
| Brush teeth 3 | 22.06219 | 4.69704 | 0.594 | 6.098 | 0.09740 |
| Drink glass 1 | 22.30684 | 4.72301 | 2.8406 | 0.1649 | 17.2213 |
| Drink glass 2 | 23.04667 | 4.80069 | 2.7862 | 0.0694 | 40.1102 |
| Drink glass 3 | 32.30555 | 5.68379 | 3.4808 | 0.0727 | 47.8618 |

ensembles is repetitive. Decision trees are formed using each ensemble by choosing the best suitable feature to split training set on the node without pruning. Thus, largest possible trees are grown by selecting appropriate feature and binary classification at every node. In programming, this is done using nested if else loop. Here, some of the feature data samples may remain out of the ensembles which are known as out of bag samples. At every iteration of a tree grown using a selected ensemble of samples, the out of bag classification error is calculated which plays an important role in determination of appropriate number of trees to be grown in the random forest to get best classification result. Thus, once the training is completed, the testing can be done using a feature set of any unknown class. Different class prediction may be obtained by different decision trees, and the final prediction result is achieved by maximum class voting.

Here from the feature table among the 25 datasets of a particular activity, 15 randomly selected datasets are taken for training the classifier model, while 10 datasets are kept for testing purpose. Thus, the input to the classifier is the feature dataset ($X$) and a class variable ($Y$) which contains the classes of activity to be trained for its corresponding features. Here, five different types activities are to be classified, and in the class variable, they are denoted as follows; descend stairs as '1', climb stairs as '2', comb hair as '3', brush teeth as '4', and drink glass as '5'. The random forest classifier then generates number of decision trees as specified by user, here in this case 10.

The number of grown trees is taken as 10 by viewing the number of grown trees versus out of bag classification error plot as shown in Fig. 5. From the graph, it is

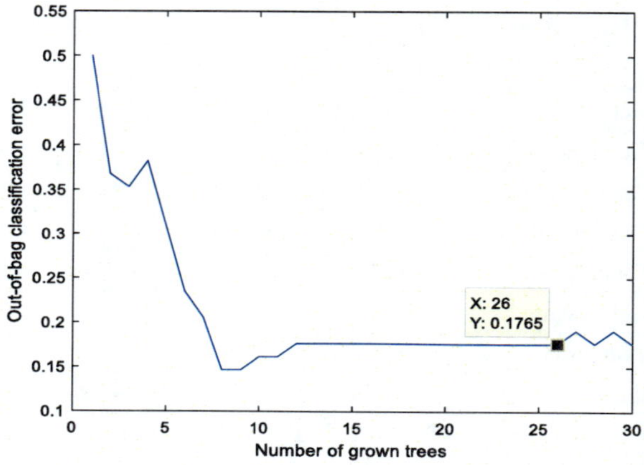


**Fig. 5** Plot of number of grown trees versus out of bag classification error

observed that the classification error is minimum for approximately 10 grown trees, which increases further and remains almost constant for up to 30 grown trees. In detail experimentation, it was observed that this error increases further with number of grown trees due to overtraining. Therefore, 10 decision trees are finally grown during training. One such grown decision tree is shown in Fig. 6, where at every node, a binary split is occurring depending on a threshold value of a certain feature denoted as $x1$, $x2$ … $x5$. Such binary classification at multiple stage gives rise to a complete tree capable to decide class of all the five activities denoted as 1–5.

Thus, the trained classifier model when fed by a test feature set of a unknown activity 10 decisions emerges from 10 classification trees among these which decision gets majority votes that comes out as final class prediction. Similarly by testing on 10 feature sets of each activity, 96% accuracy is achieved. The test results are shown in Table 3.

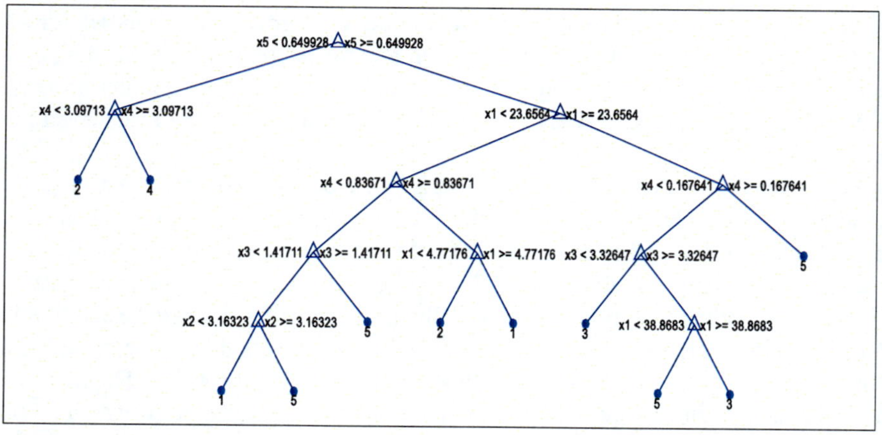


**Fig. 6** Grown classification tree

# 5 Result

The developed human activity detection system is found to be capable of detecting five activities of daily life with an average accuracy of 96%. The confusion matrix is formed with the classification results of test data which are 40% of the total available dataset. Remaining 60% data are used for training. The confusion matrix in Table 2 shows four activities climb stairs, comb hair, brush teeth, and drink something from a glass are truly detected with 100% accuracy. Where only 20% of descend stairs is detected as climb stairs which is false detection. The accuracy of a particular activity detection is calculated as (true positive/number of instances) × 100%. Table 3 shows classification accuracy of five different activities. The average of those 96% is considered to be the overall accuracy of the proposed model (Table 4).

Around 95–98% classification accuracy is achieved by a SVM-based HAR system [9] which uses camera as sensor node. The BiLSTM RNN system [16] registered average accuracy of 97.64% using three accelerometer, two gyroscope, two magnetometer, and one ECG sensor. In two other random forest-based multi-sensor HAR systems, overall accuracy of 93.44 [22] and 90% [23] is reported. In comparison with these state-of-the-art works, this proposed system ensures 96% detection accuracy with only single wrist worn accelerometer, which makes the system simple, feasible, and cost effective.

**Table 2** Confusion matrix in terms of percentage of classified activities

| | Prediction | | | | |
|---|---|---|---|---|---|
| Activity | Descend stairs | Climb stairs | Comb hair | Brush teeth | Drink glass |
| Descend stairs | 80 | 20 | 0 | 0 | 0 |
| Climb stairs | 0 | 100 | 0 | 0 | 0 |
| Comb hair | 0 | 0 | 100 | 0 | 0 |
| Brush teeth | 0 | 0 | 0 | 100 | 0 |
| Drink glass | 0 | 0 | 0 | 0 | 100 |

**Table 3** Classification accuracy of different activities

| Activity | Number of instances | True positive | False positive | Accuracy (%) |
|---|---|---|---|---|
| Descend stairs | 10 | 8 | 2 | 80 |
| Climb stairs | 10 | 10 | 0 | 100 |
| Comb hair | 10 | 10 | 0 | 100 |
| Brush teeth | 4 | 4 | 0 | 100 |
| Drink glass | 10 | 10 | 0 | 100 |

**Table 4** Performance comparison with other related works

| Author | Number of activity detected | Overall accuracy (%) |
|---|---|---|
| L. V. Shiripova et al. | 2 | 98 |
| Amir A. Aljarrah et al. | 12 | 97.6 |
| Zengtao Feng et al. | 19 | 93.4 |
| Lu Xu et al. | 6 | 90 |
| Barbara Bruno et al. | 8 | 66 |
| Pubali De et al. | 7 | 82.5 |
| Proposed work | 5 | 96 |

## 6 Discussion

This proposed method is found to be capable of recognizing human activity with a reasonably good accuracy using limited resource, i.e., only one tri-axial accelerometer. Application of PCA on the raw sensor data reduces the dimension of data keeping all important information. This leads to less time complexity and faster response of the system reducing amount of data handling requirement. The MATLAB program execution time in a i3 7th Gen PC with 8 GB RAM is approximately 4.5 s. Application of random forest classifier gives 100% detection accuracy for four activity and 80% for descend stair only. Hence, it is worth saying that the proposed method can be adopted to develop small, wearable, and cost-effective HAR systems. Application of this HAR method for more number of ADL detection is future plan of this work.

## References

1. Amina E, Anouar A, Abdellah T, Abderahim T (2018) Ambient assisted living system's models and architectures: a survey ofthe state of the art. J King Saud Univ Comput Inf Sci. https://doi.org/10.1016/j.jksuci.2018.04.009
2. Costa A, Castillo JC, Novais P, Fernndez-Caballero A, Simoes R (2012) Sensor-driven agenda for intelligent home care of the elderly. Expert Syst Appl 39:12192–12204
3. Fleury A, Vacher M, Noury N (2010) Svm-based multimodal classification of active ities of daily living in health smart homes: sensors, algorithms, and first experimental results. IEEE Trans Inform Technol Biomed 14:274–283. ID: 1
4. Chernbumroong S, Cang S, Atkins A, Hongnian Y (2013) Elderly activities recognition and classification for applications in assisted living. Expert Syst Appl 40:1662–1674
5. Bruno B, Mastrogiovanni F, Sgorbissa A, Vernazza T, Zaccaria. Analysis of human behaviour recognition algorithms based on acceleration data. In: 2013 IEEE international conference on robotics and automation (ICRA) Karlsruhe, Germany, May 6–10, 2013, pp 1602–1607
6. Bruno B, Mastrogiovanni F, Sgorbissa A, Vernazza T, Zaccaria R Human motion modelling and recognition: a computational approach. In 8th IEEE international conference on automation science and engineering, Seoul, Korea, August 20–24, 2012, pp 156–161
7. De P, Chatterjee A, Rakshit A Recognition of human behaviour for assisted living using dictionary learning approach. IEEE Sens J. https://doi.org/10.1109/JSEN.2017.2787616

8. Walse KH, Dharaskar RV, Thakare VM PCA based optimal ANN classifiers for human activity recognition using mobile sensors data. In: First international conference on information and communication technology for intelligent systems: volume 1, smart innovation, systems and technologies, p 50. https://doi.org/10.1007/978-3-319-30933-0_43
9. Shiripova LV, Myasnikov EV Human action recognition using dimensionality reduction and support vector machine. In: 5th international conference on "information technology and nanotechnology" (ITNT-2018)
10. Huang P-C, Lee S-S, Kuo Y-H, Lee K-R (2010) A flexible sequence alignment approach on pattern mining and matching for human activity recognition. Expert Syst Appl 37:298–306
11. Kasteren TV, Noulas A, Englebienne G, Krose B (2008) Accurate activity recognition in a home setting. In: Proceedings the 10th international conference ubiquitous computing, Tokyo, Sept. 11–14, pp 1–8
12. Manosha Chathuramali KG, Rodrigo R (2012 ) Faster human activity recognition with SVM. In: The international conference on advances in ICT for emerging regions—ICTer, pp 197–203
13. Jain A, Kanhangad V (2018) Human Activity classification in smartphones using accelerometer and gyroscope sensors. IEEE Sens J 18(3):1169–1177
14. Huang YC, Yi CW, Peng WC, Lin HC, Huang CY (2017) A study on multiple wearable sensors for activity recognition. In: 2017 IEEE conference dependable security computing, pp 449–452
15. Brezmes T, Gorricho JL, Cotrina J (2009) Activity recognition from accelerometer data on a mobile phone. In: Lecture notes computer science (including subseries on lecture notes artificial intelligence lecture notes bioinformatics), vol 5518 LNCS, no. PART 2, pp 796–799
16. Aljarrah AA, Ali AH (2019) Human activity recognition using PCA and BiLSTM recurrent neural networks In: 2nd international conference on engineering technology and their applications-IICET2019, IEEE
17. Abidine MB, Fergani B (2013) Evaluating a new classification method using PCA to human activity recognition. In: International conference on computer medical applications, ICCMA, pp 1–4
18. Chokka A, Sandhya Rani K (2019) PCA based regression decision tree classification for somatic mutations. Int J Eng Adv Technol (IJEAT) 8(6S3). ISSN: 2249–8958
19. Sasikala S, Appavu Alias Balamurugan S Data classification using PCA based on effective variance coverage (EVC). In: IEEE international conference on emerging trends in computing, communication and nanotechnology (ICECCN 2013)
20. Sijo Antony NJ, Kavitha R (2017) Human activity recognition from sensor data using random forest algorithm. Int J Adv Res Comput Sci [S.l.] 8(3):334–337. ISSN 0976-5697
21. Mandha P, Lavanya Devi G, Viziananda Row S (2017) A random forest based classification model for human activity recognition. Int J Adv Sci Technol Eng Manage Sci 3(1) (IJASTEMS-ISSN: 2454-356X)
22. Feng Z, Mo L, Li M (2015) A random forest-based ensemble method for activity recognition. In: 37th annual international conference of the IEEE engineering in medicine and biology society
23. Xu L, Yang W, Cao Y, Li Q (2017) Human activity recognition based on random forests. In: 2017 13th international conference on natural computation, fuzzy systems and knowledge discovery (ICNC-FSKD), pp 548–553. https://doi.org/10.1109/FSKD.2017.8393329
24. UCI Machine Learning Repository. Dataset for ADL recognition with wrist-worn accelerometer data set. https://archive.ics.uci.edu/ml/datasets/Dataset+for+ADL+Recognition+with+Wrist-worn+Accelerometer

# Plant Growth Optimization Using Internet of Things for Self-Sustaining Indoor Farms

**Karthick Nanmaran and Arun Kumar Ramaiyan**

## 1 Introduction

**Background** India is the second most populous country in the world and one of the fastest growing economies, next only to China and USA. Agriculture, with its allied sectors, accounts for the livelihood of over 70% of its rural population. Although India has achieved self-sufficiency in food production, agriculture's contribution to the gross domestic product (GDP) of India has steadily declined over the years and so is the total land area under agriculture [1]. Whilst one reason for this may be the diversification of the economy, as it grew, from being an agriculture-based economy to an urbanised industrial and service oriented economy, the resource-intensive unsustainable farming practices pose major threats to the agriculture in this country. Desertification, land degradation and climate change also expedite the problem, and there is a need to shift from a conventional to a more sustainable farming practice that is resilient to the above-mentioned threats.

**Problems and Solutions** Transforming from a soil-dependant farming practice to a soilless farming practice aided by modern technology and precision farming techniques can help mitigate most of the problems that affect conventional agriculture. A variety of interconnected sensors can be used to monitor and control crop growth, in an indoor environment. Alternative and smart farming technologies like hydroponics [2–4], vertical farming [5–7] make use of Internet of Things (IoT) to monitor the crop environment and optimise plant growth. IoT consists of devices with built-in sensors to collect and transfer data over a network without manual intervention. IoT can be used to monitor all the plant parameters to make the environment conducive for plant growth. Also, an extensive review of related works can be found in [8–15].

K. Nanmaran (✉) · A. K. Ramaiyan
Department of Computer Science and Engineering, Annamalai University, Chidambaram, Tamil Nadu, India
e-mail: karthicknanmaran@gmail.com



P. H. J. Chong et al. (eds.), *Emerging Electronics and Automation*, Lecture Notes in Electrical Engineering 937, https://doi.org/10.1007/978-981-19-4300-3_18

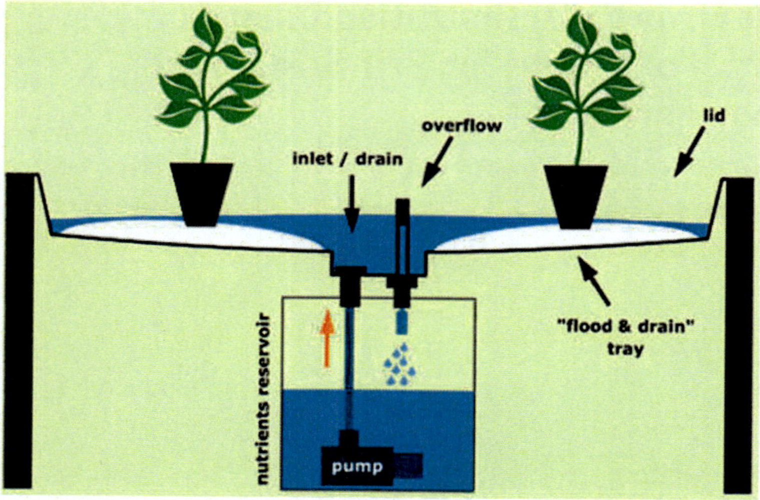


**Fig. 1** An example hydroponic system [16]

**Proposed Solution** Hydroponic farming is a farming method that involves growing plants without using traditional soil medium, using mineral nutrient solutions mixed in a water substrate, that acts as the growth medium. Plants need food, water, light, air, temperature and when broken down to these five things, it becomes simple to give plants only what they need. The hydroponic farmer controls the growing environment of plants by using a network of interconnected sensors. The plants thrive on the nutrient solution alone—the medium merely acts as a support for the plants and their root systems. We discuss such a prototype, of a smart soilless farming setup in an indoor environment monitored and controlled by IoT sensors for optimising plant growth. And thus, the main idea of this paper is to explore the possibility of controlling plant growth parameters to achieve desired results in terms of fruit yield, stem growth, leaf density, etc. (Fig. 1).

## 2 Methodology

### *2.1 IoT Prototype System Architecture*

The IoT indoor prototype system comprises of a Raspberry Pi connected to a temperate/moisture sensor, hydro system containing a water pump and to the LED Lights, through relays. The system was programmed in Python and the code measured the temperature/moisture levels and to turn on and off, the LED Lights and water pump at periodic intervals (Fig. 2).

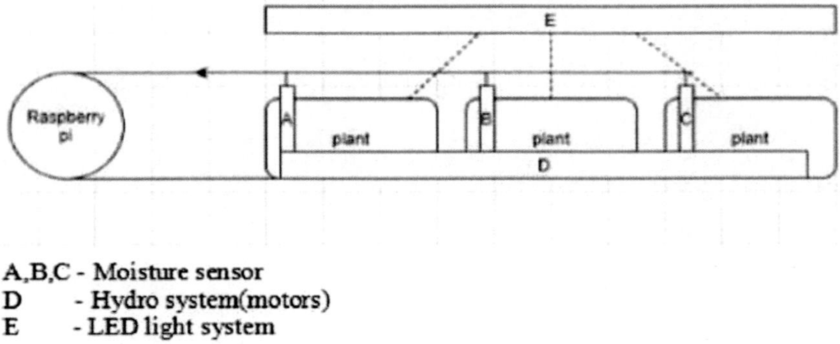


**Fig. 2** Plant growth in indoor environment

## 2.2 *The Experiment*

Coriander was grown in a nutrient mixed water solution as shown in the experimental setup given in figure in a nutrient mixed water medium. Temperature, humidity, and the level of lighting were controlled by sensors. Nutrient mixed water medium was pumped using a motor through a series of pipes.

The implementation was carried out as two cases—outdoor and indoor under natural sunlight and under artificial LED Light. The two cases were to compare the variance in the growth parameters of the plants under sunlight and artificial light.

Plants in the outdoor environmental setup were grown in a well ventilated space with adequate sunlight and wind. They were manually watered. In the indoor environment, the plants were setup in a dark space with artificial lighting, and the temperature was maintained at 25°. The seeds were placed in three support mediums—cocopeat, sponge and rockwool. The LED light in the room was turned on for 12 h of a day, and darkness was maintained during the other 12 h of the day. Nutrient solution was pumped through the pipes in the indoor environment for every 3 h with a gap of 1 h in between.

Both the indoor and outdoor setups were monitored continuously for a distinct comparison on the differences in growth and quality. The proposed prototype system was carried out on the indoor plants. Coriander was selected because of the lower harvest time and because it was easier to grow without much manual intervention. The plants were observed consistently at regular intervals on a daily basis so as to identify and pinpoint their phase of growth. In both the setups, ideal conditions were attempted, and the seeds were sown on the same day (Fig. 3).

The coriander plants showed no sign of crowning in the outdoors until day 4, and whereas in indoors, there was a single plant that crowned towards the end of day 3. On day 4, the outdoors and the indoors showed multiple crowning although there were more crowns in the indoor environment than outdoors.

**Fig. 3** Plant growth in indoor environment

**Fig. 4** Indoor plants tilting towards the direction of LED light

By day 10, the outdoor plants reached a growth of 3 cm whilst this growth was achieved by day 7 in the indoors. On the 10th day of growth, we observed that the plants in indoor environment started tilting towards the direction of LED Light placed in the room. The shoot bearing the first leaves (plumule) started getting elongated in indoors, whilst in the outdoors, the growth pattern was normal (Fig. 4).

On day 15, the indoor plant has reached a growth of 8 cm with long plumule leaves mostly, whereas in the outdoors, the length of the plants was 5 cm in average but had more branches with small leaves. Some of the long elongated shoots in the indoor plants started to fall on the sides due to lack of support.

On day 20, the plants kept in outdoors had branched out significantly well and leaves dense and healthy. The edges of the plants in the indoor environment showed signs of browning on the edges. Some of the leaves also started to turn light yellow. This cell damage in plants was due to the lack of nutrients in the water medium because the plants had used up the nutrients for growing. Hence, on this day, the nutrient solution was topped up with additional nutrients.

On addition of nutrients and after providing adequate support for the indoor plants, they started to grow and pick up pace. By day 25, the plants in the indoor setup had as much growth as the plants kept in outdoors.

The plants in the indoor environment started to grow considerably faster than the ones in the outdoors and on day 30, the day of the harvest we observed that the growth of indoor plants was around 10% greater than the plants in the outdoor environment.

From the final observation, the plants grown indoors with the proposed system provided much distinctive and satisfactory results. The results obtained on the health of the leaves and yield were greater than the actual estimate.

## 3 Results and Discussion

As observed from the above experiment, the plants grown with the help of the indoor prototype system grew at a pace 10% faster than the plants grown externally. There were some drawbacks and flaws in the system initially and were overcome as the learning process was gradual. The plants started tilting towards the direction of light, and hence, we had to reassemble the lights and place them closer and directly on top of the plants. The length of the stem was also elongated. This could have been overcome with the additional of nutrients more frequently and with a closer placement of LED lights. Signs of yellowing and browning could have been overcome with calculating the exact value of nutrients that have to be mixed in the water medium. Previous knowledge of this would have helped us achieve much higher growth rates.

## 4 Conclusion and Future Enhancements

The system was successfully implemented with a functionality equivalent to a typical to how a farmer grows it traditionally. By this method, the plants grown indoors grow at a pace 10% faster than that of the same plants grown externally. This system can be further enhanced by adding and modifying more parameters to stimulate growth such as sounds, changing LED colours [17]. As maintaining the desired temperature can seem economically inefficient, in the long term, it might seem as a smart investment, and focussing on the frequency of LED lights [18] can also give promising results.

## References


1. FAO in India—India at a glance. https://www.fao.org/india/fao-in-india/india-at-a-glance/en/
2. Jensen Merle H (1997) Hydroponics. HortScience 32(6):1018–1021
3. Roberto K (2003) How-to hydroponics. Futuregarden, Inc
4. Jones Jr, Benton J (2016) Hydroponics: a practical guide for the soilless grower. CRC press
5. Kozai T, Niu G, Takagaki M (eds) (2019) Plant factory: an indoor vertical farming system for efficient quality food production. Academic press
6. Chin YS, Audah L (2017) Vertical farming monitoring system using the internet of things (IoT). In: AIP conference proceedings, vol 1883, No 1. AIP Publishing LLC
7. Saad MHM, Hamdan NM, Sarker MR (2021) State of the art of urban smart vertical farming automation system: advanced topologies, issues and recommendations. Electronics 10(12):1422
8. Al-Kodmany K (2018) The vertical farm: a review of developments and implications for the vertical city. Buildings 8(24). https://doi.org/10.3390/buildings8020024
9. Eigenbrod C, Gruda N (2015) Urban vegetable for food security in cities. A review. Agron Sustain Devel 35(2):483–498. https://doi.org/10.1007/s13593-014-0273-y
10. Kang M, Fei-Yue Wang (2017) From parallel plants to smart plants: intelligent control and management for plant growth. IEEE/CAA J Automatica Sinica 4(2):161–166
11. Ahmed N, De D, Hussain I (2018) Internet of things (IoT) for smart precision agriculture and farming in rural areas. IEEE IoT J 5(6):4890–4899
12. Elijah O et al (2018) An overview of internet of things (IoT) and data analytics in agriculture: benefits and challenges. IEEE IoT J 5(5):3758–3773
13. Aygün S et al (2019) Sensor Fusion for IoT-based intelligent agriculture system. In: 2019 8th international conference on agro-geoinformatics (agro-geoinformatics). IEEE
14. Liu S et al (2019) Internet of things monitoring system of modern eco-agriculture based on cloud computing. IEEE Access 7:37050–37058
15. Sushanth G, Sujatha S (2018) IOT based smart agriculture system. In: 2018 International conference on wireless communications, signal processing and networking (WiSPNET). IEEE
16. Hydroponic system (Wikimedia Commons). https://upload.wikimedia.org/wikipedia/commons/b/b1/Systeme_FLOOD&DRAIN_573px.jpg
17. Namgyel T et al (2018) IoT based hydroponic system with supplementary LED light for smart home farming of lettuce. In: 2018 15th international conference on electrical engineering/electronics, computer, telecommunications and information technology (ECTI-CON). IEEE
18. Luechai Promratrak (2017) The effect of using LED lighting in the growth of crops hydroponics. Int J Smart Grid Clean Energy 6(2):133–140

# A Blockchain Point of View of Supply Chain Problems in India

**Amitesh and Divya Kumar**

## 1 Introduction

Supply chain is a process of managing a product from the stage of raw materials from a supplier, to become processed at the manufacturer, then transportation through a distributor, then sales at a retailer to reach in hands or home of the consumer. Each stake holders maintains their own accord of storing their transactional records in a conventional supply chain. The opaqueness caused enormous losses for the organisation in the past. Visibility, transparency, and traceability are fundamental problems within the supply chain. Here, blockchain technology will pave the path for the solution. Blockchain technology came into existence in 2008, after a pseudonym Satoshi Nakamoto, a person/group of persons, published a white paper on the very first cryptocurrency Bitcoin [1]. Blockchain is the heart of Bitcoin. It allows transactions directly between any two distrustful parties without any third party (like a bank) using consensus protocol. The consensus protocol is the blood pumped using blockchain into the bitcoin. It enforces the rules for the validation of any transaction/block by solving a computational puzzle using an authorised party [2]. Blockchain technology is not a disruptive technology instead a foundational technology [3].

The supply chain can be explained with the help of an example of a product such as a pen. Its life cycle begins from raw material suppliers, who send metal/glass/plastic (depending upon the material used to make the physical body of the pen) to the manufacturer of the pen. They processed the body with their proprietary ink and produced the fresh new pens, The manufacturer exports to other countries via transporter and ships to the distributors. The distributor places the order to the manufacturer based on retailers' and inventory storage demand. The distributor then sends it to the retailers,

Amitesh (✉) · D. Kumar
Motilal Nehru National Institute of Technology Allahabad, Prayagraj, India
e-mail: Amitesh@mnnit.ac.in

D. Kumar
e-mail: DivyaK@mnnit.ac.in



P. H. J. Chong et al. (eds.), *Emerging Electronics and Automation*, Lecture Notes in Electrical Engineering 937, https://doi.org/10.1007/978-981-19-4300-3_19

who earlier placed orders to the distributor for pens. The retailers sell it to the agents, who then sell it to the end customers.

Supply chain has following research problems: visibility [4], traceability [5], delayed delivery latency [6], complicated transportation network [7], stakeholder identification (accountability/custodianship) [8], demand forecasting [9], inventory management [10], etc.

Supply and demand are needed to be in synchronisation as an imbalance can decrease the profit margin, here inventory management comes into the picture [11]. Inventory as a term means stock, which are items or goods available for sale or raw materials that can come into use to manufacture products for sale. Inventory management is all about ensuring that the required quantities of the item should be in the right location at the correct time. The domain researchers introduced inventory models to improve the efficiency of inventory management. Economic Ordered Quantity or EOQ is one of the most popular inventory models ever sought. EOQ can be defined as the ideal order quantity a company should purchase to minimise inventory costs such as holding, shortage, and order costs. Inventory management plays an essential role in the supply chain management process, as it acts as the bridge between retailers and manufacturers.

In this paper, we propose a proof of concept framework for supply chain problems mentioned in the previous paragraphs. We will be using distributed ledger technologies like blockchain to resolve supply chain nuances.

## 2 Theoretical Background

### 2.1 *An Impending Need of Fixing Supply Chain Issues*

Supply chain is as old as share markets. The concepts within this area are still evolving from coarse to fine granularity. There are some problems that play essential roles in this chain; researchers are working for a long time to improve its efficiency to prevent potential losses and increase profit/revenue. The globalisation of the Supply chain opened a pandora box of fresh possibilities in terms of doing business in different regions of the world. As within one country, the supply chain is less complex, but when this transcends into another geopolitical arena, the countries cons also follow along with its pros due to various factors, like availability of resources, transportation medium, administration, and allocation of the region available for relevant industrial usage.

### 2.2 *Concepts and Characteristics of Associated Problems and a Problem Resolution System*

We have divided the associated supply chain problems into primary and derived categories. Primary problems have been there for a long time due to the unavailability of any efficient solution. Derived problems originate as a result of cascading effects of primary issues. Examples of primary problems are communication gap, less trust in sharing of data, environmental factors like lockdown, natural disasters, etc. Examples of derived problems are problems delay (due to communication gap), predictive analytics accuracy (due to distrust in data sharing), and disruption in the supply chain (due to environmental factors).

### 2.3 *Blockchain Technology for Associated Problems (Literature Survey)*

Inspite of not being a mature technology, blockchain has been in the picture on various platforms. Researchers and technocrats from various research labs are consistently striving to improve their academic and application research domain with the help of constantly improving existing blockchain platforms.

Satoshi Nakamoto first discussed this technique in 2008, where he built bitcoin (virtual currency), which internally uses the blockchain network [1]. Bitcoin was the first currency that didn't need any central server to authorise the transactions. It is maintained on the untrusted network using a decentralised replicated ledger across all nodes that are part of the network, which verifies the asset/transactions, and generates the currency based on the proof of work algorithm, which requires mathematical computation validation. It works on the P2P network, which monitors and verifies user transfers.

A set of interdependent companies that work together to manage the flow of goods and services along with the value-added chain of agricultural and food products [12]. When compared to other industries, food production takes place in more vulnerable value chains, which requires more attention over handling processes, such as producing and storing.

Everledger [13] is currently using blockchain to identify diamond's source of origin, allotment information, details of it's rough state, planning, cutting process, polishing information, and certification of genuinity of diamond based on their real-time tracing its supply chain.

Ambrosus [14] is an organisation that is leveraging blockchain secured IoT network infrastructure for the pharma and food industry, using secure communication between sensors, blockchain, and databases to optimise quality assurance and supply chain visibility.

AgriBlockIoT [15] provides a traceability solution using blockchain for "Agri-Food supply chain management". They implemented this over Ethereum and Hyperledger Fabric framework and then compared them after evaluation based on latency, network usage, CPU, and described their pros and cons.

Tradelens [16] is an organisation formed due to the collaboration of IBM and Maeserk in 2018. It provides a blockchain-based shipping transport solution for efficient, secure, transparency, and information sharing between more than 94 organisations leveraging this platform for supply chain businesses.

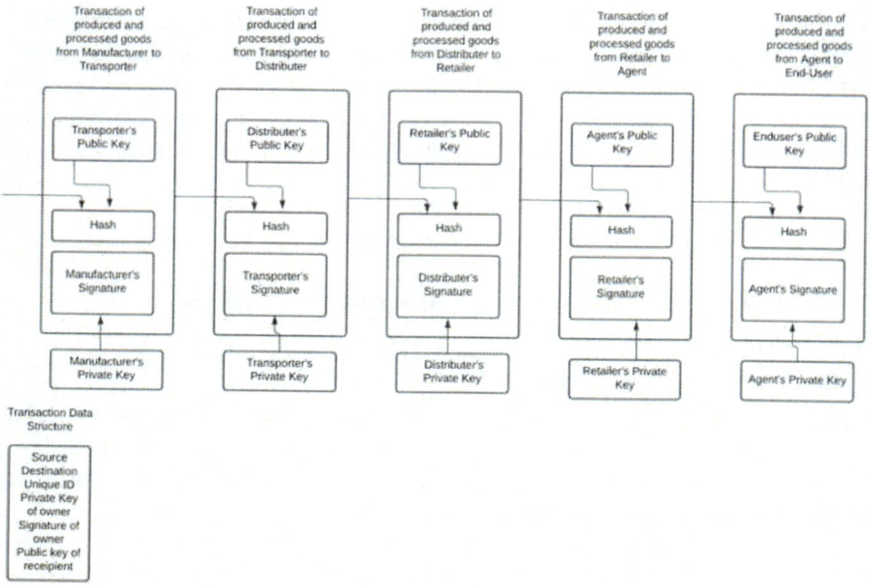


**Fig. 1** Transactions blockchain architecture

Waltonchain [17] helps in monitoring "production of items, logistics requirements, inventory warehousing, and retail distribution of all products" with the help of blockchain, RFID, and Internet of Things (IoT). On this platform, merchants create child chains for the tasks mentioned above.

Walmart [18] leverages the Hyperledger Fabric blockchain platform to trace two products, namely Mangoes and Pork, using two proof of concept projects, and the provenance time was reduced considerably from 7 days to mere 2.2 seconds.

Figorilli et al. [19] leverages the Azure blockchain workbench to store the transaction information related to marking timber, cutting the wood, sawmill processing, and final production based on RFID usage (Fig. 1).

# 3 Methodology

In this proposed method, the blockchain architecture for supply chain is depicted as follows:

## *3.1 Preliminary*

For any system to achieve traceability, a system must record and follow the product trail. The interconnected nature of the supply chain makes it difficult to introduce a centralised system in control of a third party, requiring a high level of trust such

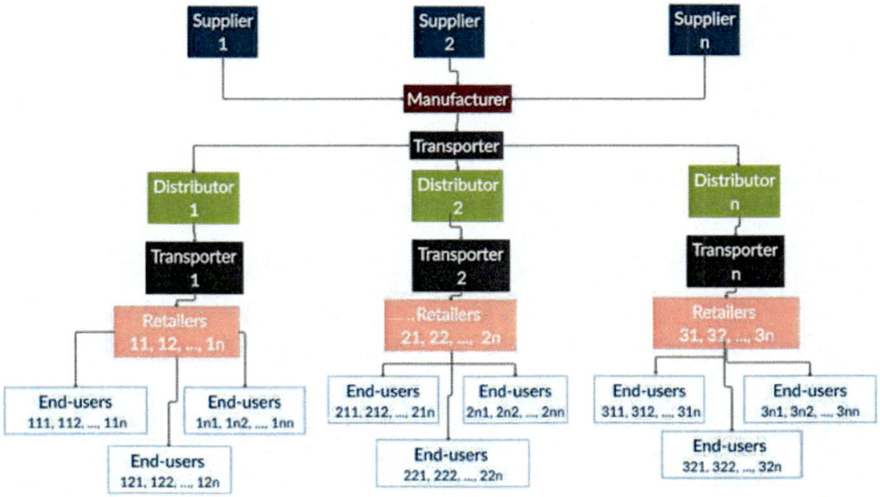


**Fig. 2** Schematic flow chart of system

that one can verify and the other as well. The limited amount of trust resulted in separate systems, limiting the possibility of achieving traceability throughout the entire supply chain and taking the distributed system to achieve traceability. We need a robust system to achieve trust between the parties (Fig. 2).

## *3.2 System Proposed*

Transaction validation is done on each actor such that only the valid transaction can be added and invalid transactions cannot be added into the block.

A manufacturer creates a product with a unique name added into the dictionary prod, which maps a list of the places where a product can be. Afterwards, the manufacturer creates the information source, a destination with the product name, where the recipient is the distributor.

Transporter is there as a mediator.

Distributor collects the product and distributes it to the retailers. This actor takes the product from the distributor and transfers it to the retailers.

Retailer collects the product and distributes it to the agents.

Agents collect the product and distribute it to the End.

End-user are the consumer of the product. We have worked on the following problems: stakeholder identification with valid ordered quantity, the average delay of the product ordered, and traceability.

A set of attributes are assumed based on the data available in the public domain regarding any product produced in the supply chain.

- unique ID
- ordered date
- product description
- universal product code
- high resolution image
- product specifications
- product current location.

### 3.3 Stakeholder Identification with Valid Ordered Quantity

**Algorithm 1:** procedure Is_Actor_Valid(S, R, K, D, h)

```
Result: Returns TRUE if actors are validated and their inventory EOQ satisfies
        or FALSE
initialization: S = Source, R = Recipient, K = Setup Cost, D = Demand Cost,
h = Holding cost ;
if (S = 'Supplier' and R = 'Manufacturer' and flag['Manufacturer']=True
and product['Supplier']== product['Manufacturer'])
OR
(S = 'Manufacturer' and R = 'Distributor' and flag['Distributor']=True and
product['Manufacturer'] == product['Distributor'])
OR
(S = 'Distributor' and R = 'Retailer' and flag['Retailer']=True and
product['Distributor']== product['Retailer'])
OR
(S = 'Retailer' and R = 'Agent' and flag['Agent']=True and
product['Retailer']== product['Agent']) then
  | EOQ = evaluate_EOQ(K, D, h) ;
  | if Order_Quantity[R]<EOQ
  | OR Order_Quantity[R]=EOQ then
  |   | return TRUE;
  | else
  |   | return FALSE;
  | end
else
  | return FALSE;
end
```

The above algorithm proposed a solution for stakeholders' validation during the supply chain process. Each OR separated conditions within the if statement demarcates the phases of product transportation, e.g. the item requested by the receiver should be the same as the item sent by the source. The source and receiver should be the genuine or relevant owners of the item. Then we evaluated the Economic Ordered

Quantity [20] using the evaluate_EOQ(). This function returns the computed result of $\sqrt{2KD/h}$, derived from the Basic Inventory model, where $K$ = inventory setup cost, $D$ = current demand, and $h$ = holding cost. If the ordered quantity from the receiver to the sender is less than or equal to Economic Ordered Quantity, the procedure will return TRUE, i.e. a valid actor with a valid ordered quantity. Else in any other possibility, the procedure will return FALSE.

### 3.4 *Traceability Information of Every Single Component Used in Any Product*

**Algorithm 2:** Procedure Traceability (Product P, index i)

```
Result: Returns "Traceability_info of Pi"
initialization: ;
for each component 'c' do
    c.supplier = fetch_supplier_details(c) ;
    c.transporter = fetch_transporter_details(c);
    c.factory_details = fetch_supplier_factory_details(c.supplier);
    c.origin_land_details = fetch_origin_details(c) ;
end
return *c;
```

In some supply chains, provenance details of each component used within the product become important depending upon the impact of the product as a whole on a consumer, e.g. food products or pharma medicine products could affect the health of the consumer, so to analyse or to contain its wrong impact, the products falling in the category of similar provenance details could be recalled to prevent it's widespread or mass impact. In the above algorithm, traceability info—like supplier details (address, location, ratings, historical details, etc.), transporter details(similar as supplier details), factory details (similar as previous two), and origin details(origin address, origin ratings, and historic origin records) are fetched to analyse the point of origin of the problem in any diverse supply chain.

## 4 Conclusion and Future Work

This article explores the challenges and opportunities in the designing of the supply chain model with blockchain. The effect of the blockchain in the supply chain model has been also discussed. Our supply chain model explains the traceability in the model using the blockchain, which continuously tracks the product. We also maintain the flow of the chain such that no product can skip the necessary step, and in case of any product transaction tries to do so it invalidates at the same time. Our model is successfully finding out the complete flow of the product, like what are the different actors from where the product was coming from. The effect of the Economic Order Quantity in the supply chain has also been discussed, which helps to make the order feasible, and it is included in the transaction to make the system. The proposed algorithms are part of the Proof of Concept model we have proposed. So, the performance parameters will only be assesed after the implementation of these algorithms, which are currently the future scope of this paper. So, another future work can be to make the system N:N in the sense that one instance of an actor can interact with multiple instances of the other actor. We will take other parameters of the supply chain's inventory model like the different costs as well as the order cycle, which help us to make our supply chain more robust and favourable to real-world applications as well.

## References

1. Nakamoto S, Bitcoin A A peer-to-peer electronic cash system, Bitcoin. https://bitcoin.org/bitcoin.pdf4
2. Kaur S, Chaturvedi S, Sharma A, Kar J (2021) A research survey on applications of consensus protocols in blockchain. Secur Commun Netw
3. Khudnev E Blockchain: foundational technology to change the world
4. Wu H, Li Z, King B, Ben Miled Z, Wassick J, Tazelaar J (2017) A distributed ledger for supply chain physical distribution visibility. Information 8(4):137
5. Aung MM, Chang YS (2014) Traceability in a food supply chain: safety and quality perspectives. Food Control 39:172–184
6. Truong Quang H, Hara Y (2018) Risks and performance in supply chain: the push effect. Int J Prod Res 56(4):1369–1388
7. Viau MA, Trepanier M, Baptiste P (2009) Integration of inventory and transportation decisions in decentralised supply chains. Int J Logist Syst Manage 5(3–4):249–272
8. Gregory AJ, Atkins JP, Midgley G, Hodgson AM (2020) Stakeholder identification and engagement in problem structuring interventions. Eur J Oper Res 283(1):321–340
9. Abolghasemi M, Beh E, Tarr G, Gerlach R (2020) Demand forecasting in supply chain: the impact of demand volatility in the presence of promotion. Comput Ind Eng 142:106380
10. Singh D, Verma A (2018) Inventory management in supply chain. Mat Today Proc 5(2):3867–3872
11. Law A (2017) Smart contracts and their application in supply chain management, Ph.D. thesis, Massachusetts Institute of Technology
12. George RV, Harsh HO, Ray P, Babu AK (2019) Food quality traceability prototype for restaurants using blockchain and food quality data index. J Clean Prod 240:118021
13. Teo C Everledger announces the industry diamond time-lapse protocol, everledger (cited 2018 Feb 22)
14. Kirejczyk M, Kędracki A, Rukhavets I, Trifa V (2017) Ambrosus white paper
15. Caro MP, Ali MM, Vecchio M, Giaffreda R (2018) Blockchain-based trace ability in agri-food supply chain management: a practical implementation,in: 2018 IoT vertical and topical summit on agriculture—Tuscany (IOTTuscany), pp 1–4. https://doi.org/10.1109/IOT-TUSCANY.2018.8373021
16. Scott T Tradelens: how ibm and maersk are sharing blockchain to build a global trade platform, Ibm.Com
17. Team W (2017) Waltonchain white paper

18. Hyperledger, How walmart brought unprecedented transparency to the food supply chain with hyperledger fabric
19. Figorilli S, Antonucci F, Costa C, Pallottino F, Raso L, Castiglione M, Pinci E, Del Vecchio D, Colle G, Proto AR et al (2018) A blockchain implementation prototype for the electronic open source traceability of wood along the whole supply chain. Sensors 18(9):3133
20. Erlenkotter D (1990) Ford Whitman Harris and the economic order quantity model. Oper Res 38(6):937–946

# Development of Smart Chatbot Using Artificial Intelligence in the Fight Against the COVID-19 Pandemic

Twarita Mahato, Aditi, Koyna Chakravorty, and Himadri Lala

## 1 Introduction

In the wake of COVID-19 pandemic, correct and precise information regarding the supplies and availability of hospital beds, medicine, oxygen cylinder and other medical facilities is not available, due to which people are misguided. A lot of people were afraid of getting tested for COVID-19 learning about the cruel ways in which they would be treated if tested positive that were constantly being posted on social media. So here we came up with the idea of CHATBOT, which will contain all updated and precise information regarding COVID-19 related issues [1, 2]. In this digital era, as we all know that almost everybody is comfortable in working with Internet as well as speech engines, so a chatbot which answers the issues verbally will prove out to be a time saver as well as it will be helpful for the people who cannot read properly.

We have created a chatbot with the help of python programming language that uses different libraries and toolkits such as NLTK, pyttsx3, gTTs, random, io. Natural Language toolkit is used in python to build a program that works with human language data for applying in statistical natural language processing.

CHATBOTs in the previous few years have proved themselves as an important part of our digital lives [1]. It gives us access to answers to our problems in a friendly way

T. Mahato · Aditi · K. Chakravorty · H. Lala (✉)
Vellore Institute of Technology, Vellore 632014, India
e-mail: lalacemk@gmail.com

T. Mahato
e-mail: twarita.mahato2020@vitstudent.ac.in

Aditi
e-mail: aditi.2020@vitstudent.ac.in

K. Chakravorty
e-mail: koyna.chakravorty2020@vitstudent.ac.in



P. H. J. Chong et al. (eds.), *Emerging Electronics and Automation*, Lecture Notes in Electrical Engineering 937, https://doi.org/10.1007/978-981-19-4300-3_20

and that makes the interaction more comfortable. Secondly privacy of our conversation with the chatbots, is also maintained. In our busy lifestyle we are in urgent need of precise and updated information and therefore any information related to COVID-19 should be correct and not misguiding, due to the cruciality of the alarming situation. The main objective for this work is to provide a chatbot for citizens to use in times of crisis. The motivations for this work are the following.

- Chatbots, unlike a newspaper, can hear and respond in a natural language.
- They can close the space between information and action through constant repetition, step wise instructions.
- Properly designed and operated, chatbots can reduce the long-term damage of isolation associated with epidemics, trauma and depression.
- Chatbots have different payloads which may reduce short- or long-term disease burden during an infectious disease epidemic, in future mental health problems.

With these motivations, the chatbot is expected fulfill the major objectives. The expected outcomes are as follows.

- Designed bot can handle user requests and identify message patterns. Information can be tailored to the needs and characteristics of the person. Answers to specific questions can be provided in a collaborative way, much faster than conventional online search methods.
- Provide the latest information and available at any time of the day to answer questions.
- Deliver time-consuming but important ethical support and instructions from health professionals.
- Translating medical information into community advice.

# 2 Proposed Methodology

## *2.1 Dataset Description*

The chatbot is required to be very clear and simple with its answers so that the user is comfortable while using it. It requires a clear datasheet and sustainable backend logic for output generation. We have developed a chatbot which answers the queries regarding COVID-19 pandemic verbally and through texts as well. The question asked by any user is understood by the use of NLTK toolkit and therefore the respective output is given [3]. For the greeting part we have used random function so that the chatbot gives any random greeting and it is not the same for everyone. This will make the conversation a bit exciting (Tables 1 and 2).

Once the engine receives any of the above given questions, it uses NLTK to understand the given question and then gives the respective output. Our chatbot assistant here uses the concept of Naïve Bayes decision tree to analyze the question and give the output based on that.

**Table 1** The greeting part

| S. no | User input | Output |
|---|---|---|
| 1 | ("hello", "hi", "greetings", "sup", "what's up", "hey",) | ["hi", "hey", "nods", "hi there", "hello", "I am glad! You are talking to me"] |

**Table 2** Questions and their answers

| S. no | Questions | Answer |
|---|---|---|
| 1 | What are the symptoms of COVID-19? | "Fever", "cold & cough", "headache", "nausea", "diarrhea |
| 2 | What if I'm exposed? | "Check your body temperature", "Maintain distance till the results arrive", "get tested", "get quarantined" |
| 3 | What are the home remedies for a COVID positive patient? | "Drink herbal tea on a regular basis", "Check your body temperature on a regular basis", "Get quarantined in a separate room", "Take medicines prescribed by the doctor |
| 4 | How can I get tested for COVID-19? | "You can get a swab test", "You can get a drive through test", "You can get blood sample test for antibodies" |
| 5 | How long do COVID symptoms last ? | "Mild symptoms long for 1–2 weeks, however, it depends on your immunity", Acute symptoms last for 1 month, however, it depends on your immunity" |
| 6 | What is the duration gap for 2nd dose of vaccine? | "If you are vaccinated with Covishield—3 to 4 months", If you are vaccinated with Covaxin—28 days", If you are vaccinated with Sputnik—21 days" |
| 7 | Whether both doses of the vaccine should be same? | "Yes, both the doses should be of the same vaccine o else you may face some health related issues." |

## 2.2 *COVID-Chatbot Architecture*

As most citizens all over the world are unaware of the contagiousness of COVID-19, therefore to convince the people about the danger of COVID-19 communication is very essential. Since we have very limited interactions among people, so we purpose for COVID19 chatbot. The architecture of the COVID-chatbot is presented in Fig. 1.

At first some of the libraries like NLTK, NumPy, etc. are imported. NumPy is the library for Python programming language that supports large and multidimensional arrays and matrices [3]. It also supports a large collection of high level mathematical functions that are being operated on these arrays.

Next, we call warnings. Filter warnings with action as "ignore" and category set to Deprecation Warning to ignore any deprecation warnings that may rise. Web scraping

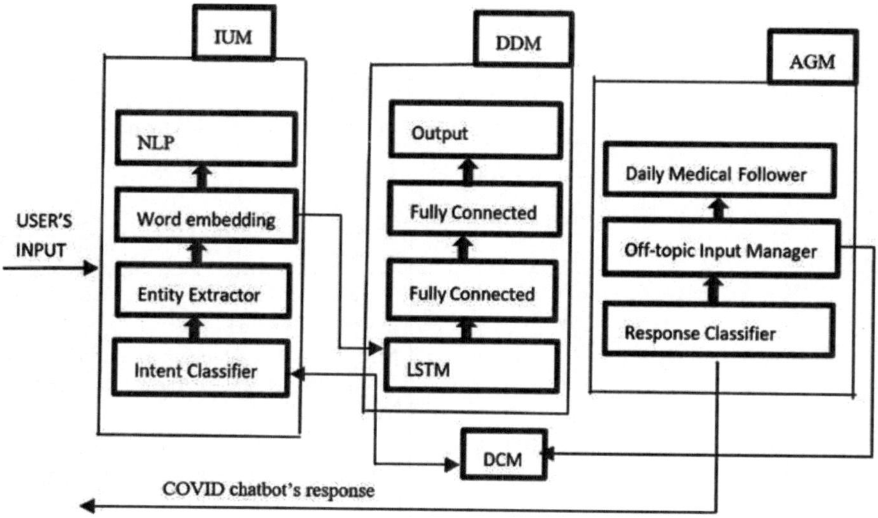


**Fig. 1** Architecture of COVID-chatbot

method is used to download the article to create a corpus of letter. This corpus is then fed into NLP task tokenization in order to remove the punctuation and some other similar things. A response function is then created using T Phi D vector which is taken from machine learning library scikit-learn and finally a chatbot is created about this response.

Input message sent by the user, COVID-chatbot converts that informal text into a business-oriented representation and purpose called natural language processing (NLP), with a few consecutive steps such as Tokenization, The Speech Marking Section. In order to increase interaction between COVID-chatbot and the user, there are a few suggestions suggested by the user to control the conversation in order to minimize errors as much as possible. The goal of the Data Collection Module is to collect non-confidential user data and create a database containing user information such as:

- Status
- Symptoms
- Contact with the Infected Person
- Recent Trips.

Depression Detector Model (DDM): We solve stress detection as a text editing problem for COVID-chatbot. Therefore, we have designed the Emotional Analysis model which is a computerized screening process and isolate the emotions expressed in the text section, mainly to determine the user's attitude to a particular topic. We have used Long Short-Term Memory Network (LSTM) as standard Recurrent Neural (RNN) networks that may not be easy to train due to translator growth by multiplying gradients in long sequences. LSTM is preferred, as it has done better than Gated

Recurrent Units (GRU) for us [4]. Therefore, a consecutive neural network model is developed that includes four layers: Embedding layer and LSTM layer followed by two fully connected layers as shown. In our example, we want to predict the user's status when communicating with our chatbot. Our model does not learn the answer to immediate dependence, but rather to long-term dependence. We have used the Sigmoid Activation function as you are able to convert output values between 0 and 1. As a result of this output, the AGM will know the status of the user during the ongoing discussion. Therefore, if a user sends three consecutive negative messages, the AGM will decide to release the user by sending him or her confirmation or happy messages or even sending questions asking the cause of the bad situation.

## 3 Implementation of Proposed Methodology

The bot that we made can cope up with user requests and it can identify the message styles with the help of artificial intelligence markup language (AIML). AIML is an XML-based markup language for creating herbal language software program marketing that provides customers with a genuine human interactive experience based on their reactions [5]. To gain access to existing consumer clinical situations, AIML common sense retrieves symptomatic key terms. Finally, we want the person to appear as if they're having a conversation with a health professional. A chatbot's functionality can be characterized in two ways:

- Request evaluation
- Provide an answer.

Initially, the chatbot uses a response from a pre-defined questionnaire to determine the virus's severity. Similarly, if the human refuses to accept perfect solutions, the bot will be unable to provide a precise solution. After assessing an affected person's situation, the chatbot responds with a clear response in the form of well-known text or text produced from the information-based response [3]. The offer question can help the chatbot interpret a user's request correctly. Because NLP is a useful generation to attract the interaction between natural language comprehensions and decide consistent response consequences [3] integrated AI and herbal language (NLP) technologies were developed to achieve the intended reaction.

The information base is centered at the reaction of the user's message on the starting of the communique—it ought to be herbal to reply within the proper way lower back. The concept of this is to perceive the first signs of Ncovid-19 in the person place, and then it will suggest whether or not the user is possibly to be inflamed or now not [5].

When a user initiates a chat conversation, the bot engine begins by inquiring about cognitive data. Once it has all of the relevant data, it calculates the percentage of viral infections the user has had and takes appropriate action, such as consulting a doctor or offering information on emergency preventive measures.

We have used NumPy which is a basic computer science package for python. It is a python library that offers a wide variety of materials of different sizes, different artifacts and a variety of fast-paced application systems. NumPy makes it easy to perform complex number tasks.

We have used NLTK, the natural language toolkit is a platform used for building python programs that work with human language data for applying in statistical natural language processing. It contains text processing libraries for tokenization, parsing, cementing reasoning and classification.

We have used pyttx3 which is a text to speech conversion library in python. This module supports two voices one is male and the other one is female. One thing about this module is that it can work offline after the installation. Here our chatbot "Timtim" which uses the female voice can also be operated offline, which makes it better than the other chatbots available.

So, after that we have made user defined functions one of which is a greeting function and the others are the question-based response functions, all these functions can be called by the main function depending on the input given by the user [6]. "Timtim", unlike other "chatbots" converses with the user verbally as well using the pyttsx3 library. The architecture of Depression Detector Model is presented in Fig. 2. Figure 3 represents the flowchart for the proposed chatbot.

## 4 Experiment

The flowchart for the CHATBOT is presented in Fig. 3. In Fig. 4, the user interface of the CHATBOT is depicted with a demo conversation when executed in Jupyter notebook [6]. This is a demo conversation between the user and the chatbot.

## 5 Comparative Analysis

The chatbot present in the market is much less user friendly than our chatbot. We are using the speech engine in our chatbot which makes our chatbot more convenient to use in comparison with the present chatbot. Speech engine makes the use of our chatbot more easy for the physically disabled people. As we all are aware that not only our country but the whole world is divided into smaller groups of linguistic and cultural diversities, a lot of people are there who can only speak English but are unable to read them. For such customers our chatbot will prove to be a savior.

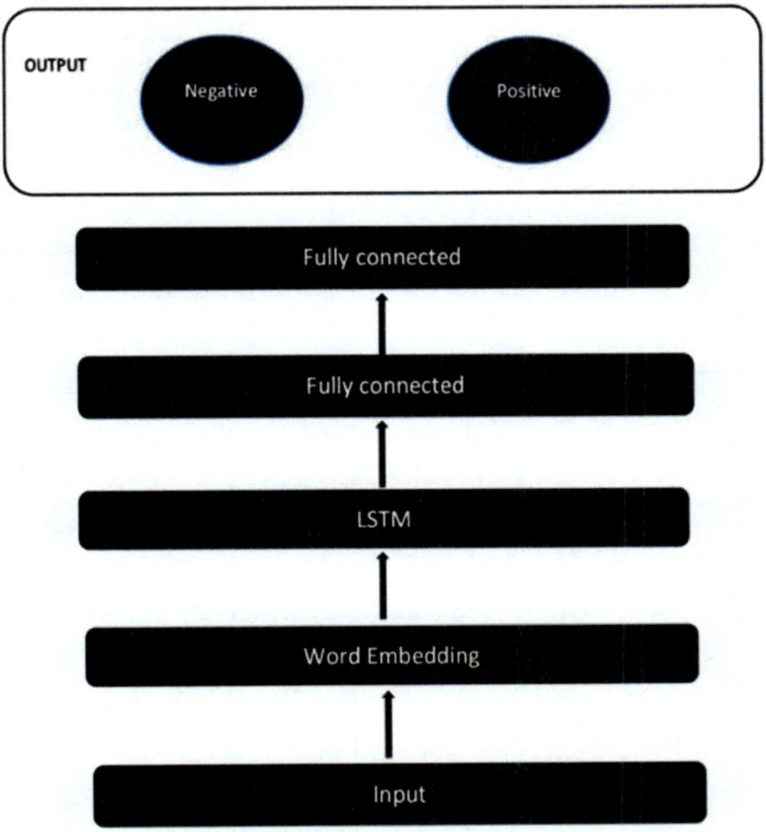


**Fig. 2** Depression detector model architecture

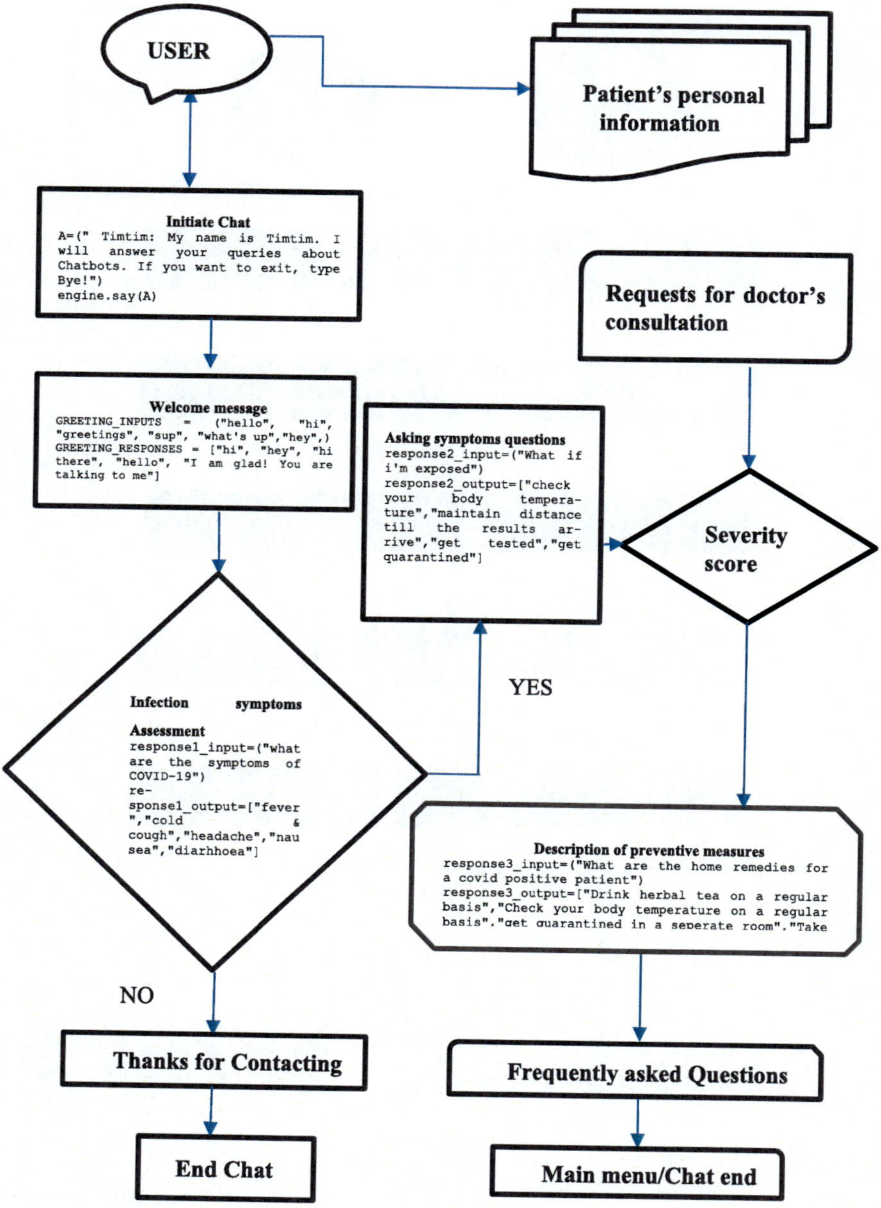


**Fig. 3** Flow chart of COVID-19 chatbot

```
Timtim: My name is Timtim. I will answer your queries about Chatbots. If you want to exit, type Bye!
hi
Timtim: hello
what are the symptoms of COVID
['fever', 'cold & cough', 'headache', 'nausea', 'diarhhoea']
what if I'm exposed
['check your body temperature', 'maintain distance till the results arrive', 'get tested', 'get quarantined']
where can i get tested
Do allow us to know your current location:    yes
There is ner covid center in the R street which is 0.5 kms from your location
what are the remedies for a COVID positive patient
['Drink herbal tea on a regular basis', 'Check your body temperature on a regular basis', 'get quarantined in a seperate room',
'Take medicenes prescribed by the doctor']
how can i get tested for COVID19
['YOU CAN GET A SWAB TEST', 'YOU CAN GET ADRIVE THROUGH TEST', 'YOU CAN GET BLOOD SAMPLE TEST FOR ANTI BODIES']
HOW long does the symptoms of COVID last
['MILD SYMPTOMS LONG FOR 1-2 WEEKS, however ,it depends on your immunity ', 'ACUTE SYMPTOMS LAST FOR 1 MONTH , however ,it depe
nds on your immunity']
what is the duration gap between the 2nd dose of the vaccine
['IF YOU ARE VACCINATED WITH COVISHIELD - 3 to 4 months', 'IF YOU ARE VACCINATED WITH COVAXIN- 28 DAYS', 'IF YOU ARE VACCINATED
WITH SPUTNIK V- 21 DAYS']
whether both the doses should be of the same vaccine
['YES, both the doses should be of the same vaccine o else you may face some health related issues']
bye
Timtim: Bye!! Take care
```

**Fig. 4** User interface of the chatbot

# 6 Conclusions

We developed a chatbot for COVID-19 assistance with Deep Sentiment Analysis Model during and after confinement. COVID-19 is introduced in its four modules: Comprehension Information Module (IUM), Action Generator Module (AGM), Data Collection Module (DCM) and Stress Diagnostic Model (DDM). The LSTM neural network based on the emotional analysis model was used to detect user referrals during continuous discussion. More work is needed to improve our emotional analysis model and make it work with the human voice and aims to add a decision support module to allow users to get an idea of the potential for infection with COVID-19. We aim to add a local tractor so that we can provide accurate patient distance from our current location to the nearest COVID location. We also want to add different languages to the speech engine so that more people can access it comfortably in a friendly environment.

# References


1. Jadhav KP, Thorat SA (2020) Towards designing conversational agent systems. In Advances in intelligent systems and computing. Springer, Berlin, Germany
2. Luo X, Tong S, Fang Z, Qu Z (2019) Frontiers: machines vs. humans: the impact of artificial intelligence chatbot disclosure on customer purchases. Mark Sci
3. Collobert R, Weston J, Bottou L, Karlen M, Kavukcuoglu K, Kuksa P (2011) Natural language processing (almost) from scratch. J Mach Learn Res 12:2493–2537
4. Dhankhar P, RNN and LSTM based Chatbot using NLP, Department of Computer Science and Engineering. MSIT, New Delhi, India.
5. Satu MS, Parvez MH (2015) Review of integrated applications with AIML based chatbot. In: Proceedings of the 1st International Conference on Computer and Information Engineering, ICCIE 2015, Rajshahi, Bangladesh, 26–27 November
6. How we created an open-source COVID-19 chatbot|by Schreurs | Towards data science

7. WHO Health Alert brings COVID-19 facts to billions via WhatsApp
8. www.messengerpeople.com | 523: Origin is unreachable
9. SAJIDA Foundation and Renata Ltd. team up to tackle the COVID-19 pandemic|Dhaka Tribune

# Communication Track—1

# Circular Microstrip Antenna with Shorting Walls for Improved Radiation Performance

**Zonunmawii, L. Lolit Kumar Singh, Sudipta Chattopadhyay, and Abhijyoti Ghosh**

## 1 Introduction

Microstrip antenna is one of the most useful and common low-profile antennas in modern wireless communication. In the present wireless world, small, compatible and affordable antennas are required. A circular microstrip antenna (CMA) is a good contender in this field due to its tiny size, light weight and less cost. CMA are easy to fabricate and assemble. The inherent advantage of a microstrip antenna (MA) have made it very popular as an attractive radiator for modern wireless communication. However, conventional microstrip antennas have several shortcomings such as narrow impedance bandwidth, poor polarization purity (co-polar (CP) radiation to cross-polar (XP) isolation), poor gain and less efficiency.

A handful number of literatures have been reported in last several years to increase the gain of MA [1–5] by inserting composite microstrip-monopole structure or modifying dielectric material as air substrate or air dielectric composite substrate. However, these all efforts are given in rectangular patch geometry rather than circular. Around 7.8 dBi gain in circular patches have been investigated using different techniques like aperture coupling [6], slot and short loaded stacked ground plane [7] without any improvement in polarization purity (PP). The employment of shorting vias with branch line couplers in annular ring antenna [8], numerous shorting vias beneath the circular patch [9] have been employed to achieve maximum 7.2 to 9 dBi gain with polarization purity of around 18 dB only. Hence, the investigations [7–9] fail to address all the main antenna parameters concurrently such as gain and polarization purity. Although the gain is good in [10], still it suffers from poor polarization purity of 17 dB. Furthermore, the radiation pattern obtained in [10] suffers from much distortion with high side lobe level. The use of multi-layered circular

Zonunmawii · L. L. K. Singh · S. Chattopadhyay · A. Ghosh (✉)
Department of Electronics and Communication Engineering, Mizoram University, Tanhril, Aizawl, Mizoram 796004, India
e-mail: abhijyoti_engineer@yahoo.co.in



P. H. J. Chong et al. (eds.), *Emerging Electronics and Automation*, Lecture Notes in Electrical Engineering 937, https://doi.org/10.1007/978-981-19-4300-3_21

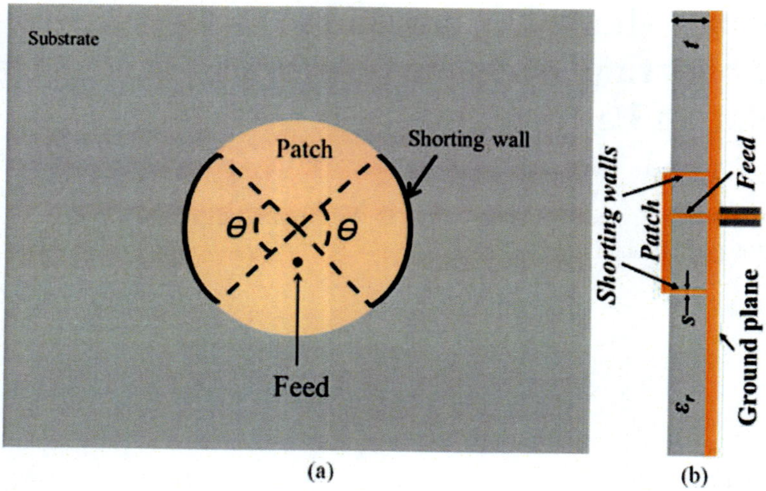


**Fig. 1** Shorting walls integrated circular patch **a** view from top **b** cross sectional view

substrate patch [11], hexadecagon circular geometry [12] or patch with graphene-based materials [13] are investigated for high gain and around 6.8 to 8 dBi gain has been achieved without any enhancement in polarization purity.

Nevertheless, the upgrading in polarization purity is very much important for modern wireless applications. Numerous efforts have been given by modifying patch geometry [14, 15], use of shorting metal patch [16] or using of DGS [17] to improve polarization purity of MA. Maximum 22–23 dB of polarization purity is achieved with no enhancement in gain.

Therefore, to discourse the limitations of previous structures and for the simultaneous enhancement in co-polarized gain and PP, four CMA, two with FR4 substrate and another two with glass substrate with different substrate thickness and a pair of shorting walls (Fig. 1) in the non-radiating periphery of CMA has been proposed which provides simultaneous enhancement of gain as well as the polarization purity. This proposed structure is very simple and provides co-polarization gain of 4.25 dBi, 6.05 dBi and polarization purity of 29.49 dB, 25.2 dB for the substrate thickness of 0.787 mm and 1.575 mm, respectively, for FR4 substrate while with glass as a substrate 7.03 dBi, 7.31 dBi of co-polarization gain and 29.95 dB, 33.62 dB of polarization purity has been achieved using substrate thickness of 0.787 mm and 1.58 mm, respectively.

## 2 Design Technique

The conventional CMA has been designed on two types of substrates having different thickness. The radius of the patch has been determined by the equation as,

$$f_r = \frac{1.84c}{2\pi a\sqrt{\epsilon_r}} \tag{1}$$

Then, shorting walls are placed along non-radiating sides of the patch with an angle $\theta = 100°$. As, we increase the shorting angle frequency increases. Along with the same, radiation performance has also been improved.

## 3 Parametric Studies

To find out the structure that provides best possible outcome in terms antenna performance a robust parametric study has been done with the support of [18]. At first, a conventional CMA with FR4 and glass substrates with different thicknesses of 0.787 and 1.58 mm has been designed for X-band frequencies. The parametric study has been started by placing shorting strips of copper at the non-radiating edge along the periphery of CMA for both antennas with FR4 and glass substrate.

Figure 2 shows the gain of the proposed model for different shorting angles ($\theta$). It is detected that the co-polar gain increases as the shorting angle ($\theta$) is increased for all four structures and attains maximum peak at $\theta = 100°$. Further increase of the short angle reduces the CP gain.

For further confirmation about the best possible structure the parametric studies on polarization purity have also been done and presented in Fig. 3. For the conventional structure, i.e., $\theta = 0°$ the polarization purity is 18 dB and 22 dB for FR4 substrate while it is around 26 dB and 29 dB in case of glass substrate with substrate thickness of 0.787 mm and 1.58 mm, respectively. As soon as the shorting strips have been introduced at the non-radiating sides of the patch the polarization purity

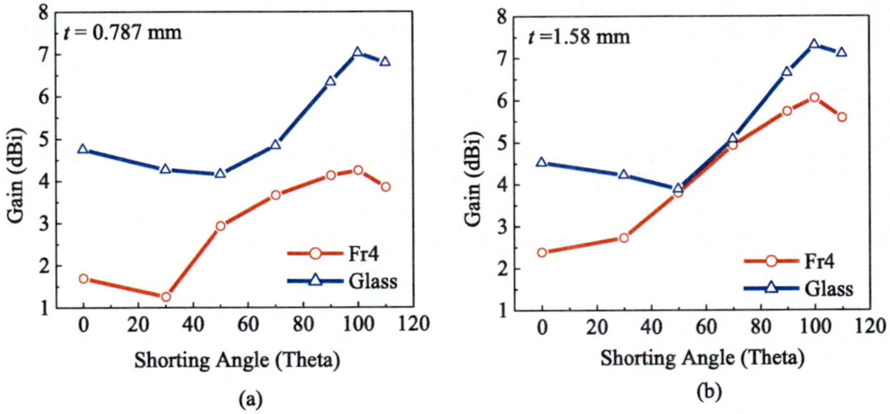


**Fig. 2** Co-polar peak gain variation with respect to the short angle for different substrate width **a** 0.787 mm **b** 1.58 mm

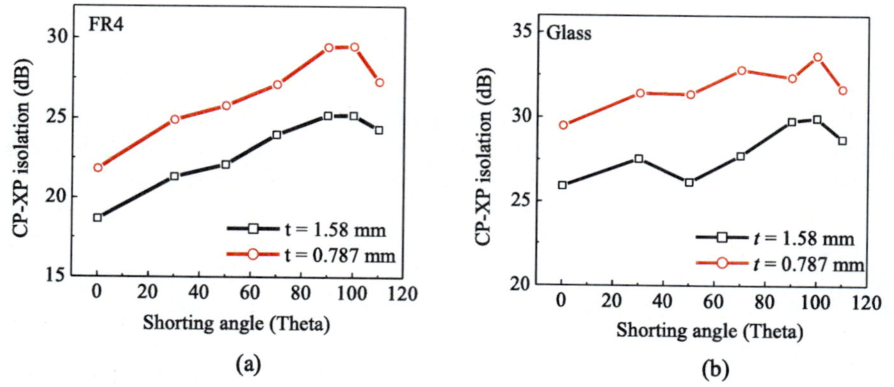


**Fig. 3** Polarization purity variation with respect to the short angle for different substrate **a** FR4 **b** glass with different substrate thickness

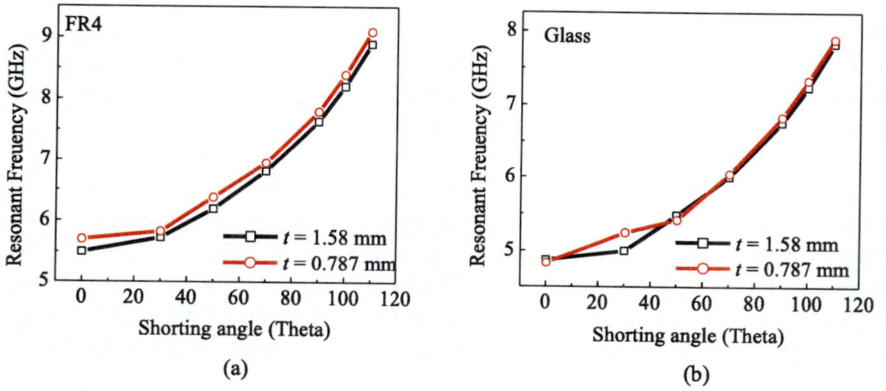


**Fig. 4** Resonant frequency variation of the proposed structure for different shorting angles ($\varphi$) **a** FR4 substrate **b** Glass substrate

starts increasing in all the structure as seen from Fig. 4. The CP-XP isolation keep on increasing as the short angle rises and attains a maximum value at $\theta = 100°$.

Figure 4 shows the resonance frequency profile of the structures under study. In all the cases for different substrate with different substrate thickness the resonance frequency increases linearly as the shorting angle increases.

## 4 Proposed Structure

At the beginning four conventional circular patch antenna of radius 7 mm has been designed on the top of FR4 ($\varepsilon_r = 4.4$) and glass ($\varepsilon_r = 5$) substrate out of which two structures are with substrate thickness of 0.787 mm and other two are with 1.58 mm.

The dimensions of the ground plane is $60 \times 60$ mm$^2$. After that a pair of grooves with thickness ($s$) 0.1 mm have been made at both non-radiating periphery of the CMA and a metal strip of same width are inserted in the grooves. In this way the final proposed structure (Fig. 1), i.e., shorted circular microstrip antenna has been designed. The investigation has been done with two different substrate thickness ($h$) of 1.58 mm and 0.787 mm. The shorting angle ($\theta$) has been varied from 30° to 110° gradually. For conventional circular patch the shorting angle has been considered as 0° as there is no short in case of conventional patch antenna.

## 5 Simulated Results

The results achieved with the optimum structure, i.e., with shorting angle ($\theta$) $100^0$ with the help of [18] is documented in this section. The $S_{11}$ profile of the conventional CMA and present optimum structure with two different substrate and thickness is shown in Fig. 5. For both the proposed structure the resonant frequency shifted toward right as compare to traditional CMA. Figure 5 also shows that all the structure is having good impedance matching.

The simulated radiation pattern of the proposed antenna ($\theta = 100^0$) with FR4 and glass substrate is shown in the Fig. 6. All the figures show improved polarization purity as compared to the conventional circular microstrip antenna. For the proposed antenna with FR4 substrate with thickness 0.787 mm and 1.58 mm a polarization purity of 27 dB is achieved (Fig. 6a and b) whereas the proposed antenna with glass substrate provides a polarization purity over 30 dB for both t = 0.787 mm and 1.58 mm (Fig. 6c and d). It may be noted that the antenna is purely linear polarized and cannot provide circular polarization. The surface current distribution of the proposed antenna is shown in Fig. 7. It confirms that the antenna is purely linear polarized with

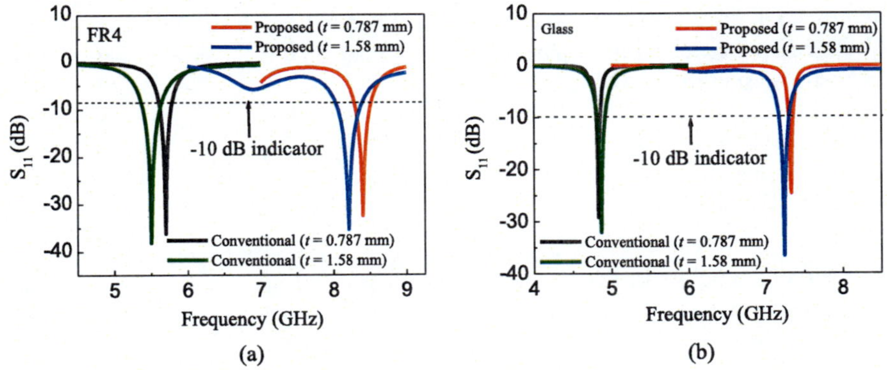


**Fig. 5** Simulated $S_{11}$ versus frequency of the CMA and present antenna with $\theta = 100°$ **a** FR4 **b** Glass substrate

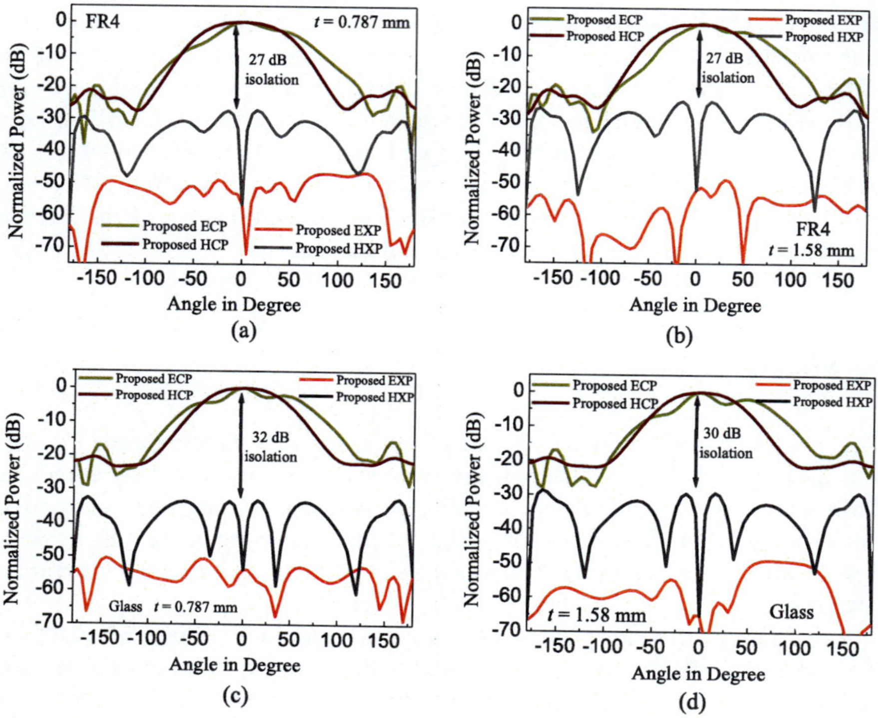


**Fig. 6** Simulated radiation pattern of the proposed antenna ($\theta = 100°$)**a** FR4 substrate with thickness 0.787 mm **b** FR4 substrate with thickness 1.58 mm **c** Glass substrate with thickness 0.787 mm **d** Glass substrate with thickness 1.58 mm

high CP-XP isolation. For all the cases the polarization purity is quite higher than the conventional circular microstrip structure.

The surface current distribution of the proposed antenna is shown in Fig. 7. It confirms that the antenna is purely linear polarized with high CP-XP isolation. For all the cases the polarization purity is quite higher than the conventional circular microstrip structure.

## 6 Conclusion

Four circular microstrip patch antennas (with shorted non-radiating sides) with different substrate (FR4 and glass) and different substrate thickness (0.787 mm and 1.575 mm) have been investigated to improve the CP-XP isolation (polarization purity) without hampering basic radiation pattern. In the present design, the patch is slightly bigger than the conventional patch. In fact, because of incorporation of shorting strips, dielectric constant of the substrate becomes modified and it eliminates surface wave. This improves overall performance of the present antenna. The

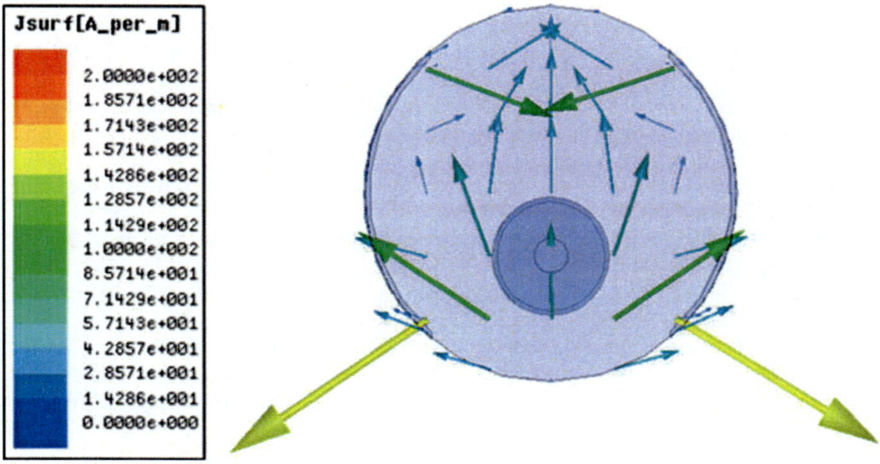


**Fig. 7** Surface current distribution over patch for $\theta = 100^0$

parametric studies over all the structures show that the proposed structures with shorting angle ($\theta$) $100^0$ provides the best results in terms of gain and polarization purity. The proposed structure ($\theta = 100^0$) with FR4 substrate provides a CP-XP isolation of 27 dB while with glass substrate the polarization purity is more than 30 dB. The polarization purity obtained from all the structures are quite higher than the conventional structures with FR4 and glass substrates.

## References


1. Garg R, Bhartia I, Ittipiboon A (2001) Microstrip antenna design handbook. Artech House, Norwood, OH, USA
2. Chattopadhyay S (Ed) (2017) Trends in research on microstrip antennas. Intech Open, Rijeka, Croatia. https://doi.org/10.5772/65580
3. Pawar UA, Chakraborty S, Sarkar T, Ghosh A, Singh LLK, Chattopadhyay S (2019) Quasi-planar composite microstrip antenna: Symmetrical flat-top radiation with high gain and low cross polarization. IEEE Access 7:68917–68929
4. Chattopadhyay S, Siddiqui JY, Guha D (2009) Rectangular microstrip patch on a composite dielectric substrate for high gain wide beam radiation patterns. IEEE Trans Antenna Prop 57(10):3324–3327
5. Guha D, Chattopadhyay S, Siddiqui JY (2010) Estimation of gain enhancement replacing PTFE by air substrate in a microstrip patch antenna. IEEE Antenna Propagation Magazine 52(3):92–95
6. Padhi SK, Karmakar NC, Law CL, Aditya S (2003) A dual polarized aperture coupled circular patch antenna using a c-shaped coupling slot. IEEE Trans Antenna Prop 51(12):3295–3298
7. Yusuf Y, Cheng H, Gong X (2013) Co-designed substrate-integrated waveguide filters with patch antennas. IET Microwaves, Antenna & Prop 7(7):493–501
8. Kojima N, Shinohara MT (2020) Integration of a via-loaded annular-ring reduced surface wave antenna and a branch-line coupler. IEEE Access 8:133645–133653

9. Rathod SM, Awale RN, Ray KP, Kakatkar SS (2019) Directivity enhancement of a circular microstrip antenna with shorting post. IETE J Res. https://doi.org/10.1080/03772063.2019.1612285
10. Ou JH, Huang J, Liu J, Tang J, Zhang XY (2020) High-gain circular patch antenna and array with introduction of multiple shorting pins. IEEE Trans Antenna Prop 68(9):6506–6515. IEEE Press
11. Xu K, Shi J (2018) High-efficiency circular dense dielectric patch antenna with frequency selectivity. Electronics Lett 54(14):861–862
12. Naik KK, Vijaya Sri PA (2018) Design of hexadecagon circular patch antenna with DGS at Ku band for satellite communications. Progress Electromagnet Res M 63:163–173
13. Khan MAK, Shaem TA, Alim MA (2020) Graphene patch antenna with different substrate shapes and materials. Optik 202. https://doi.org/10.1016/j.ijleo.2019.163700
14. Ghosh A, Ghosh SK, Ghosh D, Chattopadhyay S (2016) Improved polarization purity for circular microstrip antenna with defected patch surface. International J Microwave Wireless Technologies 8(1):89–94. Cambridge University Press and the European Microwave Association
15. Chattopadhyay S, Chakraborty S (2018) A physical insight into the influence of dominant mode of rectangular microstrip antenna on its cross-polarization characteristics and its improvement with t-shaped microstrip antenna. IEEE Access 6(1):3594–3602
16. Ghosh D, Ghosh SK, Chattopadhyay S (2014) Physical and quantitative analysis of compact rectangular microstrip antenna with shorted non radiating edges for reduced cross polarized radiation using modified cavity model. IEEE Antennas & Prop Magazine 56(4):61–72
17. Guha D, Biswas M, Antar YMM (2005) Microstrip patch antenna with defected ground structure for cross polarization suppression. IEEE Antenna Wireless Propagation Letter 4:455–458
18. HFSS: High frequency structure simulator, Ver. 14. Ansoft Corp., USA

# Modified Patch Geometry with Reduced Resonant Frequency

**Satyabrata Maiti and Khan Masood Parvez**

## 1 Introduction

The miniaturized antennas are extensively used worldwide for mobile devices, military applications, long-term evaluation (LTE), radio frequency identification (RFID), global navigation satellite systems (GNSS), etc. Due to the significant growth of wireless applications, current advancements in the modern wireless communication market to developed efficient, low-cost, light-weight, reliable antennas. Therefore, efficient miniaturized antennas are major design attractions in the current scenario to prescribe the new heights of performances of those modern wireless devices. Patch antenna is the simplest and most widely used antenna. These antennas are conducting plate printed on top of the grounded dielectric substrate, and a 50 Ω microstrip line or coaxial feed is used to excite the conducting plate or patch. The patch can be planar rectangular, circular, triangular, or any other geometrical shape. And implementation of miniaturization techniques on patch antennas is the most suitable option to fulfill the requirement due to its low profile, simple structure, and easy of fabrication and integrate facility with portable devices. A novel miniaturized techniques are implemented on microstrip patch antenna using reactive impedance surface (RIS) [1]. In addition, miniaturized antenna also improves the radiation characteristics as well as the bandwidth comparison with conventional patch antenna. Moreover, patch antenna for tunable miniaturization factor has been reported with poor efficiency in [2]. Analysis of patch antenna pasted on magneto-dielectric embedded circuit metasubstrate has been described in [3]. Furthermore, hybrid optimization technique [4] and metasurface [5] has been described for a significant size reduction. Proximity effect on antenna miniaturization has been shown in [6]. In [7], high-contrast, low-loss ceramic materials can be utilized for patch antenna miniaturization. Another

S. Maiti · K. M. Parvez (✉)
Antenna Research Laboratory, Department of Electronics & Communication Engineering, Aliah University, Newtown, Kolkata 700160, India
e-mail: masoodrph@gmail.com

 
P. H. J. Chong et al. (eds.), *Emerging Electronics and Automation*, Lecture Notes in Electrical Engineering 937, https://doi.org/10.1007/978-981-19-4300-3_22

miniaturization technique using a metamaterial-inspired is reported in [8]. Many different techniques for microstrip antenna miniaturization has been illustrated in [9]. Recent developments in patch antenna miniaturization using metamaterial unit cell [10] and inter-embedded metasurface structure [11] are reported, indicating the miniaturization toward the higher degree. Henceforth, several useful techniques for slot antenna miniaturization have been described in [12–16]. A compact rectangular patch antenna loaded with a half- U-slot and a rectangular slot on the edges of the patch for broadband applications has been shown in [17]. Due to the two resonant slots on air, substrate, the antenna gives more bandwidth compared with conventional patch. The comparative study on the effect of different slot shape on size reduction of triangular patch antenna has been investigated in [18]. Compact microstrip patch antenna with T shaped slits antenna has been shown in [19] for portable GPS handsets. The bandwidth enhancement technique has been shown in [20] on a dielectric substrate. It is claimed in [21] that shorting pin in reading patch is also an effective approach for miniaturization.

However, miniaturization techniques are well established concept for last two decades. Therefore, it is imperative to further explore the techniques for realizing higher degree of miniaturization with good antenna performances. This higher degree of miniaturized antenna will allow for integration with a package or platform of small wireless devices that are emerging. It is known that resonance of any antenna is directly linked with the antenna size. Any effort to reduce the size of the antenna may have a negative impact on its return loss, radiation pattern, and efficiency in such a way that it is unable to adequately radiate. Then it would be difficult work of antenna engineers to design, a higher degree of miniaturization with minimum effect on antenna performances. In this paper, a rectangular microstrip patch antenna has been designed and fabricated on FR glass epoxy substrate for resonate at 2.4 GHz. Then the rectangular patch is modified by removing an arc on the edges of that patch antenna. For this modification proposed antenna resonate at 37.22% reduced resonate frequency of 1.45 GHz. After that, spiral slot is integrated with modified patch and as a results resonant frequency reduced 48.48% compared with normal patch antenna keeping the overall dimension unchanged. The paper has been prepared in the following way.

- Design and analysis of a simple rectangular microstrip antenna.
- Modification of a rectangular microstrip patch by removing the arc on the edges of the patch.
- Further modification of rectangular microstip patch with a spiral slot.

## 2 Design and Analysis of Simple Rectangular Patch with Arc

It is well known that low profile rectangular fed microstrip patch antenna has attractive advantages of extremely high frequency, good impedance matching, ease of fabrication and capability for full integration with active components. The patch antenna is a very useful candidate for mobile, satellite communication systems. One of the most effective techniques for of size reduction is to introduce the shorting pins on the edge of the patch. In this way, the excited surface current paths of the patch are increased and it causes significant reduction in resonant frequency compared with a conventional patch antenna. Generally, metalized via holes is used to realize the shorting pin very accurately. Another effective approach is to load with three dimensional geometry like cavities in the substrate or folded and double folded structures. The simple microstrip patch configuration is shown in Fig. 1. The rectangular microstrip patch is connected with 50Ω microstrip fed line on the top surface of the substrate and the ground is printed on the bottom surface. The thickness of dielectric material is 1.58 mm. This antenna is considered as a reference antenna to better understand the reduction of resonant frequency. The antenna dimensions, namely, $L_S$, $W_S$, $L$, $W$, $M_L$ $M_W$, $W_{\text{ML}}$ listed in Table 1. The FR _4 glass epoxy substrate is used to increase the robustness of the antenna in adverse environmental condition compared with the air substrate used in [17, 18]. The permittivity of FR_4 glass epoxy is 4.4. Ansys High frequency structure simulator [22] based on the FEM method (finite element method), is used to evaluate the simulative response of reference and proposed antennas.

The rectangular patch antenna is modified by the arc which is illustrated in Fig. 2. The arc is removed from the non-radiating edges of the patch. So that the fringing fields are not disturbed and can effectively contribute to radiation. We are trying to

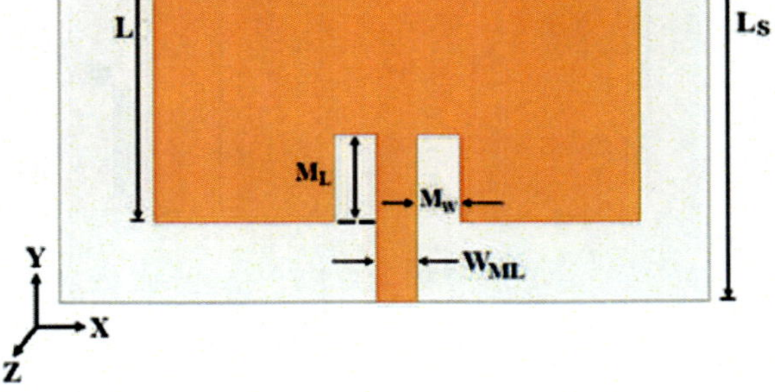


**Fig. 1** Geometry of simple rectangular microstrip antenna

**Table 1** Dimensions of antenna geometry (mm)

| $L_s$ | $W_s$ | $L$ | $W$ | $W_{ML}$ | $M_W$ | $M_L$ |
|---|---|---|---|---|---|---|
| 51 | 60 | 28 | 36 | 2.95 | 3.20 | 6.44 |

integrate the arc in radiating patch in such a way that does not require an additional patch or coupling strip to reduce the size of the antenna. The dimensions and dielectric properties are same as the simple rectangular patch antenna shown in Fig. 1.The main motivation behind the modification is to archive lower resonance frequency comparison with conventional patch. The radius (*R*) of arc is 10 mm. The minor arc angle is 125°. The center of patch is 2.80 mm far away from the arc in *X* axis. The simulated and measured return loss of rectangular microstrip antenna presented in Fig. 3. The simulated value is 2.31 GHz where as measured value is 2.40 GHz. This is the fundamental resonant frequency of rectangular microstrip antenna. Now, we try to analysis the effect of arc on radiating patch keeping the antenna design parameter and dielectric properties unchanged.

It is clearly shown from Fig. 3 that due to the modification of microstrip patch, the return loss characteristics have shifted from 2.31 GHz to 1.45 GHz. The percentage of reduced resonant frequency is 37.22. The measured resonance frequency for proposed antenna is 1.49 GHz. From the curve, it is observed that measurement results have excellent agreement with simulation ones. The small discrepancy between the measurement results can be related to lossey FR_4 dielectric materials and addition ohmic loss normally generated while soldering the SMA connector on fabrication. Due to the introduction of an arc on the rectangular patch, the area of effective radiating surface is rearranged comparison with the conventional patch. Hence, the

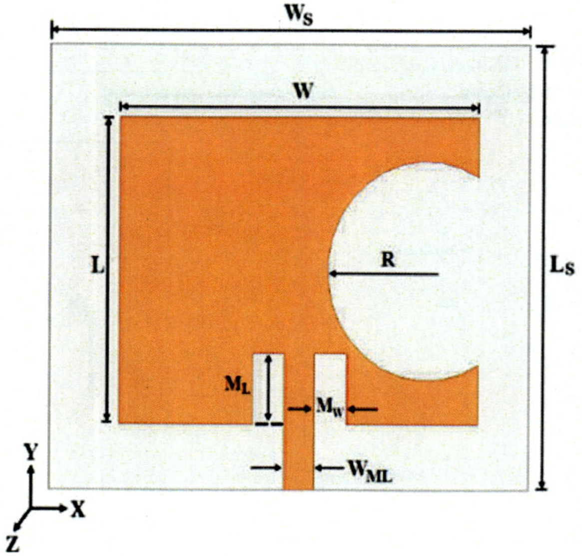


**Fig. 2** Patch antenna with removing the arc on the edges of the patch

Fig. 3 Return loss of simple patch antenna and modified with arc

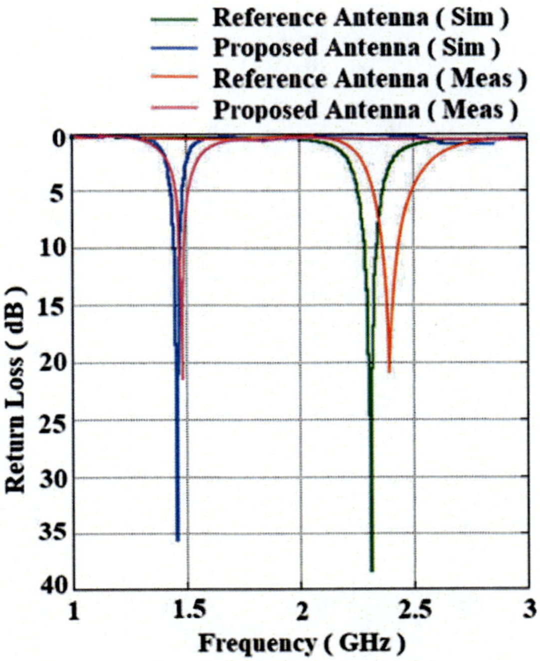


excited surface current path is made longer and therefore the significant reduction in resonant frequency. It is known to all that antenna is the largest component of any wireless system.

Therefore, it is imperative to give emphasis on simple and easy to implement miniaturization techniques. It is also arguable that the proposed antenna is very simple and easy to implement. The radiation pattern is defined as a graphical illustration of radiation characteristics of proposed antenna in a space coordinate system. Simulated H- and E-plane radiation pattern at 1.45 GHz resonant frequency is shown in Fig. 4a and b. The measurement results are also included to validate the simulation outcome. The measured Co-pol for further modified patch has shown very good agreement with simulation results. The current distribution at 1.45 GHz is shown in Fig. 5.The path length of surface current is increased due to introduction of arc on patch surface. The longer path length on the patch surface causes a significant reduction in resonant frequency compared with fundamental resonance frequency of 2.31 GHz.

## 3 Design and Analysis of Simple Rectangular Patch with Arc and Rectangular Spiral Slot

To achieve further reduction in resonant frequency, we have introduce rectangular spiral slot on the modified rectangular microstrip patch keeping the dimensions and

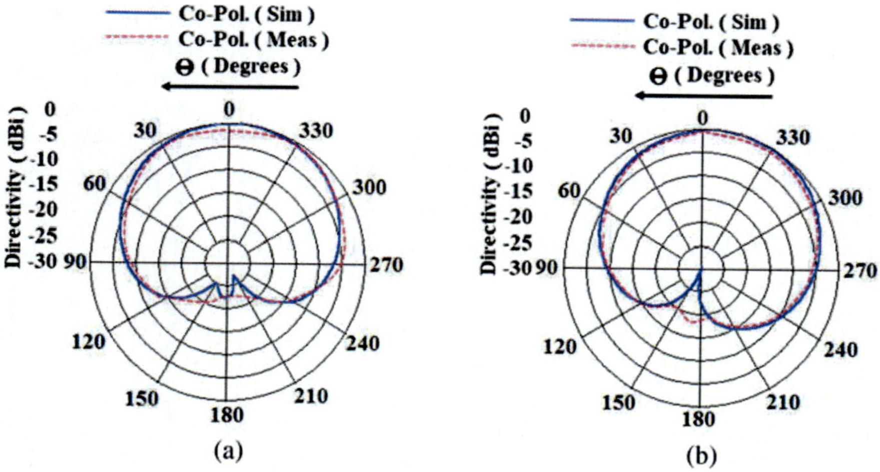


**Fig. 4** **a** H-plane radiation pattern **b** E-plane radiation pattern for modified rectangular patch antenna with arc

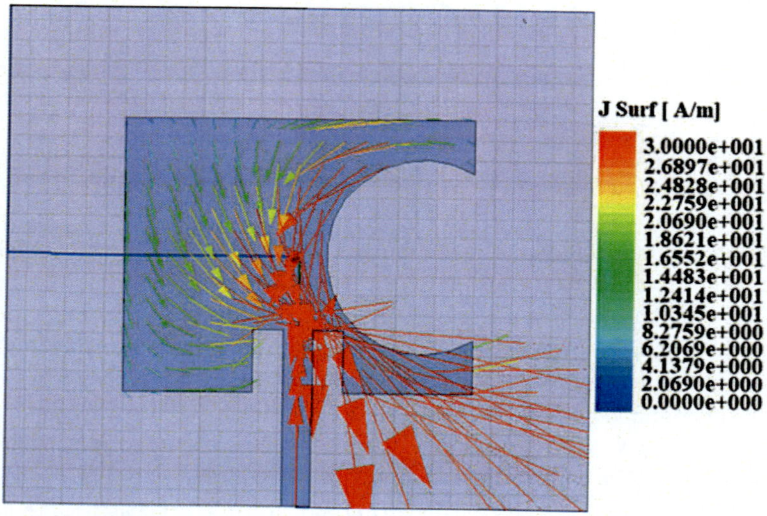


**Fig. 5** Current distribution of modified with arc at resonance frequency 1.45 GHz

dielectric properties are unchanged. The proposed antenna configuration has been shown in Fig. 6a. The spiral slot configuration shown in Fig. 6b. The rectangular spiral slot values are shown in Table 2. The rectangular split ring is 0.40 mm always in *X* axis and 1.40 mm away in *Y* axis from the edges of the patch. The simulated return loss (S11 parameter) of the further modified patch antenna comparison with simple microstrip antenna is shown in Fig. 7.

**Fig. 6** Further modified rectangular microstrip patch antenna with spiral slot

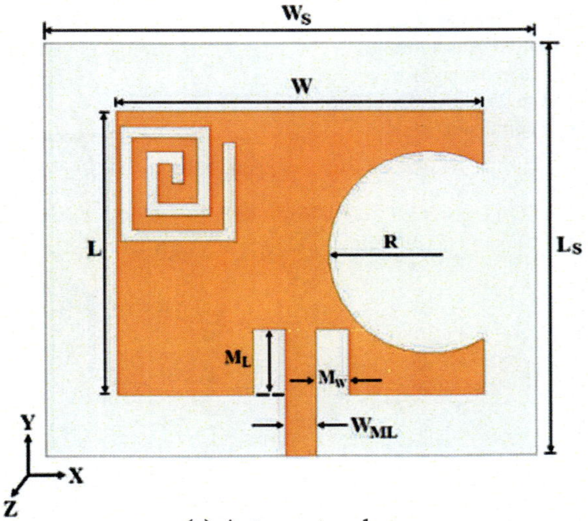


(a) Antenna topology

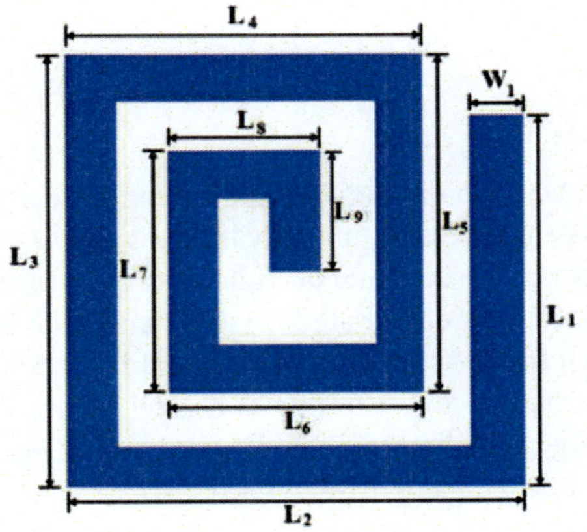


(b) Dimession of sprial slot

**Table 2** Dimensions of rectangular spiral slot (mm)

| $L_1$ | $L_2$ | $L_3$ | $L_4$ | $L_5$ | $L_6$ | $L_7$ | $L_8$ | $L_9$ | $W_1$ |
|---|---|---|---|---|---|---|---|---|---|
| 9.50 | 11.10 | 11.10 | 8.70 | 8.70 | 6.20 | 6.20 | 3.70 | 3.10 | 1.20 |

Antenna dimensions are key parameters for control over the resonant frequency. It is one of the challenging tasks to further reduce the resonant frequency in higher degree while keeping the antenna dimension unchanged. So we also need to be careful about the antenna performance parameters like return loss, radiation pattern, current distribution, etc.

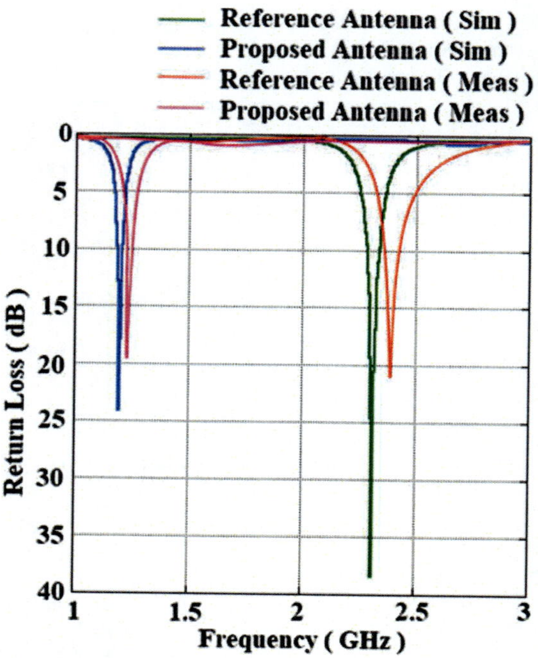


**Fig. 7** Simulated and measured return loss for simple patch and further modified rectangular patch with spiral slot

As shown in the Fig. 7, good agreement can be seen between measurement and simulation prediction, not only in terms of resonance but also with respect to miniaturization. The simple rectangular microstrip antenna and further modified simulated resonance frequencies are 2.31 GHz and 1.19 GHz, respectively. The reduction of resonant frequency is 48.48% compared with normal patch geometry. It is important to note that the higher degree of reduction can be achieved by simple modifying the radiating patch. The measured resonant frequency is 1.22 GHz. The resonant dip of simulated frequency is −24.03 dB whereas the measured dip is only −18.81 dB. The 10 dB bandwidth for modified patch antenna geometry is 1.68%. The proposed antenna produced better miniaturization comparison with [18, 19]. The simulated radiation pattern for Fig. 6 at resonant frequency 1.19 GHz is illustrated in Fig. 8. The corresponding measured value is also depicted in Fig. 8 and the measured results agree well with the simulated ones. The current distribution of proposed antenna at 1.19 GHz is shown in Fig. 9.

After the analysis of the performance of the further modified patch, we are able to achieve 48.48% reduced resonant frequency comparison with 2.31 GHz fundamental resonant frequency. The return loss and radiation pattern have no negative impact on higher degree miniaturization techniques.

It can be noted that the antennas are realized the highest degree of miniaturization in a very simple methodology that is easy to implement in any modern wireless device like cell phones and GPS systems. The fabricated prototypes are illustrated in Fig. 10.

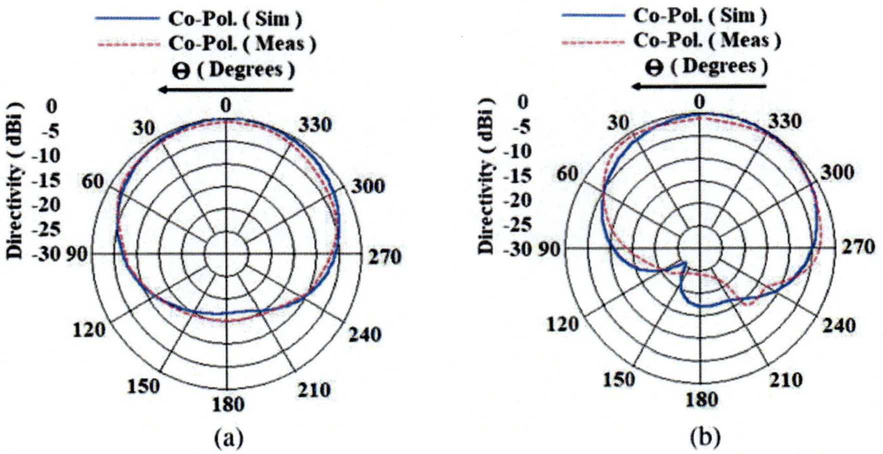


Fig. 8 a H-plane radiation pattern b E-plane radiation pattern for further modified rectangular patch antenna with split ring at resonance

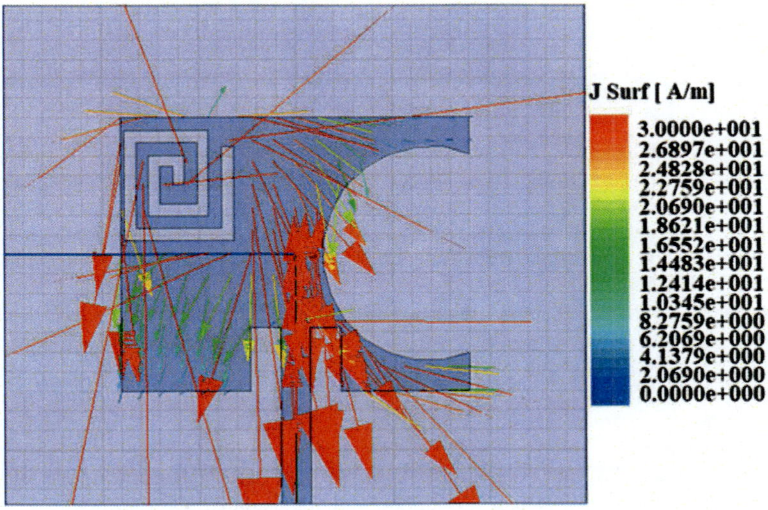


Fig. 9 Current distribution of modified with arc at resonance frequency 1.19 GHz

## 4 Conclusions

The rectangular microstrip antennas are designed to resonate at the fundamental resonance frequency. Then we modified the patch to achieve reduced resonant frequency, keeping over dimensions unchanged. In this case, the modified patch antenna with arc produced a 37.22% reduced resonant frequency comparison with fundament resonance frequency. Then we further modified the patch using the rectangular spiral slot

(a) (b)

**Fig. 10** Fabricated photography of the proposed antennas **a** Patch antenna modified with arc **b** Further modified patch antenna with rectangular spiral slot

to achieve higher degree of reduction of resonant frequency. In this case, the proposed antenna resonates at a reduced resonant frequency of 48.48%. The proposed antenna is suitable for wireless communication system.

## References


1. Mosallaei H, Sarabandi K (2004) Antenna miniaturization and bandwidth enhancement using a reactive impedance substrate. IEEE Trans Antennas Propag 52:2403–2414. https://doi.org/10.1109/TAP.2004.834135
2. Buell K, Mosallaei H, Sarabandi K (2006) A substrate for small patch antennas providing tunable miniaturization factors. IEEE Trans Microw Theory Tech 54:135–145. https://doi.org/10.1109/TMTT.2005.860329
3. Mosallaei H, Sarabandi K (2007) Design and modeling of patch antenna printed on magneto-dielectric embedded-circuit metasubstrate. IEEE Trans Antennas Propag 55:45–52. https://doi.org/10.1109/TAP.2006.886566
4. Zhu SH, Yang XS, Wang J, Wang BZ (2020) Miniaturization of patch antenna based on hybrid topology optimization. Int J RF Microw Comput Eng 30:e22308. https://doi.org/10.1002/MMCE.22308
5. Zhu HL, Cheung SW, Yuk TI (2015) Miniaturization of patch antenna using metasurface. Microw Opt Technol Lett 57:2050–2056. https://doi.org/10.1002/MOP.29275
6. Jarchi S, Rashed-Mohassel J, Faraji-Dana R (2013) Proximity effects of a layered periodic structure on miniaturization of patch antennas. Int J RF Microw Comput Eng 23:549–558. https://doi.org/10.1002/MMCE.20689
7. Kula JS, Psychoudakis D, Liao WJ, Chen CC, Volakis JL, Halloran JW (2006) Patch-antenna miniaturization using recently available ceramic substrates. IEEE Antennas Propag Mag 48:13–20. https://doi.org/10.1109/MAP.2006.323335
8. Ouedraogo RO, Rothwell EJ, Diaz AR, Fuchi K, Temme A (2012) Miniaturization of patch antennas using a metamaterial-inspired technique. IEEE Trans Antennas Propag 60:2175–2182. https://doi.org/10.1109/TAP.2012.2189699
9. Khan MU, Sharawi MS, Mittra R (2015) Microstrip patch antenna miniaturisation techniques: A review. IET Microwaves Antennas Propag 9:913–922. https://doi.org/10.1049/iet-map.2014.0602

10. Singh AK, Abegaonkar MP, Koul SK (2017) Highly miniaturized dual band patch antenna loaded with metamaterial unit cell. Microw Opt Technol Lett 59:2027–2033. https://doi.org/10.1002/MOP.30683
11. Chen D, Yang W, Xue Q, Che W (2021) Miniaturized wideband planar antenna using interembedded metasurface structure. IEEE Trans Antennas Propag 69:3021–3026. https://doi.org/10.1109/TAP.2020.3028245
12. Haque SKM, Parvez KM (2017) Slot antenna miniaturization using slit, strip, and loop loading techniques. IEEE Trans Antennas Propag 65:2215–2221. https://doi.org/10.1109/TAP.2017.2684191
13. Mitra D, Ghosh B, Sarkhel A, Chaudhuri SRB (2016) A miniaturized ring slot antenna design with enhanced radiation characteristics. IEEE Trans Antennas Propag 64:300–305. https://doi.org/10.1109/TAP.2015.2496628
14. Ghosh B, Haque SM, Mitra D, Ghosh S (2010) A loop loading technique for the miniaturization of non-planar and planar antennas. IEEE Trans Antennas Propag 58:2116–2121. https://doi.org/10.1109/TAP.2010.2046842
15. Ghosh B, Moinul Haque SK, Mitra D (2011) Miniaturization of slot antennas using slit and strip loading. IEEE Trans Antennas Propag 59:3922–3927. https://doi.org/10.1109/TAP.2011.2163754
16. Parvez KM, Haque SM (2021) Bandwidth enhancement and cross-polarization suppression of slot antenna. Electromagnetics 41:119–130. https://doi.org/10.1080/02726343.2021.1879358
17. Deshmukh AA, Ray KP (2009) Compact broadband slotted rectangular microstrip antenna. IEEE Antennas Wirel Propag Lett 8:1410–1413. https://doi.org/10.1109/LAWP.2010.2040061
18. Dasgupta S, Gupta B, Saha H (2014) Compact equilateral triangular patch antenna with slot loading. Microw Opt Technol Lett 56:268–274. https://doi.org/10.1002/mop.28073
19. Won C, Kim YD, Lee HM (2003) A compact micro strip patch antenna with T-Shaped slits for portable GPS handsets. ISAPE 2003—2003 6th Int Symp Antennas, Propag EM Theory, Proc, 335–338. https://doi.org/10.1109/ISAPE.2003.1276696
20. Sze JY, Wong KL (2000) Slotted rectangular microstrip antenna for bandwidth enhancement. IEEE Trans Antennas Propag 48:1149–1152. https://doi.org/10.1109/8.884481
21. Shackelford AK, Lee KF, Luk KM, Chair RC (2001) U-slot patch antenna with shorting pin. Electron Lett 37:729–730
22. Ansys HFSS Ver. 19.2

# Ultra-Wideband Inverted T-Shape Slot Antenna

**Sayanti Sinha, Khan Masood Parvez, and SK. Moinul Haque**

## 1 Introduction

Recently, low profile microstrip antennas are highly attractive for advanced communication systems due to their planar geometry and easily integrate facility. Actually, the rapid growth of 5th generation mobile communication systems requires wideband, ultra-wideband, and multiband antennas to cover mobile and wireless services and to minimize the system complexity as well as the overall dimensions and costs. Therefore, a great interest among the researcher community on microstrip line fed / coplanar waveguide (CPW) slot antennas has been observed in [1–7]. Simple rectangular slot antenna is considered as reference, and miniaturization techniques are applied to produce lower frequency. These antennas produce narrow bandwidth. In [8], a compact slot antennas for dual band and wide band frequency operation is described. The 10 dB bandwidth for 900 MHz and 1800 MHz are 28% and 31%, respectively. Literature on wideband antenna is noted for 4G application in [9]. Slot antenna topology based on fictitious short-circuit model for wide band frequency operation has been reported in [10]. Dual-polarized slot-coupled stacked patch antenna for wideband application is presented in [11]. A miniaturized antenna for wideband applications is illustrated in [12].

Based on the above analysis, an inverted T-shaped low profile and easily implementable slot antenna is proposed in this communication for ultra-wideband frequency operations. A 50 Ω microstrip line fed on bottom surface is used to excite an inverted T-shaped geometry etched in the middle of the top surface of the dielectric substrate. The detailed simulation on proposed antenna is conducted by finite element method (FEM) to understand its behavior and optimize for ultra-wideband operation. Return loss, radiation characteristics and efficiency are also analyzed and verified

S. Sinha · K. M. Parvez (✉) · SK. Moinul Haque
Antenna Research Laboratory, Department of Electronics and Communication Engineering, Aliah University, Newtown, Kolkata 700160, India
e-mail: masoodrph@gmail.com



P. H. J. Chong et al. (eds.), *Emerging Electronics and Automation*, Lecture Notes in Electrical Engineering 937, https://doi.org/10.1007/978-981-19-4300-3_23

with measurement results. The calculated operating frequency of this topology is 3.22 GHz, whereas the measured value is 3.10 GHz. The 10 dB bandwidth is 52.92% covering from 2.89 to 4.97 GHz (2.08 GHz). The inverted T-shaped microstrip fed slot antenna is highly efficient and suitable for ultra-wideband applications.

## 2 Antenna Design

The design procedure for the proposed inverted T-shaped microstrip fed slot antenna architecture is detailed here. In Fig. 1, the top and side views of the ultra-wideband antenna configuration are delineated. Slot antennas are broadly used in modern communication scheme because of their attractive characteristics like as low cost light weight, compactness, and simplicity of integration with other equipment. When compared to coaxial and Coplanar Waveguide (CPW) fed, microstrip fed gives higher stability and impedance matching capabilities. The proposed topology is fabricated on FR_4 glass epoxy dielectric material which is inexpensive and easily available and has $\varepsilon_r$ (relative permittivity) = 4.4 and tanδ (loss tangent) = 0.002. The antenna ground plane size is 100 × 100 mm$^2$ ($L_S \times W_S$) with dielectric substrate height of 1.60 mm. As shown in Fig. 1, an inverted T-shaped copper is etched in middle of the top surface of FR_4 glass epoxy dielectric material. The top surface is also known as ground plane.

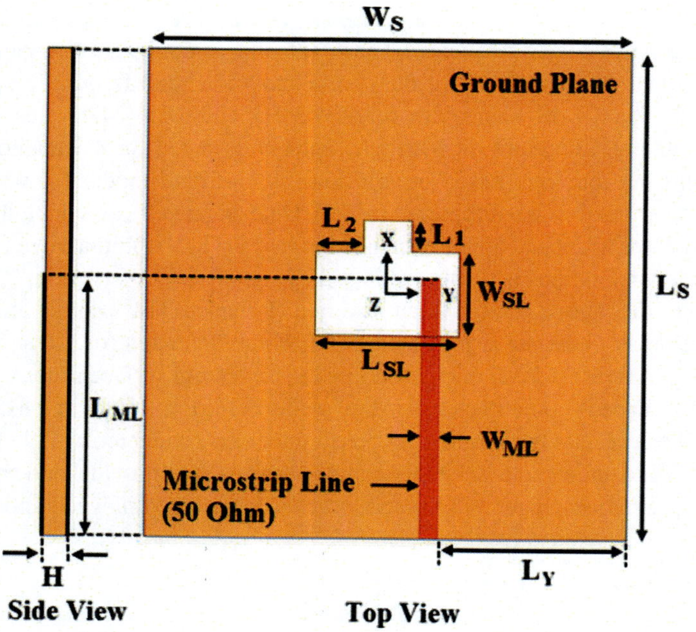


Fig. 1 Inverted T-shaped microstrip fed slot antenna geometry

**Fig. 2** Fabricated prototype of proposed inverted T-shaped microstrip antenna

The slot width ($W_{SL}$) and slot length ($L_{SL}$) are 17.20 mm and 30 mm, respectively. The antennas others parameters: $L_1 = 6.4$ mm and $L_2 = 10$ mm. To excite the inverted T-shaped slot on the top surface of FR_4 glass epoxy dielectric substrate, 50 Ω microstrip line fed on bottom surface with a length ($L_{ML}$) of 52.2 mm and width ($W_{ML}$) of 2.95 mm is applied. The fed line width of 2.95 mm is chosen such that the input impedance matches to 50 Ω of the SMA probe. The microstrip line is placed ($L_Y$) 39.20 mm away as displayed in Fig. 1. The Fig. 2 depicts the fabricated prototype of proposed inverted T-shaped microstrip fed slot antenna. The topology is designed and scrutinized using the FEM-based Ansys HFSS 19.2 simulation code [13].

## 3 Results and Discussion

In Fig. 3, the return loss ($S_{11}$) versus frequency response of the inverted T-shaped microstrip fed slot antenna topology has been presented. The measured results are used to verify the simulation response. The simulated operating frequency of this antenna (Fig. 2) is 3.22 GHz, whereas the measured frequency is 3.10 GHz. The finite conductivity of copper (ground plane), loss of FR_4 glass epoxy dielectric and surface wave all contribute to the mismatches between HFSS calculated and measured results.

The measured dip is −26.55 dB only, whereas the calculated operating frequency dip is −34.65 dB. The bandwidth (10 dB) for antenna topology is 52.92% covering from 2.89 GHz to 4.97 GHz. The bandwidth is 2.08 GHz.

In comparison to the loop loaded slot antenna [2], the inverted T-shaped slot microstrip fed antenna has a superior 10 dB bandwidth. The HFSS calculated

**Fig. 3** Measured and HFSS calculated $S_{11}$ characteristics of inverted T-shape microstrip fed slot antenna

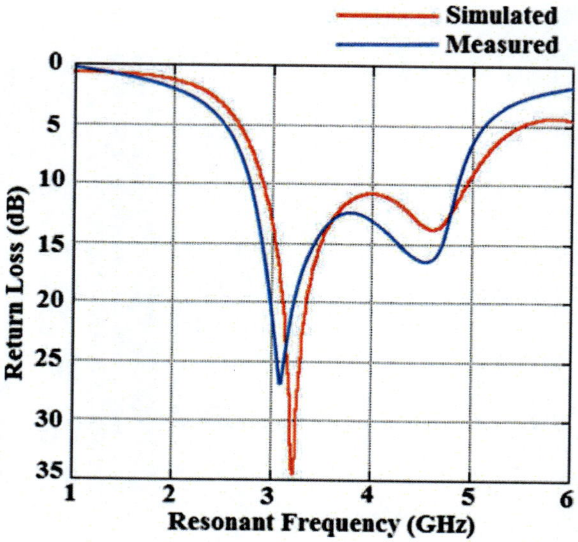


impedance characteristic of proposed antenna is presented in Fig. 4. At frequency 3.22 GHz, the input resistance is 50 Ω, which is the real part of the impedance plot, and the imaginary part reactance is on the zero axis. It is noted that the real part resistance indicates power that is either emitted or absorbed within the antenna. And imaginary part reactance signifies non-radiated power that is kept in near field region.

In spherical coordinate system, the E-plane is clarified as $\varphi = 0$ and H-plane is clarified as $\varphi = 90$. At the operating frequency of 3.22 GHz, the simulated E- and

**Fig. 4** Input impedance plot of proposed inverted T-shaped slot antenna

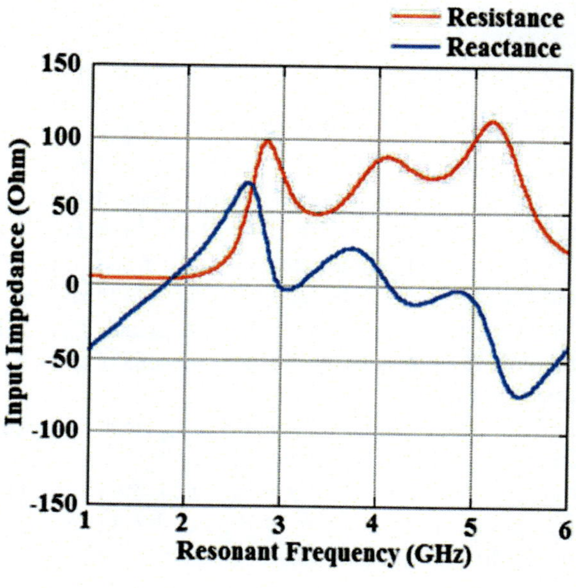

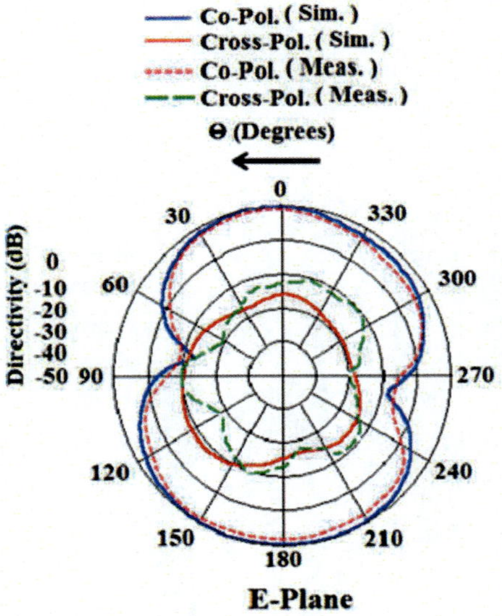


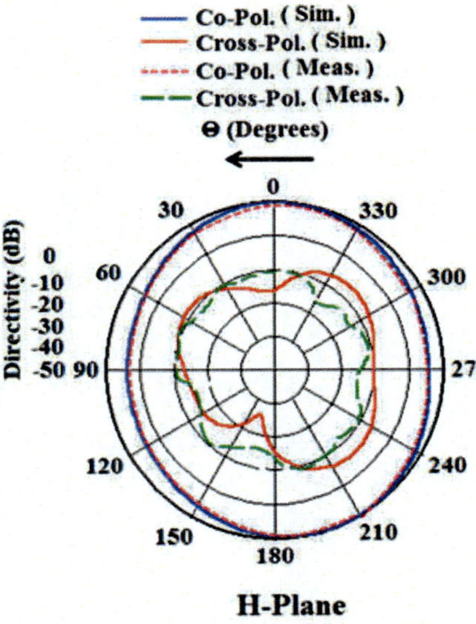


**Fig. 5** Measured and HFSS calculated radiation pattern of proposed inverted T-shaped microstrip fed slot antenna at 3.22 GHz operating frequency **a** E-Plane, **b** H-Plane

H-plane are illustrated in Fig. 5a and b. The measured radiation patterns are also includes with the simulation to verify the results and the measured results closely followed the HFSS calculated ones. The patterns show that the margin between co-polarization and cross-polarization is less than what is considered acceptable.

The distribution of surface current of inverted T-shaped slot antenna at 3.22 GHz has exhibited in Fig. 6. The slot has more current concentration on the positive Y axis as delineated in Fig. 1 due the presence of microstrip fed on the bottom surface of FR_4 glass epoxy dielectric material. The proposed inverted T-shaped slot antenna is highly efficient and its simulated efficiency is 97.66%. The corresponding measured [14] value is 95.38%. This antenna is working in between 3.1 and 10.6 GHz unlicensed spectrum assigned by U.S. FCC [15] for wide band applications.

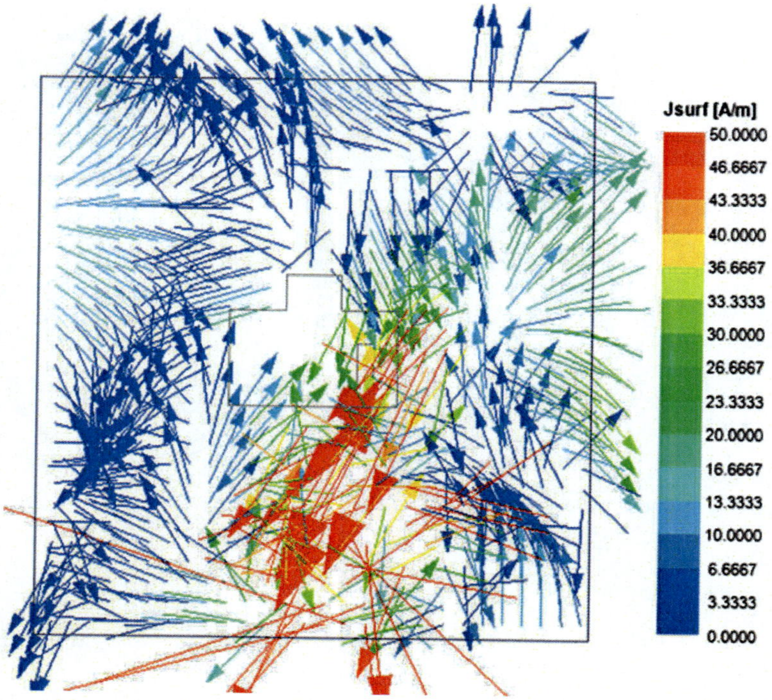


**Fig. 6** Distribution of surface current of inverted T-shaped microstrip fed slot antenna at operating frequency 3.22 GHz

## 4 Conclusions

In this paper, we have present an inverted T-shaped low profile and easily implementable slot antenna. The simulated operating frequency of this antenna is 3.22 GHz whereas measured frequency is 3.10 GHz. The bandwidth is 52.92% covering from 2.89 GHz to 4.97 GHz. At operating frequency, the separation between co-polarization and cross-polarization of radiation patterns is also less than the accepted value. Ultra-wideband applications system can benefit from the proposed inverted T-shaped microstrip fed slot antenna.

## References


1. Haque SKM, Parvez KM (2017) Slot antenna miniaturization using slit, strip, and loop loading techniques. IEEE Trans Antennas Propag 65:2215–2221
2. Parvez KM, Haque SM (2021) Bandwidth enhancement and cross-polarization suppression of slot antenna. Electromagnetics 41:119–130

3. Khan E, Parvez KM, Haque SKM (2018) Copper coin loaded miniaturized slot antenna. In: 2018 International Conference on Signal Processing and Communications (SPCOM), pp 148–151, Bangalore, India
4. Parvez KM, Sinha S, Haque SKM (2018) Miniaturization of slot antenna using meander slits. In: 2018 IEEE 88th Vehicular Technology Conference (VTC-Fall), pp 1–5, Chicago, USA
5. Ghosh B, Moinul SK, Mitra D (2011) Miniaturization of slot antennas using slit and strip loading. IEEE Trans Antennas Propag 59:3922–3927
6. Azadegan R, Sarabandi K (2003) A novel approach for miniaturization of slot antennas. IEEE Trans Antennas Propag 51:421–429
7. Ghosh B, Haque SM, Mitra D, Ghosh S (2010) A loop loading technique for the miniaturi-zation of non-planar and planar antennas. IEEE Trans Antennas Propag 58:2116–2121
8. Liu Y, Shen Z, Law CL (2006) A compact dual-band cavity-backed slot antenna. IEEE Antennas Wirel Propag Lett 5:4–6
9. Chagharvand S, Hamid MR, Kamarudin MR, Khalily M (2014) Wideband slot antenna for 4G applications. In: 2014 IEEE Asia-Pacific Conference on Applied Electromagnetics (APACE), pp 279–281, Johor Bahru, Malaysia
10. Behdad N, Sarabandi K (2005) A wide-band slot antenna design employing a fictitious short circuit concept. IEEE Trans Antennas Propag 53:475–482
11. Serra AA, Nepa P, Manara G, Tribellini G, Cioci S (2007) A wide-band dual-polarized stacked patch antenna. IEEE Antennas Wirel Propag Lett 6:141–143
12. Thomas KG, Lenin N, Sreenivasan M (2006) Wide-band dual sleeve antenna. IEEE Trans Antennas Propag 54:1034–1037
13. Ansys HFSS ver 19.2, Ansys Corp., Pittsburgh, PA, USA, 2018
14. Pozar DM, Kaufman B (1988) Comparison of three methods for the measurement of printed antenna efficiency. IEEE Trans Antennas Propag 36:136–139
15. Ma TG, Jeng SK (2005) Planar miniature tapered-slot-fed annular slot antennas for ultra-wideband radios. IEEE Trans Antennas Propag 53:1194–1202

# Fusion Rule Optimisation for Energy Efficient Cluster-Based Cooperative Spectrum Sensing

**Girraj Sharma, Vivek Upadhyaya, Ashish Kumar, Sandeep Vyas, and Ritu Sharma**

## 1 Introduction

Spectrum sensing is used to identify licenced users, also known as primary users (PUs). In CR, the efficacy of non-cooperative spectrum sensing is severely hampered by shadowing and concealed terminal issues. CSS can solve these issues, but it adds to the system's detection time and energy, reducing the total EE, particularly when the SUs are powered by batteries. Energy efficiency is calculated by dividing average throughput by total energy consumption (EE). It is a complete measure for evaluating system performance since it considers detection accuracy, energy consumption, and system efficiency all at once. Authors optimise the EE in certain works by employing the concept of total energy consumption minimisation. The authors proposed a censorship strategy in which SUs with reliable PU presence information should communicate it to the base station [1-8]. This will reduce the amount of energy used to transmit data from SUs to base stations. To enhance throughput [9] computes the best sensing time and fusion rule. It's a more realistic situation that considers PU traffic.

G. Sharma (✉) · A. Kumar · S. Vyas
Jaipur Engineering College and Research Center, Jaipur, India
e-mail: 2014rec9529@mnit.ac.in

V. Upadhyaya · R. Sharma
Malaviya National Institute of Technology, Jaipur, India

A. Kumar
Department of Electronics Engineering, Banasthali Vidyapith, Tonk, India

V. Upadhyaya
Department of Electronics Engineering, Poornima University, Jaipur, India



P. H. J. Chong et al. (eds.), *Emerging Electronics and Automation*, Lecture Notes in Electrical Engineering 937, https://doi.org/10.1007/978-981-19-4300-3_24

## 2 Related Work

EE has already been utilised as an output metric in some studies [11] discusses the trade-offs between EE and spectrum efficiency, and EE and SE are optimised by combining detector threshold modification and sensing duration. The work in [11] does not say whether the best settings meet the false alarm and detection accuracy requirements. Gao et al. [12] recommends that the sensing duration be optimised via AND fusion. Fusion, on the other hand, has a low possibility of detection. To enhance the length and transmission capacity of sensing, the authors introduced the dinkelbach technique in [13]. The double detector threshold approach is used to develop a soft fusion technique in this article. Wu and Tsang [14] describes the relationship between transmission power and sensing time. The sensing time under various situations is used to improve EE for a single CR. To increase EE, Shi et al. [15] recommends combining sensing and transmission time optimisation, whilst Zhang et al. [16] recommends combining sensing time and transmission power optimisation. Qiu et al. [17] describes the mixed spectrum syndrome. The best transmission capacity is estimated using the golden search technique to maximise EE. Wang et al. [18] proposes an OFDM-based CR system with transmission power limits and a maximum amount of interruption. The optimal power allocation is obtained using the golden search technique in this research. The optimum power allocation that optimises the EE is computed using an iterative approach in [19]. None of the research in [17–23] found a means to determine the appropriate fusion rule threshold for the Energy Efficient CSS.

The impact of changing the intra and inter fusion rule thresholds on EE is investigated in this paper. To obtain the best settings, a bisection method was applied. For a cluster-based CSS, an approach has been described for discovering the intra and inter fusion rule, as well as optimising the fusion rule to maximise the EE whilst satisfying the false alarm requirement. Detection probability limited technique is used to assess the goal function. According to the data, the best configuration options adhere to the false alarm limitations and improve EE.

The remainder of the paper is organised as follows: Sect. 3 explains the fundamental energy detector-based CSS device paradigm. The goal purpose and constraints are examined in Sect. 4. In Sect. 5, the numerical outputs are provided, followed by a conclusion.

## 3 System Model

In the CR network, it is assumed that there are '$N$' SUs and one PU. SUs needs sensing duration $\tau_s$ to detect the presence of PU. With a reporting time $\tau_r$ the sensing details must be delivered to the base station. The fusion centre (FC) is located at the base station. The FC makes a final determination on the existence or absence of PU based on the facts supplied. If the PU is recognised as absent with data transmission time $\tau_d$, one of the SU is enabled to transfer data. The whole frame duration is $T = \tau_s + N\tau_r + \tau_d$.

In compared to the distance between SUs and PUs, the distance between SUs is known to be small. As a result, the average SNR for each SU is the same. The SUs employ energy detection (ED) to look for PUs. Equation 1 may be used to calculate the false alarm and detection probability.

$$P_f^j = Q\left(\left(\frac{\varepsilon_j}{\sigma_n^2} - 1\right)\sqrt{\tau_s f_s}\right) \tag{1}$$

where $f_s$ is the sampling frequency. $Q(x) = \frac{1}{\sqrt{2\pi}}\int_x^{\infty} e^{-\frac{t^2}{2}}dt$. The likelihood of detection at the $j$th CR user is determined as [15]

$$P_d^j = Q\left(\left(\frac{\varepsilon_j}{\sigma_n^2} - \gamma - 1\right)\sqrt{\frac{\tau_s f_s}{2\gamma + 1}}\right) \tag{2}$$

For all SUs, a common threshold is utilised. Hence $P_f^j = P_f$ and $P_d^j = P_d$. Each SU will make a choice based on its local sensing and convey one bit of that decision to the base station via the reporting channel. This fusion's total detection and false alarm probability are provided by

$$Q_f(\tau_s) = \sum_{i=k}^{N}\binom{N}{i}P_f^i(1 - P_f)^{N-i} \tag{3}$$

$$Q_d(\tau_s) = \sum_{i=k}^{N}\binom{N}{i}P_d^i(1 - P_d)^{N-i} \tag{4}$$

## 3.1 Inter and Intra Fusion Rule

A complete '$L$' clusters are believed to be formed and there are '$M$' SUs in each cluster. Every SU detects PU presence in a cluster by local sensing. This local sensing outcome is shared by the SUs with CH using k out of M fusion. Detection and false alarm probability of this intra fusion is given by

$$Q_f(k) = \sum_{i=k}^{M}\binom{M}{i}P_f^i(1 - P_f)^{M-i} \tag{5}$$

$$Q_d(k) = \sum_{i=k}^{M}\binom{M}{i}P_d^i(1 - P_d)^{M-i} \tag{6}$$

here $k$ is threshold for the intra fusion rule.

The $q$ out of $L$ fusion rule is then used by all CHs to fuse their local clusters sensing data at a common base station. Here $q$ is the threshold for the inter-fusion rule. Detection and false alarm and probability are calculated as

$$\overline{Q}_f(k, q) = \sum_{i=q}^{L} \binom{L}{i} Q_f^i (1 - Q_f)^{L-i} \tag{7}$$

$$\overline{Q}_d(k, q) = \sum_{i=q}^{L} \binom{L}{i} Q_d^i (1 - Q_d)^{L-i} \tag{8}$$

## 3.2 Energy Efficiency in CSS

If $P_1$ and $P_0$ are the chances of PU being active and idle, respectively, there are four alternative circumstances based on the current status of PU and result of sensing.

Scenario 1: SU recognises PUs idle condition with probability $P_0(1 - P_f)$.

Scenario 2: SU is properly recognising PUs busy status with the probability of $P_1 P_d$. SU will not communicate its data because channel is occupied by PU.

Scenario 3: SU wrongly detects PUs status as idle with probability $P_1(1 - P_d)$. As a result, the PU's data would clash with the data of CR user, rendering the CR user data meaningless in this case.

Scenario 4: SU incorrectly interprets PUs status as active with probability $P_0 P_f$. In the same way as in example 2, there will be no data transfer.

There are two aspects to the total power consumption: spectrum sensing $\theta_s$ and data transfer $\theta_t$. The energy usage for each of the four instances inside a frame is calculated as follows:

$$\text{Case 1: } E_1 = N\tau_s\theta_s + \tau_d\theta_t \tag{9}$$

$$\text{Case2 : } E_2 = N(\tau_s\theta_s + \tau_r\theta_t) \tag{10}$$

$$\text{Case 3: } E_3 = N\tau_s\theta_s + \tau_d\theta_t \tag{11}$$

$$\text{Case4 : } E_4 = N(\tau_s\theta_s + \tau_r\theta_t) \tag{12}$$

where $N = \text{MXL}$. The total energy consumption $\mathbb{E}$ can be determined as

$$\mathbb{E} = N\tau_s\theta_s + N\tau_r\theta_t + \left(P_0\left(1 - \overline{Q_f}(k)\right) + P_1\left(1 - \overline{Q_d(k)}\right)\right)\tau_d\theta_t \tag{13}$$

Only if a whole frame is successfully sent is data believed to be accurate in the suggested model. Otherwise, data is considered worthless. As a result, only those data bits broadcast for which the PU is idle are found true, and the SU correctly detects the PU's condition. The average throughput R of the system, also known as the number of relevant data bits transferred every frame, may thus be written as

$$\mathbb{R} = P_0 \tau_d C\left(1 - \overline{Q}_f(k)\right) \tag{14}$$

where channel capacity is $C$, it may be calculated as $C = B\log_2\left(1 + \frac{\theta_t}{\Gamma}\right)$, where $\Gamma$ is the overall noise power.

To fulfil the PU interference limit, the thresholds (ε) can be set as constant. The parameters $\theta_t$ and $k$ effects EE at the fixed sensing duration. EE($\xi$) can be defined as the amount of effective transmission of information per energy consumption unit that can be given by

$$\xi(k, \theta_t) = \frac{\mathbb{R}(k, \theta_t)}{\mathbb{E}(k, \theta_t)} \tag{15}$$

By adjusting fusion rule and transmission power, the purpose of the design challenge is to optimise EE. The optimisation issue may be expressed mathematically as.

$$\begin{aligned}&\text{To find: } (k_0, \theta_{t0})\\&\text{Max.}: \ \xi(k, \theta_t)\\&\text{S.t.}: N \geq k \geq 1, \theta_{\max} \geq \theta_t \geq \theta_{\min}, P_d \geq \overline{P}_d\end{aligned} \tag{16}$$

where $k_0\theta_{t0}$ are the maximum fusion rule threshold value and transmission power level, respectively. $\theta_{\min}$ and $\theta_{\max}$ are the lowest and highest transmission power levels.

## 4 Problem Formulation and Constraints

We can define normalised sensing time as $(\tau) = \frac{\tau_s}{T}$ and transmission time $(\tau_d) = (1-\tau)T$ if we assume reporting time $\tau_r \ll T$. The complexity of the optimisation problem can be minimised by applying this assumption. Equation (18) of the optimisation problem's detection probability restriction may be satisfied by defining the false alarm probability as

$$P_f(\tau) = Q\left(\sqrt{2\gamma + 1}Q^{-1}\left(\overline{P}_d\right) + \gamma\sqrt{\tau f_s T}\right) \tag{17}$$

We can always assume $P_d(\tau) = \overline{P}_d$ according to IEEE802.22 standard. The energy consumption $\hat{\mathbb{E}}$ can be defined in terms of $\tau$ and $\theta_t$ as follows.

$$\hat{\mathbb{E}}(\tau, \theta_t) = N\tau T\theta_s + \left(P_0\left(1 - \overline{Q}_f(k)\right) + P_1\left(1 - \overline{Q}_d(k)\right)\right)(1-\tau)T\theta_t \tag{18}$$

We can determine the throughput as follows:

$$\hat{\mathbb{R}}(\tau, \theta_t) = P_0(1-\tau)T \times \left(1 - \overline{Q}_f(k)\right) \times \exp\left(\frac{-((1-\tau)T)}{a_0}\right) \times \beta\log_2\left(1 + \frac{\theta_t}{\Gamma}\right) \tag{19}$$

The modified EE ($\hat{\xi}$) is given by $\hat{\xi}(k, \theta_t) = \frac{\hat{\mathbb{R}}(k,\theta_t)}{\hat{\mathbb{E}}(k,\theta_t)}$ and optimisation problem can be rewritten as.

$$\begin{aligned} &\text{Find:}(k_0, \theta_{t0}) \\ &\text{Max. : } \hat{\xi}(k, \theta_t) \\ &\text{S.t. : } N > k_0 \geq 1, \theta_{\max} \geq \theta_t \geq \theta_{\min} \end{aligned} \tag{20}$$

## 5 Numerical Results and Discussion

To perform the simulation analysis, the parameters are taken as $f_s = 6\text{MHz}$, $\theta_s = 0.2W$, $\sigma_n^2 = 1$, $\tau_r = 10\mu s$,, $P_0 = 0.5$, $T = 45$ ms, $L = 4$, $M = 6$,$\theta_t = 3W$, $\overline{P}_d = 0.9$

At SNR = −18 dB, Fig. 1 shows an EE vs normalised sensing time curve. The OR fusion ($k = 1$) is employed in intra fusion. The inter fusion threshold ($q$) can be adjusted. EE is at its greatest when $q = 6$, and at its weakest when $q = 1$. 3.068 Mbits/Hz/joule is the maximum EE in this situation.

Figure 2 shows the effect of intra AND fusion ($k = 6$). The results reveal that when $q = 1$, EE is lowest, and when $q = 6$, EE is highest. At a sensing time of 0.8 ms, the maximum EE is determined to be 3.292 Mbits/Hz/joule. Overall, the EE is higher in this situation than in the previous one. The maximal EE changes as the inter and intra fusion changes, as seen in Figs. 1 and 2.

The EE is compared to the intra fusion rule's threshold at 1.25 ms sensing period. Because the likelihood of detection is low at higher k values and the possibility of false alarm is larger at lower k values, the concavity of EE with respect to k is depicted in Fig. 3. when $k = 3$ and $q = 4$, the highest EE of 3.391 Mbits/Hz/joule is achieved.

The effect of adjusting the intra and inter fusion rule thresholds is shown in Fig. 4. The result also reveals that the maximum EE for a 4X6 sized cluster is achieved when the intra fusion rule's threshold is 3 and the inter fusion rule's threshold is 4. With a sensing time of 0.65 ms, the highest EE = 3.17 Mbits/Hz/joule is detected.

Figure 5 shows the graph of EE vs. normalised sensing time. The graph was created using the bisection approach's optimum threshold for intra and inter fusion rules. Because energy consumption is larger in the worst radio scenario, the maximum EE drops as SNR lowers. Furthermore, the SUs require additional sensing time in order

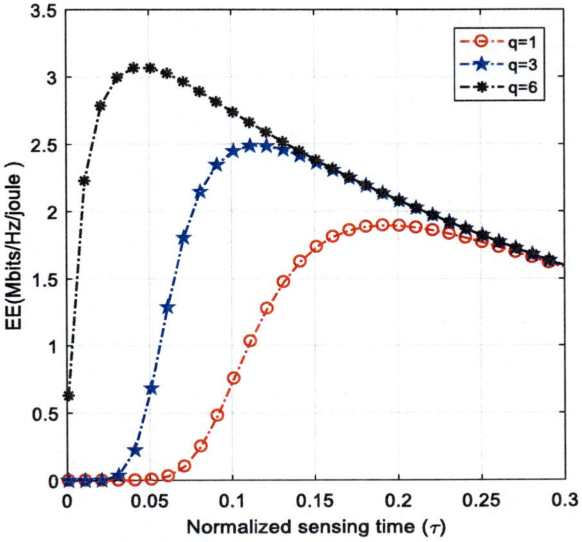


**Fig. 1** EE versus Normalised sensing time ($\tau$) graph with OR intra fusion rule

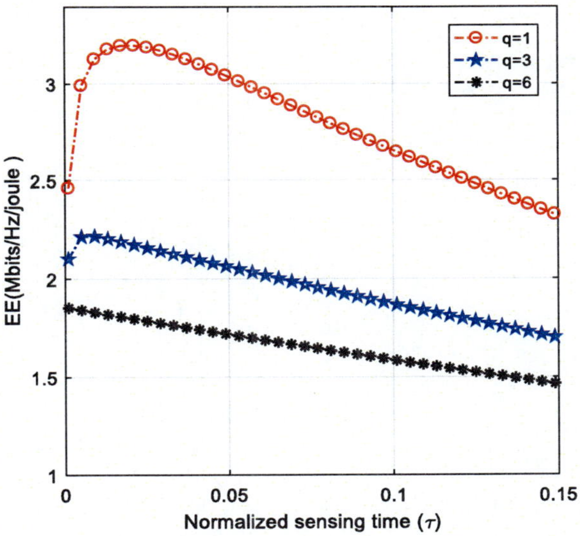


**Fig. 2** Variation of EE with respect to normalised sensing time for AND intra fusion

to meet the target detection probability. The EE is maximal = 3.38 Mbits/Hz/joule at a sensing time of 0.45 ms and an SNR of −18 dB.

**Fig. 3** Variation of EE with respect to intra fusion rule for different inter fusion rule

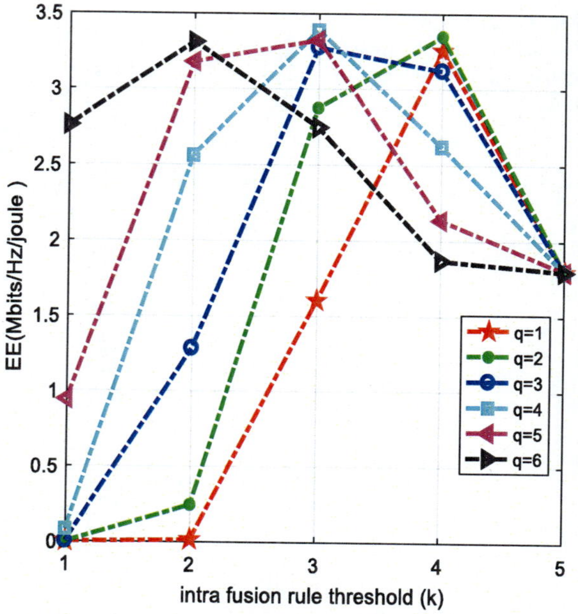


**Fig. 4** Variation of EE with respect to normalised sensing duration for variable inter and intra fusion rule

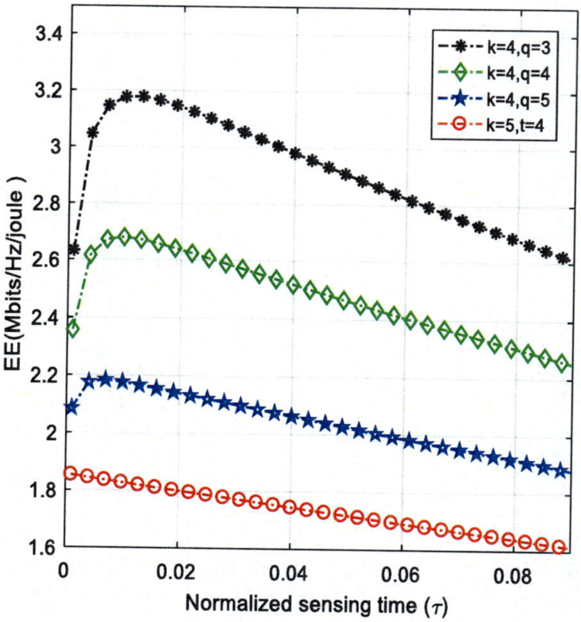

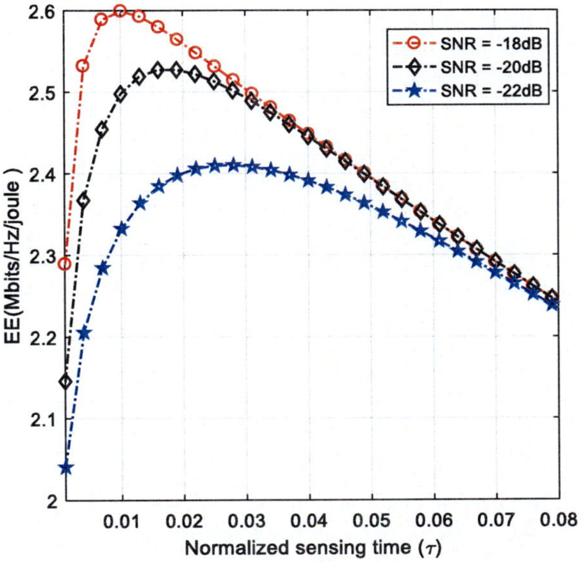


**Fig. 5** Variation of EE with respect to normalised sensing duration at optimum inter and intra fusion rule

# 6 Conclusion

Based on intra fusion and inter fusion criteria, this study proposes a CB-CSS. There is an optimal intra and inter fusion rule threshold value that optimises the EE, according to the system model and simulation research. The maximum EE is 3.391Mbits/Hz/Joule when the intra fusion rule threshold is 3 and the inter fusion rule threshold is 4. The bisection method was used to find the best settings. Both the likelihood of detection and the false alarm limitations are satisfied by the optimum design parameters.

# References


1. Lunden J, Koivunen V, Huttunen A, Poor HV (2007) Censoring for collaborative spectrum sensing in cognitive radios. In: 2007 Conference Record of the Forty-First Asilomar Conference on Signals, Systems and Computers, Pacific Grove, CA, pp 772–776
2. Sun C, Zhang W, Letaief KB (2007) Cooperative spectrum sensing for cognitive radios under bandwidth constraints. In: 2007 IEEE Wireless Communications and Networking Conference, Kowloon, pp 1–5
3. Sun C, Zhang W, Letaief KB (2007) Cluster-based cooperative spectrum sensing in cognitive radio systems. In: 2007 IEEE International Conference on Communications, Glasgow, pp 2511–2515
4. Wang Y, Feng C, Zeng Z, Guo C (2009) A robust and energy efficient cooperative spectrum sensing scheme in cognitive radio networks. In: 2009 11th International Conference on Advanced Communication Technology, Phoenix Park, pp 640–645

5. Lee C, Wolf W (2008) Energy efficient techniques for cooperative spectrum sensing in cognitive radios. In: 2008 5th IEEE Consumer Communications and Networking Conference, Las Vegas, NV, pp 968–972
6. Liu X, Jia M, Xuemai G, Tan X (2013) Optimal periodic cooperative spectrum sensing based on weight fusion in cognitive radio networks. Sensors 13(4):5251–5272
7. Liu X, Li F, Na Z (2017) Optimal resource allocation in simultaneous cooperative spectrum sensing and energy harvesting for multichannel cognitive radio. IEEE Access 5:3801–3812
8. Liang Y-C, Zeng Y, Peh ECY, Hoang AT (2008) Sensing throughput tradeoff for cognitive radio networks. IEEE Trans Wireless Commun 7(4):1326–1337
9. Peh ECY, Liang Y-C, Guan YL, Zeng Y (2009) Optimization of cooperative sensing in cognitive radio networks: A sensing-throughput tradeoff view. IEEE Trans Vehicular Tech 58(9):5294–5299
10. Tang L, Chen Y, Hines EL, Alouini M-S (2011) Effect of primary user traffic on sensing-throughput tradeoff for cognitive radios. IEEE Trans Wireless Commun 10(4):1063–1068
11. Hu H, Zhang H, Liang Y-C (2016) On the spectrum-and energy-efficiency tradeoff in cognitive radio networks. IEEE Trans Commun 64(2):490–501
12. Gao Y, Xu W, Yang K, Niu K, Lin J (2013) Energy-efficient transmission with cooperative spectrum sensing in cognitive radio networks. In: 2013 IEEE Wireless Communications and Networking Conference (WCNC), Shanghai, pp 7–12
13. Das D, Das S (2017) A novel approach for energy-efficient resource allocation in double threshold-based cognitive radio network. Int J Commun Syst 30(9):e3198
14. Wu Y, Tsang DHK (2011) Energy-efficient spectrum sensing and transmission for cognitive radio system. IEEE Commun Lett 15(5):545–547
15. Shi Z, Teh KC, Li KH (2013) Energy-efficient joint design of sensing and transmission durations for protection of primary user in cognitive radio systems. IEEE Commun Lett 17(3):565–568
16. Zhang L, Xiao M, Wu G, Li S, Liang Y-C (2016) Energy-efficient cognitive transmission with imperfect spectrum sensing. IEEE J Sel Areas Commun 34(5):1320–1335
17. Qiu T, Xu W, Song T, He Z, Tian B (2011) Energy-efficient transmission for hybrid spectrum sharing in cognitive radio networks. In: 2011 IEEE 73rd Vehicular Technology Conference (VTC Spring), Yokohama, pp 1–5
18. Wang Y, Xu W, Yang K, Lin J (2012) Optimal energy-efficient power allocation for OFDM-based cognitive radio networks. IEEE Commun Lett 16(9):1420–1423
19. Ozcan G, Gursoy MC (2014) Energy-efficient power adaptation for cognitive radio systems under imperfect channel sensing. In: 2014 IEEE Conference on Computer Communications Workshops (INFOCOM WKSHPS), Toronto, ON, pp 706–711
20. Sharma G, Sharma R (2019) Energy efficient collaborative spectrum sensing with clustering of secondary users in cognitive radio networks. IET Commun 13(8):1101–1109
21. Sharma G, Sharma R (2019) Optimised fusion rule in cluster-based energy-efficient CSS for cognitive radio networks. Int J Electron 106(5):741–755. https://doi.org/10.1080/00207217.2018.1553248
22. Sharma G, Sharma R (2018) Performance evaluation of distributed CSS with clustering of secondary users over fading channels. Int J Elect Lett 6(3):288–301. https://doi.org/10.1080/21681724.2017.1357762
23. Sharma G, Sharma R (2019) Cluster-based distributed cooperative spectrum sensing over Nakagami fading using diversity reception. IET Net 8(3):211–217. https://doi.org/10.1049/iet-net.2018.5002

# Communication Track—2

# Design of Dispersion Compensated with NRZ Modulation-Based 8- and 16-Channels WDM System for Long-Haul Communication

**Mijanur Rahim, Anjumanara Begam, and Md. Asraful Sekh**

## 1 Introduction

In fiber optic communication system, the organized and structured data are transmitted by the transmission of light through the optical fibers. Here, the carrier wave is EM wave in the form of light, which is modulated to carry the information. In a higher data rates communication system, there are some essential things, i.e., huge bandwidth and the low cost. When light propagates through a single-mode fiber (SMF) of a fiber optic communication system, the chromatic dispersion or group velocity dispersion (GVD) unfavorably affects the high data rate of the system [1, 2]. Due to the dispersed signals, the broadening pulses overlap each other and create ambiguity of data retrieval; it is called inter-symbol interference (ISI) [2–4].

Earlier in our previous work, we discussed different types of dispersion compensation methods, i.e., post-, pre-, and symmetry for different data rates where we found the performance of symmetry dispersion compensation method is best [5–7]. There are three schemes of symmetry dispersion compensation method using dispersion compensation fiber (DCF), using fiber Bragg grating (FBG) and using a combination of DCF + FBG together. It is also found that this combination of DCF + FBG performance is superior to the other two schemes in the symmetry dispersion compensation method, especially when there is a requirement of high data rates in long-distance communication [7, 8]. In this paper, we have discussed the comparison of these three schemes of symmetry dispersion compensation in the 8- and 16-channel DWDM system with high data rates designed for long-distance optical communication. We have made this comparison in terms of eye diagrams, Q factors, O/P gains, and O/P powers which we have obtained from the received signals. We have also compared the obtained results with the dispersion compensation scheme

M. Rahim · A. Begam · Md. A. Sekh (✉)
Optical Communication and Networks Lab, Department of ECE, Aliah University, II-A/27 Newtown, Kolkata 700160, India
e-mail: asraful.ece@aliah.ac.in



P. H. J. Chong et al. (eds.), *Emerging Electronics and Automation*, Lecture Notes in Electrical Engineering 937, https://doi.org/10.1007/978-981-19-4300-3_25

using DCF individually in the symmetry dispersion compensation method. The main challenging task in such a long-haul optical communication system is to maintain the good quality factor and gain at moderate or high data rates. Generally, we use erbium-doped fiber amplifier (EDFA) to reduce the attenuation of the signal in such DCF-based symmetrical dispersion-compensating system, but the output signal is distorted enough to retrieve it [7]. Therefore, using DCF + FBG combination in the symmetry dispersion compensation method, the performance of this DWDM system is being enhanced; we have FBG in this DWDM system for achieving high gain [8, 9]. A combinational dispersion compensation scheme is designed for long-haul communication in 8-channels and 16-channels DWDM systems at the data rates of 10 Gbps and 20 Gbps, respectively.

## 2 Mathematical Modeling

The Fresnel reflection is followed in FBG, where light travels between media of different refractive indices may both reflect and refract at the interface. The Bragg wavelength is the reflected wavelength which is denoted by $\lambda_b$ [1, 2].

$$\lambda_b = 2m_e\Lambda \tag{1}$$

where $m_e$is effective refractive index of the grating in the fiber core and $\Lambda$ is grating period [1, 2].

$$\Delta\lambda = \left[\frac{2\delta n_0 n}{\pi}\right]\lambda b \tag{2}$$

where $\Delta\lambda$ is bandwidth, $\delta n_0$ is difference between the two refractive index $(n_2 - n_1)$, n is the fractional power in the fiber core [1, 2].

$$P_B(\lambda_b) \approx \tanh^2\left[\frac{Nn(V\delta n_0)}{n}\right] \tag{3}$$

where $P_B(\lambda_B)$ is the peak reflection power and $N$ is the number of periodic variations.

Usually, we use DCF to compensate for the chromatic dispersion. It is used after the single-mode fiber (SMF), and therefore, the equivalent wavelength-dependent dispersion is denoted as $D_{eq}(\lambda)$ [10, 11].

$$D_{eq}(\lambda) = [RD_{\mathrm{SMF}}(\lambda) + D_{\mathrm{DCF}}(\lambda)]/(1 + R) \tag{4}$$

where $R$ is the length ratio between SMF and DCF, $D_{\mathrm{SMF}}$ and $D_{\mathrm{DCF}}$ are the dispersion of the SMF and the DCF, respectively. From Eq. (5), given $D_{eq}$ as zero fulfills the ideal compensation condition, and $R = L_{\mathrm{SMF}}/L_{\mathrm{DCF}}$, the formula would be [8–11],

$$D_{\mathrm{SMF}} L_{\mathrm{SMF}} + D_{\mathrm{DCF}} L_{\mathrm{DCF}} = 0 \tag{5}$$

For externally modulated sources, transmission distance limited by chromatic dispersion is [8–11]

$$L < \frac{2\pi c}{16|D|\lambda^2 B_t^2} \tag{6}$$

where $L$ is length of fiber in km, c is speed of light (m/s), λ is wavelength in meter, $B_T$ is $B$ it rate in Gbps.

## 3 Simulation Layout and Parameters Setting

In our previous works, we have found that the DCF–FBG combinational scheme is the best dispersion compensation scheme [7, 8]. Therefore, Fig. 1 is the simulation layout of the proposed model of 8-channel DWDM system for 500 km. dispersion compensated long-haul communication at data rates of 10 and 20 Gbps. We have taken the CW laser power of 1mW with channel spacing of 0.8 nm; the extinction ratio of Mach–Zehnder modulator (MZM) is 100 dB. This optical transmission is done over 100–500 km length SMF; attenuation of SMF is 0.25 dB/km, SMF dispersion 16 ps/nm/km, and dispersion slope 0.08 ps/nm/km. The length of DCF is 5 km, DCF dispersion—80 ps/nm/km, and the effective index of 1–5 mm length FBG is 1.45. The length of EDFA is 5 m with a numerical aperture of 0.24. The same things are set for 16-channel WDM system with a channel spacing of 0.4 nm.

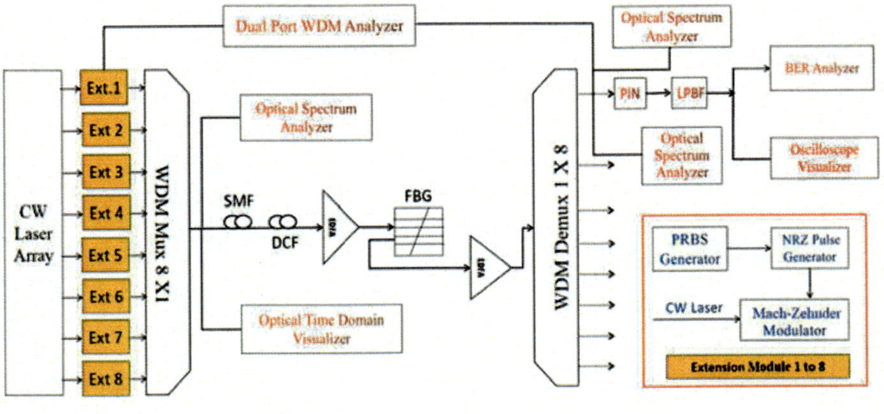


**Fig. 1** Simulation layout of the proposed model

## 4 Results and Discussions

Figure 3 is the comparison of Q factor (QF) vs transmission distance (km) for 8-channel WDM system using different dispersion compensation techniques. We have achieved the QF of 42.57 at 10 Gbps and QF of 35.29 at 20 Gbps using the combinational scheme of dispersion compensation (DCF + FBG) for the transmission distance of 500 km. Figure 3 is the comparison of Q factor (QF) vs transmission distance (km) for 16-channel WDM system using different dispersion compensation techniques. We have achieved the QF of 35.86 at 10 Gbps and QF of 29.77 at 20 Gbps using the combinational scheme of dispersion compensation (DCF + FBG) for the transmission distance of 500 km.

Both Figs. 2 and 3 are showing that using this combinational dispersion compensation scheme (DCF + FBG) is giving the best Q factors for the application of above 500 km long-haul communication both based on 8- and 16-channels DWDM systems at higher data rates.

Figure 4a, b is the eye diagrams at 500 km transmission distance for 8-channels DWDM system. It is clearly showing that there is no inter-symbol interference in

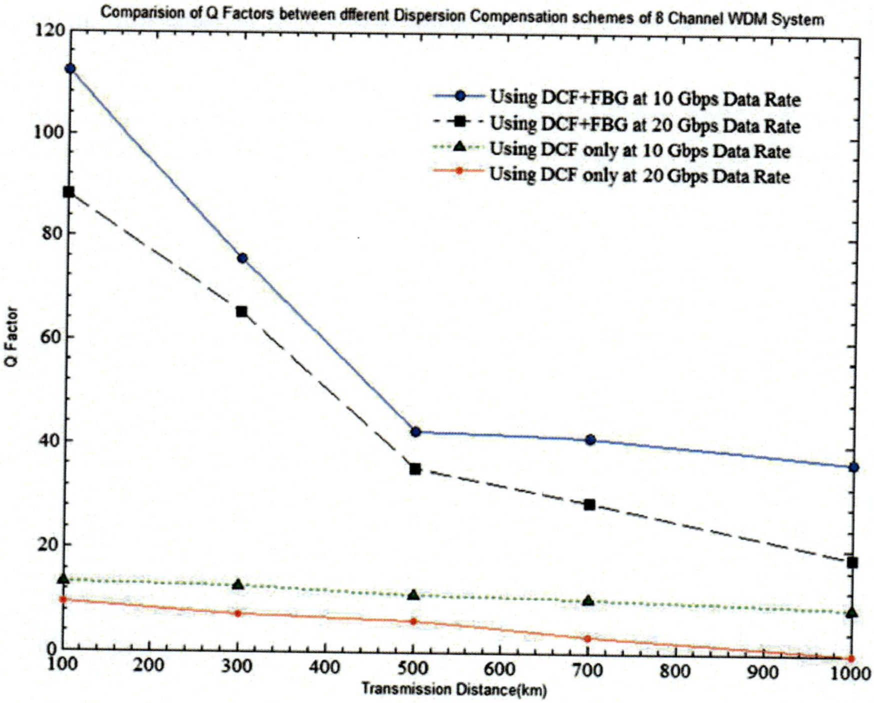


**Fig. 2** Comparison of Q factor versus transmission distance (km) for 8-channel WDM system using different dispersion compensation techniques

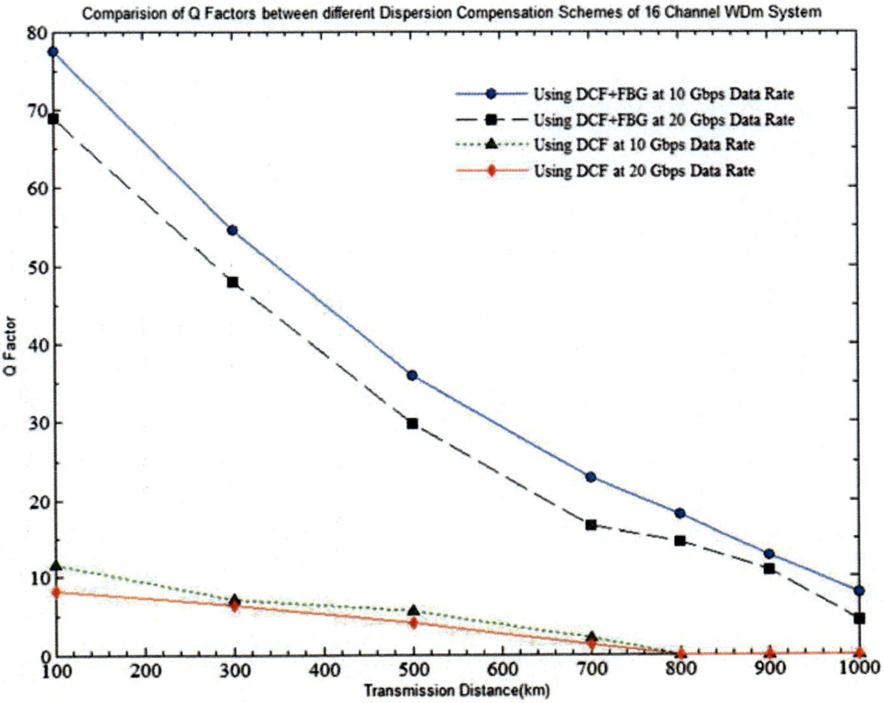


**Fig. 3** Comparison of Q factor versus transmission distance (km) for 16-channel WDM system using different dispersion compensation techniques

the retrieved data as the chromatic dispersion is well managed by this combinational dispersion compensation scheme.

Figure 5 is the comparison graph of O/P gain (dB) versus transmission distance (km) for 8-channels WDM. We have achieved the O/P gain of 38.1 dB and 25.2 dB at 10 Gbps and 20 Gbps, respectively, in 8-channels dispersion compensated (DCF + FBG) DWDM system for 500 km transmission. We have also achieved the O/P gain of 24.11 dB and 22.14 dB at 10 Gbps and 20 Gbps, respectively, in the 16-channels dispersion compensated (DCF + FBG) DWDM system for 500 km length optical transmission because of this combinational scheme in symmetry dispersion compensation method where the chromatic dispersion very well managed.

In Fig. 6, it is clearly showing that we are getting a better Q factor in the higher wavelength region. In 1552.52 nm wavelength at 10 Gbps data rate, we have achieved the highest Q factor 42.57 and 35.86 for 8-channels and 16-channels DWDM system, respectively. When the data rate is 20 Gbps in 1552.52 nm wavelength, the Q factors become 35.29 and 29.77 for 8-channels and 16-channels DWDM system, respectively. Here, we have applied the combinational dispersion compensation scheme in those DWDM systems.

Tables 1 and 2 show the performance of the 8-and 16-channels DWDM systems, respectively, for different data rates.

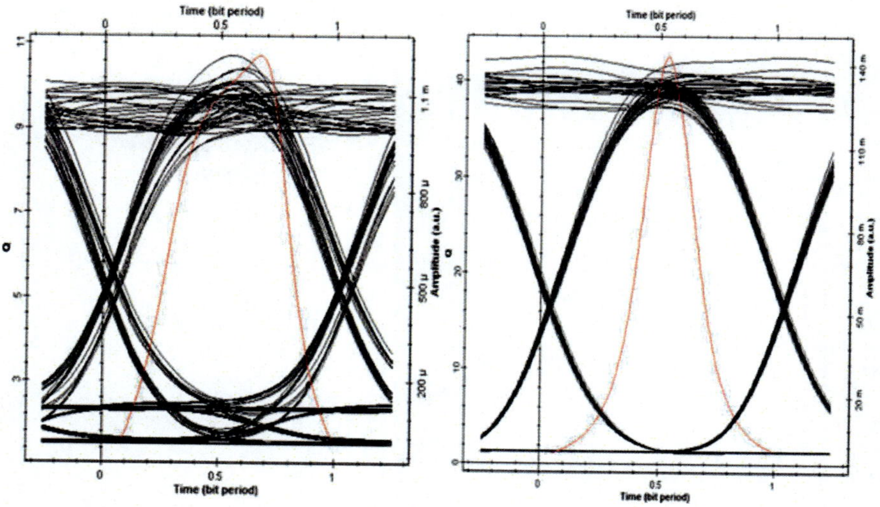


**Fig. 4** **a** Eye diagram at 500 km transmission distance for normal 8-channels DWDM system **b** Eye diagram at 500 km transmission distance for dispersion compensated WDM system using combinational dispersion compensation scheme (DCF + FBG)

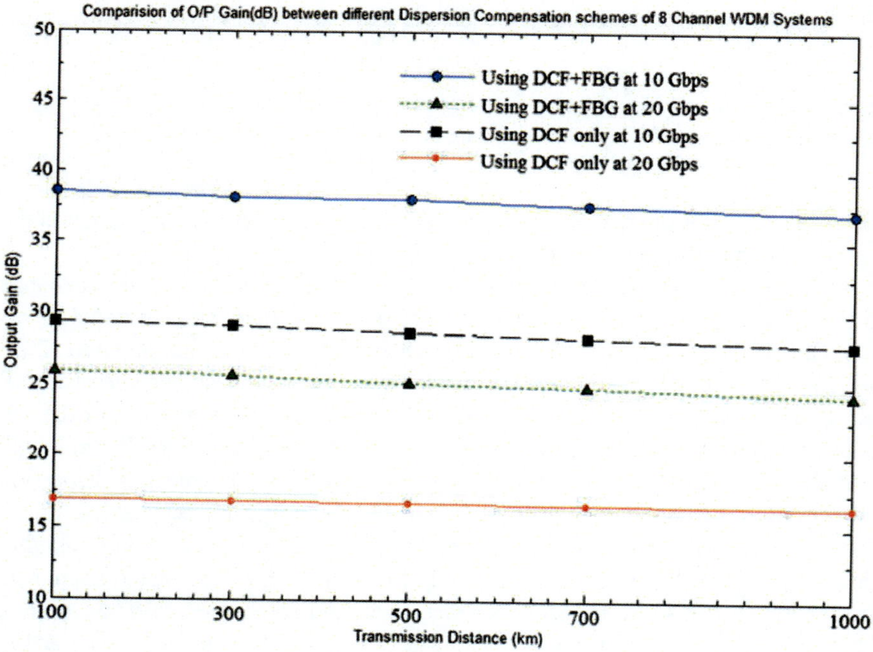


**Fig. 5** Comparison of O/P gain (dB) versus transmission distance (km) for 8-channels WDM

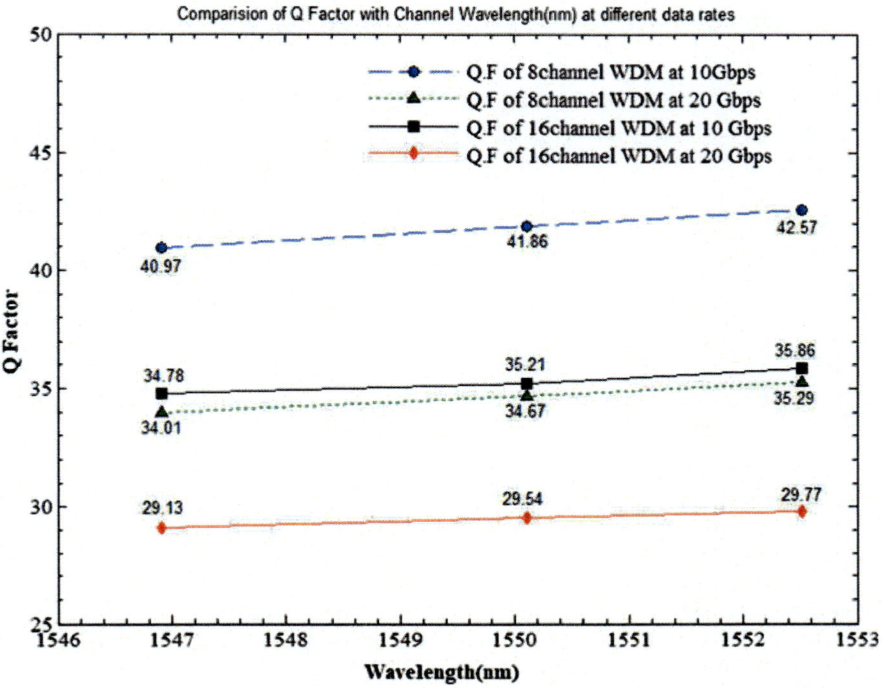


**Fig. 6** Q factor versus wavelength (nm) for 8- & 16-channels DWDM system at the data rate of 10 and 20 Gbps

**Table 1** Performance of the 8-channels DWDM system for different data rates

| Distance (km) | 8-channels DWDM system using combinational schemes (DCF + FBG) | | | | | |
|---|---|---|---|---|---|---|
| | Q factor | | O/P gain (dB) | | O/P power (dBm) | |
| | 10 Gbps | 20 Gbps | 10 Gbps | 20 Gbps | 10 Gbps | 20 Gbps |
| 100 | 112 | 88.1 | 38.6 | 25.9 | 17.1 | 14.2 |
| 300 | 75.4 | 65.2 | 38.2 | 25.6 | 17.1 | 13 |
| 500 | 42.6 | 35.3 | 38.1 | 25.2 | 17.1 | 12.8 |

**Table 2** Performance of the 16-channels DWDM system for different data rates

| Distance (km) | 16-channels DWDM system using combinational schemes (DCF + FBG) | | | | | |
|---|---|---|---|---|---|---|
| | Q factor | | O/P gain (dB) | | O/P power (dBm) | |
| | 10 Gbps | 20 Gbps | 10 Gbps | 20 Gbps | 10 Gbps | 20 Gbps |
| 100 | 77.56 | 68.86 | 24.45 | 22.37 | 16.2 | 13.6 |
| 300 | 54.67 | 47.95 | 24.34 | 22.25 | 16.2 | 12.9 |
| 500 | 35.86 | 29.77 | 24.11 | 22.14 | 16.1 | 12.6 |

## 5 Conclusions

The simulated transmission system has been analyzed on the basis of better eye diagram, higher quality factor, gain, output power. It is found that the DCF–FBG combinational scheme shows better Q factor and O/P gain compared to dispersion compensation scheme using DCF of 8- & 16-channel WDM system at 10 & 20 Gbps data rate. It is also found that using DCF dispersion compensation scheme shows better O/P power in 16-channel WDM system at 20 Gbps data rate. This work can be extended for more high data rates in such long-haul optical communication.

## References


1. Agrawal G (2001) Applications of nonlinear fiber optics. Academic Press, Cambridge
2. Keiser G (2003) Optical fiber communications. Wiley Online Library, Hoboken
3. Kashyap R (2010) Fiber bragg gratings, 2nd edn. Academic Press, Cambridge
4. Ning HB, Jing W, Wei W, Zhao RM (July 2010) Analysis on dispersion compensation with DCF based on optisystem. In: 2010 2nd international conference on industrial and information systems, pp 40–43
5. Hayee MI, Willner AE (1997) Pre- and postcompensation of dispersion and nonlinearities in 10-Gb/s WDM systems. IEEE Photon Tech Lett 9:1271
6. Liu Z, Xu T, Saavedra G, Bayvel P (2018) 448-Gb/s PAM4 transmission over 300-km SMF-28 without dispersion compensation fiber. In: Optical fiber communication conference. Optical Society of America, San Diego, California, p W1 J.6
7. Rahim M, Bar AT, Begam A, Sekh MA (2019) Investigation of dispersion compensation methods for the data rates 2.5 and 10 Gbps using standard and dispersion compensated fibers. J Emerg Technol Innov Res (JETIR) 6(5):731–734
8. Sekh MA, Rahim M, Bar AT. Investigating combinational dispersion compensation schemes using DCF and FBG at data rate of 10 and 20 Gbps. In: Saini HS, Singh RK, Beg MT, Sahambi JS (eds) Innovations in electronics and communication engineering, lecture notes in networks and systems, vol 107. Springer, Singapore. https://doi.org/10.1007/978-981-15-3172-9_28
9. Weiner M (1996) Broadband fiber dispersion compensation for sub-100-fs pulses with a compression ratio of 300. Opt Lett 21(15):1141–1143
10. Xu T, Jacobsen G, Popov S, Li J, Sergeyev S, Friberg AT (2017) Analysis of chromatic dispersion compensation and carrier phase recovery in long-haul optical transmission system influenced by equalization enhanced phase noise. Optik-Int J Light Electron Opt 138:494–508
11. Hussein TF, Rizk MRM, Aly MH (2019) Opt Quant Electron 51:103. https://doi.org/10.1007/s11082-019-1823-y

# D2D Communication for Next Generation Cellular Systems: A Review

**Radhika Gour**

## 1 Introduction

The massive demand for high data rate applications and enormous use of smart devices is contributing to heavy data traffic in the wireless communication domain. Handling the considerable number of users, high data rate requirements, and heterogeneous applications require modifications in the existing cellular communication. Integration of D2D communication into the existing cellular system can fulfill these expectations and improve the system performances. Unlike traditional two-hop uplink/downlink transmission via a base station (BS), D2D communication allows a direct transmission between users in close proximity.

Over the last decade, the volume of mobile data traffic and demand for high data rate applications have been growing up remarkably. Traffic growth is being driven by both the rising number of smartphone subscriptions and an increasing average data volume per subscription, fueled primarily by more viewing of video content. Mobile data traffic grew 82% between 2018 and 2019. As reported in [1], there will be 7.2 billion smartphone subscriptions by the end of 2024. Data traffic per smartphone in 2024 will be 18 GB per month that is more than two fold as compared to the data traffic in 2018. According to the CISCO report [2], mobile data traffic will increase seven times as shown in Fig. 1a, and the number of smart devices is expected to grow two times between 2017 and 2022. The bar graph in Fig. 1b shows the proliferation rate of smart devices subscriptions from 2017 to 2022 [1]. According to CISCO's annual visual network index (VNI) reports [3], the amount of data that must be transported will continue to grow dramatically. By 2020 [4], a huge number of applications other than phone, texts, and videos are predicted to emerge, putting existing cellular networks under strain. Video streaming, online gaming, online social networks, file sharing, and other heterogeneous applications are examples of high

R. Gour (✉)
Indian Institute of Information Technology, Allahabad, India
e-mail: radhika@iiita.ac.in



P. H. J. Chong et al. (eds.), *Emerging Electronics and Automation*, Lecture Notes in Electrical Engineering 937, https://doi.org/10.1007/978-981-19-4300-3_26

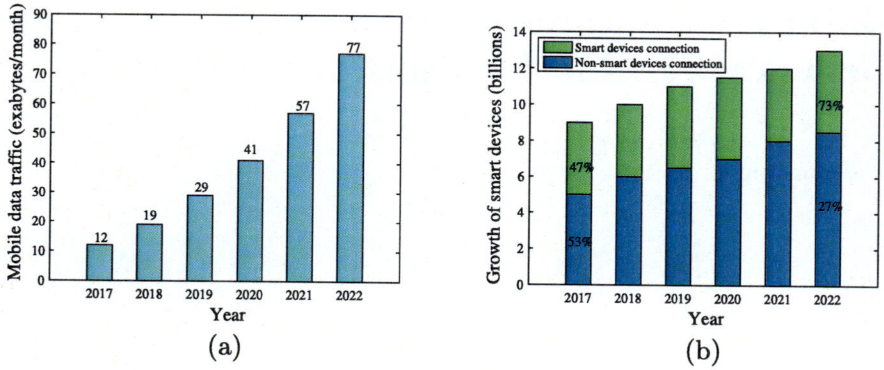


**Fig. 1** **a** Mobile data traffic growth. **b** Global growth of smart devices and connections. These figures are taken from [2]

data rate applications. Existing 4G technologies aren't up to the task of meeting these future demands. As a result, the Third Generation Partnership Project (3GPP) has proposed LTE-Advanced, an upgraded radio interface for Long Term Evolution (LTE) known as LTE-Advanced (LTE-A). (LTE-A). Although, 3GPP proposed LTE-A, to cater the high data rate applications like video in cellular networks, 5G is expected to avail 1000 times per unit area data rate and 100 times peak user data rate than 4G proposed in LTE-A. In terms of latency, 5G is needed to support the delay of 1 ms that is 15 ms for the current 4G technology [4]. While satisfying these requirements with the consideration of cost, performance, and complexity, the resource allocation is a challenging task and seeking the attention of both academia as well as industry.

The motivation behind D2D communication came from the existing technologies such as a mobile ad hoc network (MANET), Wi-Fi, and bluetooth. Bluetooth and Wi-Fi allow peer to peer communication, but they operate on the unlicensed band. The use of an unlicensed band restricts the performance because of uncontrolled interference and lack of security. In addition to this, manual pairing is required for connecting two terminals. On the other hand, MANET is also a distributed network, where devices communicate directly with each other by creating an autonomous network.

Despite this, there is no central controller in place to oversee terminal operations. D2D communication, on the other hand, allows devices to communicate directly on the licensed band while being controlled by BS [5]. In this way, D2D communication is capable of resolving the challenges associated with MANET, Bluetooth, and Wi-Fi while taking advantage of peer to peer communication from them [6]. As the number of devices increases, the possibility that two devices will come at a shorter distance increases. It will be advantageous to enable D2D communication for nearby devices to be benefited from the advantage of D2D technology.

## 2 Variants and Applications of D2D Communication

D2D communication can be classified as inband or outband D2D based on what the D2D users are using a licensed or unlicensed band [7]. Outband D2D has been inferred in the form of Bluetooth (based on IEEE 802.15) and Wi-Fi Direct (based on IEEE 802.11), but manual pairing is required, and these technologies are security-sensitive because they operate on an unlicensed band with no central entity to supervise their operations [7]. However, inband D2D is operated on the licensed cellular band under the control of BS. Furthermore, inband D2D is categorized into underlay (non-orthogonal) and overlay (orthogonal) D2D. In underlay D2D, a portion of the underlying CU's spectrum band is reused by D2D pairs, but in overlay D2D, a portion of the licensed band is assigned explicitly to D2D users. Increased spectrum and energy efficiency, lower latency, and offloading the BS all seem to be benefits of the underlying D2D connection.

Some foreseen applications that support D2D for proximity-based services are shown in Fig. 2.

Resource allocation can be grouped into two categories based on the channel sharing criterion: (1) number of channels per D2D user and (2) number of D2D users per channel. Four alternative scenarios have been identified based on the above classification: (1) One-on-one interaction: Each D2D has a maximum of one channel and a maximum of one D2D per channel. (2) One-to-many: Each D2D has several channels, with a maximum of one D2D per channel. (3) One-to-many relationship: Each D2D gets one channel at most, with several D2Ds on each channel. (4) Many-to-many communication: Each D2D has numerous channels, and each channel has multiple D2Ds.

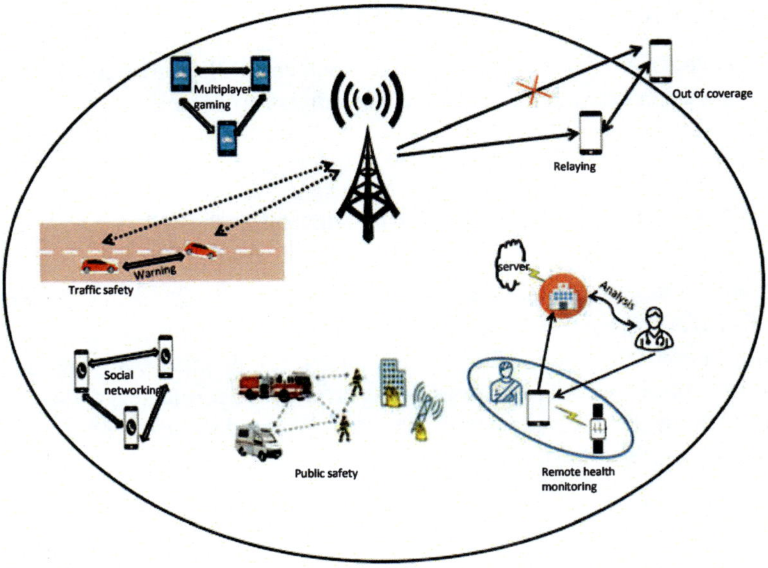


**Fig. 2** D2D applications

## 3 Design Challenges

In spite of numerous features of D2D technology, this field is immature in its real-world implementation. There are many design challenges associated with D2D deployment in 5G. Some of the major challenges in D2D implementation are as follows:.

**Interference Management**: In reuse mode of D2D, the D2D user equipment uses the same spectrum as the existing CU(s), which leads to interference from D2D to the cellular network (vice-versa) and among D2Ds. It is a major issue to be resolved for the practical implementation of D2D.

**Power allocation**: An optimal power allocation is required to maximize the performance of each user without causing severe interference to the others. Thus, optimal power allocation is another design challenge to maximize each user's throughput and minimize interference on other users.

**Mode Selection**: An adequate selection of D2D mode in which a device should be operated can improve the throughput. Mode selection is a problem of finding whether a device is performing better in D2D mode or cellular mode [7]. Sometimes it may be more beneficial for user equipments (UEs) to communicate through BS rather than direct communication or vice-versa. Hence, the decision regarding the selection of the mode is crucial.

**Channel Measurement**: D2D communication requires not only channel gain information between D2D transmitter and receiver but also channel gain between D2D transmitter and CU, and between cellular transmitter and D2D receiver. That introduces additional signaling overhead on the network, so channel measurement with less complexity is an issue to be resolved in D2D.

**Admission control**: It is necessary to judge whether a D2D pair should be allowed to communicate or not. This strategy is called admission control.

## 4 Performance Gain of D2D Communication

A great deal of work has been done in the field of resource allocation for underlay and overlay D2D with various objectives. These schemes include power and channel allocation, mode selection, admission control, etc. A detailed overview on D2D technology is presented in [7]. Various D2D performance parameters and objectives are considered by the researchers in their work. We provide a survey of literature for each category in the following sections.

## *4.1 Sum Rate Maximization*

In the literature, a substantial amount of work has been proposed to maximize the data rate in D2D networks using diverse characteristics. This section presents a brief overview of the existing sum rate maximization schemes.

In [8], interference is identified for uplink and downlink reuse cases separately for a multi-cell scenario. The authors have formulated the resource allocation problem to maximize the overall sum rate with the minimum signal to interference plus noise ratio (SINR) constraint for each user considering one-to-one pairing. The authors have designed a semi-distributed resource allocation scheme by exploiting the centralized and distributed schemes in [9].

Existing work in [10] optimizes the overall sum rate by allocating resources while taking into account the QoS of CUs in terms of lowest feasible data rate. Authors in [11] dealt with several D2D pairs running on the same channel while assuming a fixed power allocation. In [10], multiple D2D pairs are allowed to share the same channel. To resolve the MINLP problem, a 'alternating optimization' strategy is used to adapt power and channel allocation. Wang et al. [12] proposes a channel allocation technique that maximizes the weighted sum rate while adhering to QoS restrictions for all users. Same as [10], one D2D can use at most one channel, but reuse of channel among D2Dis permitted. D2D can initiate its communication, either sharing uplink or downlink channel from existing CU. An optimal dynamic programming and suboptimal cluster-based channel assignment greedy algorithm is proposed.

Overall sum rate maximization for D2D uplink reuse mode was presented by the authors of [13]. Two D2D users can only use one CU's channel at a time. The goal of an optimal power allocation is to maximize the cumulative rate. Following that, a greedy channel allocation algorithm, which is a suboptimal solution, is presented. The authors of [14] take into account numerous CUs and D2D pairs, as well as QoS for both users. Then, to enhance overall system throughput and D2D access rate while achieving the goal SINR for both D2Ds and CUs, a channel and power allocation approach for underlay D2D is presented. This research is limited to D2D and CU pairings that are one-to-one. In contrast to [14], channel uncertainty is considered in [15]. The work in [16] differs from existing work in terms of channel sharing criterion. Authors in [14] used bipartite matching to find the best reuse partner, but multiple channels are not allowed to use by D2D users. Here a D2D pair can use multiple CU's channels, and a CU can be reused by multiple D2D pairs. This work is proposed for the single cell where all CUs transmit their data with the fixed power.

Authors in [17] focused on resource allocation and criterion for the mode selection jointly. Power control is applied to mitigate the interference while satisfying the minimum SINR constraint. This proposal is limited to the scenario where only one D2D link is restricted to use only one CU's channel. The overall sum rate is optimized in [18] by proper pairing of a D2D pair and CU as its reuse partner. In [18], the authors introduced an inverse popularity pairing (IPPO) algorithm, which is a one-to-one pairing scheme. The technique is not perfect, but it ensures that D2D couples are treated equally when assigning the channel. Authors in [19] proposed a D2D link to communicate in multiple modes by using resource multiplexing. A resource utilization scheme with a one-to-many reuse scenario is given in [20] where multiple D2D pairs can reuse channels among them, but one D2D pair can use at most one channel. When allocating powers on the same uplink channel, mutual interference between D2D couples is ignored. The authors of [21] focused on maximizing network throughput by performing channel and power allocation. In this study, the authors propose a resource allocation approach for relay-assisted D2D. Idle D2D users (who don't have anything to send) operate as a relay for the active D2D pair. Authors in [22] proposed a power allocation technique to maximize the D2D sum rate in a many-to-many channel reuse scenario. The authors in [23, 24] proposed D2D technology for relaying the traffic with the aid of overlay mode of D2D transmission.

## *4.2 Interference Mitigation Techniques*

Due to the sharing of the channel among multiple users, there is a variety of additional interference in underlay D2D. There are many schemes proposed in the literature to deal with this design challenge, such as power control and fractional frequency reuse (FFR). Power control is used to determine the best power level for minimizing interference to other users while staying within the tolerance limit. On the other hand, FFR assigns the frequency band optimally to avoid the interference.

**Power Control Techniques**: In [25], cellular link is allowed to transmit at its maximum power level, while power control is applied on D2D users to attain the QoS of prioritized cellular system. In a single cell environment, underlay D2D is considered with one CU and one D2D pair. A direct link between D2D terminals can share or reuse either uplink or downlink channel of CU. It is believed that perfect/instantaneous CSI is available for all the links at the BS. However, to reduce the signaling overhead, statistics of SINR for both kinds of users are formulated with the assumption of average CSI available instead of perfect CSI.

Authors in [26] have optimized the overall cell throughput and outage probability of CU with average CSI in contrast to work done in [25], where only D2D power is optimized for interference coordination under the same scenario. Interference from cellular network on the performance of D2D is addressed in [27]. Multiple antenna BS is chosen so that it may serve M CUs and one D2D pair simultaneously on the same channel. To manage the interference from M CUs to the D2D receiver, interference limited area (ILA) is defined. ILA is the area where interference to signal ratio (ISR) at D2D receiver is less than a threshold value. CUs and D2D pair are prohibited to use same channel if the CU is within ILA. Coverage area for ILA is derived by assuming average CSI at the BS. Semi-orthogonal user selection algorithm is used to select M out of K users that can share the same channel.

SINR and data rate distributions are derived in closed or semi-closed form analytically [28]. Fully loaded network where D2D may operate in dedicated or reuse mode is assumed. D2D link operates during downlink period of cellular communication.

Work in [29] selects the mode between overlay and underlay in addition to cellular or D2D mode. Moreover, mode selection criterion is based on the distance between D2D transmitter and receiver. Rate expression in overlay and underlay case is derived analytically by using stochastic geometry. This work includes mode selection, power control, and scheduling of CU within a single cell for uplink reuse case. Authors of [30] specified that spectrum efficiency is improved by mode selection scheme with power control. Network topology and user distribution are modeled as PPPs with the help of stochastic geometry. Underlay access of D2D with uplink sharing is taken in a single cell. Mode selection and power control are modeled analytically. Another analytical modeling of interference is done in [31] where one time frequency resource block is shared by one CU and multiple D2D pairs. Exclusion regions are formed to protect the CU and hence spectrum efficiency is improved. This framework considers underlay and overlay D2D where uplink is shared in a multi-cell environment.

**Fractional Frequency Reuse**: In general, BS is deployed at the center of the cell. Users close to the BS have strong signals, while edge users have more probability of outage due to less received power and interference from adjacent cell and co-channel interference. It is desirable to set a high frequency reuse factor (FRF) to reduce the effect of interference. But this will reduce spectral efficiency. High spectral efficiency demands the FRF equal to unity. Both requirements are contradictory, and proposed FFR is a recommended solution in the conventional cellular system [32].

Authors in [33] have used FFR-based resource allocation in D2D-enabled cellular networks. FFR decides which frequency band should be used by any D2D user and CU, based on the location of D2D pairs so that throughput is maximized, and interference from D2D to CU is relaxed. FFR is explained for the case when D2D transmission held during the downlink period of cellular communication.

An example of FFR in a 7-cell network is shown in Fig. 3. FRF for inner region is one and greater than one for outer areas, that is 3 in this particular case. Authors in [33] do not consider interference on the D2D receiver from CU. However, the aim of [34] is to explain the frequency reuse method and performance evaluation of D2D in FFR-based OFDMA cellular system. Same as [33], the cell is divided into inner and central areas, and spectral efficiency is determined analytically for the central zone and edge area. The multi-cell scenario is taken, which is less explored in the literature. Analytical expression for interference between CU and D2D is derived by considering downlink D2D.

The resource allocation scheme proposed in [35] is different from [33, 34] in the sense that it applies FFR in D2D-enabled communication system where D2D occur during the uplink period. Each cell is divided into the inner and outer regions, and the outer region is then further divided into sectors. Also, the frequency allocation is performed based on the location of the D2D pair. The proposed scheme is compared with [33], and significant gain is obtained. Similar to the [35], the authors in [36] proposed interference mitigation technique for uplink D2D. The objective of this technique is to minimize interference by employing FFR in the D2D system. Authors in [33] assumed fixed transmit power for all the users. However, the boundaries of accessible and reusable regions are calculated in [36]. The D2D pair within the

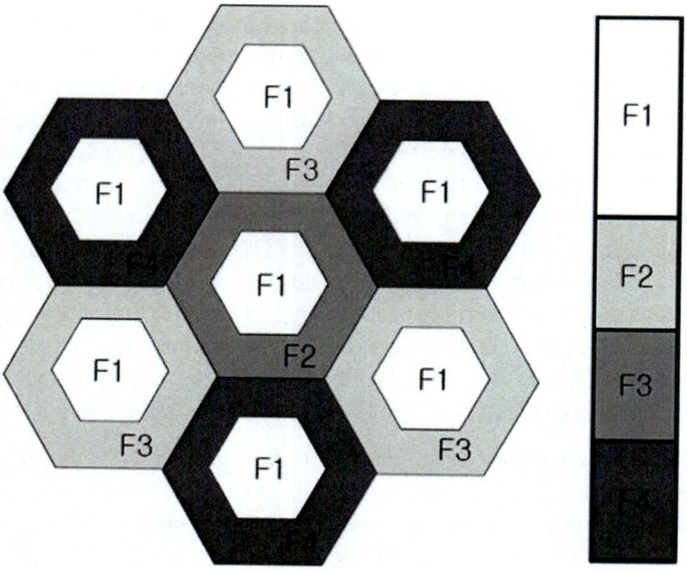


**Fig. 3** Fractional frequency reuse in D2D [33]

accessible region only can reuse the CU's channel. Outage probability is the QoS criterion for each user.

**Energy Efficiency**: Since handheld devices are battery-driven, thus in addition to the sum rate, EE is a key performance parameter for assessing the potential of D2D technology. There are two ways to achieve EE, (1) Maximizing EE, which is defined as the ratio of total sum rate to total power consumption and (2) decreasing total power consumption. Two channel and power allocation techniques were developed in [37] to maximize system and total individual EE of D2D users while preserving the minimum necessary rate for shared uplink CUs. Only a one-to-one channel reuse scenario is considered in this study. In [38], a power control strategy is presented to enhance the D2D system's EE while maintaining the QoS of the CU and D2D pairings. By considering full and average CSI for the interfering channel gains, the problem is formulated as a multi-objective optimization problem and addressed using a d.c. programming approach. To maximize the EE of D2D lines, the authors of [39] investigated the joint channel allocation and power control problem. The problem of EE maximization is nonconvex. Because of the nonconvexity, the original problem is broken into two subproblems. To address it, the authors presented a low-complexity iterative technique.

Article [40] proposes a dual-based resource allocation technique that maximizes the least weighted EE of D2D users while maintaining cellular link QoS. Bhardwaj *et al.* investigated the spectrum-EE trade-off in multicast D2D groups. A power allocation algorithm solves the problem of EE maximization with constraint on spectrum efficiency in [41]. The authors of [42] formulated the resource and EE challenges in order to maximize the system EE for a multiuser downlink D2D network based on OFDMA. The proposed approach allows D2D communications to operate in an overlaying mode that prevents cellular and D2D users from interfering. The EE was improved by the authors in [43], who proposed D2D communication as a network supplement to replace the coverage hole left by turning off the BS.

In [44], Feng et al. proposed a mode selection strategy for a D2D integrated cellular system. The goal is to optimize EE for all users while maintaining QoS and power restrictions. To reduce total power consumption by users, Kai et al. developed a joint channel and power allocation for the uplink D2D system in [45]. The QoS in terms of minimum required data rate is considered for cellular as well as D2D users. The channel assignment is based on many-to-many pairing scheme. Authors in [46] proposed an energy efficient scheme by employing channel and power algorithm for downlink D2D. Each CU can share its channel with at most one D2D pair while a D2D can have multiple channels (many-to-one matching). The optimization problem is to maximize the EE of D2D users subject to QoS constraints for CU. By using Dinkelbach and convex optimization, an iterative algorithm is given. In [47], a combinatorial auction algorithm is proposed to solve the joint channel and power allocation problem. The objective is to maximize the EE of all users in uplink D2D with one-to-many channel reuse strategy. In [48], channel assignment and power control scheme are proposed for uplink D2D in cellular networks to maximize the EE of D2D users. The formulated problem is nonconvex and solved using fractional programming.

## 5 Conclusion

This paper provides an overview of the resource allocation schemes for D2D communication. In the literature, researchers have focused on D2D with different objectives and optimization parameters such as sum rate, energy efficiency, channel and power allocation, etc. Underlay D2D allows to reuse a channel and thus various channel reuse strategies adopted by the researchers are also summarized in this paper. By employing intelligent resource utilization, D2D technology can be chosen as a next generation cellular technology to provide seamless connectivity. D2D communication using Non-orthogonal multiple access (NOMA) schemes can be studied in the future for further improvement in frequency reuse factor.

## References

1. Ericsson (2019) Ericsson mobility report. Technical report
2. Index CVN (2019) Cisco visual networking index: forecast and trends, 2017–2022. CA, USA, Cisco, San Jose
3. Cisco CVNI (2014) Global mobile data traffic forecast update, 2013–2018. White paper (2014)

4. Andrews JG, Buzzi S, Choi W, Hanly SV, Lozano A, Soong ACK, Zhang JC (2014) What will 5G be? IEEE J Sel Areas Commun 32(6):1065–1082. https://doi.org/10.1109/JSAC.2014.2328098
5. Liu J, Kato N, Ma J, Kadowaki N (2015) Device-to-device communication in LTE-Advanced networks: a survey. IEEE Commun Surveys Tutor 17(4):1923–1940. https://doi.org/10.1109/COMST.2014.2375934
6. Mach P, Becvar Z, Vanek T (2015) Inband device-to-device communication in OFDMA cellular networks: a survey and challenges. IEEE Commun Surveys Tutorials 17(4):1885–1922. https://doi.org/10.1109/COMST.2015.2447036
7. Asadi A, Wang Q, Mancuso V (2014) A survey on device-to-device communication in cellular networks. IEEE Commun Surveys Tutorials 16(4):1801–1819. https://doi.org/10.1109/COMST.2014.2319555
8. Zulhasnine M, Huang C, Srinivasan A (2010) Efficient resource allocation for device-to-device communication underlaying LTE network. In: 2010 IEEE 6th international conference on wireless and mobile computing, networking and communications, pp 368–375. https://doi.org/10.1109/WIMOB.2010.5645039
9. Lee DH, Choi KW, Jeon WS, Jeong DG (2013) Resource allocation scheme for device-to-device communication for maximizing spatial reuse. In: 2013 IEEE wireless communications and networking conference (WCNC), pp 112–117. https://doi.org/10.1109/WCNC.2013.6554548
10. Zhao W, Wang S (2015) Resource allocation for device-to-device communication underlaying cellular networks: an alternating optimization method. IEEE Commun Lett 19(8):1398–1401. https://doi.org/10.1109/LCOMM.2015.2444403
11. Lee DH, Choi KW, Jeon WS, Jeong DG (2014) Two stage semi-distributed resource management for device-to-device communication in cellular networks. IEEE Trans Wireless Commun 13(4):1908–1920. https://doi.org/10.1109/TWC.2014.022014.130480
12. Wang R, Zhang J, Song SH, Letaief KB (2015) QoS aware channel assignment for weighted sum rate maximization in D2D communications. In: 2015 IEEE global communications conference (GLOBECOM), pp 1–6. https://doi.org/10.1109/GLOCOM.2015.7417341
13. Cheng Y, Gu Y, Lin X (2014) Power and channel allocation for device-to-device enabled cellular networks. J Comput Inf Syst 10:463–472. https://doi.org/10.12733/jcis8765
14. Feng D, Lu L, Yuan-Wu Y, Li GY, Feng G, Li S (2013) Device-to-device communications underlaying cellular networks. IEEE Trans Commun 61(8):3541–3551. https://doi.org/10.1109/TCOMM.2013.071013.120787
15. Feng D, Lu L, Yi Y, Li GY, Feng G, Li S (2016) QoS-aware resource allocation for device-to-device communications with channel uncertainty. IEEE Trans Vehic Technol 65(8):6051–6062. https://doi.org/10.1109/TVT.2015.2479258
16. Esmat HH, Elmesalawy MM, Ibrahim II (2016) Adaptive resource sharing algorithm for device-to-device communications underlaying cellular networks. IEEE Commun Lett 20(3):530–533. https://doi.org/10.1109/LCOMM.2016.2517012
17. Yu G, Xu L, Feng D, Yin R, Li GY, Jiang Y (2014) Joint mode selection and resource allocation for device-to-device communications. IEEE Trans Commun 62(11):3814–3824. https://doi.org/10.1109/TCOMM.2014.2363092
18. Wang L, Wu H (2014) Fast pairing of device-to-device link underlay for spectrum sharing with cellular users. IEEE Commun Lett 18(10):1803–1806. https://doi.org/10.1109/LCOMM.2014.2351400
19. Tang H, Ding Z (2016) Mixed mode transmission and resource allocation for D2D communication. IEEE Trans Wireless Commun 15(1):162–175. https://doi.org/10.1109/TWC.2015.2468725
20. Zhao W, Wang S (2015) Resource sharing scheme for device-to-device communication underlaying cellular networks. IEEE Trans Commun 63(12):4838–4848. https://doi.org/10.1109/TCOMM.2015.2495217
21. Kim T, Dong M (2014) An iterative hungarian method to joint relay selection and resource allocation for D2D communications. IEEE Wireless Commun Lett 3(6):625–628. https://doi.org/10.1109/LWC.2014.2338318

22. Ma L, Deng X, Wang J, Huang Y, Shi F (2019) Downlink resource sharing in multichannel device-to-device communication. IEEE Wireless Commun Lett 8(3):741–744
23. Gurjar DS, Upadhyay PK (2017) Overlay device-to-device communications in asymmetric two-way cellular systems with hybrid relaying. IEEE Syst J 12(4):3713–3724
24. Gurjar DS, Upadhyay PK (2017) Overlay spectrum sharing for device-to-device communications in two-way cellular networks with nodes mobility. Trans Emerg Telecommun Technol 28(10):e3164
25. Yu C, Tirkkonen O, Doppler K, Ribeiro C (2009) On the performance of device-to-device underlay communication with simple power control. In: VTC spring 2009 IEEE 69th vehicular technology conference, pp 1–5. https://doi.org/10.1109/VETECS.2009.5073734
26. Yu C, Doppler K, Ribeiro C, Tirkkonen O (2009) Performance impact of fading interference to device-to-device communication underlaying cellular networks. In: 2009 IEEE 20th international symposium on personal, indoor and mobile radio communications, pp 858–862. https://doi.org/10.1109/PIMRC.2009.5450264
27. Min H, Lee J, Park S, Hong D (2011) Capacity enhancement using an interference limited area for device-to-device uplink underlaying cellular networks. IEEE Trans Wireless Commun 10(12):3995–4000. https://doi.org/10.1109/TWC.2011.100611.101684
28. Ye Q, Al-Shalash M, Caramanis C, Andrews JG (2014) Resource optimization in device-to-device cellular systems using time-frequency hopping. IEEE Trans Wireless Commun 13(10):5467–5480. https://doi.org/10.1109/TWC.2014.2340879
29. Lin X, Andrews JG, Ghosh A (2014) Spectrum sharing for device-to-device communication in cellular networks. IEEE Trans Wireless Commun 13(12):6727–6740. https://doi.org/10.1109/TWC.2014.2360202
30. ElSawy H, Hossain E, Alouini M (2014) Analytical modeling of mode selection and power control for underlay D2D communication in cellular networks. IEEE Trans Commun 62(11):4147–4161. https://doi.org/10.1109/TCOMM.2014.2363849
31. George G, Mungara RK, Lozano A (2015) An analytical framework for device-to-device communication in cellular networks. IEEE Trans Wireless Commun 14(11):6297–6310. https://doi.org/10.1109/TWC.2015.2452264
32. Darshi S, Bhattacharjee R (2014) Interference analysis of subband allocation for femtocells in fractional frequency reuse based OFDMA networks. In: 2014 international conference on signal processing and communications (SPCOM), pp 1–6. https://doi.org/10.1109/SPCOM.2014.6983997
33. Chae HS, Gu J, Choi BG, Chung MY (2011) Radio resource allocation scheme for device-to-device communication in cellular networks using fractional frequency reuse. In: 2011 IEEE 17th Asia Pacific conference on communications, pp 58–62
34. Zhu H, Wang J (2014) Device-to-device communication in cellular networks with fractional frequency reuse. In: 2014 IEEE international conference on communications (ICC), pp 5503–5507. https://doi.org/10.1109/ICC.2014.6884197
35. Shah ST, Gu J, Chung MY, Hasan SF (2014) FFR based resource allocation scheme for device-to-device communication. In: 2014 IEEE 3rd global conference on consumer electronics (GCCE), pp 622–623. https://doi.org/10.1109/GCCE.2014.7031096
36. Bao P, Yu G, Yin R (2013) Novel frequency reusing scheme for interference mitigation in D2D uplink underlaying networks. In: 2013 9th international wireless communications and mobile computing conference (IWCMC), pp 491–496. https://doi.org/10.1109/IWCMC.2013.6583607
37. Hoang TD, Le LB, Le-Ngoc T (2015) Energy-efficient resource allocation for D2D communications in cellular networks. In: 2015 IEEE international conference on communications (ICC), pp 2251–2256. https://doi.org/10.1109/ICC.2015.7248660
38. Robat Mili M, Tehrani P, Bennis M (2016) Energy-efficient power allocation in OFDMA D2D communication by multiobjective optimization. IEEE Wireless Commun Lett 5(6):668–671. https://doi.org/10.1109/LWC.2016.2614507
39. Xu H, Xu W, Yang Z, Pan Y, Shi J, Chen M (2017) Energy-efficient resource allocation in D2D underlaid cellular uplinks. IEEE Commun Lett 21(3):560–563. https://doi.org/10.1109/LCOMM.2016.2633338

40. Hoang TD, Le LB, Le-Ngoc T (2015) Dual decomposition method for energy-efficient resource allocation in D2D communications underlying cellular networks. In: 2015 IEEE global communications conference (GLOBECOM), pp 1–6. https://doi.org/10.1109/GLOCOM.2015.7417275
41. Bhardwaj A, Agnihotri S (2018) Energy and spectral efficiency trade-off for D2D multicasts in underlay cellular networks. IEEE Wireless Commun Lett 7(4):546–549. https://doi.org/10.1109/LWC.2018.2794353
42. Idris F, Tang J, So DKC (2018) Resource and energy efficient device-to-device communications in downlink cellular system. In: 2018 IEEE wireless communications and networking conference (WCNC), pp 1–6. https://doi.org/10.1109/WCNC.2018.8377382
43. Panahi FH, Panahi FH, Hattab G, Ohtsuki T, Cabric D (2018) Green heterogeneous networks via an intelligent sleep/wake-up mechanism and D2D communications. IEEE Trans Green Commun Netw 2(4):915–931. https://doi.org/10.1109/TGCN.2018.2844301
44. Feng D, Yu G, Xiong C, Yuan-Wu Y, Li GY, Feng G, Li S (2015) Mode switching for energy-efficient device-to-device communications in cellular networks. IEEE Trans Wireless Commun 14(12):6993–7003. https://doi.org/10.1109/TWC.2015.2463280
45. Kai C, Li H, Xu L, Li Y, Jiang T (2018) Energy-efficient device-to-device communications for green smart cities. IEEE Trans Industrial Inf 14(4):1542–1551. https://doi.org/10.1109/TII.2017.2789304
46. Hu J, Heng W, Li X, Wu J (2017) Energy-efficient resource reuse scheme for D2D communications underlaying cellular networks. IEEE Commun Lett 21(9):2097–2100. https://doi.org/10.1109/LCOMM.2017.2711490
47. Wang F, Xu C, Song L, Han Z (2015) Energy-efficient resource allocation for device-to-device underlay communication. IEEE Trans Wireless Commun 14(4):2082–2092. https://doi.org/10.1109/TWC.2014.2379653
48. Jiang Y, Liu Q, Zheng F, Gao X, You X (2016) Energy-efficient joint resource allocation and power control for D2D communications. IEEE Trans Vehic Technol 65(8):6119–6127. https://doi.org/10.1109/TVT.2015.2472995

# Intelligent Reflecting Surface at Mm-Wave Band for D2D Communication: An Insight

**Subhra Sankha Sarma and Ranjay Hazra**

## 1 Introduction

Recent advancements in Internet technology prognosticates the arrival of Internet of everything (IoE). It is expected that the number of IoE devices to be connected would touch the 50 billion mark within a few years. The fifth generation (5G) network has to deal with a massive increase in number of devices thereby introducing a number of issues to tackle [1, 2]. The ever increasing urge for resources makes the current multiple access scheme such as, time division multiple access (TDMA), frequency division multiple access (FDMA), code division multiple access (CDMA), obsolete. Low spectral efficiency of these techniques is assumed to be the main reason for this insufficiency. Basically, IoE is an amalgamation of people, processes, data and things. For enabling IoE in next generation networks, 5G and B5G (beyond 5G), device-to-device (D2D) communication may prove to be a viable candidate as its architecture comes under the IoE networks [3]. The basic components of an IoE network are summarized as follows,

- People: Artificial intelligence (AI) analyzes the personal insights provided by humans through various platforms such as the websites, applications etc.
- Things: A variety of physical objects acquire the data generated from sensors and actuators and transmits them to the designated user.
- Data: The acquired data from the devices is analyzed and then the information is used to propel various systems and thus empower smart technologies.

S. S. Sarma (✉) · R. Hazra
Department of Electronics and Instrumentation Engineering,
National Institute of Technology Silchar, Silchar, India 788010
e-mail: subhra3s@gmail.com



P. H. J. Chong et al. (eds.), *Emerging Electronics and Automation*, Lecture Notes in Electrical Engineering 937, https://doi.org/10.1007/978-981-19-4300-3_27

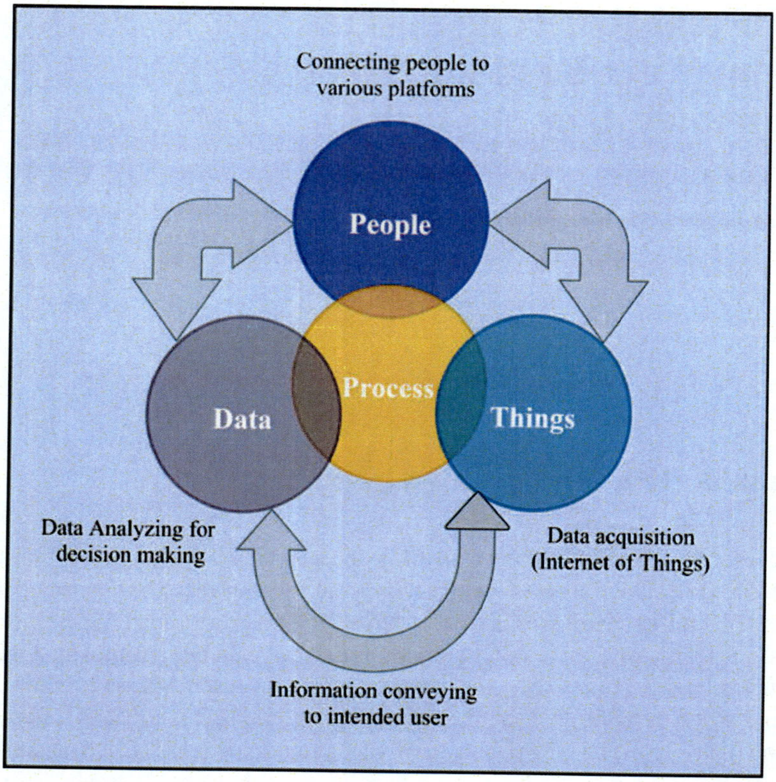


**Fig. 1** Internet of everything [4]

- Process: Machine learning paradigm based processes are used for conveying the information to the intended user and also enhance the quality of service (QoS) (Fig. 1).

## 2 D2D Communication

D2D communication may be defined as the communication between two devices which are in propinquity to each other without the information traversing the base station (BS) [5, 6]. The exchange of information can be done either directly or through hopping. Among all of the applications of IoE, the fastest evolving is the vehicle-to-vehicle (V2V) communication. It is garnering widespread application as well as prominence among the researchers recently. As D2D communication provides higher data rate for small cell, thus the symbiosis between local D2D and V2V communication is gaining importance [7]. Teaming up with mm-Wave 5G cellular network, D2D communication has the potential which could be used for IoE. D2D communication also can offload cellular communication resulting in communication

**Table 1** List of acronyms

| S. No. | Acronyms | Meanings |
|---|---|---|
| 1 | IRS | Intelligent reflecting surface |
| 2 | IoE | Internet of everything |
| 3 | TDMA | Time division multiple access |
| 4 | FDMA | Frequency division multiple access |
| 5 | CDMA | Code division multiple access |
| 6 | B5G | Beyond fifth generation |
| 7 | D2D | Device-to-device communication |
| 8 | AI | Artificial intelligence |
| 9 | QoS | Quality of service |
| 10 | BS | Base station |
| 11 | V2V | Vehicle-to-vehicle communication |
| 12 | LOS | Line-of-sight |
| 13 | UAV | Unmanned aerial vehicle |
| 14 | CSI | Channel state information |
| 15 | 3-D | Three dimensional |
| 16 | SINR | Signal to interference plus noise ratio |

among higher number of users. The list of acronyms along with their meanings are shown in Table 1.

## 3 Mm-Wave Propagation

Mm-Wave band frequency ranges from (30–300) GHz with wavelength ranging from (10–1) mm [8]. The propagation features of mm-Wave band are as follows: (a) The transmitted signal at mm-Wave band undergoes higher propagation loss as compared to the GSM frequency band which is why directional antennas are used for compensating the losses. (b) Signal at mm-Wave band has trouble diffracting around edges or obstacles. This compels the system to use Line-of-Sight transmission (LOS) of signals. But this makes the system vulnerable to different types of obstacles. The signals may get blocked by these obstacles and as a result, the efficiency gets reduced. (c) Signals have difficulty in penetrating through concrete walls, solid structures, trees, humans, etc. Due to this problem, several losses arise such as, blockage effect, shadowing effect, rain attenuation, atmospheric absorption, foliage loss. [9, 10] These challenging issues appear to be a hindrance for the successful implementation of mm-Wave band for reliable and efficient D2D communication. Dynamic obstacles also hamper the system by attenuating the signal strength in major urban cities due to the movement of various transportation vehicles. Apparently, the issue of dynamic obstacles could be solved through the use of unmanned aerial vehicle (UAV). UAVs

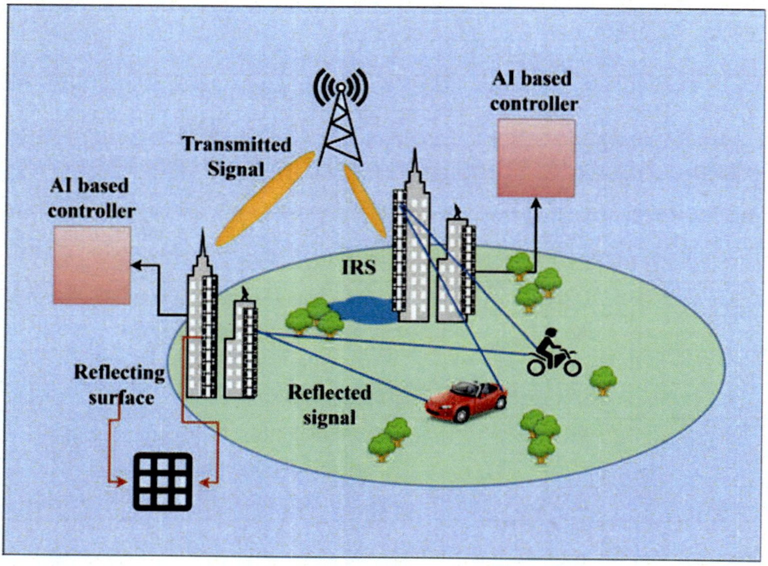


**Fig. 2** Application of intelligent reflective surface in mm-Wave network based on AI controller

can be flown to the defined position for providing LOS signal path to the desired user effectively. But there is an issue with the UAV. Since, it is an aerial device, it requires a large amount of power backup for staying in air for a longer time period. But the battery could only provide back up for a few hours of flight time. This flaw proves to be fatal for continuous usage of signal transmission to the desired user. Thus, there is another technique known as intelligent reflecting surface (IRS) through which the problem of dynamic obstacle can be tackled.

The issue of dynamic obstacles [11] can be partially solved by the use of intelligent reflecting surface. IRS is basically a planar surface which consists of numerous passive reflecting elements as shown in Fig. 2. Each of these reflecting elements has the capability of inducing a change independently in phase and amplitude of the incident transmitted signal. In a 5G mm-Wave cellular network, IRS could be densely installed and through smart controlling of the reflected transmitted signal, the propagation channel can be reconfigured to yield intended optimal configurations between the transmitter and receiver. As a result, it has the potential to yield highly optimal channel and thereby provide a better way to manage interference. Thus, we can get a reliable system which enhances the overall system capacity.

# 4 IRS: The Key to Dynamic Obstacle

We have assumed D2D communication in a 5G mm-Wave underlaying cellular network with a carrier frequency of 28 GHz. The base station (BS) is situated at the center of the dense cell. The cellular and D2D users are randomly situated in the cell following the poisson point distribution. It has been assumed that the BS has complete knowledge of the channel state information (CSI). It also has knowledge about the positions of the cellular and D2D users in the cell along with their respective pathloss exponents. An urban scenario is considered for the purpose. In this article, three urban scenarios are considered as shown in Fig. 3 which are then compared amidst dynamic obstacles for better efficiency and reliability. Signal propagation and resource allocation are the two factors which demands attention. Some key features and insights are laid down in the following subsections as follows,

## *4.1 Signal Propagation*

Figure 3a gives us the scenario of signal transmission in an urban environment using directional antenna. Rician channel is considered for signal propagation which signifies that only the LOS component of the transmitted beam is taken into account as it is the strongest of all the beams. The transmitted mm-Wave signal propagation undergoes several attenuation factors which degrades the signal strength as described in the earlier section. Thus, the received signal is of poor signal strength. The intended user is assumed to be shadowed by large vehicles on the road. In the figure, it can be seen that the user is behind a concrete structure which attenuates the signal strength. And if the user is behind a moving large vehicle then the strength of the received signal is further attenuated for a time period of few milliseconds. Thus, dynamic obstacles also cater to the problem and poses to be an alarming factor in the signal reception for the intended user (in this case, a vehicle). And the penetration loss for the transmitted signal at 28 GHz is very high as compared to 2.4 GHz carrier frequency. Again, in Fig. 3b scenario, we can observe that the signal transmission takes place through the help of an UAV (also popularly known as drone). The UAV acts as a relay in this scenario. The UAV helps in maintaining LOS propagation of transmitted signal. The intended user which is shadowed by a large vehicle in an urban environment tends to have a better signal strength as compared to the previous scenario depicted in Fig. 3a. The UAV can be reconfigured to a new position at a certain height with new coordinates to provide a better signal strength and enhanced data rate of the D2D user. But there is a trade off attached to it. For relocating an UAV to its new three-dimensional (3-D) coordinate, it would require a substantial amount of energy consumption. Moreover, the UAV has a flight time of only a few hours. So, the implementation of an UAV may not be cost effective. Thus, we look into the third scenario depicted in Fig. 3c. The figure shows a scenario wherein the transmitted signal from the BS is incident on an IRS [12]. The amplitude and/ or phase of the signal

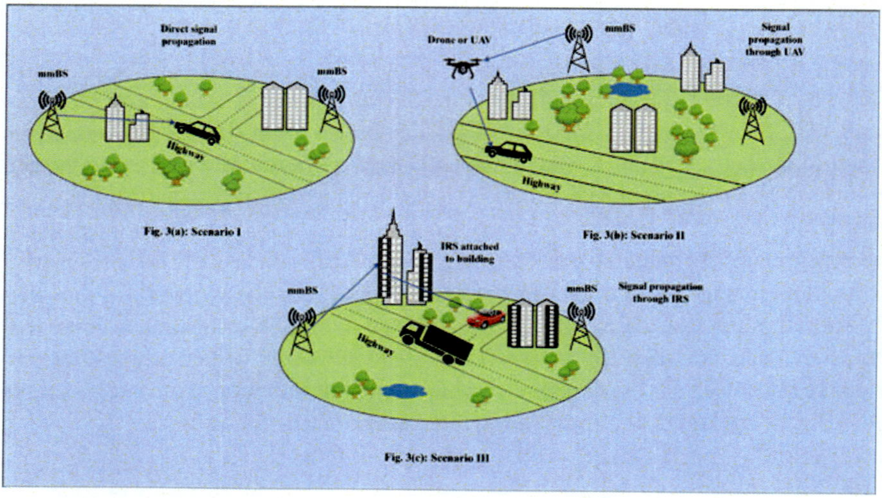


**Fig. 3** Three different scenarios for signal propagation in mm-Wave network. **a** The signal propagation in the presence of various blockages, **b** the signal propagation through UAV (drone), and **c** the signal propagation through IRS when obstructed by moving vehicles (dynamic obstacles)

is then altered and reconfigured as per the needs of the desired configuration. The IRS supports passive beamforming [13]. A controller based on AI is attached to the passive reflecting elements which dynamically changes the amplitude and phase of the incident signal beam. This is basically a speculative architecture which requires extensive work and formidable research. This dynamic reconfiguration of incident signal would help in minimizing the interference due to the dynamic obstacle that blocks the signal from reaching the intended D2D user. In [14], the authors have continuously changed the phases of the incident signal. The IRS acts as a frequency mixer which introduces new frequencies and makes the propagation environment non-linear. The phases of each of these signal paths are aligned by choosing the phases of the IRS. This would help in increasing the SINR.

## *4.2 Resource Allocation*

For an underlaying mm-Wave 5G cellular network, mitigation of interference is of foremost priority. Thus, resource allocation plays a pivotal part in the modeling of the communication system. We have considered a system where the D2D pairs use the same resources as that of the cellular users. Partial resource multiplexing [15] is a viable solution for the interference minimization problem. This scheme tells us the number of cellular resources to be multiplexed with the D2D users. Additionally, it also lets us know about the particular cellular user which is about to share its resources with the D2D user. Firstly, the BS will sort the D2D according to its priority status.

The resources would be allocated to the intended D2D user accordingly through the application of a Hungarian algorithm. This stands true if the number of D2D users are lesser in number to that of the cellular users. But if the number of the D2D users are greater than that of the cellular users, partial resource multiplexing scheme will be applied. According to this scheme, partial resource multiplexing would be applied to only the rest of the D2D users who were not allocated any resources due to the insufficiency of cellular resources. At any time instant, cellular resource is shared with only one of the D2D user but a particular D2D user can utilize the resource of more than one cellular user. A multiplexing ratio is also considered in this case so that the D2D user can be sorted out. The D2D user which has lower signal-to-interference-plus-noise ratio (SINR) would showcase lower priority level status.

## 5 Conclusion

Signal propagation in the mm-Wave band suffers extensive degradation due to the presence of a variety of interference. Dynamic obstacles also cater to the signal attenuation. Thus, we have focused our work on this parameter so that interference can be minimized for urban environment. Dynamic obstacle refers to the obstructions in urban environments when signal strength gets degraded abruptly for a few milliseconds due to movement of large vehicles as the signal having carrier frequency of 28 GHz cannot penetrate the vehicle. The signals at 28 GHz frequency has the problem of diffraction which restricts them from bending around the obstacles. UAV can solve this pressing issue but suffers from the issue of energy saving and limited time of flight. Thus, this signal degradation can be controlled through the application of intelligent reflecting elements. These IRS are connected to AI-based controllers which dynamically alters the amplitude and/or phase of the incident beam on the passive reflecting elements. Furthermore, resource allocation is done based on partial resource multiplexing which enables the cellular users to share the resources with the D2D users. Hungarian algorithm can be used for sorting of the D2D users based on priority status. The proposed scheme has the potential to uphold the minimum SINR required for the D2D communication against dynamic obstacles.

## References

1. Minoli D (2019) Occhiogrosso B (2019) Practical aspects for the integration of 5G networks and IoT applications in smart cities environments. Wireless Commun Mobile Comput 5710834:30. https://doi.org/10.1155/2019/5710834
2. Sarma SS, Hazra R (2020) Interference management for D2D communication in mmWave 5G network: an alternate offer bargaining game theory approach. In: 2020 7th international conference on signal processing and integrated networks (SPIN). IEEE, pp 202–207. https://doi.org/10.1109/SPIN48934.2020.9071419

3. Sahoo BPS, Chou C, Weng C, Wei H (2019) Enabling millimeter-wave 5G networks for massive IoT applications: a closer look at the issues impacting millimeter-waves in consumer devices under the 5G framework. IEEE Consumer Electron Mag 8(1):49–54. https://doi.org/10.1109/MCE.2018.2868111
4. Miraz MH, Ali M, Excell PS, Picking R (2015) A review on Internet of Things (IoT), Internet of Everything (IoE) and Internet of Nano Things (IoNT). In: 2015 internet technologies and applications (ITA), Wrexham, UK, pp 219-224. https://doi.org/10.1109/ITechA.2015.7317398
5. Cai Y, Ni Y, Zhang J, Zhao S, Zhu H (2019) Energy efficiency and spectrum efficiency in underlay device-to-device communications enabled cellular networks. China Commun 16(4):16–34. https://doi.org/10.12676/j.cc.2019.04.002
6. Sarma SS, Khuntia P, Hazra R (2021) Power control scheme for device-to-device communication using uplink channel in 5G mm-Wave network. Trans Emerg Telecommun Technol. https://doi.org/10.1002/ett.4267
7. Mousam M, Chakarbaty R, Kumari N, Mukherjee A (2018) Millimeter wave vehicular communications to generate high data rates. In: 2018 international conference on communication, computing and Internet of Things (IC3IoT), pp 191–194
8. Ju S, Kanhere O, Xing Y, Rappaport TS (2019) A millimeter-wave channel simulator NYUSIM with spatial consistency and human blockage. In: 2019 IEEE global communications conference (GLOBECOM), pp 1–6. https://doi.org/10.1109/GLOBECOM38437.2019.9013273
9. Wei L, Hu RQ, Qian Y, Wu G (2014) Key elements to enable millimeter wave communications for 5G wireless systems. IEEE Wireless Commun 21(6):136–143. https://doi.org/10.1109/MWC.2014.7000981
10. Rappaport TS et al (2013) Millimeter wave mobile communications for 5G cellular: it will work! IEEE Access 1:335–349. https://doi.org/10.1109/ACCESS.2013.2260813
11. Singh D, Chattopadhyay A, Ghosh SC (2020) Distributed relay selection in presence of dynamic obstacles in millimeter wave D2D communication. In: ICC 2020—2020 IEEE international conference on communications (ICC), Dublin, Ireland, pp 1–6. https://doi.org/10.1109/ICC40277.2020.9148816.
12. Wu Q, Zhang S, Zheng B, You C, Zhang R (2021) Intelligent reflecting surface aided wireless communications: a tutorial. IEEE Trans Commun 69(5):3313–3351. https://doi.org/10.1109/TCOMM.2021.3051897
13. Björnson E, Özdogan O, Larsson EG (2020) Reconfigurable intelligent surfaces: three myths and two critical questions. IEEE Commun Mag 58(12):90–96. https://doi.org/10.1109/MCOM.001.2000407
14. Yuan J, Carvalho ED, Williams RJ, Bjornson E, Popovski P (2021) Frequency-mixing intelligent reflecting surfaces for nonlinear wireless propagation. IEEE Wireless Commun Lett 10(8):1672–1676. https://doi.org/10.1109/LWC.2021.3077085
15. Li Y, Liang Y, Liu Q, Wang H (2018) Resources allocation in multicell D2D communications for Internet of Things. IEEE Internet Things J 5(5):4100–4108. https://doi.org/10.1109/JIOT.2018.2870614

# Signal Processing Track—1

# Pre-trained EfficientNet-B0 with Adjusted Optimizer, Learning Rate and Image Size to Improve Diabetic Foot Ulcers Diagnosis

Sujit Kumar Das, Pinki Roy, and Arnab Kumar Mishra

## 1 Introduction

The DFU can result in costly treatment, lower feet, or limb amputation if proper care is not taken. Facts indicate that among all diabetes patients, 15% suffer from DFU complications later stage [1]. It is the most common cause of non-traumatic lower extremity amputations in the world. About 85% of the lower limb amputations are caused by DFU [2]. Almost a million diabetic patients go through necessary surgery of their legs due to late detection and lack of proper treatment of DFU every year [3]. So, to avoid such severe situations, a patient having diabetes is required to go for a check-up of their foot by medical experts (specifically DFU experts). It is equally important to maintain proper medication and self-foot care suggested by DFU experts. Most of the time, all these processes bring a huge financial burden and physical and mental stress to the patients and their families.

In current medical practices evaluating DFU undergoes several subtasks like early diagnosis and management. These are achieved by several expensive clinical tests such as CT scans, X-Ray and MRIs. Based on these test results, DFU experts have their visual expertise identify DFU cases' presence and make a treatment plan. However, in general, DFU has irregular structures and uncertain outer boundaries, making it difficult for the DFU experts to appear into a conclusive decision and hence correct treatment plan. The evolution of ML in the medical imaging fields [4–9], are very promising. Using these techniques in solving various problems suggests using these techniques in designing automatic, cost-effective, and reliable DFU identification systems to help medical experts provide better patient care. However, designing such systems is a challenging task, and it becomes more challenging in the case of DFU for the following reasons:

S. K. Das (✉) · P. Roy · A. K. Mishra
Department of Computer Science and Engineering, National Institute of Technology Silchar, Silchar, Assam 788010, India
e-mail: dassujit88@gmail.com



P. H. J. Chong et al. (eds.), *Emerging Electronics and Automation*, Lecture Notes in Electrical Engineering 937, https://doi.org/10.1007/978-981-19-4300-3_28

- Unavailability of publicly available training dataset.
- Data collection requires a massive amount of time.
- Heterogeneous image qualities due to multiple capturing devices.
- The visual appearances of DFU wounds can have different characteristics and low contrast between object and background makes it more difficult to analyze them.

In this work, DL-based automatic DFU identification system is proposed. The Imagenet-based Pre-Trained EfficentNet-B0 [10] model has been used as a bottleneck network combined with VGG-based Fully connected layers for automatic DFU identification. The primary objectives of the proposed work are as follows:

- To evaluate the performances of Pre-Trained EfficientNet-B0 on various input image sizes.
- To find the best Optimizer and Learning rate settings to enhance the model performance.

The rest of the article is organized as follows: Sect. 2 includes a discussion of related work, Sect. 3 includes material and method. The experimental results and discussion are in Sect. 4 and 5, respectively. At last, Sect. 6 includes conclusions and future works.

## 2 Related Works

Automatic classification of DFU to make the diagnosis process faster and cost-effective is an important aspect to deal with it. The application of ML and DL approaches to do automatic DFU classification made tremendous contributions in recent years. Goyal et al. [11] have proposed a parallel convolution layer-based CNN (DFUNet). In the same work, the authors explored traditional ML-based features to identify DFU cases. The standard CNNs namely, LeNet [12], AlexNet, and GoogleNet [9] architectures are also used to extract features from image patches. They have observed that DFUNet outperformed all unique feature set-based classification results. Though the DFUNet has achieved promising results, the absence of automatic annotators to delineate and classify images is one of the major limitations. In a similar work, Alzubaidi et al. [14] have proposed DFU_QUTNet to extract multi-level features, and fed into support vector machine (SVM) and K-Nearest Neighbors (KNN) classifiers. The novel DFU_QUTNet architecture is a width network with less computational cost. The results of DFU_QUTNet are compared to standard CNN models (GoogleNet, AlexNet, and VGG16 ). It has been found that the proposed model with the combination of support vector machine (SVM) classifier outperformed state-of-the-art CNN architectures. Zhao et al. [15], have proposed a Bilinear CNN (Bi-CNN) model to diabetic wound grading tasks based on the idea of fine-grained classification. They have performed various preprocessing like deblurring, image super-resolution, and illumination correction before training VGG16 based Bi-CNN model. The evaluation of the proposed models has outperformed state-of-the-art works results.

In The Medical Image Computing and Computer-Assisted Intervention Society (MICCAI), DFUC 2020 challenge, Cassidy et al. [16] have used Faster-RCNN [17] and Inception-v2-Resnet101, FRCNN Inception-v2-ResNet101, EfficientDet [18], and YOLOv5 [19] to identify DFU cases. They have observed that these networks are significant in providing promising results.

Goyal et al. [20] in one of their works used Faster R-CNN [17] with InceptionV2 [13] architecture and a two-stages transfer learning approach to perform real-time DFU localization. Faster R-CNN produced the highest accuracy results, and this approach has been implemented in a mobile application to provide better health service. The model proposed by Claudia et al. [21] consisted of three stages, namely, data normalization, competitive neural layer (CNL)-based clustering, and the at last design of the proposed method consisted of defining which of the obtained groups presented the higher risk of developing diabetic foot. In a work, Yap et al. [22] used Faster R-CNN [17], EfficientDet [18] and YOLOv5 [19] for DFU identification. The result shows that Faster R-CNN is useful in reducing false positives and hence improving the identification results.

# 3 Material and Methods

## 3.1 Dataset Description

The DFU image dataset consists of 292 DFU abnormal (positive) and 105 normal (negative) full foot images collected from Lancashire Teaching Hospital, United Kingdom, by a research group at Manchester Metropolitan University. Also, images are captured in room light condition and parallel orientation from a distance of 30–40 cm from the ulcer area. The dataset creator also extracted Region-of-Interests (ROIs) and annotated them with the help of DFU experts. Once the annotation is carried out, the complete dataset consists of 641 normal and 1038 abnormal skin patches (Total 1679). The dataset[1] is not publicly accessible but can acquire it, following the proper guidelines given in the dataset release agreement. A few examples of both the classes are shown in Fig. 1.

## 3.2 Method

The proposed model used EfficientNet-B0 [25], which is the first version of the EfficientNet family. The family of EfficientNet from B0-B7 is proved to be computationally efficient compared to other CNN models proposed on the Imagenet

[1] http://www2.docm.mmu.ac.uk/STAFF/M.Yap/dataset.php

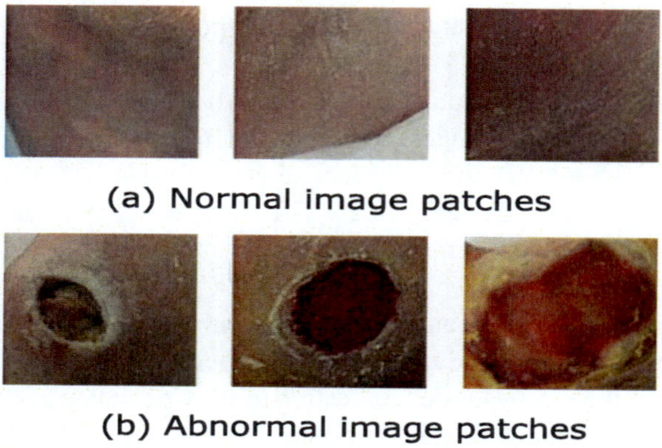


**Fig. 1** DFU dataset examples

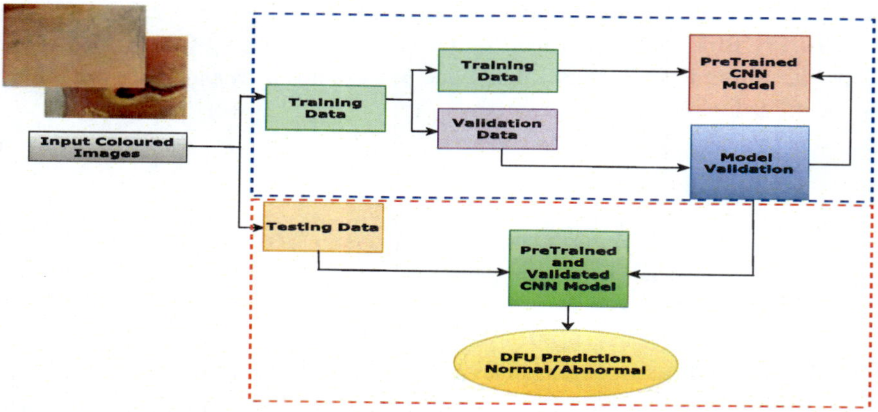


**Fig. 2** A higher level view of proposed model diagram

challenge. A higher-level view of the proposed model is shown in Fig. 2. The idea of EfficientNet-B0 is a combination of inverted residual blocks with different settings. The Squeeze Excitation block and Swish activation are two important operations in these inverted residual blocks. The details of each of these are discussed in the following subsection.

**Inverted Residual Block** The idea of the inverted residual block was first introduced in MobileNet [23]. In inverted residual blocks (MBconv), the skip connections are made between narrower layers instead of wider layers. This skip connection reduces the number of parameters. Thereby reduces computations without compromising with accuracy. A typical inverted residual block is shown in Fig. 3.

**Squeeze and Excitation Block (SE)** The basic idea of using the SE block is to give weightage to each channel instead of equal weightage to the channels of the convolution operation. The network itself learns these individual weights. For input

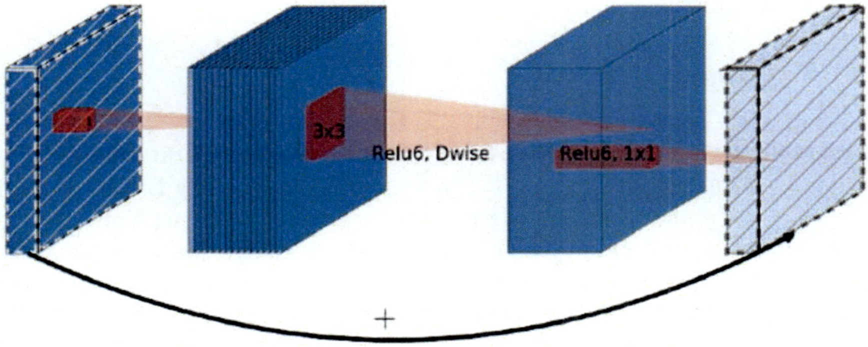


**Fig. 3** Inverted residual block

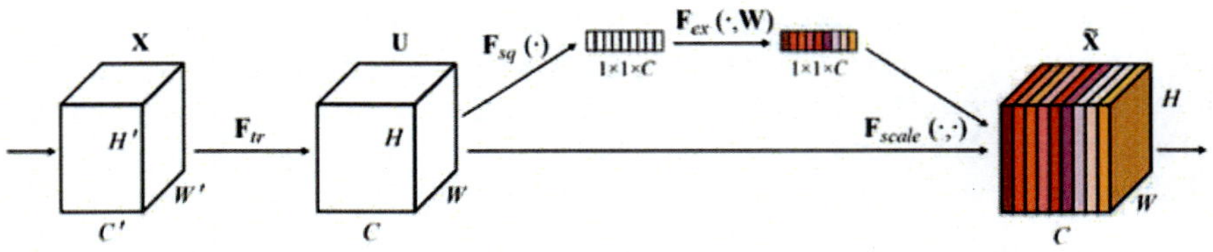


**Fig. 4** Squeeze and excitation block

*X*, a mapping transformation ($F_{tr}$) with feature map *U*, a Squeeze Excitation block (SE), is constructed to make feature improvement. At first, a channel descriptor is created by passing features *U* through a squeeze operator ($F_{sq}$). This step helps in producing an embedded global distribution of channel-wise feature responses [24]. Then, it goes through an excitation operation ($F_{ex}$) to generate a collection of pre-channel improved weights. These improved weights are applied to *U* and produce the output of SE block and feed into the subsequent layers. A pictorial representation of squeeze and excitation block is shown in Fig. 4.

**Swish Activation** To tackle the problem of ReLU, where it nullifies all negative values, a new activation function has been proposed. The swish activation function proved to be more useful in deeper networks. It ensures one sided boundedness at 0, smoothness, and non-monotonicity. The swish activation function is a multiplication of a linear value and sigmoid activation function. The readers can have more insight on the Swish activation function at [25] published work.

# 4 Experimental Results

## 4.1 Experimental Setup

In the proposed model Imagenet-based Pre-Trained EfficientNet-B0 has been used as transfer learning from Keras API. The reason behind choosing EfficientNet-B0 is that it is one of the latest CNNs frameworks capable of achieving high accuracy and efficiency over the existing CNNs. Additionally, including the top argument as False, a sequence of 3 fully connected (FC) layers with a 'ReLU' activation function following standard VGG network architecture has been added after the flatten vector. Also, a dropout layer with 30% random dropout has been used for intermediate FC layers. Finally, to get the prediction, the output layer contains a 'sigmoid' activation function. For experimental evaluation, the dataset was split into 80:20 ratio for training and testing purposes, respectively. Further, from the training 10% of the data are kept for model validation. The random state has been kept the same for every optimizer_learning rate (Opt_Lr) setting and input image size for a fair comparison. All the experiments have been performed using a freely available google colaboratory GPU environment. The experiments are carried out in a step by step manner as:

- ImageNet Pre-Trained EfficientNet-B0 with modified fully connected are evaluated in the same random split with optimizers Adaptive Moment Estimation (Adam), Stochastic Gradient Descent (SGD) and Adaptive Gradient Algorithm (Adagrad), learning rate (0.001, 0.0001, 0.00001) and input image sizes ($32 \times 32$, $64 \times 64$, and $128 \times 128$).
- The performance of the proposed model is evaluated using multiple evaluation metrics like Accuracy, Sensitivity (or Recall), Specificity, Precision, $F1$-Score, AUC(area under the curve value).
- After finding the best setting among considered, Imagenet Pre-Trained VGG16, ResNet50, and DenseNet121 models are evaluated in a found dominant setting to compare the results with the Pre-Trained EfficientNet-B0.

## 4.2 Results

The identification of DFU normal versus abnormal classes has been evaluated using different important evaluation metrics. In Table 1 Accuracy of the proposed model has been reported with different optimizer_learning rates (Opt_lr) and input size among considered cases. The accuracy table has indicated that the Adagrad optimizer with 0.01 learning rate and $128(H) \times 128(W)$ input image size proved to the best setting with the highest accuracy of 94.05%.

Another important evaluation metric to evaluate the proposed model is $F$-measures ($F1$-Score). Table 2 includes the $F1$-Score with different Opt_lr and input image sizes. It is clear from the table that in $F1$-Score also the proposed model shows similar characteristics as accuracy.

**Table 1** Accuracy on different learning rate and optimizer setting for multiple dimension input

| Opt_lr | $32(H) \times 32(W)$ | $64(H) \times 64(W)$ | $128(H) \times 128(W)$ |
|---|---|---|---|
| Adam_0.01 | 86.90 | 91.37 | 90.77 |
| Adam_0.001 | 92.26 | 89.58 | 92.86 |
| Adam_0.0001 | 91.96 | 91.67 | 92.86 |
| SGD_0.01 | 90.77 | 92.86 | 91.96 |
| SGD_0.001 | 92.86 | 91.96 | 89.58 |
| SGD_0.0001 | 89.58 | 90.77 | 91.07 |
| Adagrad_0.01 | 90.18 | 90.48 | **94.05** |
| Adagrad_0.001 | 90.18 | 91.67 | 92.56 |
| Adagrad_0.0001 | 83.93 | 87.20 | 89.27 |

The boldface values represents the metric-wise best performances, as compared to the others present in the table

**Table 2** $F1$-score(%) on different learning rate and optimizer setting for multiple dimension input

| Opt_lr | $32(H) \times 32(W)$ | $64(H) \times 64(W)$ | $128(H) \times 128(W)$ |
|---|---|---|---|
| Adam_0.01 | 82.94 | 88.71 | 87.84 |
| Adam_0.001 | 90.07 | 85.83 | 90.77 |
| Adam_0.0001 | 89.81 | 88.52 | 90.77 |
| SGD_0.01 | 88.12 | 91.04 | 90.11 |
| SGD_0.001 | 90.84 | 88.98 | 87.08 |
| SGD_0.0001 | 86.17 | 88.47 | 88.46 |
| Adagrad_0.01 | 87.45 | 87.09 | **92.42** |
| Adagrad_0.001 | 86.95 | 88.88 | 90.63 |
| Adagrad_0.0001 | 77.87 | 83.52 | 86.15 |

The boldface values represents the metric-wise best performances, as compared to the others present in the table

The AUC values in similar settings as Accuracy and $F1$-Score are reported in Table 3. The observation from the AUC values has given more confidence to conclude that Adagrad_0.01 with $128(H) \times 128(W)$ input image size is one of the best settings among all considered with achieving the highest AUC value 94.15%.

The precision results are shown in Table 4. The table shows that the proposed model could not achieve the highest precision values among all settings, which is quite obvious. The high value of recall (or sensitivity) comes with compromising of precision value. Moreover, in automatic medical systems, sensitivity is a more important metric than precision. Next, in Figs. 5 and 6, sensitivity and specificity are shown using bar graphs. The results from these graphs have shown an impressive performance in the dominant setting of the proposed model.

**Table 3** AUC on different learning rate and optimizer setting for multiple dimension input

| Opt_lr | $32(H) \times 32(W)$ | $64(H) \times 64(W)$ | $128(H) \times 128(W)$ |
|---|---|---|---|
| Adam_0.01 | 86.16 | 90.80 | 90.03 |
| Adam_0.001 | 92.11 | 88.19 | 92.60 |
| Adam_0.0001 | 92.02 | 90.17 | 92.57 |
| SGD_0.01 | 90.47 | 93.18 | 92.60 |
| SGD_0.001 | 92.74 | 90.56 | 89.94 |
| SGD_0.0001 | 88.62 | 91.05 | 90.71 |
| Adagrad_0.01 | 89.98 | 89.20 | **94.15** |
| Adagrad_0.001 | 89.25 | 90.75 | 92.79 |
| Adagrad_0.0001 | 81.99 | 86.69 | 88.82 |

The boldface values represents the metric-wise best performances, as compared to the others present in the table

**Table 4** Precision on different learning rate and optimizer setting for multiple dimension input

| Opt_lr | $32(H) \times 32(W)$ | $64(H) \times 64(W)$ | $128(H) \times 128(W)$ |
|---|---|---|---|
| Adam_0.01 | 82.94 | 89.06 | 88.88 |
| Adam_0.001 | 88.72 | 89.83 | 90.07 |
| Adam_0.0001 | 87.50 | 93.91 | 90.08 |
| SGD_0.01 | 87.12 | 87.77 | 85.42 |
| SGD_0.001 | 89.47 | **93.96** | 83.10 |
| SGD_0.0001 | 87.90 | 85.00 | 87.79 |
| Adagrad_0.01 | 85.82 | 90.76 | 90.37 |
| Adagrad_0.001 | 88.71 | 91.05 | 87.68 |
| Adagrad_0.0001 | 82.61 | 82.57 | 85.50 |

The boldface values represents the metric-wise best performances, as compared to the others present in the table

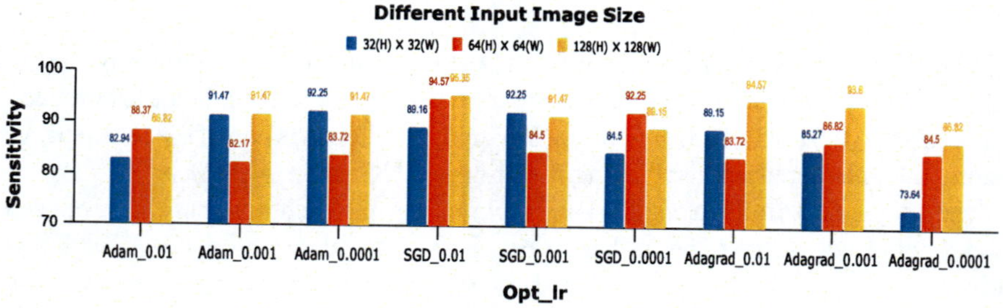


**Fig. 5** Sensitivity based on different optimizer_learning rate and input image size

# 5 Discussion

In DFU identification using Pre-Trained EfficientNet-B0 with multiple optimizers, learning rate and input image sizes are explored. Various important evaluation met-

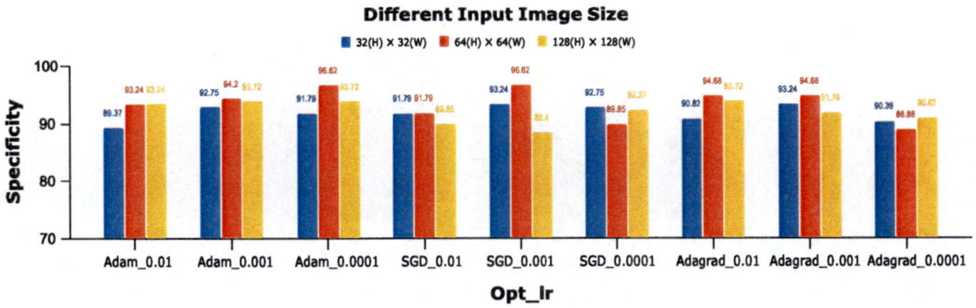


**Fig. 6** Specificity based on different optimizer_learning rate and input image sizes

rics results are reported to find the best Optimizer_Learning rate (Opt_lr) setting and input image size among considered ones. Based on the most important evaluation metrics Accuracy, $F1$-Score, and AUC value, the dominant setting is Adagrad_0.01 and $128(H) \times 128(W)$. Although the models' performance in other metrics is not the highest with dominant settings, they are promising with more than 90% except precision. Further, to compare the results with other Pre-Trained models on similar settings, VGG16, DenseNet121, and ResNet50 have been considered. The comparison is included in Table 5.

**Table 5** Comparison with different pretrained model with Adagrad_0.01 and $128(H) \times 128(W)$

| Metrics | VGG16 | ResNet50 | DenseNet121 | EfficientNetB0 |
|---|---|---|---|---|
| Accuracy | 91.07 | 89.58 | 92.262 | **94.05** |
| Sensitivity | 87.57 | 89.92 | 83.72 | **94.57** |
| Specificity | 93.24 | 89.37 | **97.58** | 93.72 |
| Precision | 88.98 | 84.06 | **95.57** | 90.37 |
| Recall | 87.60 | 89.92 | 83.72 | **94.57** |
| $F1$-score | 88.28 | 86.89 | 89.25 | **92.42** |
| AUC | 90.42 | 89.64 | 90.65 | **94.15** |

**Table 6** Comparison with SOTA

| Metrics | DFUNet [11] | DFUQUT_Net[14] | Pre-Trained EfficientB0 |
|---|---|---|---|
| Accuracy | 92.50 | – | **94.05** |
| Sensitivity | 93.40 | – | **94.57** |
| Specificity | 91.10 | – | **93.72** |
| Precision | 94.50 | **95.40** | 90.37 |
| Recall | – | 93.60 | **94.57** |
| $F1$-score | 93.90 | **94.50** | 92.42 |
| AUC | **96.10** | – | 94.15 |

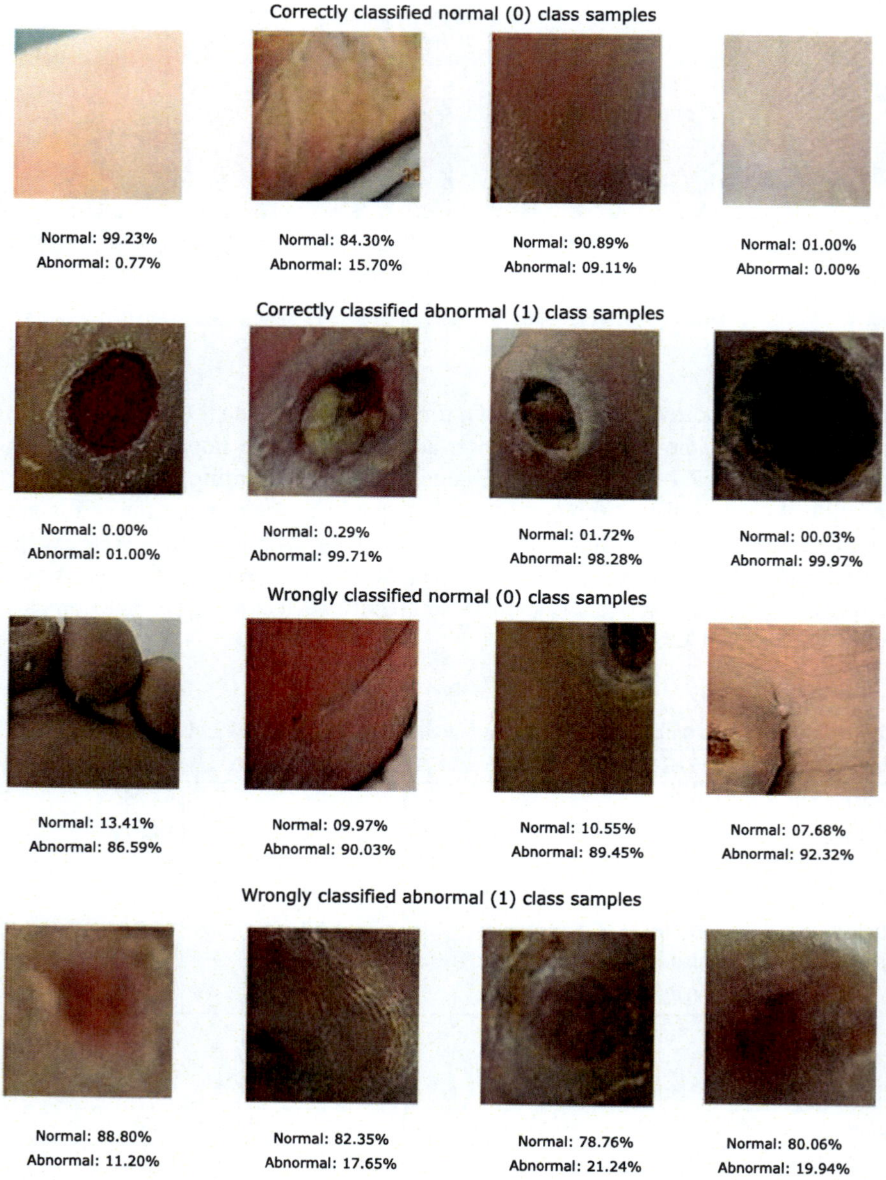


**Fig. 7** A few examples of correctly and wrongly classified samples by the proposed approach

The proposed model's result is compared with state-of-art work in Table 6. It is shown in the comparison table that the performance of the proposed model has not outperformed SOTA in all the evaluation metrics except accuracy, specificity, and recall. However, it is important to note that the input image size was $224(H) \times 224(W)$ in SOTA models. Also, data augmentation had been performed to increase the number of training samples. Further, investigations have been made by looking at some of the correctly and wrongly classified examples in Fig. 7. The first and second rows represent correctly classified samples of normal (0) and abnormal (1) classes. From the correctly classified samples, it has been observed that the proposed model was quite confident in classifying samples correctly where clear skin tone was present for both normal and abnormal classes. In the last two rows of Fig. 7 wrongly classified samples of normal and abnormal classes are shown. The wrongly classified normal classes, where sharp skin tones were present, the model got confused and labeled them as an abnormal class. However, in the case of wrongly classified abnormal cases, the image patches were blurred in nature.

## 6 Conclusion

DFU identification is one of the trending research areas where deep learning algorithms proved to be an effective way to provide faster, cost-effective solutions to support clinician and patient care. This work aimed, having fewer training examples, how optimization algorithms, learning rate, and input shape enhance the performance of the Pre-Trained DL model. The use of ImageNet-based Pre-Trained model when we have fewer training samples in hand is one of the best approaches. Therefore, ImageNet-based Pre-Trained EfficientNet-B0, one of the latest DL architectures with modified FC parts, has been picked to solve the DFU identification problem. Also, the most frequently used optimization algorithms Adam, SGD, and Adagrad with learning rate 0.001, 0.0001, 0.00001 have been explored in different input image sizes as $32(H) \times 32(W)$, $64(H) \times 64(W)$, and $128(H) \times 128(W)$. We have observed that the Adagrad optimizer with an adjusted learning rate of 0.01 in $128(H) \times 128(W)$ input image size successfully achieved promising results in most evaluation metrics. From the observation and assumption, we are hoping that a further increase of the input image size of the DFU examples could improve the performance of the proposed model. However, due to the unavailability of computing power, we could not explore that part. We consider this as one of the future tasks in DFU identification. Further, data augmentation to increase the number of training examples and designing new CNN architecture in DFU identification are other research scopes which we plan to explore in our later work.

## References


1. Leone S, Pascale R, Vitale M, Esposito S (2012) Epidemiologia del piede diabetico [Epidemiology of diabetic foot]. Infez Med 20(Suppl 1):8–13 (in Italian) PMID: 22982692
2. Ramsey SD et al (1999) Incidence outcomes and cost of foot ulcers in patients with diabetes. Diabetes Care 22(3):382–387. https://doi.org/10.2337/diacare.22.3.382
3. Armstrong David G, Lavery Lawrence A, Harkless Lawrence B (1998) Validation of a diabetic wound classification system: the contribution of depth, infection, and ischemia to risk of amputation. Diabetes Care 21(5):855–859. https://doi.org/10.2337/diacare.21.5.855
4. Das SK, Roy P, Mishra AK (2021) Recognition of ischaemia and infection in diabetic foot ulcer: a deep convolutional neural network based approach. Int J Imaging Syst Technol. https://doi.org/10.1002/ima.22598
5. Das SK, Roy P, Mishra AK (2021) DFU_SPNet: a stacked parallel convolution layers based CNN to improve Diabetic Foot Ulcer classification. ICT Express. https://doi.org/10.1016/j.icte.2021.08.022
6. Das SK, Roy P, Mishra AK (2021) Deep learning techniques dealing with diabetes mellitus: a comprehensive study. Health informatics: a computational perspective in healthcare. Springer, Singapore, pp 295–323
7. Mishra AK et al (2020) Identifying COVID19 from chest CT images: a deep convolutional neural networks based approach. J Healthcare Eng. https://doi.org/10.1155/2020/8843664
8. Singh PP et al (2019) Classification of effusion and cartilage erosion affects in osteoarthritis knee mri images using deep learning model. In: International conference on computer vision and image processing. Springer, Singapore
9. Das S, Mishra A, Roy P (2018) Automatic diabetes prediction using tree based ensemble learners. In: Proceedings of international conference on computational intelligence and IoT (ICCIIoT)
10. Tan M, Le Q (2019) Efficientnet: rethinking model scaling for convolutional neural networks. In: International conference on machine learning. PMLR
11. Goyal M, Reeves ND et al (2020) DFUNet: CNNs for diabetic foot ulcer classification. IEEE Trans Emerging Topics Comput Intell 4(5):728–739. https://doi.org/10.1109/TETCI.2018.2866254
12. Yann L, Bengio Y (1995) Convolutional networks for images, speech, and time series. Handbook Brain Theory Neural Netw 3361(10)
13. Szegedy C et al (2015) Going deeper with convolutions. In: Proceedings of the IEEE conference on computer vision and pattern recognition
14. Alzubaidi L et al (2020) DFU_QUTNet: diabetic foot ulcer classification using novel deep convolutional neural network. Multimedia Tools Appl 79(21):15655–15677. https://doi.org/10.1007/s11042-019-07820-w
15. Zhao X et al (2019) Fine-grained diabetic wound depth and granulation tissue amount assessment using bilinear convolutional neural network. IEEE Access 7:179151–179162. https://doi.org/10.1109/ACCESS.2019.2959027
16. Cassidy B et al (2020) Dfuc2020: analysis towards diabetic foot ulcer detection. arXiv:2004.11853
17. Ren S et al (2015) Faster R-CNN: towards real-time object detection with region proposal networks. arXiv:1506.01497, https://doi.org/10.1109/TPAMI.2016.2577031
18. Tan M, Pang R, Le QV (2020) Efficientdet: scalable and efficient object detection. In: Proceedings of the IEEE/CVF conference on computer vision and pattern recognition
19. Jocher G et al (2020) ultralytics/yolov5: v3.0. Zenodo
20. Goyal M et al (2018) Robust methods for real-time diabetic foot ulcer detection and localization on mobile devices. IEEE J Biomed Health Inform 23(4):1730–1741. https://doi.org/10.1109/JBHI.2018.2868656
21. Ferreira ACBH et al (2020) Competitive neural layer-based method to identify people with high risk for diabetic foot. Comput Biol Med 120:103744. https://doi.org/10.1016/j.compbiomed.2020.103744

22. Yap MH et al (2020) Deep learning in diabetic foot ulcers detection: a comprehensive evaluation. arXiv:2010.03341, https://doi.org/10.1016/j.compbiomed.2021.104596
23. Howard AG et al (2017) Mobilenets: efficient convolutional neural networks for mobile vision applications. arXiv:1704.04861
24. Hu J, Shen L, Sun G (2018) Squeeze-and-excitation networks. In: Proceedings of the IEEE conference on computer vision and pattern recognition
25. Ramachandran P, Zoph B, Le QV (2017) Swish: a self-gated activation function. 7:1. arXiv:1710.05941

# Performance Analysis of Image Enhancement Techniques for MRI Brain Images

**Jyoti Krayla, Upendra Kumar Acharya, and Sandeep Kumar**

## 1 Introduction

An image is a data or visual information, and it is defined in terms of gray levels or intensity values. During image acquisition & transmission, noise present in the channel alters the pixel of image and thus makes the image noisy. So, the pixels of image show different intensity value instead of actual ones. Thus, it is important to process & enhance image to make it suitable for desired application [1]. Images like medical images play very significant role to examine patient's health condition and to provide effective diagnosis. These images are very complicated because of overlapped objects, tiny cells, and tissues [2].

For enhancing the image quality, image enhancement techniques are used. The main purpose of image enhancement is to enhance minute details of image having low luminance for human or machine interpretation. For enhancing image quality, we will change brightness, contrast of image [3]. Image enhancement is useful in many areas like medical industry, satellite communication, forensic departments, and underwater operations.

Image enhancement technique is broadly classified as, spatial domain methods & frequency domain methods [4]. In spatial domain, most widely used technique is HE because of its effectiveness and simplicity [5]. The overall contrast of the image is improved by HE by increasing dynamic range resulting in flat profile of histogram. Main disadvantage of HE is over brightness and reduced information content. This drawback is removed by another technique called as DSIHE [6]. This method improves contrast and preserves information content. To maximize entropy and preserve the brightness of image, another technique named as ESIHE has been developed [7].

J. Krayla · U. K. Acharya · S. Kumar (✉)
National Institute of Technology, Delhi, India
e-mail: sandeep@nitdelhi.ac.in



P. H. J. Chong et al. (eds.), *Emerging Electronics and Automation*, Lecture Notes in Electrical Engineering 937, https://doi.org/10.1007/978-981-19-4300-3_29

Many other techniques have been developed for image enhancement; one is to employ the concept of fuzzy logic in which fuzzification, membership function modification, and defuzzification are done to enhance image quality [8]. Contrast enhancement on the basis of layered difference representation of 2D histogram is also developed in which contrast of image is improved by amplifying the gray-level differences between adjacent pixels [9]. Brightness of MRI brain images is preserved along with contrast enhancement by a method in which concept of fuzzy logic is combined with adaptive histogram equalization [10].

Another technique is introduced, in which image enhancement is carried out based on fuzzy theory, & after that image segmentation is done using new level-set method for visualizing & measuring the brain's anatomical structures [11]. Another approach of image enhancement, named as automatic contrast enhancement of brain MR images, has been developed which was applied to flair images on the basis of contrast mapping technique & intensity adjustment [12].

This research paper is structured as; Sect. 2 explains the HE technique. The detail about DSIHE technique is explained in Sect. 3. Section 4 explains the ESIHE technique. Section 5 explains about the image parameters on the basis of which these techniques are compared. Section 6 shows experimental results, & Sect. 7 concludes the research paper.

## 2 HE Technique

HE is a technique of image processing in which image contrast is improved. Through this technique, intensities are distributed over entire range of image. Image containing lower contrast areas gains higher contrast region. Its main advantage is that it is simple & straightforward technique. The method to perform histogram equalization is described in following steps:

1. Read the original image.
2. Calculate number of pixels corresponding to gray level of image.
3. Cumulative histogram is formed.
4. Values are divided by the total number of pixels for normalization.
5. Multiply these values by maximum gray level & then round off values.
6. At last, one to one mapping of original value to the outcome of step 5.

On comparing input with output graph in Fig. 1, it has been observed that, the histogram values are at the center, and in the output graph, the histogram values are distributed. So, we are moving toward flat profile of histogram which indicates high-contrast image. It is a simple technique, but it is having disadvantage as it create loss of contrast and information in some regions of image. After simulation, it has also been observed that, the resultant image becomes more bright compared to the original brain images. The gap among the bins of the histogram is large which indicates the noise amplification in the enhanced image. So, to preserve the brightness, contrast,

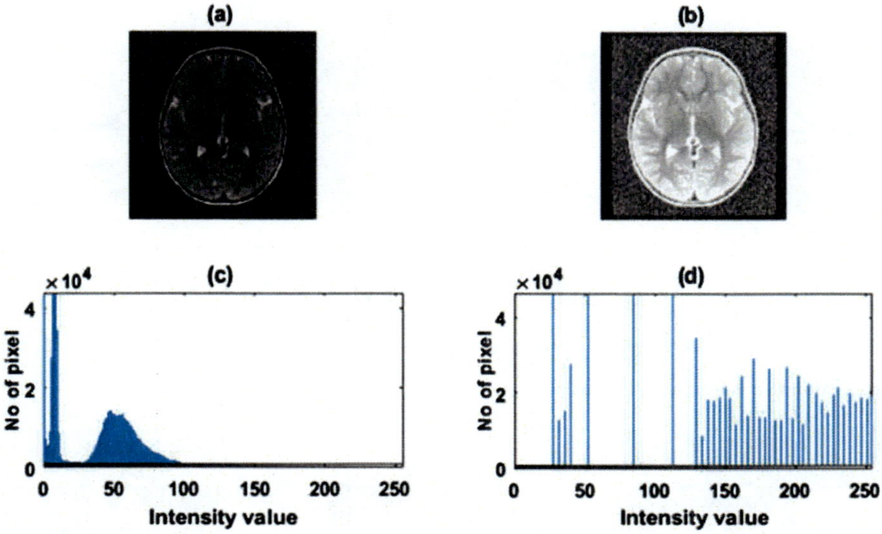


**Fig. 1** **a** Original image; **b** HE image; **c** Histogram of original image; **d** Histogram of HE image

and minimize information loss, the HE technique is modified which is discussed in Sect. 3.

## 3 DSIHE Technique

HE is very simple & the most effective method of image enhancement. But, this method after equalization alters the luminance of an image so significantly as compared to original one that it becomes difficult to use in video system. An alternative to HE is DSIHE [6]. The DSIHE technique not only improves or enhance image effectively but also preserves brightness which is shown in Fig. 3 & makes it possible to be used in video system. This method is comprised of three steps. Firstly, on the basis of original probability density function, original image is segmented into two sub-images of equal area. The image division is carried out the basis of median value. Then, HE is performed on both sub-mages, and at last, the processed equal area sub-images are integrated to form the resultant enhanced image. This method can be described in three main steps, and its block diagram is shown in Fig. 2.

1. Input image is segmented into two equal area sub-images on the basis of median value.
2. HE is performed individually on both the sub-images.
3. Individually processed equal area sub-images are finally integrated to get the enhanced output image.

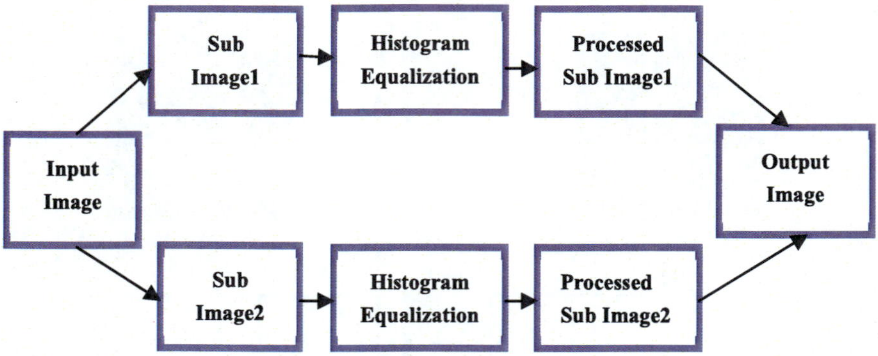


**Fig. 2** Block diagram of DSIHE technique

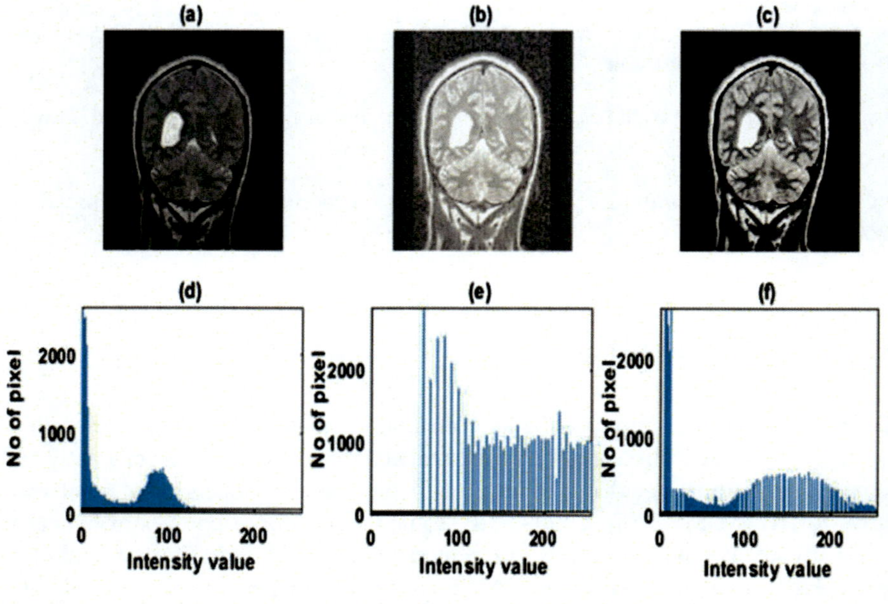


**Fig. 3** **a** Original image; **b** HE image; **c** DSIHE image; **d** Histogram of original image; **e** Histogram of HE image; **f** Histogram of DSIHE image

## 4 ESIHE Technique

Based on intensity exposure, an image is categorized as under exposed image & over exposed image. Images whose histogram bins are present toward the lower part then image possesses low intensity exposure. In similar way, images whose histogram bins are present toward the higher part or brighter part then image possesses high intensity value. Exposure-based sub-image histogram equalization technique focuses on contrast improvement of low-exposure grayscale image [7]. In this method, exposure

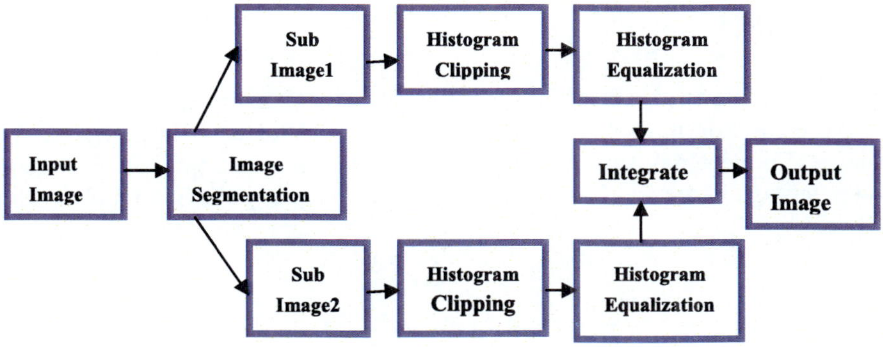


**Fig. 4** Block diagram of ESIHE technique

threshold is calculated to segment the input image into sub-images of different intensity levels. Then, histogram clipping is carried out based on the clipping threshold. Histogram equalization is implemented individually on both the sub-images, and finally, we will integrate the processed sub-images to get final image for analysis.

ESIHE technique involves following steps, & its block diagram is shown in Fig. 4.

1. Original image histogram (*H(k)*) is computed first.
2. Calculation of exposure parameter and threshold parameter.
3. Clipping threshold (*Tc*) is calculated, and histogram *H(k)* is clipped.
4. Clipped histogram is divided into two sub-histograms using the threshold parameter (*Xa*).
5. HE is performed on both the sub-histograms.
6. At last, the sub-images are combined to get output image.

## 5 Image Parameters

There are different parameters on the basis of which we can evaluate the quality of processed image.

### *5.1 Visual Quality*

Visual quality indicates that just by looking at the image how much information can be perceived, as done by medical professionals for early detecting of diseases.

## 5.2 Entropy

Entropy also known as average information content (AIC) indicates the information value present in the image [10]. It can be calculated by histogram of the image which indicates different gray-level probabilities. Image with high entropy value indicates that it has high information content. Entropy of an image is given by formula,

$$\text{Entropy} = -\sum_{k} P_k \log_2 P_k \tag{1}$$

whereas, $P_k$ = normalized histogram value of the image.

## 5.3 PSNR

The ratio of maximum possible signal power to the noise power is defined as PSNR. It is performance parameter used for evaluating the quality of enhanced image [13, 14]. Image with high PSNR value is considered as good quality image because it contains more signal power and less noise power which indicates the image is less noisy. It is given as,

$$\text{PSNR} = 20\log_{10} \frac{255}{\text{Root Mean Square Error}} \tag{2}$$

Images with high PSNR value indicate the low mean square error and thus high image quality. However, human perspective about image can be different.

## 5.4 AMBE

The difference between the mean intensity values of input and output image is called as AMBE [15, 16]. It is given by formula,

$$\text{AMBE} = |m_1 - m_2| \tag{3}$$

where $m_1 and m_2$ are the average intensity of the input and enhanced image. Image with smaller AMBE value shows that the average intensity of original image is well preserved in the enhanced output image.

## 6 Results

In this section, performance of HE, DSIHE, & ESIHE is analyzed based on visual quality, entropy, PSNR, and AMBE. Figures 5, 6, 7, 8, and 9 show different brain images which are taken from publicly available database [17]. From visual quality analysis, it is evident that, ESIHE technique results better images as compared to DSIHE and HE techniques. In HE, the contrast improvement is significant, but there is more information loss. DSIHE shows improvement in terms of contrast improvement and information contents than HE technique. Originality of brain image is maintained in DSIHE and better than HE technique. But, ESIHE shows significant improvement in visual quality along with contrast enhancement and less information loss as compared to both the techniques. Noise amplification in the enhanced image is also reduced by ESIHE technique.

Figure 10 represents the histograms of the brain image 1. In HE, histogram is shifted to higher intensity value which indicates the image is getting brighter, but here, histogram bins are distributed and separated which indicates the information loss in image. In DSIHE, histogram is shifted toward higher intensity value. It also follows the shape of histogram of original image which indicates that information is preserved in this technique along with contrast enhancement. Also, the histogram is

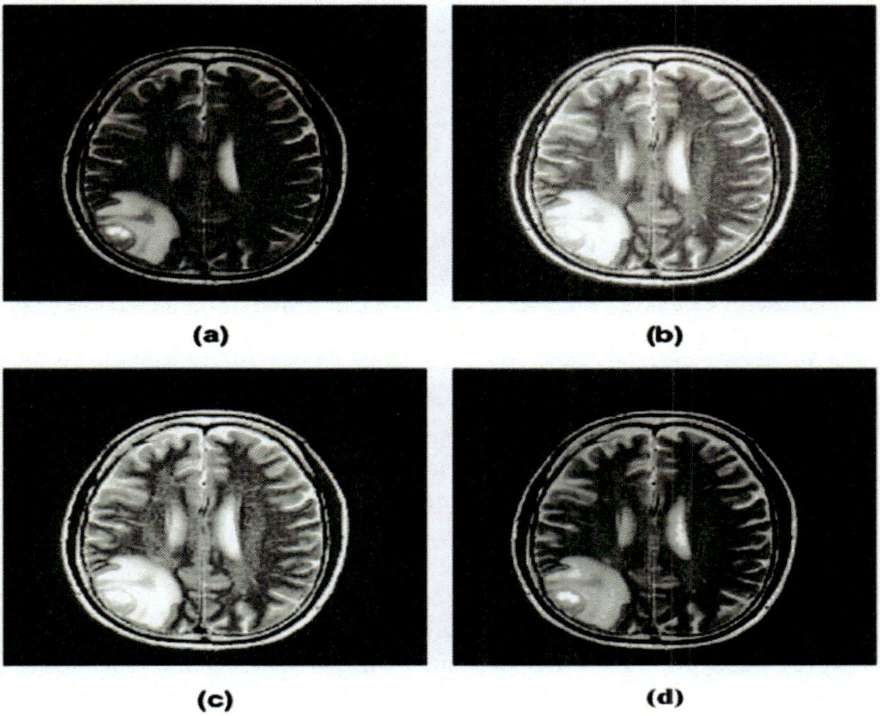

**Fig. 5** **a** Image-1; **b** HE; **c** DSIHE; **d** ESIHE

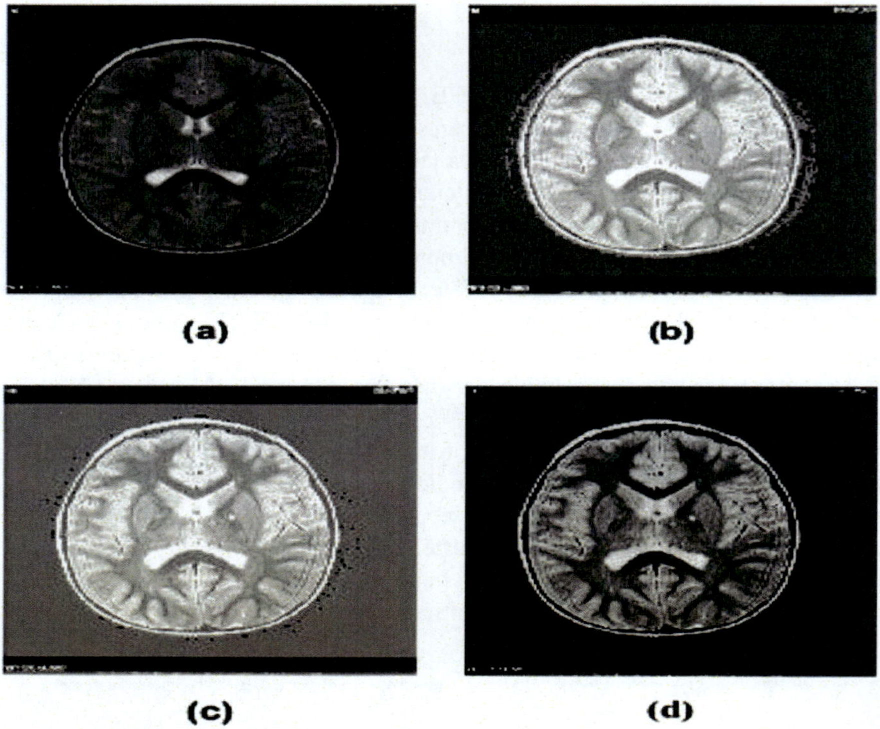

Fig. 6 **a** Image-2; **b** HE; **c** DSIHE; **d** ESIHE

distributed over entire range giving the flat profile which means significant contrast enhancement. In ESIHE, the performance is highly improved in comparison to both the techniques. The shape of histogram in ESIHE is similar to the histogram of original image which indicates that the originality of image is preserved here along with contrast improvement. So, we can conclude that ESIHE technique is superior to both HE & DSIHE in terms of contrast enhancement and information loss.

Average information content (AIC) of image is given by entropy. Table 1 shows entropy values of different brain images after applying all three techniques. The average entropy of original image is 5.389 which reduce to 4.365 after HE. It means that the image obtained by HE technique results very less information content compared to original image. DSIHE technique shows improved result over HE as its average entropy is 5.030. ESIHE technique shows significant improvement with its average entropy value 5.289. So, it can be said that, less information loss is occurring in ESIHE technique as compared to the discussed techniques.

Table 2 shows AMBE values of different brain images after applying all three techniques. Image with smaller AMBE value shows that intensity of original image is well preserved in the resulting output image. It is evident by looking at average AMBE of three techniques that ESIHE gives minimum AMBE value; thus, in this technique, intensity of image is preserved better than HE and DSIHE. Table 3 shows

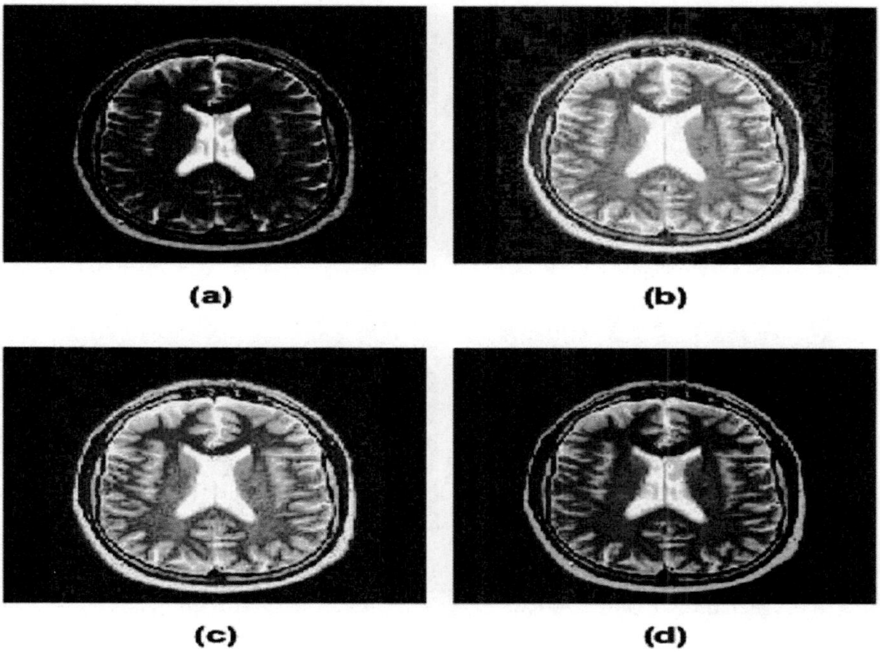


**Fig. 7** **a** Image-3; **b** HE; **c** DSIHE; **d** ESIHE

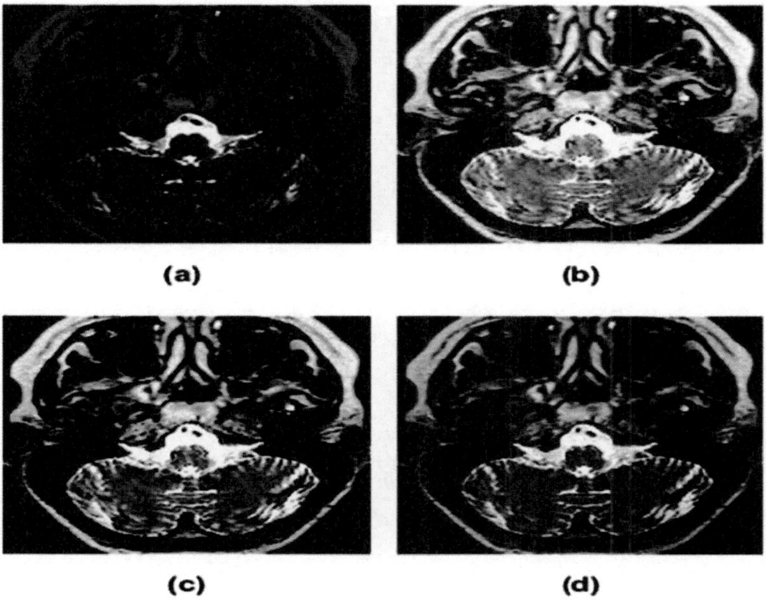


**Fig. 8** **a** Image-4; **b** HE; **c** DSIHE; **d** ESIHE

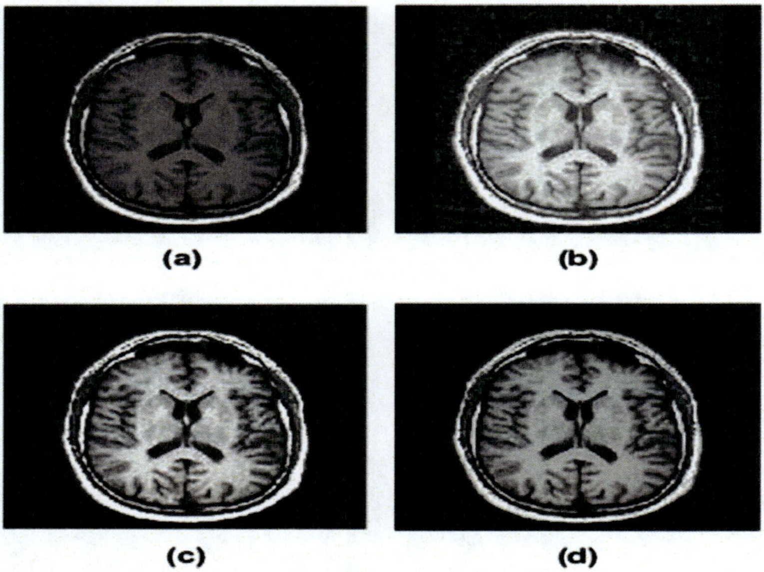


**Fig. 9** **a** Image-5; **b** HE; **c** DSIHE; **d** ESIHE

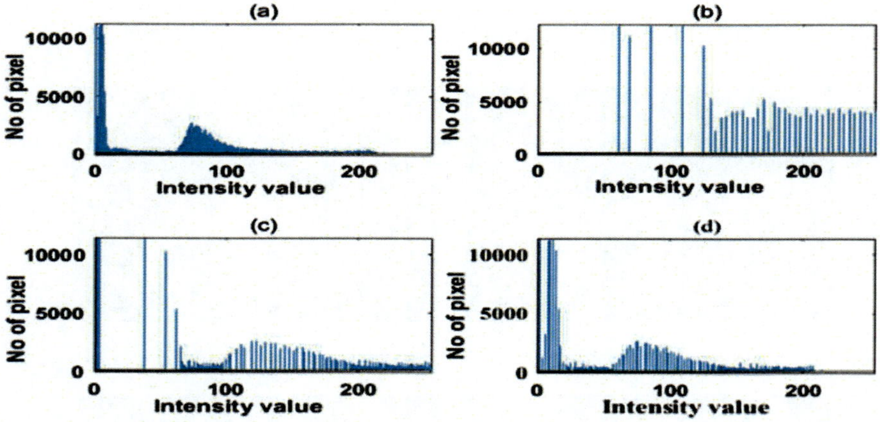


**Fig. 10** Histogram of image 1 **a** Original image; **b** HE; **c** DSIHE; **d** ESIHE

**Table 1** Comparison of entropy values

| Image/Techniques | Original | HE | DSIHE | ESIHE |
|---|---|---|---|---|
| Image-1 | 5.652 | 4.450 | 5.374 | 5.530 |
| Image-2 | 4.449 | 3.564 | 4.222 | 4.353 |
| Image-3 | 5.466 | 4.300 | 4.899 | 5.359 |
| Image-4 | 5.888 | 5.096 | 5.599 | 5.811 |
| Image-5 | 5.491 | 4.417 | 5.054 | 5.390 |
| Average entropy | 5.389 | 4.365 | 5.030 | 5.289 |

**Table 2** Comparison of AMBE values

| Image/Techniques | HE | DSIHE | ESIHE |
|---|---|---|---|
| Image-1 | 88.646 | 36.668 | 8.873 |
| Image-2 | 96.661 | 68.513 | 21.109 |
| Image-3 | 94.560 | 45.158 | 17.063 |
| Image-4 | 104.068 | 45.462 | 32.979 |
| Image-5 | 85.811 | 23.951 | 16.068 |
| Average AMBE | 93.949 | 43.950 | 19.219 |

**Table 3** Comparison of PSNR values

| Image/Techniques | HE | DSIHE | ESIHE |
|---|---|---|---|
| Image-1 | 24.06 | 26.14 | 29.13 |
| Image-2 | 24.06 | 24.06 | 28.10 |
| Image-3 | 24.12 | 26.14 | 27.68 |
| Image-4 | 24.10 | 26.84 | 26.91 |
| Image-5 | 24.11 | 26.95 | 28.42 |
| Average PSNR | 24.09 | 26.03 | 27.87 |

the comparison of PSNR values of all three techniques. High PSNR means that image is having good signal strength and minimum noise. The average PSNR value resulted by ESIHE technique is 27.87 which is higher than HE and DSIHE. So, ESIHE technique is much superior to HE and DSIHE on the basis of PSNR.

## 7 Conclusion

In this paper, HE, DSIHE, and ESIHE techniques are implemented to improve the overall quality of the brain images. Then, performance of each technique is analyzed & compared on the basis of visual quality, entropy, PSNR, & AMBE. From the simulation results and measured parameter values, it has been concluded that the ESIHE technique gives improved performance as compared to other two techniques. As the information loss and noise amplification is less in ESIHE technique, so it can be used as a preprocessing technique for scientific evaluation and early detection of diseases. Every technique has its advantages & disadvantages. Some technique suitable for one application might not be suitable for other application areas. We can modify the existing image enhancement techniques to get improvised results.

## References


1. Reddy PS, Kommuri SR, Singh H, Kumar A, Balyan LK (2019) Gamma corrected fusion framework of dominant orientation and edge based textural flistogram equalization for image enhancement. In: 2019 international conference on communication and signal processing (ICCSP). IEEE, pp 0475–0479
2. Acharya UK, Kumar S (2020) Particle swarm optimized texture based histogram equalization (PSOTHE) for MRI brain image enhancement. Optik 224:165760
3. Acharya UK, Kumar S (2021) Directed searching optimized mean-exposure based sub-image histogram equalization for grayscale image enhancement. Multimedia Tools Appl 1–21
4. Maini R, Aggarwal H (2010) A comprehensive review of image enhancement techniques. arXiv preprint arXiv:1003.4053
5. Gonzalez RC, Woods RE (2002) Digital image processing, 3rd edn. Prentice Hall Press, Upper Saddle River, NJ, USA
6. Wang Y, Chen Q, Zhang B (1999) Image enhancement based on equal area dualistic sub-image histogram equalization method. IEEE Trans Consum Electron 45(1):68–75
7. Singh K, Kapoor R (2014) Image enhancement using exposure based sub image histogram equalization. Pattern Recogn Lett 36:10–14
8. Mahashwari T, Asthana A (2013) Image enhancement using fuzzy technique. Int J Res Eng Sci Technol 2(2):1–4
9. Lee C, Lee C, Kim CS (2013) Contrast enhancement based on layered difference representation of 2D histograms. IEEE Trans Image Process 22(12):5372–5384
10. Subramani B, Veluchamy M (2018) MRI brain image enhancement using brightness preserving adaptive fuzzy histogram equalization. Int J Imaging Syst Technol 28(3):217–222
11. Fenshia Singh J, Magudeeswaran V (2017) A machine learning approach for brain image enhancement and segmentation. Int J Imaging Syst Technol 27(4):311–316
12. Isa IS, Sulaiman SN, Mustapha M, Karim NKA (2017) Automatic contrast enhancement of brain MR images using average intensity replacement based on adaptive histogram equalization (AIR-AHE). Biocybernetics Biomed Eng 37(1):24–34
13. Acharya UK, Kumar S (2021) Genetic algorithm based adaptive histogram equalization (GAAHE) technique for medical image enhancement. Optik 230:166273
14. Acharya UK, Kumar S (2021) Swarm intelligence based adaptive gamma corrected (SIAGC) retinal image enhancement technique for early detection of diabetic retinopathy. Optik 247:167904
15. Ali QM, Yan Z, Li H (2015) Iterative thresholded bi-histogram equalization for medical image enhancement. Int J Comput Appl 114(8)
16. Acharya UK, Kumar S (2021) Image enhancement using exposure and standard deviation-based sub-image histogram equalization for night-time images. In: Proceedings of international conference on artificial intelligence and applications. Springer, Singapore, pp 607–615
17. Image Data Base. Magnetic resonance-technology information portal. Web link: http://www.mrtip.com/serv1.php?type5img&img5Brain%20MRI%20Images%20T1

# Effect of End Point Detection on Fixed Phrase Speaker Verification

**Nirupam Shome, Rabul Hussain Laskar, and Richik Kashyap**

## 1 Introduction

Speaker verification refers to the process of verifying a claim of a person based on his/her voice sample. In a fixed phrase speaker verification system, the verification text is fixed (3–4 s phrase) for the training and testing phase and the system has a-priori knowledge about the speaker characteristic. Here the system takes the claim speaker's voice sample and matches it with the stored speaker model of the same speaker to take the decision. Many speaker verification systems reported in the literature perform well under clean data conditions [1]. The performance of a fixed phrase speaker verification system largely depends on the various distortions like background noise, chirping noise, background speech, channel mismatch, sensor mismatch, and other environmental conditions, etc. [2, 3]. These degradations can be handled with compensation in signal level, model level, score level, feature level, individually, or all of them together. Due to these degradations, detection of speech boundaries becomes difficult and results in poor performance of the speaker verification system. When the speech boundaries are not identified properly then it leads to treating noise and other redundant information as speech, which results in degradation in system performance. In literature, several methods are available for dealing with train and test speech degradation [4, 5]. In this paper, we aimed to find an efficient method to identify the start and endpoint of the speech signal as it is very crucial for the fixed phrase speaker verification process [6–9]. In fixed phrase speaker verification, the features are evaluated for a speech sample within that speech boundary for both the enrollment and verification phase. Then the testing template is matched with

N. Shome (✉) · R. Kashyap
Assam University, Silchar, Assam, India
e-mail: nirupam_shome@yahoo.com

R. H. Laskar
National Institute Technology, Silchar, Assam, India
e-mail: rhlaskar@ece.nits.ac.in



P. H. J. Chong et al. (eds.), *Emerging Electronics and Automation*, Lecture Notes in Electrical Engineering 937, https://doi.org/10.1007/978-981-19-4300-3_30

the training template using the pattern matching technique like dynamic programming. The dynamic time warping (DTW) algorithm [10] is the most popular template matching technique and it is extensively used by researchers for pattern matching.

The rest of the paper is organized as follows: Sect. 2 illustrates different endpoint detection methods and describes three state of art techniques of endpoint detection. Section 3 describes the baseline system for fixed phrase speaker verification and detailed speaker-specific threshold and cohort speaker-based modules for speaker verification. Database used for our investigation is given in Sects. 4 and 5 describes experimental results and discuss the findings for our analysis. Lastly, Sect. 6 provides the overall conclusion of our study.

## 2 Endpoint Detection Methods

The identification of starting point and terminus point of a speech sample is referred to as endpoint detection. Speech endpoint detection can be categorized by the threshold-based approach and pattern matching approach. In a threshold-based system, acoustic features are compared with predefined thresholds to get speech frames. On basis of speech and noise models, speech frames are identified in pattern matching methods. In literature, several methods have been reported for endpoint detection that uses features like Mel Frequency Cepstral Coefficient (MFCC), Short-time Energy (STE), Zero Crossing Rate (ZCR), Entropy, Teager energy operator algorithm, and wavelet transform. Other than these methods, long-term signal features like the Long-term Spectral Flatness Measure (LSFM) [11] feature and the Long-term Signal Variability (LTSV) [12] feature has been extensively used for endpoint detection. The energy-based VAD [13, 14] is one of the most popular endpoint detection techniques. This method performs well in a clean environment but utterly fails when tested on real-life data [5]. An improvement over energy-based VAD has been achieved with periodicity-based VAD [15] for endpoint detection. Entropy-based detection proposed in [16] uses speech and noise entropy to detect endpoints of a signal. The performance of this method can be improved by changing the spectral probability density function [17]. Using one-dimensional Mahalanobis distance function [18] endpoint of speech has been detected in [13]. Endpoint detection for low SNR signal has been proposed in [19], which extracts features by Savitzky-Golay filtering, constant Q-transform, and improved sub-band energy entropy. The single parameter double threshold method has been adopted to find endpoints of the speech signal. These days machine learning approaches have become popular in endpoint detection. A neural network system has been proposed in [20] to detect endpoints of the speech signal and it uses raw waveform as its input. A deep residual network model, MarbleNet [21] has also been proposed for endpoint detection and validated in the real-world dataset.

For our analysis, three methods are chosen based on signal energy, glottal information, and a combination of suprasegmental feature and energy of the signal as these methods are very popular signal processing techniques for endpoint detection.

We have implemented the energy-based voice activity detection (VAD) [14], glottal activity detection (GAD) [22], and pitch and energy-based detection (PED) [23] for endpoint identification and are discussed in the following subsections.

## *2.1 Voice Activity Detection (VAD)*

The endpoint detection based on energy VAD [14] makes use of the instantaneous energy of the speech signal to identify the start and terminus point. The signals are processed in short frames of length 20 with 10 ms overlap. First framewise energy is calculated and is compared with the energy threshold. The threshold is calculated by 6% of the average energy of the whole speech sample. Speech and non-speech frames are identified by comparing framewise energy with a predefined threshold. If energy exceeds the threshold, then it is considered as speech frame otherwise non-speech frame. The first speech frame is taken as the start point and the last speech frame is the endpoint of the whole speech signal.

## *2.2 Glottal Activity Detection (GAD)*

The glottal activity detection technique proposed in [22] is divided into two parts. In the first part, a zero-frequency filter (ZFF) signal is obtained from a zero-frequency filter, and then using adaptive threshold glottal regions are detected. The threshold has been increased from low to high in finite steps until the number of speech regions selected by GAD is less than VAD. Finally, the obtained threshold has been applied on energy frames of the ZFF signal and the decision is taken between glottal and non-glottal regions.

## *2.3 Pitch and Energy-Based Detection (PED)*

The method proposed in [23] uses energy and pitch information to detect endpoints of the speech signal. Utilizing the prosodic information like pitch with energy gives a significant amount of improvement in endpoint detection. Here the energy and pitch information of the whole speech has been calculated to get energy and pitch thresholds. To identify the speech and non-speech frame, framewise energy and pitch have been calculated. These energy and pitch values are compared with the predefined thresholds. If a frame exceeds the threshold of energy and pitch, it is considered a speech frame otherwise non-speech frame. The first frame of the first quad of speech frames is taken as the start point and the last frame of the last quad of speech frames is considered as the terminus point of the speech sample.

## 3 Baseline System for Speaker Verification

The fixed phrase speaker verification systems are generally based on either template matching or model sequence matching techniques. These methods are based on the alignment of time axes of input speech utterance and registered speakers reference models. The alignments are done from the beginning to the end of utterance [6, 24]. A speaker verification model performs better if the linguistic information is known because it improves the system's ability to characterize the speaker-specific phonetic content of a speech signal. A speaker verification system can be implemented for practical application if it is session, language, and sensor independent and most importantly it should be robust to surrounding noise. The performance of a fixed phrase speaker verification largely depends on environmental factors like variability in the channel [25]. Accurate matching or alignment between training speaker model and test features is important for speaker verification [1].

We have developed our system using 39-dimensional MFCC features and DTW as template matching techniques. The speech samples are processed in a 20 ms frame size with a 10 ms frameshift. The MFCC templates are generated using 22 logarithmically spaced filter banks from 20 ms frames. From one to thirteen MFCC coefficients are used as the feature vector in our analysis. The velocity and acceleration features are calculated by taking first-order derivative delta ($\Delta$) and second-order derivative delta-delta ($\Delta\Delta$) from the current feature vector. Thus, the resulting complete feature vector contains 13-MFCC, 13-$\Delta$MFCC, and 13-$\Delta\Delta$MFCC coefficients. The feature normalization is carried out with the help of the cepstral mean and variance normalization (CMVN) technique [6]. The DTW algorithm is used for feature matching, which provides many to many mapping between two templates under observation by minimizing the overall accumulated distance [10]. Due to the dynamic programming of DTW, an efficient minimization can be achieved.

To carry out our study, we have designed and implemented two fixed phrase speaker verification modules. The first one is the speaker-specific threshold-based system and the second cohort speaker-based system. The complete description of these methods is given in the following subsection.

### *3.1 System Based on Speaker-Specific Threshold*

The speaker-specific threshold-based module is shown in Fig. 1. The total number of speech samples is divided into three parts for each speaker. The first part contains three thresholding samples, the second part have three training samples and the third part contains the rest of the speech samples for testing the system performance. Three speech samples of the first part are used as the reference concerning three training samples to estimate speaker-specific threshold. All of the speech samples are first passed through the endpoint detection module, here VAD, GAD, and PED are used one by one. Then the signal features are extracted and feature normalization is done

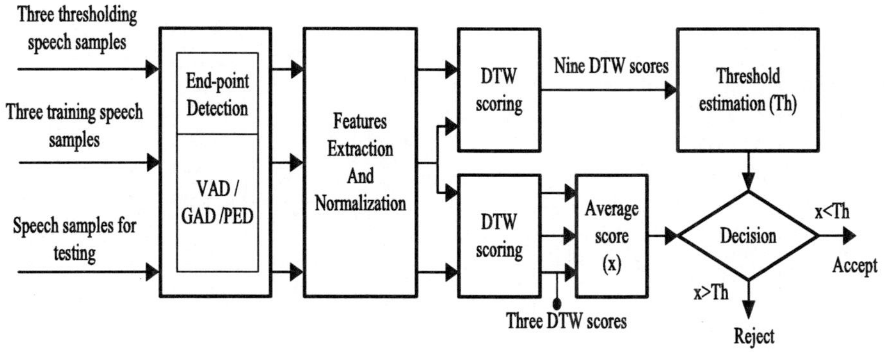


**Fig. 1** System description of speaker-specific threshold-based speaker verification

in the next module. Now, the distance between three thresholding templates and three training templates are estimated by the Euclidean distance-based DTW algorithm, which gives a total of nine distance scores. These nine scores are averaged to get the average distance, i.e., mean ($\mu$) and subsequently standard deviation ($\sigma$) of these scores are estimated. The speaker-specific threshold is calculated as threshold (Th) = $\mu + 2\sigma$. In the case of testing, the distance between the test template and three training templates are calculated and three distance scores are averaged. This averaged DTW score ($x$) is compared with the predefined threshold (Th). If the distance score ($x$) is less than the threshold (Th) then the claim is accepted, otherwise rejected. The thresholding used here is called hard thresholding.

## *3.2 System Based on Cohort Speaker*

Here we have described the methodology for a cohort speaker-based system. The complete structure for this investigation is shown in Fig. 2. We have divided the system into two phases, the first is for training the system and the second is for testing the system. A completely different dataset is being used to train and test the system. In both the phases endpoint detection, feature extraction and normalization are common. Endpoint detection is performed by VAD, GAD, and PED one by one. In the training phase, extracted features of all speakers are stored in the feature database and are used as reference templates for verification purposes. The difference in methodology with the earlier system lies in the decision-making process. In this methodology, the decision is made on basis of a set of cohort speaker scores. For a test trial, the test template is compared with 3 templates of claimed speakers and 12 templates of cohort speakers to get the respective DTW score. These 12 cohort speaker templates are obtained from four randomly selected speakers. Thus, for a particular testing trial total of 15 DTW scores are obtained which are then sorted in ascending order according to their distance scores. Now the decision of a certain test trial is made on

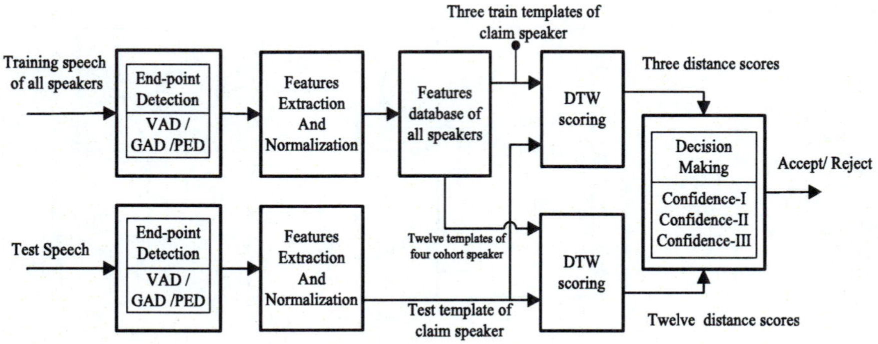


**Fig. 2** System description of cohort speaker-based speaker verification

three principles, namely, confidence-I, confidence-II, and confidence-III. The claim of a speaker is accepted by the following conditions.

- Confidence-I: If one out of three scores obtained from claimed speaker occupies a position in the top three places in the array of 15 sorted scores.
- Confidence-II: If two out of three scores obtained from claimed speaker occupies any two positions in the top three places in the array of 15 sorted scores.
- Confidence-III: If all three scores obtained from the claimed speaker occupy all three top positions in the array of 15 sorted scores.

If any one of these conditions is not satisfied then the claim is rejected. In this way, the decision is made for a test claim of a particular speaker.

## 4 Database Used for Experimentation

We have carried out our analysis on two different datasets. The first one is RSR 2015 database [25], which is a standard corpus collected under a fully controlled environment. The second dataset is the NITS speech corpus [26], which was collected in a fully uncontrolled environment. The RSR2015 database contains 300 speakers and is divided into three parts. The first part of the dataset contains 30 fixed phrases spoken by 300 speakers. And the second and third part contains 30 short commands and 13-digit sequences. The fixed phrase dataset has been chosen for our analysis, as it is appropriate for our analysis. This database has 157 male and 143 female speakers and the samples are recorded in nine different sessions. The NITS database contains 298 speakers and has divided into three parts, namely, voice password, text-dependent, and text-independent. Each part consists of 247 male and 51 female speakers. We have taken only the text-dependent part as we are interested in fixed phrase speaker verification. There are 4819 test trails for text-dependent modules in this dataset and it had recorded in 16 different sessions.

## 5 Experimental Results and Discussion

This section discusses the performance of two fixed phrase speaker verification methodology by using three endpoint detection methods. Here we analyze the effect of endpoint detection on fixed phrase speaker verification.

Here we have discussed the findings of our investigation on fixed phrase speaker verification. For this analysis, we have adopted two methods, namely, speaker-specific threshold-based module and cohort speaker-based module. The results of the two methodologies are given in Table 1. The results are expressed in terms of equal error rate (EER) and are calculated from the false rejection rate (FRR) and false acceptance rate (FAR). The imposter testing is done ten times more than the genuine testing for both methodologies. From the results, it is clear that the results for RSR 2015 database are better than the NITS database for all endpoint detection cases irrespective of analyzing methods. The reason behind it is the quality of speech samples in the respective corpus. As the RSR2015 database has been collected under a controlled environment, the signal contains minimum noise in it, but the NITS database has been collected in an uncontrolled environment, so it contains several types of noises and it affects the performance of the speaker verification system.

The comparative analysis of the effect of endpoint detection on fixed phrase speaker verification is discussed in subsequent sections. We have chosen three endpoint detection modules from the literature and compared their performance for both the speaker verification methodologies.

### *5.1 Results Analysis for Speaker-Specific Threshold-Based System*

Here we have presented the results for the speaker-specific threshold-based system in Fig. 3. From the given figure it is clear that for both the datasets, PED gives better results than VAD and GAD-based endpoint detection. The energy-based VAD performs poorly than the other two methods. In the case of the RSR 2015 corpus, an

**Table 1** Results of fixed phrase speaker verification analysis

| Speaker verification method | | Equal error rate (EER) in % | | | | | |
|---|---|---|---|---|---|---|---|
| | | RSR 2015 database | | | NITS database | | |
| | | End point detection method | | | | | |
| | | VAD | GAD | PED | VAD | GAD | PED |
| Speaker-specific threshold | | 5.15 | 4.53 | 3.08 | 11.81 | 10.41 | 8.77 |
| Cohort speaker-based approach | Confidence-I | 4.66 | 3.96 | 2.27 | 10.78 | 9.58 | 7.11 |
| | Confidence-II | 9.11 | 8.24 | 2.78 | 15.19 | 14.42 | 8.62 |
| | Confidence-III | 15.92 | 13.99 | 7.60 | 21.78 | 19.27 | 12.96 |

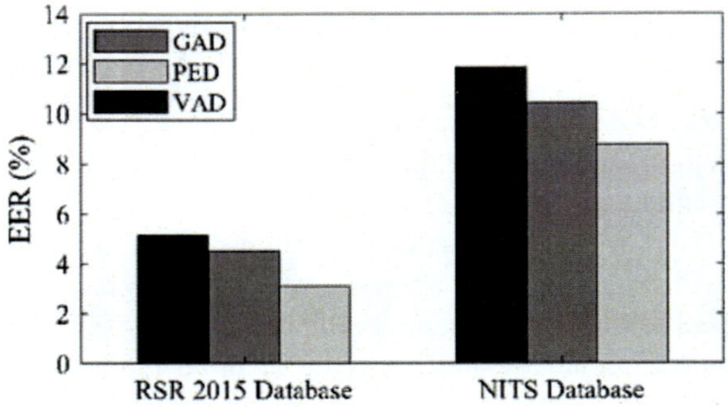


**Fig. 3** Performance of three endpoint detection methods under RSRS2015 and NITS datasets for speaker-specific threshold-based speaker verification

improvement of 2.07% and 1.45% are achieved in comparison to VAD and GAD, respectively. For the NITS database, though the error increased for each method, an improvement of 3.04 and 1.64% are achieved by PED over the other two methods.

## 5.2 *Results Analysis for Cohort Speaker-Based System*

The experimental result comparison for the cohort speaker-based system is shown in Fig. 4. The results are presented for two datasets and performance measures are divided into confidence I, II, III. From the Fig. 4, it can be observed that with the increase in the confidence measure from I to III, the performance of the cohort speaker-based system gets deteriorates for both the corpora. For the RSR database, PED gives an improvement of 2.39%, 6.33%, and 8.32% over VAD-based endpoint detection for the confidence I, II, and III, respectively. Similarly, an improvement of 3.67, 6.57, and 8.82% is achieved for the NITS database. When PED and GAD are compared for both datasets, improvement of 1.69%, 5.46%, 6.39% for the RSR 2015 database and 2.47%, 5.8%, 6.31% for the NITS database can be observed in case of confidence I, II, III, respectively, for PED-based endpoint detection. For confidence-I, improvement with PED is less in comparison to the performance of confidence II and III for both datasets.

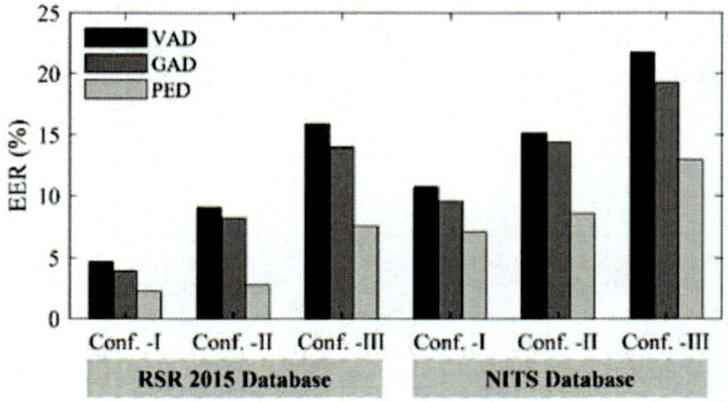


**Fig. 4** Performance of three endpoint detection methods under RSRS2015 and NITS datasets for cohort speaker-based speaker verification

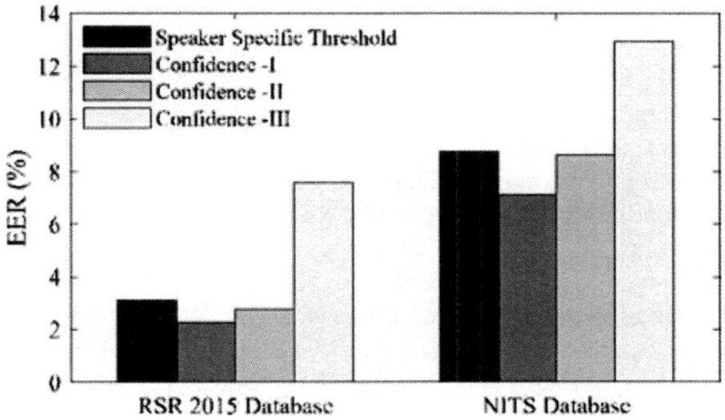


**Fig. 5** Comparison of performance of PED as an endpoint detection method for RSRS2015 and NITS datasets under different speaker verification approaches

In Fig. 5, we have shown the performance comparison of PED as an endpoint detection method for different speaker verification approaches under RSR 2015 and NITS datasets. From the figure, it is clear that the PED-based endpoint detection gives the best performance in the case of confidence-I measure of cohort speaker-based approach. The results of the confidence-II and speaker-specific threshold-based endpoint detection are quite similar. In the case of confidence-III, in both the datasets, the PED method of endpoint detection gives a large error but it is less than VAD and GAD-based methods.

## 6 Conclusion

This paper discusses the significance of speech endpoint detection on fixed phrase speaker verification. Here, the speaker verification is accomplished by speaker-specific threshold and cohort speaker-based approach. The endpoints of a speech sample are detected with VAD, GAD, and PED, to find the best performing method. For our analysis, a clean corpus RSR 2015 and a noisy corpus NITS have been used. With help of these databases, we have evaluated the performance of each endpoint detection algorithm under controlled and uncontrolled surroundings. From our analysis, it is clear that the PED as endpoint detection method outperforms the other two methods. And most importantly, it is substantiated that choice of proper endpoint detection is very much crucial for fixed phrase speaker verification. When the endpoints of the speech signals are not detected properly then the performance of the template matching-based fixed phrase verification gets affected severely.

## References


1. Kinnunen T, Li H (2010) An overview of text-independent speaker recognition: from features to supervisors. Speech Commun 52(1):12–40
2. Ming J, Hazen TJ, Glass JR, Reynolds DA (2007) Robust speaker recognition in noisy conditions. IEEE Trans Audio Speech Lang Process 15(5):1711–1723
3. Wong LP, Russell M (2001) Text-dependent speaker verification under noisy conditions using parallel model combination. In: 2001 IEEE international conference on acoustics, speech, and signal processing. Proceedings (Cat. No. 01CH37221). IEEE, New York, pp 457–460
4. Prasanna SRM, Pradhan G (2011) Significance of vowel-like regions for speaker verification under degraded conditions. IEEE Trans Audio Speech Lang Process 19(8):2552–2565
5. Pradhan G (2013) Speaker verification under degraded conditions using vowel-like and nonvowel-like regions
6. Furui S (1981) Cepstral analysis technique for automatic speaker verification. IEEE Trans Acoust Speech Sign Process 29(2):254–272
7. Prasanna SRM, Zachariah JM, Yegnanarayana B (2003) Begin-end detection using vowel onset points. In: Workshop on spoken language processing
8. Rabiner L (1993) Fundamentals of speech recognition
9. Yegnanarayana B, Prasanna SRM, Zachariah JM, Gupta CS (2005) Combining evidence from source, suprasegmental and spectral features for a fixed-text speaker verification system. IEEE Trans Speech Audio Process 13(4):575–582
10. Rabiner L, Rosenberg A, Levinson S (1978) Considerations in dynamic time warping algorithms for discrete word recognition. IEEE Trans Acoust Speech Sign Process 26(6):575–582
11. Ma Y, Nishihara A (2013) Efficient voice activity detection algorithm using long-term spectral flatness measure. EURASIP J Audio Speech Music Process 2013
12. Ghosh PK, Tsiartas A, Narayanan S (2010) Robust voice activity detection using long-term signal variability. IEEE Trans Audio Speech Lang Process 19(3):600–613
13. Saha G, Chakroborty S, Senapati S (2005) A new silence removal and endpoint detection algorithm for speech and speaker recognition applications. In: Proceedings of the NCC. Citeseer, p 5
14. Benyassine A, Shlomot E, Su H-Y, Massaloux D, Lamblin C, Petit J-P (1997) ITU-T recommendation G. 729 Annex B: a silence compression scheme for use with G. 729 optimized for V. 70 digital simultaneous voice and data applications. IEEE Commun Maga 35(9):64–73
15. Hautamäki V, Tuononen M, Niemi-Laitinen T, Fränti P (2007) Improving speaker verification by periodicity based voice activity detection. In: Proceedings of the 12th international conference speech and computer (SPECOM'2007) (2007), pp 645–650
16. Shen J, Hung J, Lee L (1998) Robust entropy-based endpoint detection for speech recognition in noisy environments. In: Fifth international conference on spoken language processing
17. Jia C, Xu B (2002) An improved entropy-based endpoint detection algorithm. In: International symposium on chinese spoken language processing
18. Hart PE, Stork DG, Duda RO (2000) Pattern classification. Wiley, Hoboken
19. Gan Z, Hou M, Hou H, Yang H (2020) Savitzky-Golay filtering and improved energy entropy for speech endpoint detection under low SNR. J Phys: Conf Ser 1617:12070
20. Sainath TN, Simko G, San Martin MCP, Candil RZ (2019) Voice activity detection
21. Jia F, Majumdar S, Ginsburg B (2021) MarbleNet: deep 1D time-channel separable convolutional neural network for voice activity detection. In: ICASSP 2021 - 2021 IEEE international conference on acoustics, speech and signal processing (ICASSP) (2021), pp 6818–6822
22. Murty KSR, Yegnanarayana B, Joseph MA (2009) Characterization of glottal activity from speech signals. IEEE Sign Process Lett 16(6):469–472
23. Shome N, Laskar RH, Kashyap R, Bandyopadhyay S (2020) A robust technique for end point detection under practical environment. In: International conference on machine learning, image processing, network security and data sciences. Springer, Berlin, pp 131–144

24. Dutta T (2008) Dynamic time warping based approach to text-dependent speaker identification using spectrograms. In: 2008 Congress on image and signal processing. IEEE, New York, pp 354–360
25. Larcher A, Lee KA, Ma B, Li H (2014) Text-dependent speaker verification: classifiers, databases and RSR2015. Speech Commun 60:56–77
26. Das RK, Jelil S, Prasanna SRM (2018) Multi-style speaker recognition database in practical conditions. Int J Speech Technol 21(3):409–419

# Signal Processing Track—2

# Detection of DME in OCT Images Based on Histogram Descriptor

**Puspita Dash and A. N. Sigappi**

## 1 Introduction

The human eye has an important coat called retina having ten different layers. It forms a window to the body as it is the only place directly accessible to visualization. Many of the changes that happen in different diseases can be visualized and thus scanned and studied directly. The most common diseases like diabetes and hypertension affect the blood vessels throughout the body but can be visualized in the retina. The macula is central part of the retina and is responsible for the visual acuity and color vision. Swelling or fluid accumulation in this portion of macula due to leaking abnormal blood vessels known as micro aneurysms is called diabetic macular edema (DME). The disease DME can cause permanent visual loss in patients with diabetes with diabetic retinopathy [1]. DME patterns are classified into Diffuse, Cystoid and Focal abnormalities with or without vitreomacular traction. A large majority of DME is due to diabetes. The OCT is a diagnostic imaging tool for the visualization of the retinal anatomy and its different layered structure. It is a fast procedure similar to the CT scan, using a coherent beam of light to acquire cross section images of the retina to show anatomy and structure of the eye [2]. This helps in the early diagnosis; treatment and follow up in various retinal disorders. Generally the DME on OCT images are classified into different types such as Focal, Diffuse retinal thickening, Cystoid macular edema and presence or absence of different types of vitreomacular traction and sub-retinal detachment and other abnormalities. OCT was quickly and easily adopted by clinicians for the assessment of patients with detection of DME.

P. Dash (✉)
Department of Information Technology, Sri Manakula Vinayagar Engineering College, Pondicherry, Pondicherry, India
e-mail: puspitadashit@smvec.ac.in

A. N. Sigappi
Department of Computer Science and Engineering, Annamalai University, Chidambaram, Tamil Nadu, India



P. H. J. Chong et al. (eds.), *Emerging Electronics and Automation*, Lecture Notes in Electrical Engineering 937, https://doi.org/10.1007/978-981-19-4300-3_31

In this study an automated method which concentrates on the detection and classification of normal and abnormal, such as the presence of DME in macular area in diabetes patients. The training and test images are filtered with three methods as Gaussian, Median and Weiner filters. The detection of the presence of DME disease from OCT images was done by using HOG descriptor method. HOG and SURF features of OCT images are extracted. The extracted features are then classified using two different SVM kernel classifiers namely SVM-Polykernel, SVM-RBF and to classify whether the image is normal or abnormal diseases DME.

## 2 Material and Methods

The block diagram of the proposed method is as shown in Fig. 1.

### 2.1 *Input OCT Images*

The Retinal OCT images containing both the normal and abnormal of diabetic patients were obtained from the Jyoti Eye Care, Puducherry using CIRRUS 5000 Spectralis B-Scan SD-OCT imaging device. Initially collected DICOM format OCT images are converted into JPEG format for further processing. An OCT scan set of normal OCT image in the right eye of 66-year-old female is shown in the Fig. 2 and abnormal with DME in the right eye of 75-year-old man is shown in Fig. 3. All the collected OCT images shows the following data in each figure such as a normal or abnormal with DME, first is the horizontal B-scan color image, second is the grayscale fundus image and third is the color coded thickness between the ILM-RPE map of the retina [3].

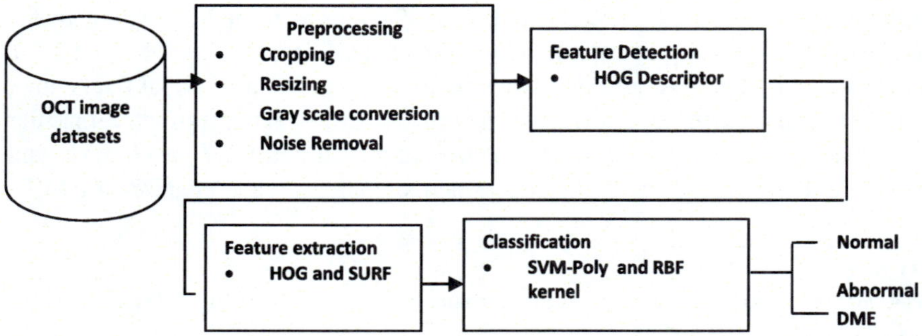


**Fig. 1** Diagram of proposed methods

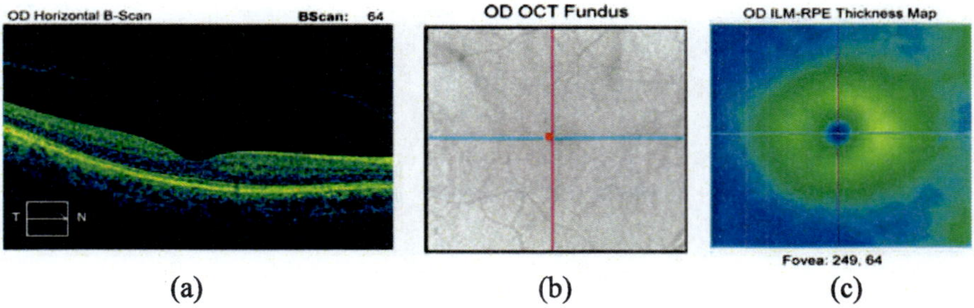


**Fig. 2** OCT scan of right eye showing **a** original Normal OCT image **b** its OCT fundus image and **c** Thickness map

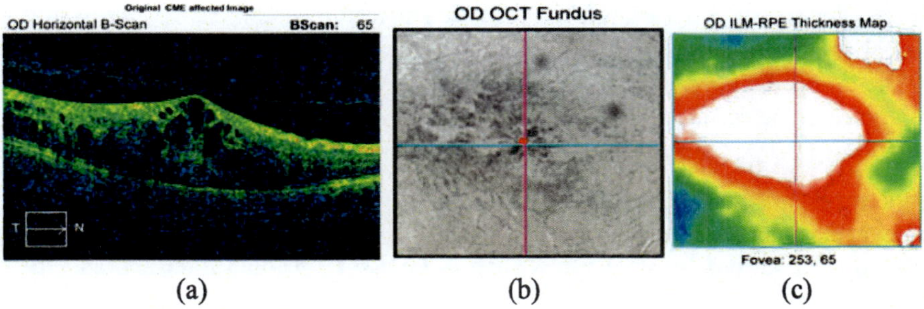


**Fig. 3** OCT of right eye showing **a** original cystoid macular edema affected image **b** its OCT fundus image and **c** Thickness map

## 2.2 *Preprocessing*

In this study, the first step of the study is the preprocessing of OCT images, this is carried out by cropping, resizing, RGB to grayscale conversion and denoising, as discussed below.

Before further processing for detection of DME and classification of OCT images, it is required to crop each OCT image. The OCT image of the retina is cropped based on smooth pixel intensity and is to focus on the region of the retinal image that contains morphological structures with sufficient variation between different patterns of OCT images. Cropping of the each OCT images is done 102 pixels from the center of fovea, i.e., on both the nasal and temporal side, thus bringing into focus the area of interest and highlighting most important changes. Here we have the OCT image with cropped up area showing only the region of interest (ROI) by eliminating the unnecessary areas to continue our focus on the areas where DME affected disease is most likely to be present. In this study, before extracting the features and to avoid excessive features as well as to increase the speed of further processing, for better result each B-scan OCT images are resized to 430 $\times$ 351 pixel dimension using imresize(img,[row, column]) method. The original input OCT images are is in RGB

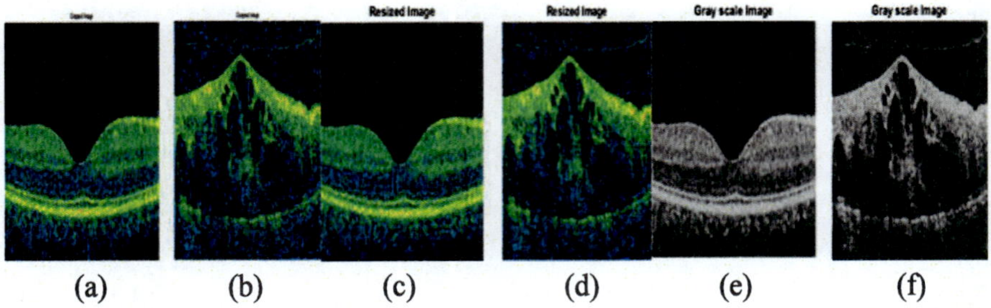


**Fig. 4** OCT images of **a** Cropped Normal image **b** Cropped DME affected image **c** Resized Normal image **d** Resized DME affected image **e** Grayscale Normal image **f** Grayscale DME affected image

format. For image processing is to get a bright feature, the RGB OCT image of normal and DME disease is converted into grayscale image.

The cropped, resized, converted RGB into grayscale OCT image of normal and DME disease performance is shown in Fig. 4a–f.

**Denoising**: All the OCT images has a lot of speckle noise due to high frequency sound waves, so it is better to denoise them. This will be help to reduce the effect of noise on the detection and classification results. In this work, Speckle noise has been reduced with various types of filters such as Gaussian, median filter and adaptive wiener filters.

**Gaussian and Median filter**: The OCT images get corrupted by speckle noise during imaging processing which can reduce the efficiency of the classification results. Firstly, this grayscale image is then filtered with a 2-D Gaussian filter method with sigma = 0.5 is used to remove the speckle noise and to blur the images shown in Fig. 5a, b. Then images are filtered using an order-statistics digital image filtering approach median filter [4] method. In this the average Peak Signal to Noise Ratio of the filtered of window size 5 × 5 in order to remove the speckle noise and smoothen the image.

**Adaptive wiener filtering**: In this work, the OCT images is then filtered using Wiener filter method to remove the blurred and additional noise present in the images by

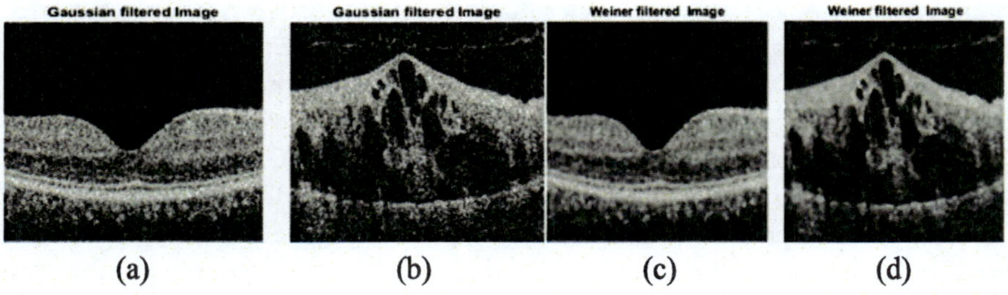


**Fig. 5** OCT image of **a** Gaussian filtered Normal image **b** Gaussian filtered DME affected image **c** Wiener filtered Normal image **d** Wiener filtered DME affected image

Gaussian and Median filter and with a compression operation it is minimized the mean square error during removal of noise filtering operation. Wiener filter is an inverse filtering and noise smoothing image restoration technique. In this type of linear adaptive filtering the edges are well preserved of an OCT images. The Wiener filtered image is shown in Fig. 5c, d.

The effect of Gaussian filtering is to remove the speckle noise and also to smoothen the image but the problem of Gaussian filter is that when it removes the noise, some of the details are also lost resulting in a blurred the image. Wiener is an inverse filtering noise smoothing method used for reducing the blurring while removing the noise when one is by using the Gaussian filter and recover the original image details using inverse filtering technique for better noise reduction as well as without the loss of vital information. Wiener filter's denoising performance is better compared than of Gaussian median filter. Wiener filter in Fourier transform [5, 6] can be expressed in the following Eq. 1,

$$W(f1, f2) = \frac{H^{*}(f1, f2)S_{xx}(f1f2)}{|H(f1, f2)|^{2}SS_{xx}(f1, f2) + S_{\eta\eta}(f1, f2)'} \tag{1}$$

where $SS_{xx}(f1, f2)$, $S_{\eta\eta}(f1, f2)'$ are the power spectra of OCT images and additive noise and $H(f1, f2)$ is the blurring filter.

In OCT medical retinal images noises are random as well as deterministic and it is not very easy to identify. In this work, Gaussian, median and Weiner various filter is used to identified the noise and thus suppress it to have a better quality of image to work on further. The noise free images are used for line, point and edge detection and classify them. So, different filtering techniques are used for smoothing and removal of the noise. High frequency parts are retained in the image, median filter can be used to remove the random noise but less cloudy than of linear smoothing filter. Different types of filters have different properties like the Gaussian filter is used to smoothen as well as removal of the noise while suppressing the high frequencies. Wiener filter is to remove the blurred and additional noise present in the images processed by Gaussian method. The Median filter is used to minimize the mean square error during removal of noise filtering operation.

The combination of three filtering methods is to successfully reduce the effect of unnecessary noise on the OCT images for better results of detection and classification images. Then the filtered result images are used for feature extraction process to detect the retinal diseases of OCT images.

## 2.3 *Feature Detection and Extraction of Images*

### Histogram of Oriented Gradients (HOG)

In this study, HOG feature vector descriptor algorithm [7] is used to extract feature vectors from the ROI portion in OCT images, as per the features extracted the images

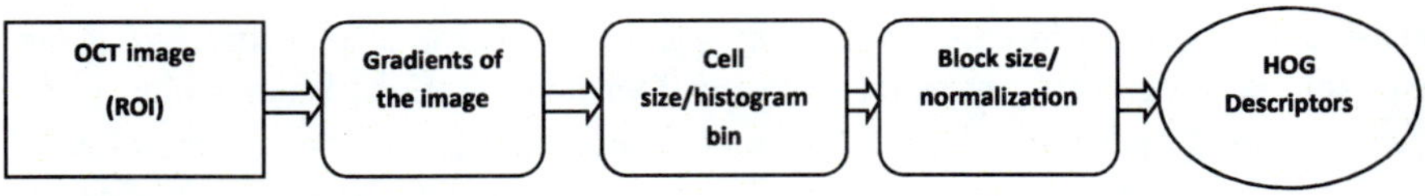


**Fig. 6** Process of HOG algorithm

can now be classified as normal or abnormal, here DME. So, the HOG feature descriptor generalizes the disease object in OCT image and produce the feature vector descriptors from the disease objects in the OCT image. The HOG is based on the evaluating well-normalized local histograms of image gradient orientations in a dense grid. HOG techniques counts occurrence gradient orientation in disease region of interest (ROI) portions of an OCT image. HOG feature descriptor converts an OCT image of any size into a feature vector and from the original image to cropped resize image is used for calculating HOG feature descriptor. Then the extracted HOG feature vector descriptors are given to the SVM kernel classifiers for training and testing to detect and classify normal and DME affected OCT images. The HOG algorithm process flow chart is as shown in Fig. 6.

The algorithm for HOG descriptor is as follows:

- Initially, The HOG is to divide the ROI of grayscale OCT image into 8 × 8 cells. A [8 × 8] cell size, [2 × 2] cells per block.
- Gradient of an image for each pixel is calculated with x and y axis individually and then the gradient mask of the image is computed by following formula,

$$h_{m=}[-1\ 0\ 1] \quad v_{n=}\begin{bmatrix}-1\\0\\1\end{bmatrix}$$

- A mask is used with the original image to obtain the $x$ and $y$ gradient by using the following formula,

$$G_{x=}(I.h_m) \quad G_{y=}(I.v_n)$$

Result of $G_x$ and $G_y$ axis gradient of an original image is shown in Fig. 7a–c as follows,

- Calculate magnitude and angle orientation of the gradient vector in 8 × 8 cells using following formula The gradient vector's angle is as follows,

$$\Theta = \tan^{-1}\left(\frac{G_x}{G_y}\right)$$

- The length of the gradient vector's magnitude is as follow,

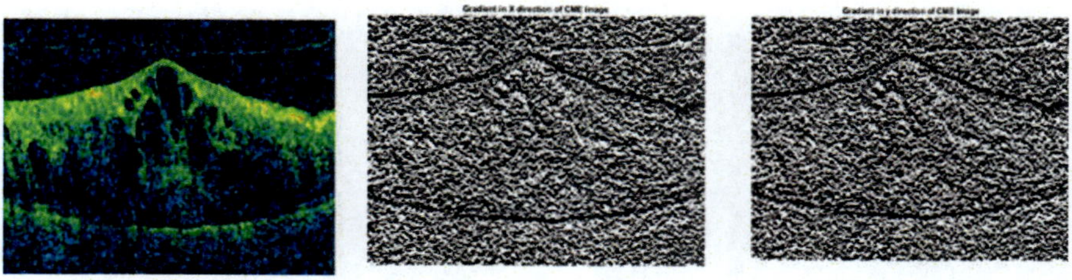

(a) (b) (c)

**Fig. 7** Gradient of OCT images **a** DME affected image **b** $G_x$ axis gradient and **c** $G_y$ axis gradient image

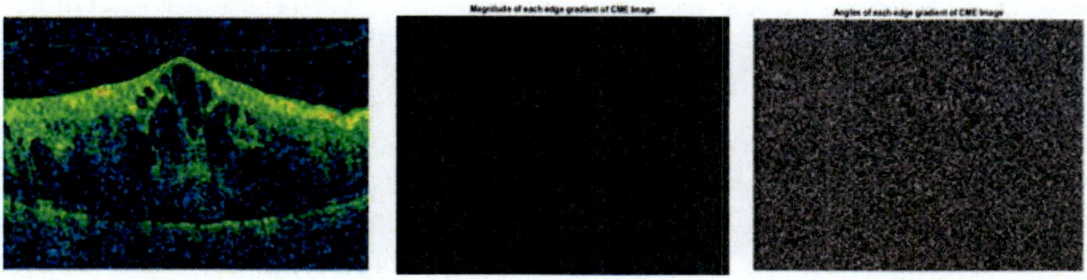

(a) (b) (c)

**Fig. 8** OCT images gradient vector orientation **a** DME affected image **b** magnitude and **c** angles of DME image

$$G = \sqrt{G_x^2 + G_y^2}$$

To calculate the angle atan () function and compute magnitude using sqrt $(x, y)$ function for each edge gradient of image and result shown in Fig. 8a–c.

- Calculate the histogram of gradient in 8 × 8 cells and extracted feature vector is calculated by comparing the magnitude and angle direction of the gradient.
- Gradient vectors with magnitude value are place them into the 9 histogram bin and in this gradient vector have an angle of 90°. The histogram's bin different angles from 0° to 90°.
- The set of normalized block of histograms represents the descriptor and normalization. Descriptor block are determined by histogram bin and after 8 × 8 cells histogram of gradient is calculated the blocks are normalized.
- The combined histogram entries are used as the HOG feature descriptor vector of the image.
- HOG algorithm elapsed time is 10.028380 s and is faster than of compared to SURF feature extraction method.

The input OCT image and its visualization of Hog feature descriptor extracted grayscale Image shown in Fig. 9a, b.

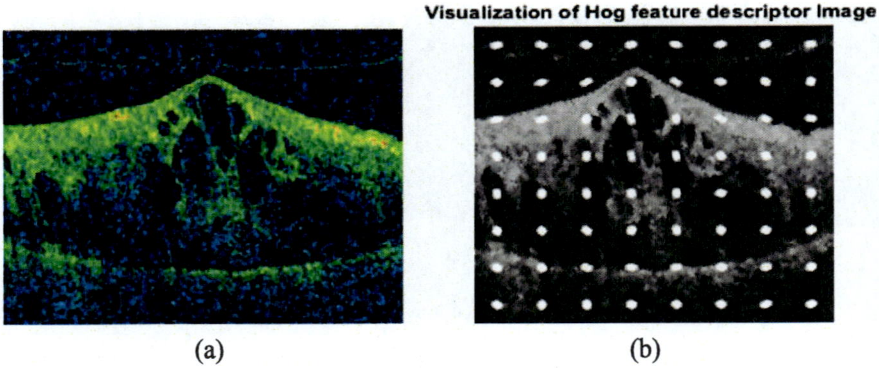


**Fig. 9** OCT images of **a** DME affected image and **b** Its visualization of Hog descriptors of the image

**Speeded-Up Robust Features (SURF)**: In this study, SURF [8] used for detection of interest key points and extract local interest key points feature descriptor elements from the retinal OCT images. The SURF method is similar to SIFT descriptor in terms of speed of extraction and feature vector comparison obtained from different images but it is invariant to scaling, rotation and translation than SIFT. SURF algorithm [9] used for the detection of DME diseases and feature extraction. SURF uses Gaussian square shaped filters of alternative integral representation of the image to speed up the calculation of features. SURF algorithm involves using the following steps:

(a) In SURF method the blob detector is used based on Hessian matrix to find and detection of interest points for selecting and filtering the scale of image of size $9 \times 9$.

Given a point $p = (x, y)$ in an image $I$, then the Hessian matrix $H(x, \sigma)$ at point $p$ and scale $\sigma$, is as follow on Eq. 2 and

$$H(x, \sigma) = \begin{bmatrix} L_{xx}(x, \sigma) \; L_{xy}(x, \sigma) \\ L_{xy}(x, \sigma) \; L_{yy}(x, \sigma) \end{bmatrix} \tag{2}$$

where $L_{xx}(x, \sigma)$, $L_{xx}(p, \sigma)$ is the convolution of the second-order derivative of the Gaussian with the image $I$.

$I(x, y)$ SURF uses the determinant of ($H_{\text{approx}}$) to find selecting the location and scale of interest points both using following Eq. 3 of Determinant of hessian matrix.

$$\det(H_{\text{approx}}) = D_{xx}D_{yy} - (0.9D_{xy})^2 \tag{3}$$

(b) After calculating the Hessian matrix SURF searches the maxima interest points of the image on different scales.
(c) The last step is the suppression of non-maxima points and calculation of feature descriptors using haar wavelet kernel filter, i.e., integral image taken from the

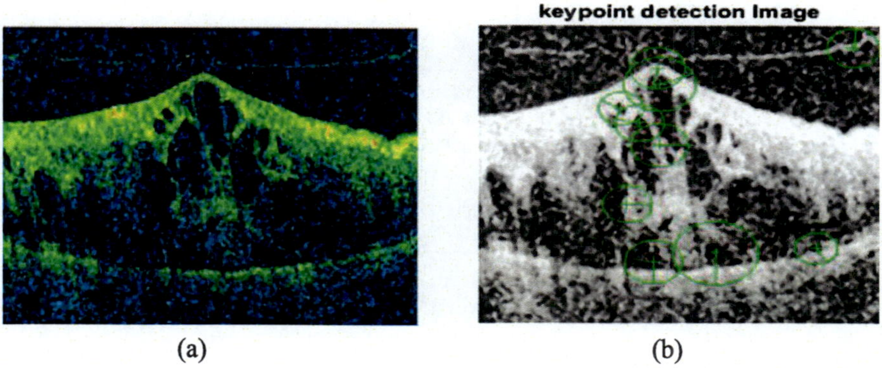


**Fig. 10** OCT images of **a** DME affected image and **b** Its SURF keypoints

input OCT image. From that image, the existing intensity values and their absolutes at both horizontal and vertical position are summed. Finally, by comparing the descriptors obtained from different images, matching pairs can be found.

(d) SURF uses haar wavelet for feature description and in this keypoint's around the circular neighborhood is selected which is divided into sub-regions and then each sub-region is detected and represented then extract SURF feature descriptor from the OCT images.

(e) Elapsed time is 23.159179 s of SURF algorithm.

Consider, for each sub-region 4 * 4 the 16 sub-regions vectors are then aggregated into a single vector of 16 * 4 = 64 which is the size of feature vector descriptor of the region of interest of the OCT image. Finally the summed value of 4 × 4 sub-region gives the 64 feature vector of DME affected disease of OCT image in our work.

In this SURF descriptor is used to detect and extract local interest key points feature descriptor elements from the images and SURF compute gradients by applying haar wavelet feature but HOG descriptor is also similar to SURF edge orientation histogram but difference is computed either horizontally or vertically grid of cells and compute the histogram of gradient in the region of cells to improve the accuracy. The input OCT image and its SURF key points of interest of grayscale image shown in the following Fig. 10a, b.

## *2.4 Classification*

Based on the feature extraction of images using SURF or HOG methods, our proposed supervising SVM-Polykernel and SVM-RBF kernel classifier methods are trained and tested with normal and DME affected OCT images and the steps of classifier algorithm are discussed in Fig. 11.

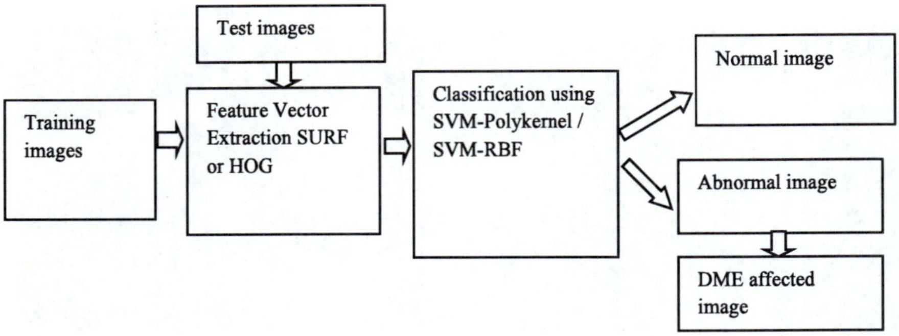


**Fig. 11** Steps of classifier algorithm

**Support Vector Machines (SVM)**: SVM are the supervised machine learning training algorithms trained on separable data [10]. SVM was introduced by Vapnik [11] are the supervised machine learning training algorithms trained on separable data is used for classification [12]. The classifiers have been used for classifying the pixels of OCT images into two groups, normal or abnormal such as DME affected image. In this work, based on the extracted HOG or SURF features type of SVM kernels techniques such as SVM-polykernel and SVM-RBF or Gaussian kernel classification is applied to classify the OCT images into two different categories such as normal and abnormal, where DME affected diseases is identified. In this SVM classification is used to separating the datasets into training and testing data. In the SVM classification the training extracted features are transformed into n-dimensional space used for feature vectors by using kernel function $K(x_i, x_j)$. SVM locates separating hyperplane in the feature space as well as classify training points in that space [13]. So, the training points are the support vectors to define the optimal hyperplane which is closest distance between support vectors shown in linear SVM architecture. SVM classifier model is to classify to find the hyperplane between linearly separate two different classes such as normal [+1] and abnormal DME [−1] in the OCT image by the margin represents distance between hyperplane and support vectors as shown in Fig. 12 with support vector + symbol denotes + 1 (normal) class and support vector – symbol denotes −1 (abnormal DME diseases) classes.

There are different kernel functions can be used in SVM method such as Linear, Polynomial, Radial-Basis Function (RBF) or Gaussian, PUK and Sigmoid. In this work, we used Polynomial and RBF two types of SVM kernels to classify the OCT images into different categories such as normal and abnormal, where DME affected abnormal diseases is affected.

**SVM-Polynomial Kernel**: Polynomial of power $p$ kernel in SVM classification as given in the Eq. (4). Polynomial kernel is defined as follows,

$$K(x_i, x_j) = \left\langle 1 + x_i^T\, x_j^p \right\rangle \quad \text{for any} \quad p > 0 \tag{4}$$

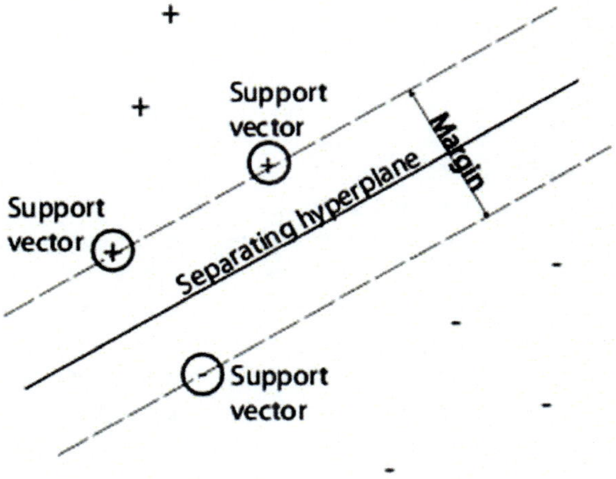


**Fig. 12** Architecture of SVM

where $x$ is the input OCT image pattern, $x_i$, $x_j$ are the training input support vectors for the $i$th and $j$th elements; $x_i$ is the input vector for $i$th elements and $x_j$ is the corresponding result. Power $p$ is the kernel parameter, $K$ is the kernel function that corresponds to the inner product $x_i^T x_j$ in feature space and used this function to find the SVM performance.

**SVM-Radial-basis function (RBF): Gaussian Kernels**

SVM-RBF classifiers can be used to learn vectors are nonlinearly mapped into a very high infinite dimensional feature space. Gaussian (Radial-Basis Function network) Kernels given in Eq. (5). RBF kernel is defined as follows,

$$K\left(x_{i,}x_{j,}\right) = \exp\left(\frac{\|x_i - x_j\|^2}{2\sigma^2}\right) \quad \text{for} \quad \sigma > 0 \tag{5}$$

$\sigma$ is the variance $x_i, x_j$ are the training input support vectors for the $i$th and $j$th elements; $x_i$ is the input vector for the $i$th elements and $x_j$ is the corresponding output, $K$ is the RBF kernel function that corresponds to the inner product $x_i x_j$ in feature space and used this function evaluated the SVM performance accuracy. RBF function also expressed as following Eq. (6),

$$K\left(x_{i,}x_{j,}\right) = \exp\left(-\gamma \frac{\|x_i - x_j\|^2}{2\sigma^2}\right) \quad \text{for} \quad \gamma > 0 \tag{6}$$

There are $C$ and gamma ($\gamma$) two parameters functions of SVM with RBF kernel increases the SVM classification performance accuracy. The gamma ($\gamma$) parameter defines distance such as 'far' or 'close' values can reach single training example. The $C$ parameters give the trade-off of training examples misclassification against decision surface simplicity. Through variation of gamma ($\gamma$) and $C$ parameters SVM classification performance accuracy is evaluated.

## 2.5 Experiment and Results

To carry this experiment initial OCT images datasets were collected from Jyoti Eye Care, Puducherry. A careful selection of the datasets were carried out to include only two categories of images, the normal images and the images with DME, all these patients were known diabetic. Care was taken to exclude the patients with DME from other causes and pictures with diabetic maculopathy without DME. This ensures that the experiment is focused on detecting the presence or absence of DME in diabetic patients. Initial experiment includes the training of the program with datasets for proper validation. This was followed by running the program again on the random datasets obtained. The results thus obtained were tabulated and experiments indices were carried out to evaluate and compared the performance of classifiers for the detection of retinal DME diseases to know which was giving better results. In this all the normal and abnormal DME OCT images were trained and tested by the SVM-polykernel and SVM-RBF two classifiers. In this work, the experiments were conducted for tenfold cross validation. Using two classifiers 65 OCT image data sets were used for training in which 31 were normal and 34 abnormal with DME images. After training was successfully completed the program was tested out of 45 new datasets of which 15 were normal and 30 abnormal with DME affected OCT images. Among all the classifiers out of 45 tested images 44 images were correctly classified by the HOG features with SVM-polykernel shown in Table 1.

The classification performance results obtained are tabulated and presented. After completion of the classification a performance validation is carried out by finding the values of True positive (TP), True negative (TN), False positive (FP) and False negative (FN). This is done by creating a confusion matrix of HOG and SURF with SVM kernel functions classifiers as shown from (a) to (d) in Table 2. After this a comparison shows that the HOG with SVM-polykernel kernel classifier is better to correctly classified the normal and abnormal images than the other kernel classifiers as shown in Table 2 (a) and using this classifier out of 45 testing datasets 1 normal image was classified as abnormal (False positive). Comparison of the overall performance measurement calculated results of both the classifiers of normal and DME affected OCT images using SVM classifier with HOG and SURF are presented in Table 3 and performance measurement Chart is shown in Fig. 13. In this the

**Table 1** Classifiers results

| Classifiers | SVM-polykernel | SVM-RBF | |
|---|---|---|---|
| No. of training images = 65 | Normal | 31 | 31 |
| | DME | 34 | 34 |
| No. of test images = 45 | Normal | 15 | 15 |
| | DME | 30 | 30 |
| No. of images correctly classified with HOG | 44 | 42 | 42 |
| No. of images correctly classified with SURF | 36 | 34 | 34 |

**Table 2** Confusion matrix of HOG and SURF with classifiers and Confusion matrix of SURF with SVM-Polynomial and SURF with SVM-RBF

| Actual class | Predicted class | |
|---|---|---|
| | DME | NOR MAL |
| *(a) HOG with SVM-polynomial* | | |
| DME | 30(TP) | 0(FN) |
| NORMAL | 1(FP) | 14(TN) |
| *(b) HOG with SVM-RBF* | | |
| DME | 30(TP) | 0(FN) |
| NORMAL | 3(FP) | 12(TN) |
| *(c) SURF with SVM-polynomial* | | |
| DME | 27(TP) | 3(FN) |
| NORMAL | 6(FP) | 9(TN) |
| *(d) SURF with SVM-RBF* | | |
| DME | 26(TP) | 4(FN) |
| NORMAL | 7(FP) | 8(TN) |

classifier SVM-Polykernel detected DME disease using HOG feature descriptor with sensitivity of 100%, specificity of 93%,accuracy of 98%, Precision of 97% and 98% of *F*-Score better than of SVM classifiers with SURF and SVM-RBF with HOG feature extraction method.

**Table 3** Overall performance measurements of the SVM classifiers with HOG and SURF

| Classifiers | Features | Sensitivity (%) | Specificity (%) | Accuracy (%) | Precision (%) | *F*-Score (%) |
|---|---|---|---|---|---|---|
| SVM-polynomial | SURF | 90 | 60 | 80 | 82 | 85 |
| SVM-RBF | | 87 | 53 | 76 | 79 | 82 |
| SVM-polynomial | HOG | 100 | 93 | 98 | 97 | 98 |
| SVM-RBF | | 100 | 80 | 93 | 91 | 95 |

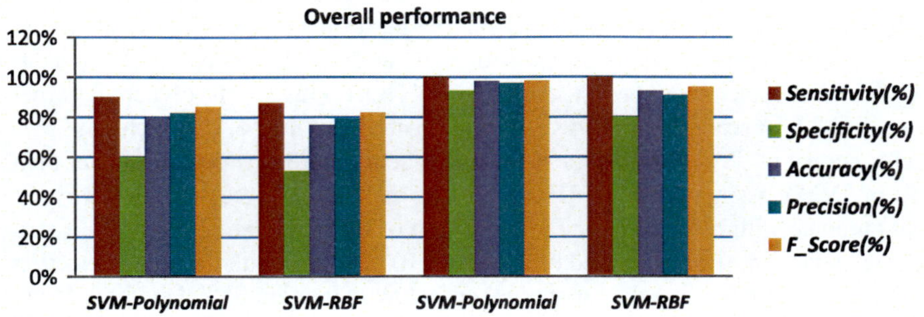


**Fig. 13** Overall performance measurement of the SURF and HOG with SVM kernel classifiers

The confusion matrix of the classification result of normal and DME affected images is as follows,

The overall performance measurement parameters are calculated by using the following equations,

$$\text{Sensitivity} = \frac{\text{TP}}{(\text{TP} + \text{FN})} \tag{7}$$

$$\text{Specificity} = \frac{\text{TN}}{(\text{TN} + \text{FP})} \tag{8}$$

$$\text{Accuracy} = \frac{(\text{TP} + \text{TN})}{(\text{TP} + \text{FP} + \text{TN} + \text{FN})} \tag{9}$$

$$\text{Precision} = \frac{(\text{TP})}{(\text{TP} + \text{FP})} \tag{10}$$

$$F - \text{Score} = \frac{(2 * \text{TP})}{(2 * \text{TP} + \text{FP} + \text{FN})} \tag{11}$$

The overall performance measurement parameters of the SVM classifiers with HOG and SURF features is shown in Table 3.

The chart of overall performance measurement for DME affected diseases classification using SURF with SVM kernel functions and HOG with SVM kernel functions classifiers is shown in Fig. 13 as follow.

## 2.6 Conclusion and Future Works

This paper deals with the development of one of the methods for the early and fast detection of DME in diabetic patients from their retinal OCT images. To undertake this task detection of DME abnormal diseases from OCT images using HOG descriptor method and successfully detects the diseases from the OCT images. Then HOG and SURF features were extracted which were then classified using SVM-polykernel and SVM-RBF two classifiers. The performance measure for classification was done using SVM classifiers with HOG and SURF feature extraction techniques. The results show best performance of HOG with SVM-polykernel kernel classifier. This particular method showed a sensitivity of 100%, specificity of 93%, and accuracy of 98%, Precision of 97% and F-Score of 98% which was better than that of SVM classifiers with SURF feature. The experimental results show an excellent outcome with better accuracy for detection of DME in diabetic patients. It also showed a fast processing elapsed time of only 10.028380 s with the HOG detection technique, which is faster than that of other feature extraction SURF method. Many more work need to be done in future using other advance features and also different classifiers such as deep learning to get faster better accuracy and get more details of disease changes in all retina layers.

**Acknowledgements** My special thanks to the ophthalmologist Dr. Vanaja, Jyoti Eye Care, Puducherry for providing the original OCT images and ophthalmologist Dr. Susil Kumar Pani, Raghu eye clinic, Puducherry for helping in the analysis and interpretation of data collected.

## References

1. Ciulla TA, Amador AG, Zinman B (2003) Diabetic retinopathy and diabetic macular edema. Diabetes Care 26(9):2653–2664
2. Schuman JS (2012) Introduction to optical coherence tomography
3. Dash P, Sigappi AN (2018) Detection and classification of retinal diseases in spectral domain optical coherence tomography images based on SURF descriptors. In: Proceedings of IEEE international conference on system, computation, automation & networking 2018, ISBN No. 978-1-5386-4865-0, pp 85–90
4. Pitas I, Venetsanopoulos AN (2012) Nonlinear digital filters: principles and applications. Kluwer Academic Publishers, Boston
5. Stela V (2000) Denoising via block wiener filtering in wavelet domain. In: European Congress of mathematics, pp 619–625
6. Choi H, Baraniuk R (1998) Analysis of wavelet domain wiener filters. In: Proceeding of the IEEE-SP international symposium on time-frequency and time-scale analysis
7. Dalal N, Triggs B (2005) Histograms of oriented gradients for human detection. In: Proceeding of the IEEE computer society conference on computer vision and pattern recognition. San Diego, California, USA, vol 1, pp 886–893
8. Bay H et al (2008) Speeded-up robust features (SURF)
9. Chihaoui T et al (2016) Personal verification system based on retina and SURF descriptors
10. Burges C (1998) A tutorial on support vector machines for pattern recognition. Data Min Knowl Disc 2(2):121–167
11. Vapnik VN (1998) Statistical learning theory. A Wiley Inter Science publication. Wiley
12. Burges JC (1998) A tutorial on support vector machines for pattern recognition. J Data Mining Knowl Discovery 2(2):121–167
13. Tan M (2004) Support vector machine & its applications. The University of British Columbia

# Pain Assessment Using Randomness and Busyness of EMG Signal

**Kausik Sen and Saurabh Pal**

## 1 Introduction

Pain is one of the most complex [1, 2] phenomenon which is not fully perceived by doctors and researchers. Chronic pain is a major problem for elderly people [3]. Existing pain assessment methods are mostly subjects' verbal response dependent. In Verbal Pain Intensity Scale (VPIS), apart from no pain, pain sensation is classified in three stages such as mild pain, moderate pain and severe pain. In Numeric Pain Rating Scale [4] (NPRS), patients are requested to scale their pain between the scale of 0–10, where 0 indicates no pain and 10 denotes maximum pain. Visual Analog Scale (VAS) [1], is a straight line with two extreme points' no pain' and 'worst possible pain'. Subjects are asked to compare their pain with these scales. Subjects unable to communicate will not be able to participate in these techniques. Infants, a senseless person or a person with cognitive disorder will not able to express their pain and compare it with the scales.

Hence, an alternative method is required to asses pain for the patients who are not able to communicate. As, pain is an emotional experience it changes facial expression [5]. Alteration of facial expression is considered as a key tool to detect pain [6–8]. Different stages of pain were classified on the basis of facial responses like Prkachin and Solomon Pain Intensity (PSPI), Subjective Rating and Optical Flow. Rare facial expression was observed during lower stimulus. In [9] Residual Generative Adversarial Network (R-GAN) approach was considered to classify between actual pain and posed pain. Most of the facial expression-based methods are individual images or video-frames-based analysis. That might eliminate the spatio-temporal information for pain recognition. Spatio-temporal information in pain assessment was introduced

K. Sen (✉) · S. Pal
Department of Applied Physics, University of Calcutta, Kolkata, India
e-mail: kausiksen1993@gmail.com

S. Pal
e-mail: spaphy@caluniv.ac.in



P. H. J. Chong et al. (eds.), *Emerging Electronics and Automation*, Lecture Notes in Electrical Engineering 937, https://doi.org/10.1007/978-981-19-4300-3_32

in [10]. Cross database pain analysis was investigated in [11] between The BioVid heat pain database and X-ITE database, where 'no pain' stage was classified with all the other pain stages together. In [5], mutual effect of pain and various emotion were investigated. However, facial expression-based pain analysis requires complex algorithm and costly setup also the results are not highly satisfactory.

Specificity theory [12] suggested the relation of pain with Autonomic Nervous System (ANS), it also specified the pathway of pain. This suggest that pain is associated with ANS. So, biosignals associated with ANS alters due to pain elicitation. In [13], 135 features were extracted from Electromyogram (EMG), Skin conductance level (SCL), Electroencephalogram (EEG) and Electrocardiogram (ECG). Gamma band of EEG was found to be affective to classify different stages of pain, in [14]. However, the results are not quite satisfactory considering complexity of the methods. Correlation of pain with EMG is established by researchers [15, 16].

In this study, an alternate subject-response independent method is proposed for pain assessment based on autonomic parameter. Here, Entropy (EN) and Zero Crossing (ZC) are extracted from EMG of Zygomaticus muscle during various stages of pain. Further various pain stages are classified using Support Vector Machine k-NN classifiers with 'No Pain' stage. Methodology of the proposed work is described in Sect. 3. Results of pain stage classification are elaborated in Sect. 4. Further performance of classifiers is investigated by alternating classifier parameters which shows the robustness of SVM. The study is concluded in Sect. 5.

## 2 BioVid Heat Pain Database

This study was performed on BioVid heat pain database [17, 18]. The database contains the data of 90 subjects who were introduced with a four pain stages such as $P_1, P_2, P_3, P_4$. First, threshold temperature ($P_1$) was induced followed by tolerance temperature ($P_4$) of the subject which never exceeded 50.5 °C to avoid any kind of injury. Further, two evenly distributed intermediate temperatures stimuli ($P_2$, $P_3$) were also used.

$$P_2 = \left(\frac{P_4 - P_1}{3}\right) + P_1 \tag{1}$$

$$P_3 = \left(\left(\frac{P_4 - P_1}{3}\right) \times 2\right) + P_1 \tag{2}$$

Biosignals (EMG, ECG, EEG and GSR) and facial video expression were recorded for four seconds during each pain stages including baseline temperature or 'No Pain' (NP), considered as 32 °C.

# 3 Methodology

Facial expression depends on the muscle movements, which can be recoded using EMG signal. Here, EMG of zygomaticus muscle is taken from random twenty-two subjects of BioVid heat pain database. Figure 1 shows the change in typical EMG signal from NP stage to $P_4$ stage. From the Fig. 1 it can be observed that the busyness and randomness are increased during $P_4$ condition compared to NP condition.

To replicate these to attributes statistically Zero Crossing (ZC) and Entropy (EN) are extracted from the signal. Further pain stages are classified using these extracted features. Diagram of overall methodology is given in Fig. 2.

## 3.1 *Normalization*

Inter and intra subject variability reduce the reliability of the actual signal amplitude for different intrinsic and extrinsic factors like subject's measurement condition, body composition and variability of measuring instrument, etc. These effects are reduced by amplitude normalization in the range of (0, 1) on EMG signal using the formula given below.

$$y' = \frac{y - y_{\min}}{y_{\max} - y_{\min}} \tag{3}$$

where $y$ is the data under test and $y'$ is normalized data.

## 3.2 *Feature Extraction*

**Zero Crossing (ZC).** Busyness of a signal can be estimated by zero crossing. It is defined as the number of times a signal crosses zero or a reference line. These feature

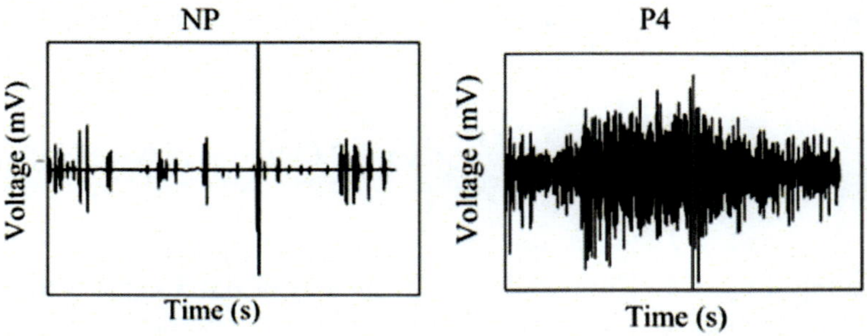


**Fig. 1** EMG signal during 'No Pain' and $P_4$ pain stages

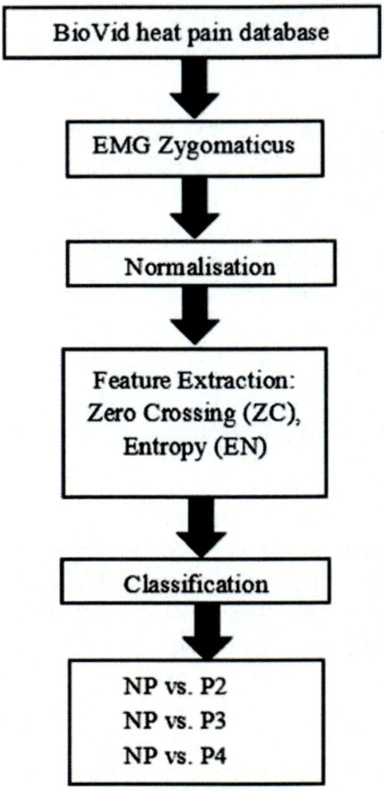


**Fig. 2** Overall methodology

can approximate the frequency of a signal [19]. Further ZC was also normalized using Eq. 3.

**Entropy (EN)** [20, 21]. EMG signals are complex in nature. It also shows a hint of randomness. This randomness increases with introduction of different stimuli such as pain. Hence, Entropy can be used to measure the randomness of the signal in different stages of pain.

Let, $X$ be an EMG signal taken from zygomaticus muscle during any pain stage. Signal $X$ consists of $x_1, x_2, x_3, \ldots, x_{n-1}, x_n$ discrete samples. Entropy, $H(X)$ of signal $X$ can be defined as,

$$H(X) = -\sum_{i=1}^{n} P(x_i) \log P(x_i) \tag{4}$$

where $P(x_i)$ is the probability distribution of $x_i$ sample of signal $X$.

From Fig. 1 it can be observed that the irregularity of EMG signal is increased with introduction of pain. Boxplots of the two features are given in Fig. 3.

Mean of ZC and EN during different stages of pain are given in Table 1.

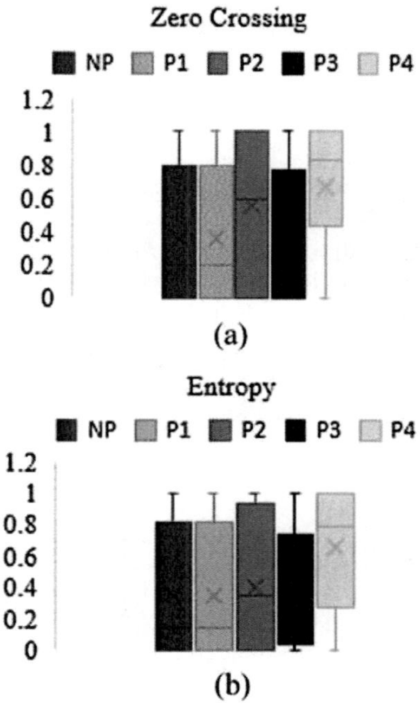


**Fig. 3** Boxplot of **a** Zero crossing **b** Entropy of ENG of Zygomaticus muscle during different stages of pain

**Table 1** Mean ZC and entropy during different stages of pain

| Features | NP | $P_1$ | $P_2$ | $P_3$ | $P_4$ |
|---|---|---|---|---|---|
| ZC | 0.20 | 0.20 | 0.60 | 0.31 | 0.82 |
| EN | 0.14 | 0.14 | 0.34 | 0.25 | 0.79 |

From Table 1 it can be observed that there is hardly any change during $P_1$ pain stage compared to NP. However, noticeable changes can be found in the higher stage of pain compared to baseline. Alteration in the features is noticeable for higher stimuli. Higher pain stages ($P_2$, $P_3$ and $P_4$) are compared with no pain condition.

## 3.3 *Classification*

Binary classification between pain stages is done using Support Vector Machine (SVM) and k-Nearest Neighbor algorithm.

**SVM**. SVM is one of the most used classifiers in biomedical engineering field. It was initially developed for classification and have been extended for regression and preference (or rank) learning. Here, maximum separation between two classes is obtained by choosing the hyperplane which offers maximum width. Support vectors are the nearest data points between two classes. The hyperplane acts as a decision function which can be linear as well as non-linear. The linear hyper plane can be defined as,

$$w.x + b = 0 \tag{5}$$

**k-NN**. One of the most popular instance-based method is the k-nearest neighbor algorithm. In this algorithm all instances are assumed to be correspond to points in the $n$-dimensional space $\mathbb{R}^n$. The nearest neighbors of an instance are defined in terms of the standard Euclidean distance. More precisely, let an arbitrary instance $x$ be described by the feature vector ($b_1(x)$, $b_2(x)$, …, $b_n(x)$) where $b_r(x)$ denotes the value of the $r$th attribute of instance $x$. Then the distanced ($x_i$, $x_j$), between two instances $x_i$ and $x_j$ is defined as given below.

$$d(x_i, x_j) = \sqrt{\sum_{r=1}^{n} (b_r(x_i) - b_r(x_r))^2} \tag{6}$$

## 4 Result and Discussion

As mentioned in Sect. 3.2 mean of ZC and EN are during NP and $P_1$ pain stages are found to be too close to discriminate. To substantiate this inference $t$-test was performed for same data. From the result it can be concluded that hardly any alteration is found during NP and $P_1$ stage. Hence, comparison between NP and $P_1$ is excluded in this study. Result of paired $t$-test on the extracted features between 'No Pain' and 'Tolerance temperature' is given in Table 2, which suggests that the two pain conditions can be discriminated.

Results of classification for NP with different stages of pain using ZC and EN extracted from EMG of zygomaticus muscle is given in Table 3.

**Table 2** Result of $t$-test

| Feature | NP versus $P_4$ |
|---|---|
| ZC | 0.02 |
| EN | 0.04 |

**Table 3** Classification results

| | SVM | K-NN |
|---|---|---|
| Np versus $P_4$ | 71.4 | 73.8 |
| Np versus $P_3$ | 55.8 | 46.5 |
| NP versus $P_2$ | 54.8 | 45.2 |

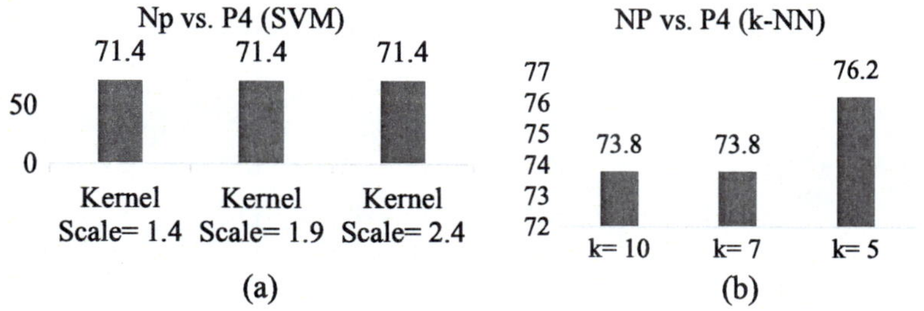


**Fig. 4** Classification accuracy for **a** different kernel scale in SVM, **b** different k value in k-NN

This study was further extended to exploit the change in accuracy by altering the parameters of SVM and k-NN. In Table 3, kernel scale and k value were chosen as 1.4 and 10, respectively. As, the pain stages considered in this study are not widely apart, kernel scale in SVM was increased by ½ in two stages. However, classification accuracy remained unchanged which indicates the robustnessof the proposed method. In k-NN, initially k value was chosen as 10. Then, ½ of the initial k value was taken, finally median of these two values which is preferably an odd number was chosen. From the results it can be seen that decreased in k value offers increase in classification accuracy. Graph chart of the classification accuracy for different parameters of SVM and k-NN is given in Fig. 4.

The proposed work is compared with some the recent relevant work in Table 4, where efficacy of this method can be observed over the existing techniques.

## 5 Conclusion

In this study, biosignal-based pain assessment technique proposed. It is seen that randomness and busyness of EMG signal is increased with introduction of pain stimulus. Zero Crossing and Entropy of EMG signal of zygomaticus muscle are obtained to classify different stages of pain from 'No Pain' condition. Study was further extended by changing the kernel scale and k value of SVM and k-NN algorithm, respectively. From the results robustness of SVM classifier and the enhancement of classification accuracy by reducing the k value which reduced overfitting problem, can be observed. Finally, to substantiate the performance of the proposed work it is compared with some recently reported study. Where the superiority of the proposed work is noticeable. From this study it can be also be seen that the proposed method offers superior accuracy compared to a complex and costly video signal-based pain analysis. However, other autonomic parameters and features can be incorporated to achieve enhanced pain recognition. Also, performance of the classifiers can be enhanced by using optimizing the classifier parameters.

**Table 4** Comparison of the proposed work

| | Proposed method | | Kächele et al. [22] | | | | Othman et al. [11] | Yang et al. [10] | Singh et al. [23] |
|---|---|---|---|---|---|---|---|---|---|
| *Parameters extracted* | | | | | | | | | |
| | EMG (zygomaticus) | | EMG (zygomaticus) | EMG (trapezius) | ECG | GSR | Video signal | Facial expression | Facial expression |
| *Classifier* | | | | | | | | | |
| | SVM | K-NN ($k = 5$) | SVM | | | | Random forest | SVM | Random forest |
| Np versus $P_4$ | 71.4 | 76.2 | 67.8 | 61.3 | 63.5 | 67.0 | 65.8 | 71.0.0 | 50.0 |
| Np versus $P_3$ | 55.8 | 57.1 | – | – | – | – | | | |
| NP versus $P_2$ | 54.8 | 47.6 | – | – | – | – | | – | |

**Acknowledgements** Authors acknowledge their gratitude to BioVid heat pain database (http://www.iikt.ovgu.de/BioVid) administrators for providing persmission to use the database.

## References

1. Haefeli M, Elfering A (2006) Eur Spine J 15:S17
2. Williams ACDC, Craig KD (2016) Pain 157:2420
3. Zwakhalen SMG, Hamers JPH, Abu-Saad HH, Berger MPF (2006) BMC Geriatr 6:1
4. Wong DL, Baker CM, Pediatr Nurs 14:9 (n.d.)
5. Reicherts P, Gerdes ABM, Pauli P, Wieser MJ (2013) Pain 154:793
6. Zhang X, Yin L, Cohn JF, Canavan S, Reale M, Horowitz A, Liu P, Girard JM (2014) Image Vis Comput 32:692
7. Martinez DL, Rudovic O, Picard R (2017) IEEE Comput Soc Conf Comput Vis Pattern Recognit Work 2017-July, 2318
8. Werner P, Al-Hamadi A, Niese R, Walter S, Gruss S, Traue H (2014) In: Brain Machinery Vision Conference 2013 (2014), pp 118.1–118.11
9. Tavakolian M, Cruces CGB, Hadid A (2019) In: Proceedings—14th IEEE international conference on automatic face and gesture recognition, FG 2019. IEEE, New York, pp 1–8
10. Yang R, Tong S, Bordallo M, Boutellaa E, Peng J, Feng X, Hadid A (2017) In: 2016 6th international conference on image processing theory, tools and applications, IPTA 2016 (Oulu, Finland)
11. Othman E, Werner P, Saxen F, Al-Hamadi A, Walter S (2019) In: International symposium on image and signal processing and analysis, ISPA. IEEE, New York, pp 181–186
12. Britton NF, Skevington SM (1989) J Theor Biol 137:91
13. Walter S, Gruss S, Limbrecht-Ecklundt K, Traue HC, Werner P, Al-Hamadi A, Diniz N, da Silva GM, Andrade AO (2014) Psychol Neurosci 7:363
14. Vijayakumar V, Case M, Shirinpour S, He B (2017) IEEE Trans Biomed Eng 64:2988
15. Khoshbin S, Hallett M, Lunbeck R (1987) Muscle Nerve 10:629
16. Das P, Bhattacharyya J, Sen K, Pal S (2020) In: Proceedings 2020 IEEE applied signal processing conference ASPCON 2020. Institute of Electrical and Electronics Engineers Inc., pp 349–353
17. (n.d.)
18. Werner P, Al-Hamadi A, Niese R, Walter S, Gruss S, Traue HC (2014) In: Proceedings—international conference on pattern recognition. Institute of Electrical and Electronics Engineers Inc., pp 4582–4587
19. Toledo-Perez DC, Rodriguez-Resendiz J, Gomez-Loenzo RA (2020) IEEE Access 8:8783
20. Dutta A, Chattopadhyay H, Yasmin H, Rahimi-Gorji M (2019) Comput Methods Programs Biomed 180:105010
21. Yang D, Geng D, Zheng L, Cai M, Hao W (2021) Biomed Signal Process Control 65:102391
22. Kächele M, Werner P, Al-Hamadi A, Palm G, Walter S, Schwenker F (2015) In: Lectures Notes in Computer Science (Including Subseries Lecture Notes Artificial Intelligence Lectures Notes Bioinformatics). Springer, Berlin, pp 220–230
23. Singh SK, Tiwari S, Abidi AI, Singh A (2017) Multimed Tools Appl 76:19317

# Literature Review on Brain Tumor Detection

**Rafik Ahmad and Kalyan Acharjya**

## 1 Introduction

Automated defect detection in diagnostic imaging has become a hot subject in a number of medical clinical diagnostics. Automated tumor diagnosis in MRI is critical because it offers facts related to aberrant tissues, those are vital in treatment planning. Human inspection is regular approach for detecting defects in resonance brain imaging. Due to a lack of information, this technique is impracticable. Most analysis in developed countries has exposed that the death rate of individuals suffering from tumor has exaggerated over the past three decades. Today one in every of the foremost cause of rise of fatality among kids and adults is tumor.

In our reviewed literature survey we found that Brain tumor is a disease developing in the intracranial architecture owing to inappropriate and disorganized multiplication of cells. It is an extremely aggressive and life-threatening illness that must be recognized and treated as soon as possible in order to prevent morality. There are varying sizes, regions, and placements of brain tumors. They even have sensitivities that overlap with normal tissues. Neoplasm, which can be malignant or benign, can be seen in many portions of the brain and may or may not be primary tumors. As a result, detecting cancers before they become uncontrolled is critical. MRI is the method used to locate the tumor.

R. Ahmad (✉) · K. Acharjya
Maharishi University of Information Technology, Lucknow, India



P. H. J. Chong et al. (eds.), *Emerging Electronics and Automation*, Lecture Notes in Electrical Engineering 937, https://doi.org/10.1007/978-981-19-4300-3_33

### 1.1 Classification of Brain Tumors by Machine Learning Algorithms (2019)

Çınarer and Emiroğlu [1] suggested an approach that helps in detecting brain tumor integrating Machine learning (ML) and Magnetic resonance technique. The core objective of his main goal is to learn mechanical and classification algorithms to learn automatically from training and make the right decision. In the paper, the performances of tumor classification techniques for n/a, multifocal, multicentric, and gliomatosis classification of brain MRI symptoms such as n/a, multifocal, multicentric, and gliomatosis was investigated in this work. The statistical properties of the incoming pictures were examined during the classification process, and the data was meticulously split into several groups. These statistics were put to the test, and a 90% accurateness rates was discovered to found better compared to other algorithms.

The accuracy attained is good due to the various appearances and intricacy of malignancies. Taking into consideration of a big dataset and removing density based features in adding up to texture based property will almost certainly enhance this accuracy.

### 1.2 Brain Tumor Detection by Image Processing Using MATLAB (2016)

Sudharson et al. [2] had tried a new method to detect brain tumor by using MATLAB software. Traditional imaging techniques such as pneumoencephalography and cerebral radiology were intrusive, therefore computed tomography and magnetic resonance imaging (MRI) were used to enable neurosurgeon see clearly enough to spot brain tumors.

In this, through MATLAB three major steps are involved such as preliminary processing, segmentation, and morphological operation. When the original image of brain tumor is seen after the is pre-processed by switching from black and white to grayscale using an amplifier and a medium filter for distortion reduction and quality enhancement. This is followed by an improvement phase, which produces a histogramic comparable image.

The watershed algorithm is used to segment data during the processing stage. The major goal of this article is to minimize manpower requirements for obtaining precise and efficient results in the treatment of brain tissue.

## 1.3 *Design and Implementing Brain Tumor Detection Using Machine Learning Approach (2019)*

Hemanth et al.—Nowadays [3], brain tumor detection and identification has become a great deal for the doctors. The image segmentation process is adopted for finding the crucial tumor region within the brain. In this research paper, brain tumor detection is done by applying machine learning technique. MRI report contains lots of important data about the disease but proper data mining and utilization will bring great output.

The research proposes a technique called automatic segmentation method based on Convolutional Neural Network (CNN) determining 3 * 3 Kernels. For segmentation and classification process various steps are used such as data gathering, preprocessing, baseline filtering, segmentation, extraction of features, and CNN. The output is satisfactory in determining the presence of brain tumor in early stage as well as in normal condition.

## 1.4 *Classification of Brain Tumor Types by Deep Learning with Convolutional Neural Network on Magnetic Resonance Images Using a Developed Web-Based Interface (2019)*

Ucuzal et al.—In this paper [4], the development of automated web-based software using deep learning is being discussed with abundant data, apex accuracy and defined method of classification of brain tumor. They have mentioned a particular Python library named as Keras library for the detection of unwanted cell collection formed in brain. All these detection and classification is done with the help of MRI reports.

The result shown points toward the high quality classification of types of brain tumor nearly with 98% accuracy on the training dataset. This classification of brain tumor will help the surgeons to understand and remove tumor easily.

## 1.5 *Current Trends on Deep Learning Models for Brain Tumor Segmentation and Detection—A Review (2019)*

Somasundaram and Gobinath [5]—In this paper [6], the development of an automated web-based software using deep learning is being discussed with abundant data, apex accuracy and defined method of classification of brain tumor. They have mentioned a particular Python library named as Keras library for the detection of unwanted cell collection formed in brain. All these detection and classification is done with the help of MRI reports.

The result shown points toward the high quality classification of types of brain tumor nearly with 98% accuracy on the training dataset. This classification of brain tumor will help the surgeons to understand and remove tumor easily.

## 1.6 A Deep Convolutional Neural Network Learning Transfer to SVM-Based Segmentation Method for Brain Tumor (2019)

Cui et al.—Segmentation of brain tumor [7] is useful for aiding with diagnosis, care plans, and surgical route planning. They propose a learning handover from convolutional neural networks to help the process of separating the vector machine of the brain tumor detection in this study. The algorithm is made up of two phases that are linked together. They taught CNN to learn how to map from image space to label space in the first step. They have passed the anticipated brand output from convolution of neural network, together with the trial picture, in the SVM separator for the correct partition especially during the show phase. The deep CNN-SVM separator is reversed. Experiments and tests show that the proposed design transcends SVM or CNN-based segments separately. They introduced a simple but effective MRI brain fragmentation procedure that involves transferring learning data based on a flexible network to an SVM separator. Compared to existing methods, the suggested process works better. It also beat the use of CNN or SVM alone for abdominal split.

## 1.7 Surgical Process Identification System Using Machine Learning in Awake Surgery for Brain Tumor (2019)

Tomohiro Nagai, Ikuma Sato, Yuichi Fujino, Manabu Tamura, Yoshihiro Muragakiand Ken Masamune: Qualified practitioners evaluate maximal brain tumor excision and minimal postoperative glial problems during malignancy surgery [8]. The neurosurgeon resects the brain tumor depending on their expertise and experience, and the surgical method, work contents, and operation length differ on case. Younger doctors and surgical personnel have a tough time understanding the surgery. Visual representation of the surgical procedure is an excellent technique for assisting surgical team knowledge. A postoperative diagnosis system based on intra-operative data and machine learning is presented in this study. Using the navigational system's log, MR pictures, and microscopic footage, they derive operating characteristics. The surgical procedures are then analyzed with the help of a Hierarchical Hidden Markov Model. Previous log data (navigation system, MR images, and microscope images) was used to assess the technique [9]. In 12 of the recognized surgical procedures, the accuracy was 84%. These result showed that the error of diagnosing the surgical procedure is only a few minutes long and has a high revel accuracy. Furthermore, this result provides a chance to aid new surgeons and surgical personnel in their comprehension of surgical procedures.

In awake surgery for brain tumors, we built a surgical operating model and surgical procedure diagnostic procedure. The specification of the diagnostic procedure using previous clinical data was 84%, according to their technique. This finding shown that utilizing current medical technology in the operating room, it is feasible to detect the skilled surgeon's surgical procedures [10]. As a result of implementing this method in the operating room, a young surgeon and a surgical team will no longer be required to continually watch the procedure. The navigation system's record was lost in the experiment, yet utilizing microscope footage, the precision did not deteriorate.

## 2 Conclusion

Automated tumors detection approaches are being developed in order to reduce radiologist time while also achieving a high level of accuracy. Because of the intricacy and variety of malignancies, MRI brain tumors identification is very difficult undertaking. During this research, we point out on utilizing machine learning techniques to conquer the limitations regarding conventional classifiers in the detection of tumors in brain MRI. Machine learning and image classifiers are frequently used in MRI to detect cancer cells in the brain. In addition to this project, an android app is collaborated with nearby radiologist lab that provide the medical imaginary in accurate manner also the final result would be finally verified by doctors about the status and level of severity of that tumor and treatment urgency.

## References

1. Çınarer G, Emiroğlu BG (2019) Classification of brain tumors by machine learning algorithms. In: 3rd international symposium on multidisciplinary studies and innovative technologies (ISMSIT). https://doi.org/10.1109/ISMSIT.2019.8932878
2. Sudharson M, Thangadurai Rajapandiyan SR, Ilavarasi PU (2016) Brain tumor detection by image processing using MATLAB. Middle-East J Sci Res 24(S1):143–148. ISSN 1990-9233. https://doi.org/10.5829/idosi.mejsr.2016.24.S1.30
3. Hemanth G, Janardhan M, Sujihelen L (2019) Design and implementing brain tumor detection using machine learning approach. In: 3rd international conference on trends in electronics and informatics (ICOEI). https://doi.org/10.1109/ICOEI.2019.8862553
4. Ucuzal H, Yaşar Ş, Çolak C (2019) Classification of brain tumor types by deep learning with convolutional neural network on magnetic resonance images using a developed web-based interface. In: 3rd international symposium on multidisciplinary studies and innovative technologies (ISMSIT). https://doi.org/10.1109/ISMSIT.2019.8932761
5. Somasundaram S, Gobinath R (2019) Current trends on deep learning models for brain tumor segmentation and detection—a review. In: International conference on machine learning, big data, cloud and parallel computing (COMITCon). https://doi.org/10.1109/COMITCon.2019.8862209

6. Wu W et al (2020) An intelligent diagnosis method of brain MRI tumor segmentation using deep convolutional neural network and SVM algorithm. In: Computational and mathematical methods in medicine
7. Cui B, Xie M, Wang C (2019) A deep convolution neural network learning transfer to SVM-based segmentation method for brain tumor. In: 2019 IEEE 11th international conference on advanced infocomm technology (ICAIT). https://doi.org/10.1109/ICAIT.2019.8935904
8. Nagai T, Sato I, Fujino Y, Tamura M, Muragaki Y, Masamune K (2019) Surgical process identification system using machine learning in awake surgery for brain tumor. In: IEEE 1st global conference on life sciences and technologies (LifeTech). https://doi.org/10.1109/LifeTech.2019.8884047
9. Kiranmayee BV, Rajinikanth TV, Nagini S (2017) Explorative data analytics of brain tumour data using R. In: 2017 International conference on current trends in computer, electrical, electronics and communication (CTCEEC). IEEE, New York
10. Yang G et al (2015) Discrete wavelet transform-based whole-spectral and subspectral analysis for improved brain tumor clustering using single voxel MR spectroscopy. IEEE Trans Biomed Eng 62(12)

# MEMS and VLSI Track—1

# Design and Comparative Analysis of Dynamic Comparators for SAR ADC

**Noman Ahmed Ansari, Priyansh Jaiswal, Mohit Tyagi, and Poornima Mittal**

## 1 Introduction

Signal processing is universally used in almost every field ranging from communication to medical systems. It deals with both analog and digital signals, thereby giving utmost importance to data converters [1], thereby serving as a link b/w the analog and digital world. Comparison is one of the principal actions used in data converters. Comparators play crucial part in high-speed ADCs. A comparator compares 2 analog signals and based on that comparison produces the digital output. It is also referred to as a 1-bit digital converter.

Comparators produce a logical value, indicating which one of the inputs is higher or lower. Nowadays, the demand for high-speed ADC is increasing day by day. Since it is a fundamental building block used in ADCs, high-speed, less hysteresis, and low-power specifications are praised. Generally, due to less delay and power consumption, clocked comparators are preferred in ADCs. Clocked comparators are basically referred to as dynamic comparators. The positive feedback mechanism provided by the back-to-back inverter converts the smallest voltage differences into full scale digital level output. Circuit design of high-speed comparators becomes complicated, whenever input power becomes small. Earlier, ADC architectures such as flash and pipeline employ pre-amps-based comparators which have the main drawback of offset voltage. In order to nullify this issue, dynamic ones are usually employed because of the comparison frequency of 1/CLOCK and need very low-offset voltage. In addition to technological tweaks, the establishment of new circuit structures that prevent excessive stacking of transistors between the given rails is most effective for low-voltage usage, particularly if they do not raise the complexity of the circuit. Further, some more elements are combined to the conventional circuit

N. A. Ansari · P. Jaiswal · M. Tyagi (✉) · P. Mittal
Delhi Technological University, New Delhi 110085, India
e-mail: mohit.tyagi@kiet.edu



P. H. J. Chong et al. (eds.), *Emerging Electronics and Automation*, Lecture Notes in Electrical Engineering 937, https://doi.org/10.1007/978-981-19-4300-3_34

for the improvement of speed at lower supply. In this work, designing of dynamic comparator is considered.

As compared to static comparators, dynamic comparators offer very high-noise immunity as differential input stage offers noise cancelation and also improves the sensitivity of the digital dynamic comparator circuit. For ultra-low-power SAR ADC, power consumption of designed dynamic comparator for sampling rate of few MS/s power consumption of the order of few pW is desirable.

Goll et al. [2] have worked on techniques to reduce delay of comparator in—65 nm complementary MOS for supply reduced to 0.65 V. Furthermore, Petrie et al. [3] have worked on ultra-low-supply and dynamic-bulk-biasing comparator. Then, Chan et al. [4] designed a 400 nW 19.5 fJ/conversion step with 8 effective no. of bits, 80-kiloS/S successive approximation register ADC in 0.18-μm CMOS. Hu et al. [5] designed an 8-bit single ended ultra-low-power successive approximation register ADC with a novel DAC switching-method and a counter-based digital control circuit. Thereafter, Kumar et al. [6] developed a low-power 10-bit SAR ADC using varying threshold method for the application in Biomedical field.

This paper shows the designing and comparative analysis of dynamic comparator structures. Further, Sect. 2 includes the different architecture of dynamic comparators. Results and simulation analysis are presented in Sect. 3. Lastly, conclusion is given in Sect. 4.

## 2 Different Architecture of Comparators

Clocked-regenerative comparator has observed large usage in many high-speed analog to digital converters due to their rapid decision-making capability owing to good positive feedback. With the use of dynamic CMOS latch, the comparator is designed. The working of dynamic comparators is divided in to two stages: reset phase and comparison phase. In recent times, many comprehensive analyzes have been done studying the performance of these comparators with respect to noise, delay, random decision error, and kickback noise. The presented paper does a full scale analyzes of delay as well as power with varying supply.

### *2.1 Conventional Comparator*

Conventional comparator is a simple one used in ADC because of its enormous input impedance, oscillating rail to rail output and almost zero static power. The implemented circuit diagram of the above-mentioned design commonly used in ADC is depicted in Fig. 1 with its transient behavior in Fig. 2, working mechanism of which is explained as follows. There are primarily 2 clock phases, reset and comparison, which supports the functioning of the comparator. During reset phase, CLOCK is at 0 V, and during comparison phase, CLOCK is at $V_{dd}$. INN and INP are given to the transistors M7 and M8, respectively. In reset phase, transistor M9 is off. Transistors M3 and M4 start conducting in reset phase and which pull up both the outputs $out_p$ and $out_n$ to $V_{dd}$ to have a justifiable logic state during reset phase. Then, as the time passes, the precharged outputs start to discharge to ground depending on the inputs applied. Assume INN > INP, then $out_n$ will discharge faster than $out_p$. Whenever, $out_n$ falls below $V_{dd} - V_{thn}$ before $out_p$, the corresponding PMOS transistor M2 starts conducting and initiates regeneration.

Finally, $out_p$ will be pull back to supply and output $n$ degrades to $g_{nd}$.

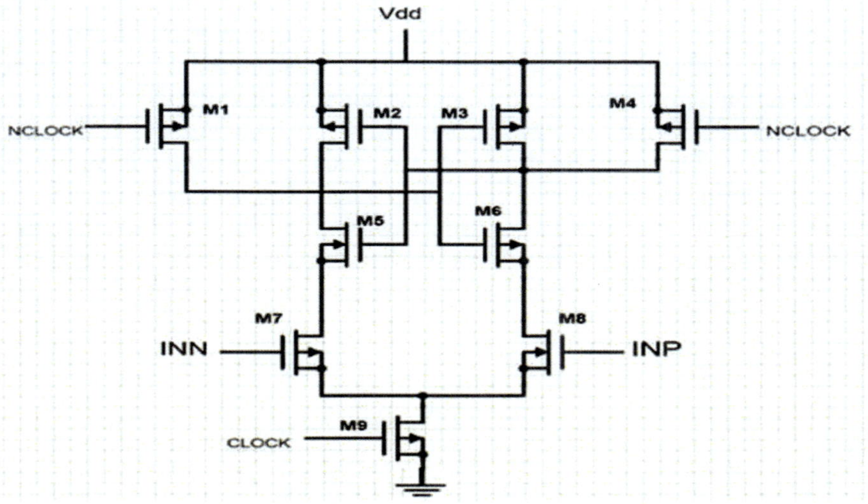


**Fig. 1** Schematic of conventional comparator

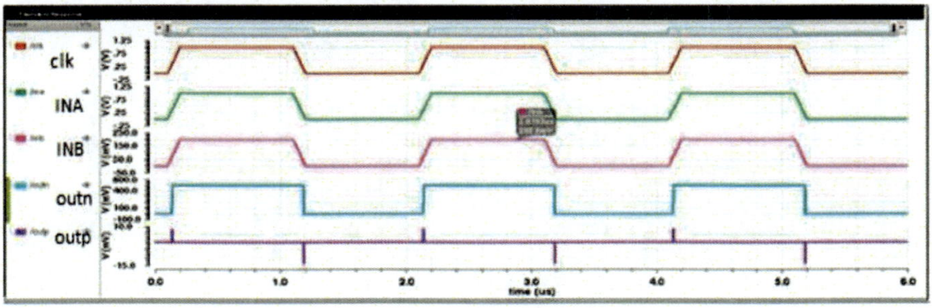


**Fig. 2** Transient behavior for INN > INP

## 2.2 *Double-Tailed-Dynamic Comparator (DTD)*

Figure 3 depicts the implemented circuitry of the DTD comparator and Fig. 4 its transient response. This layout has much less stacking for the transistors and consequently may be used at low voltages, that is, the benefit of this topology in order to offer separate cutting-edge route for regeneration route, one extra tail transistor is delivered to the circuit. Two intermediate transistors are indicated via way of means of M1 and M4, and intermediate factors are denoted via way of means of $F_n$ and $F_p$. During reset phase, each tail transistors M10 and M9 are off. The intermediate factors are pulled to $V_{dd}$ via way of means of each M7 and M8 transistors. Then, intermediate transistors turn on and which will cause both the outputs to discharge to ground. In the comparison phase, each tail transistors activate, and intermediate point voltage begins discharging relying upon the inputs applied. If INP > INN, $F_n$ voltage discharges quicker than $F_p$ voltage. So, $\text{out}_p$ will be pull up to $V_{dd}$ via M2. Finally, output will be at logic high. At the end of evaluation level, each the intermediate point voltages discharges to ground that is the principle downside of this circuit. So, those nodes should be charged from ground to $V_{dd}$ within side the subsequent reset phase, which can also additionally cause extended energy consumption. This double tail comparator is much better than conventional one in terms of delay.

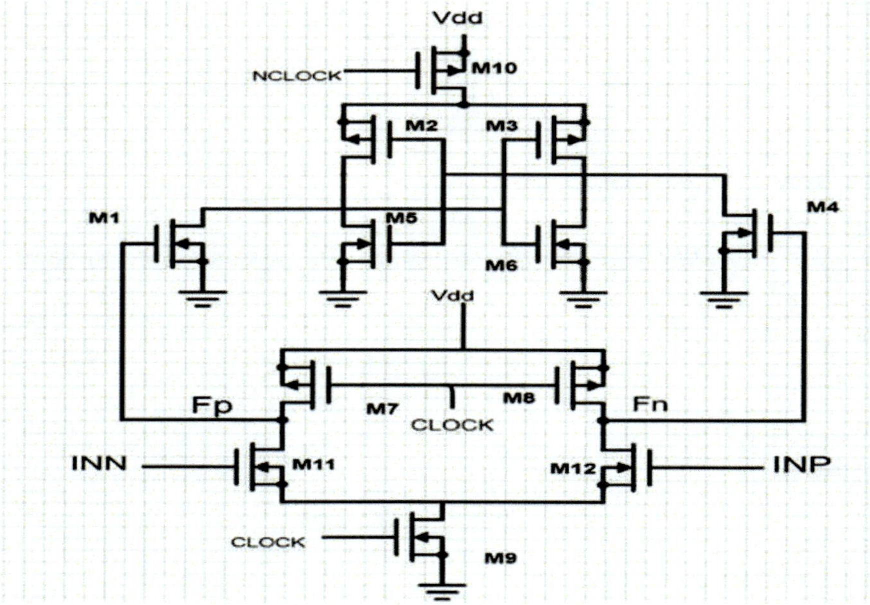


**Fig. 3** Schematic of double tail dynamic comparator

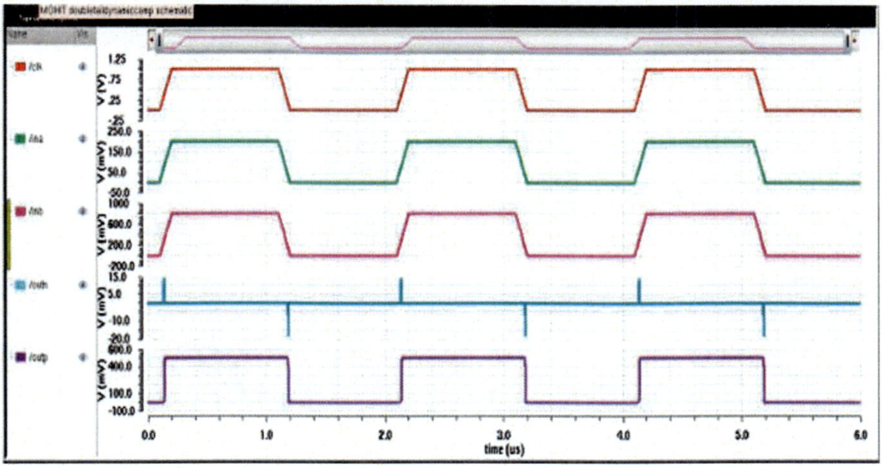


**Fig. 4** Transient behavior for INN > INP

## 2.3 *Controlled Comparator*

Controlled comparator is meant with latch regeneration abilities with the assist of cross coupled control transistors M7 and M8 positioned in parallel to M13 and M14, as depicted in Fig. 5 with transient response in Fig. 6. It additionally employs a pass coupled CMOS latch fashioned with the aid of using M2, M3, M5, and M6 with $\text{out}_n$ and $\text{out}_p$ as output nodes. Further as compared to conventional comparator, this layout makes use of tail transistors; M9 and M10 to facilitate the static power reduction abilities. Inclusion of M15 and M16 transistors offers a widespread development within side the speed than that of conventional one.

On the clock at low logic, nodes $F_p$ and $F_n$ are charged to $V_{dd}$ that in flip switches, the M1 and M4 to ON country and resultantly nodes; $\text{out}_n$ and $\text{out}_p$ stay in logic low below this situation. Additionally, change $c_{lk}$ to logic high, ends within the transistors M9, M10, M2, and M3 in ON state as a result of the pre-charging of $\text{out}_n$ and $\text{out}_p$ nodes to $V_{dd}$.

# 3 Results and Analysis

Upon simulating the above discussed dynamic comparators and observing power and delay with varying supply voltage, the following data have been obtained.

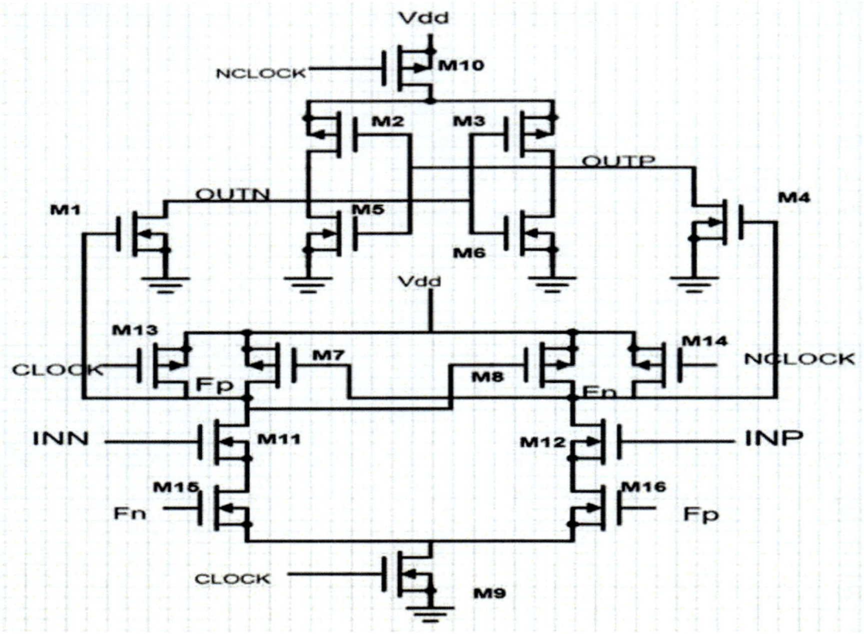


**Fig. 5** Schematic of controlled comparator

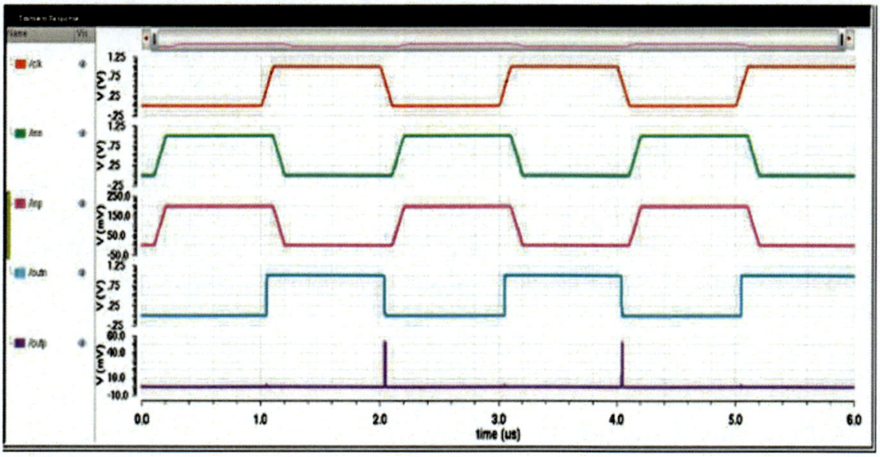


**Fig. 6** Transient behavior for INN > INP

## 3.1 Conventional Comparator

Conventional dynamic comparator is a simple comparator used in ADCs because of its some advantages like high rail to rail output swing and input impedance. Table 1 depicts delay and power variation of conventional comparator at different supply voltages ($V_{dd}$). Figures 7, 8, and 9 are a plot for variation in power dissipation with varying supply voltage ($V_{dd}$), whereas Fig. 8 is for variation of delay at varying $V_{dd}$.

**Table 1** Delay and power variations of conventional comparator at different supply voltage

| S. No. | Supply voltage (V) | Delay (ns) | Power |
|---|---|---|---|
| 1 | 0.4 | 172.227 | 13.3 pW |
| 2 | 1.0 | 167.133 | 2.07 μW |
| 3 | 1.5 | 165.282 | 58.1 μW |
| 4 | 2.0 | 162.018 | 281.3 μW |

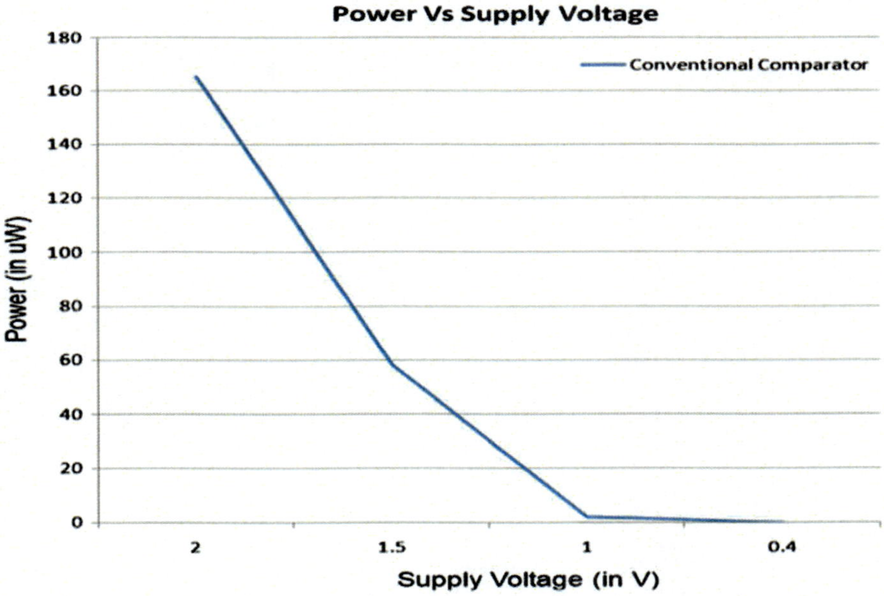


**Fig. 7** Power profile at varying $V_{dd}$

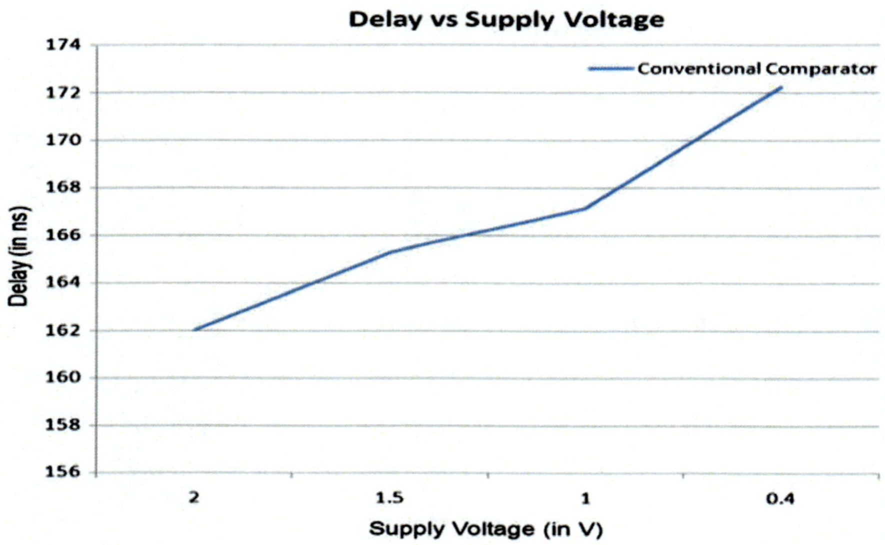


**Fig. 8** Delay profile at varying $V_{dd}$

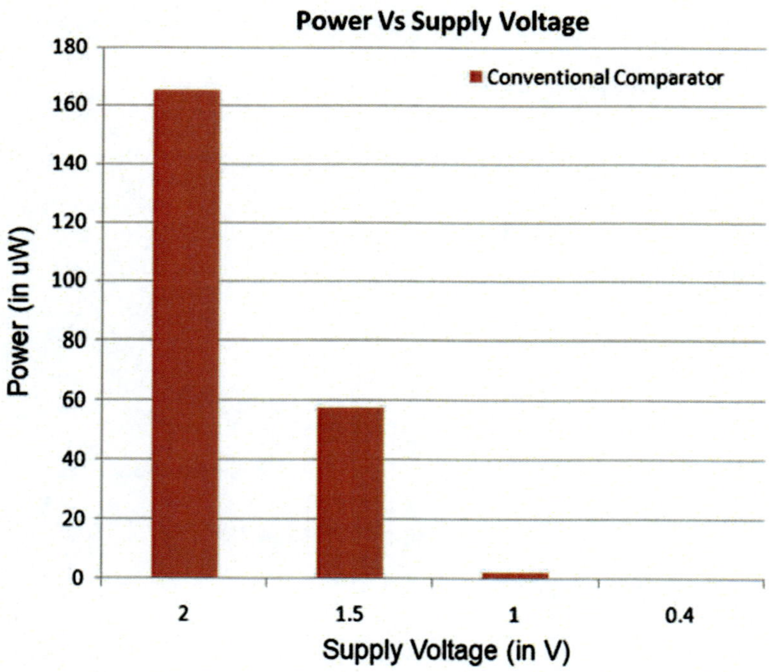


**Fig. 9** Power variation with supply voltage

**Table 2** Delay and power variations of double tail dynamic comparator at different supply voltage ($V_{dd}$)

| S. No. | Supply voltage (V) | Delay (ps) | Power (pW) |
|---|---|---|---|
| 1 | 0.4 | 249.3 | 4.71 |
| 2 | 1.0 | 166.1 | 6.22 |
| 3 | 1.5 | 138.9 | 8.73 |
| 4 | 2.0 | 130.4 | 12.2 |

## *3.2 Double Tail Dynamic Comparator*

Table 2 presents the same trend as above for power and delay variation of double tail dynamic comparator at varying supply voltage ($V_{dd}$). Figure 10 is a plot for power variation of double tail dynamic comparator against supply voltage ($V_{dd}$).

Delay and power variations of double tail dynamic comparator at different supply voltage are shown in Table 2.

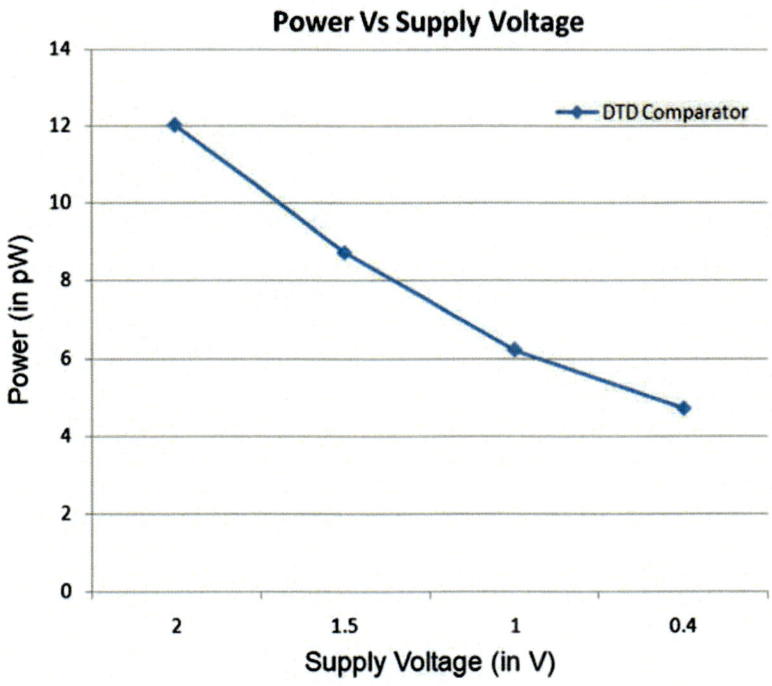


**Fig. 10** Power profile at varying $V_{dd}$

## 4 Conclusion

MOSFET-based dynamic comparators have enormous applications in the manufacture of analog to digital converters. This paper presents the design and comparative analysis of different dynamic comparators; conventional, double tail dynamic, and controlled comparator for SAR ADC. Simulation results have validated that at the minimum supply voltage of 0.4 V, the delay has shown a decreasing profile from 172 ns in conventional comparator to the reduced value of 249 ps in double tail dynamic comparator. Furthermore, rigorous power analysis of power profile for designed comparators has observed that power consumption of conventional comparator is 13.3 pW at 0.4 V, while this has been reduced to the value of 4.7 pW in double tail dynamic comparator at same operating voltage. Also, the comparative power analysis of comparators is carried out with increase in supply voltage up to maximum of 2 V. The simulation results and plots for the aforementioned traits have been shown.

## References

1. He J, Zhan S, Chen D, Geiger RL (2009) Analyses of static and dynamic random offset voltages in dynamic comparators. IEEE Trans Circuits Syst-I: Regul Papers 56(5):911–919
2. Goll B, Zimmermann H (2009) A comparator with reduced delay time in 65-nm CMOS for supply voltages down to 0.65 V. IEEE Trans Circuits Syst-II: Exp Briefs 56(11):810–814
3. Petrie A, Kinnison W, Song Y, Layton K, Chiang SHW (2020) A 0.2-V 10-bit 5-kHz SAR ADC with dynamic bulk biasing and ultra-low-supply-voltage comparator. In: IEEE custom integrated circuits conference (CICC), pp 1–4
4. Chan P, Tiew KK, Je MA (2011) 400–nW 19.5 FJ/conversion step 8 ENOB 80-KS/S SAR ADC in 0.18- μm CMOS. IEEE Trans Circuits Syst-11 Exp Briefs 58(7)
5. Hu W, Liu Y, Lie DYC, Ginsburg BP (2013) An 8-bit single-ended ultra-low-power SAR ADC with a novel DAC switching method and a counter-based digital control circuitry. IEEE Trans Circuits Syst–I: Regul Papers 60(7):1726–1738
6. Mao W, LiY, Heng C, Lian Y (2018) A low power 12 bit 1-KS/s SAR ADC for biomedical signal processing SAR-assisted time interleaved SAR (SATI-SAR) ADC. IEEE Trans Circuits Syst-1, Regul Papers pp 1549–8328
7. Shim J, Kim M, Hong S, Hwon O (2018) An ultra-low power 16 bit second order incremental ADC with SAR–based integrator for IOT sensor applications. IEEE Trans Circuits Syst-11 Exp Briefs 65(12):1899–1903
8. Nuzzo P, Bernardinis FD, Terreni P, VanderPlas G (2008) Noise analysis of regenerative comparators for reconfigurable ADC architectures. IEEE Trans Circuits Syst I Regul Papers 55(6):1441–1454
9. Rawat B, Mittal P (2021) Single bit line accessed high performance ultra low voltage operating 7-T SRAM cell with improved read stability. Int J Circuit Theory Appl 49:1435–1449
10. Kim J, Leibowits BS, Ren J, Madden CJ (2009) Simulation and analysis of random decision errors in clocked comparators. IEEE Trans Circuits Syst I, Reg Papers 56(8):1844–1857
11. Figueiredo PM, Vital JC (2006) Kickback noise reduction technique for CMOS latched comparators. IEEE Trans Circuits Syst II, Exp Briefs 53(7):541–545
12. Negi S, Mittal P, Kumar B (2019) Analytical modelling and parameter sextraction of multilayered OLED. IET Circuits Devices Syst 13:1255–1261

13. Mashhadi SB, Lotfi R (2014) Analysis and design of a low-voltage-power double-tail comparator. IEEE Trans Very Large-Scale Integr Syst 22(2):343–352
14. Shinkel D, Mensink E, Klumperink E, Van TE, Nauta B (2007) A double-tail latch-type voltage sense amplifier with 18 ps setup and hold time. In: Proceedings of IEEE international solid-state circuits conference, pp 314–315
15. Negi S, Mittal P, Kumar B (2020) Numerical modeling and parameters extraction of novel triple hole block layer based organic light emitting diode for display. J Soc Inform Display 22(12):956–964
16. Mittal P, Kumar N (2020) Comparative analysis of 90 nm MOSFET and 18 nm FinFET based different multiplexers for low power digital circuits. Int J Adv Sci Technol 29(8):4089–4096

# FinFET: A Revolution in Nanometer Regime

**Umayia Mushtaq, Md. Waseem Akram, and Dinesh Prasad**

## 1 Introduction

In the modern electronic industry, the demand for small feature size, high performance and low power consumption is of primary concern while designing efficient field-effect transistors (FETs) for very large-scale integration (VLSI) applications. Due to the continuous down scaling of metal oxide semiconductor field-effect transistor (MOSFET) devices, the performance of digital logic circuits gets improved but at the same time, it paves a way for increased dissipation of heat and power dissipation in lower technology logic circuits. The studies in the past few decades come up with ideas to overcome the challenges posed by continuous down scaling. These include increasing the channel height, changing doping polarities of source and drain, increasing number of channels, changing channel material and shape of the channels. Many devices like gallium nitride-based single gate tunnel field-effect transistor (GaN TFET), FinFET and graphene nanoribbon field-effect transistor (GNRFET) come with different features [1, 2]. In this paper, our aim is to explore FinFET device and compare with the previous existing technologies. FinFET device has some unique features like greater space for the flow of electrons on applying gate voltage, increased channel height and different design style in comparison with the already existing complementary metal oxide semiconductor (CMOS) devices [1]. In order to increase gate control over the channel charge, which overcomes the deterioration caused due to short channel effects to a large extent, multi-gate MOSFETs are used. Among multi-gate devices, dual gate devices and tri-gate devices are more preferred due to the reason that it results in lesser value of parasitic capacitances and potent nature against dopant behavior [2–5]. The rest of the paper is organized as follows. In Sect. 2, history and evolution of FinFET devices are explained. Section 3 defines the

U. Mushtaq (✉) · Md. Waseem Akram · D. Prasad
Department of Electronics and Communication Engineering, Faculty of Engineering and Technology, Jamia Millia Islamia, New Delhi 110025, India
e-mail: umaiyamushtaq1247@gmail.com



P. H. J. Chong et al. (eds.), *Emerging Electronics and Automation*, Lecture Notes in Electrical Engineering 937, https://doi.org/10.1007/978-981-19-4300-3_35

structure, operation and classification of FinFET devices in detail. Section 4 defines the different manufacturing parameters of FinFET device. In Sect. 5, FinFET device structures of different materials are explained and comparative analysis between them is also taken into consideration. Section 6 represents the reliability of FinFET device and explains the various process parameters which are important while performing reliability analysis for FinFET logic devices. In Sect. 7, different applications of FinFET logic devices are mentioned with special focus on FinFET SRAM cells. Section 8 shows the various performance parameters and important tradeoffs which should be taken into consideration while designing FinFET SRAM cells, and finally, the conclusion is presented in Sect. 9.

## 2 History and Evolution of FinFET Devices

The first FinFET device type called as DELTA transistor or fully depleted lean channel transistor was invented in 1989 [6]. This type of device has a vertical channel, which depends on height of island. Hence, results in new gate structure with silicon on insulator (SOI) effects induced. Although the channel is vertical in this case but the current direction is same as in conventional device. This DELTA transistor forms a fully isolated body of transistor by making use of selective oxidation and bulk silicon wafer. This SOI provides the full isolation of body from the remaining wafer as shown in Fig. 1 [7]. Another derivative of FinFET structure called triangular wire has been proposed as shown in Fig. 2. This device has multi-wire or multi-fin structure. It has lower source drain parasitic resistance and good short channel effects (SCE) control in comparison with the parallel double gate structure [8].

Another device called vertical gate replacement transistor (VRG) [9] has improved performance compared to its predecessors. In this device, sacrificial layer is removed to form gate dielectric and electrode. The removal of sacrificial layer results in exposing of channel area which results in formation of electrode and gate dielectric

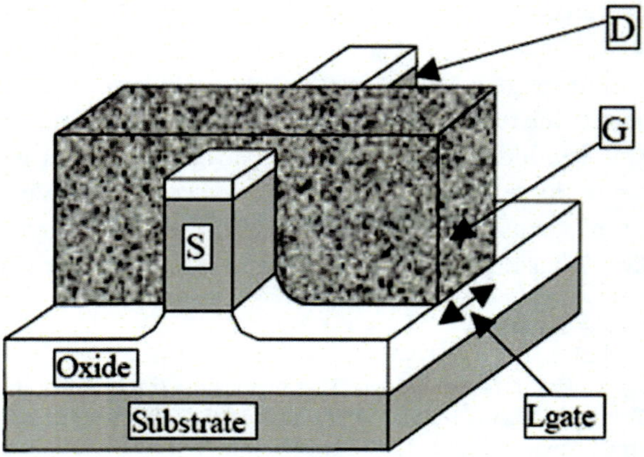


**Fig. 1** FinFET on SOI wafer [7]

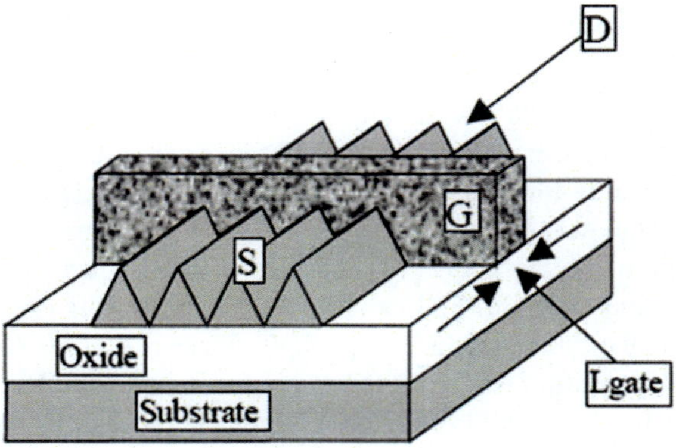


**Fig. 2** Triangular wire fin transistor [8]

of choice. Besides, it does not make use of costly SOI wafer which is used in FinFET device. But the challenging part of this device is the integration of various gate lengths. In addition to the above-mentioned devices, there exists one more device called planner double gate device or tunnel epitaxy device which has planner double gate formed by tunnel epitaxy. It has the better control on channel thickness and sacrificial gates and is formed with low budget materials [7]. This whole discussion clearly depicts multi-gate devices brought a diversity of design styles of various levels of complexity. New solutions are put-forth by different researchers all over the globe from time to time. Though, every structure has some good characteristics as well as incompetence at some level, which is then vanquished by the new design styles in the past few decades. Researchers from the past few decades are looking for devices with better electrostatic control for the flow of carriers. For this devices like planner double gate, SOI-FinFET, tri-gate, Pi-gate, omega gate and gate all around (GAA) FET come into existence. Among them, FinFET can be one of the best choices for designing of digital logic circuits up to 7 nm technology node. Electrostatic control of gate increases from planner double gate devices to GAA FET but at the same time, fabrication cost also increases [10]. While comparing the characteristics of double gate and tri-gate FinFET, it is evident that on reducing tox (oxide thickness) in bulk devices, threshold voltage increases, while in SOI-FinFET, the threshold voltage is reduced by reducing tox. Tri-gate FinFET is more reliable and efficient in subthreshold swing and threshold voltage in comparison with the SOI-FinFET. Also DIBL is lower in tri-gate FinFET devices as compared to the SOI-FinFET. Other devices include Pi-gate and omega gate devices. These devices have reduced SCE's and high drive current. These are also tri-gate devices with extended gate electrode. Due to the extended gate control, DIBL is reduced compared to conventional devices at lower technology nodes [11]. The prominent semiconductor companies have successfully developed integrated circuits at lower technology nodes like 7 nm, 5 nm, etc. FinFET has proved the best performing devices up to 7 nm. But, with technology nodes lesser than 7 nm, there exists some challenges like self-heating and severe short channel effects. These problems are reduced by introducing GAA FETs in modern semiconductor industry [12].

## 3 Structure, Operation and Classification of FinFET Device

Continuous downscaling of MOSFET transistors makes power dissipation a crucial issue at lower technology nodes. Many innovations with the aim to implement devices with ultra-low leakage power have been proposed in the past few decades. The prominent ones include fully depleted silicon on insulator transistor (FDSOI) [13], ultra-thin body transistors [14] and fin-type field-effect transistor (FinFET) devices [15]. Among them, FinFET can prove as the promising candidate to replace conventional metal oxide semiconductor (MOS) devices. Figures 3 and 4 show the three terminal (3T) and four terminal (4T) three-dimensional (3D) structure of FinFET.

In the 3T FinFET device, the thin silicon body is wrapped by gate electrode. This device bears resemblance to the conventional MOSFET device but has better performance parameters like low power dissipation and higher drive strength compared to planner MOSFET device. This FinFET device is called quasi-planner due to the reason that current flow is parallel to the wafer plane but formation of channel occurs perpendicular to the wafer plane. However, in 4T FinFET devices, the gate electrode is etched at the top of channel giving rise to independently controlled front and back gates. These gates can be controlled separately and results in better control of flow of carriers across the channel [16, 17]. In order to obtain higher ON currents, multiple number of fins are used which gives rise to wider FinFET devices. FinFETs can be categorized into 3 modes—short gate (SG) mode, independent gate mode (IG) mode and low power mode as shown in Fig. 5. In SG mode, both front gate and back gate are tied together, while as in IG mode, the gates are not connected. SG mode has few advantages like it offers high ON and OFF currents and has better electrostatic

Fig. 3 3T or SG FinFET [16]

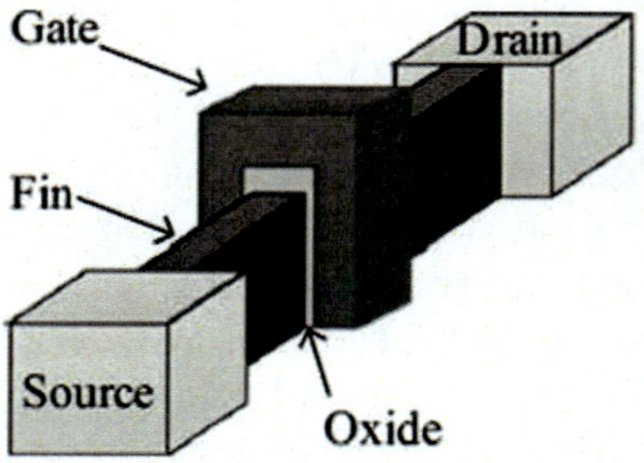


Fig. 4 4T or IG FinFET [16]

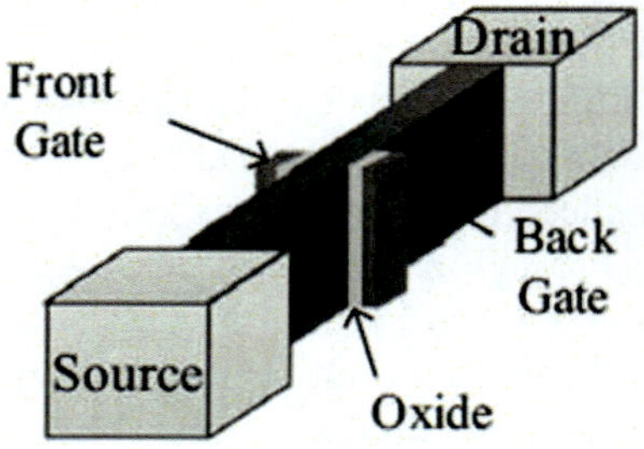

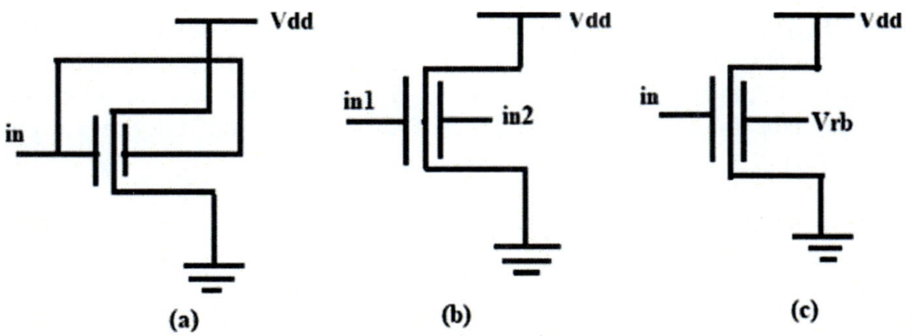


**Fig. 5** Schematic diagram of *n*-type FinFET **a** short gate mode, **b** independent gate mode and **c** low power mode

control over the channel. Besides, short gate mode of FinFET offers the best drive strength as compared to other two modes. In contrast to this, IG mode FinFETs provide better flexibility from application point of view due to the separate control of front and back gates. But, IG mode FinFET is more expensive and occupies more area in comparison to SG mode FinFET device. In addition to this, it has reduced ON current as compared to short gate mode. Another mode called low power mode has reverse bias voltage connected to back gate of the FinFET device. This reduces the subthreshold leakage in this type of device. It also decreases ON current as compared to short gate mode [10, 18].

## 4 Manufacturing Parameters in FinFET Device

Many challenges occur while making a transition from conventional planner devices to FinFET technology, i.e., many parameters are taken into consideration while designing FinFET devices. These parameters include shape of fin, fin orientation, variable fin dimensions, doping concentration, reliability of FinFET devices, parasitic capacitances and FinFET integration. It is obvious that FinFET device structure with smaller width-height ratio called aspect ratio shows more robustness compared to the one with higher aspect ratio. This is due to the reason that as fin inclination increases, short channel effects also increase. Hence, increase in slope of fin can result in serious problems on scaling of gate length [19]. Fin orientation in $\prec$100$\succ$ direction is responsible for current flow in $\prec$110$\succ$ sidewall surface. The $\prec$100$\succ$ has fast electron flow as compared to $\prec$110$\succ$ direction. The opposite case occurs for FinFET devices. The electron mobility is more desirable in $\prec$110$\succ$ direction due to quantum confinement effect [20]. Besides, fin height should also be taken care of due to the reason that width of FinFET devices depends on height of Fin, i.e.,

$$W_{\text{minimum}} = 2H_{\text{fin}} + T_{\text{fin}} \tag{1}$$

where $H_{fin}$ is the height of fin and $T_{fin}$ is the thickness of silicon body of FinFET device. It is clear from Eq. 1, $H_{fin}$ directly affects width of FinFET as compared to $T_{fin.}$ Hence, in order to increase the fin width, multiple fin structures are formed as $H_{fin}$ is fixed for any particular FinFET device. Total width for FinFET device is given in Eq. 2 [21].

$$\begin{aligned} W_{total} &= n(W_{minimum}) \\ &= n(2H_{fin} + T_{fin}) \end{aligned} \tag{2}$$

where $n$ is the number of fins. As the numbers of fins is increased, it increases the current through FinFET device but at the same time, hot carrier effect degrades the performance of the device [22]. Therefore, fin height is one of the crucial parameters of FinFET device and has to be taken care of while designing FinFET device. Scaling capabilities in double gate and tri-gate FinFET device are limited by ratio ($L_{eff}/T_{fin}$), where $L_{eff}$ is the effective width of channel. Tri-gate FinFET is more scalable than double gate FinFET because this ratio is less in tri-gate FinFET device in comparison with double gate device. In order to get better ON current per unit width, this ratio of $H_{fin}$ and $T_{fin}$ should be maximum at a given $T_{fin}$ and $L_{eff}$. For a given gate length, careful optimization of $T_{fin}$ and $H_{fin}$ is required to get the better performance because increase in $H_{fin}$ degrades SCEs and fin stability [23]. Another parameter called doping concentration of FinFET device is also of prime concern. Usually, undoped channel is used to design FinFET device moreover to have better control on leakage current, light doping is preferred. The source and drain regions are highly doped, hence increases series resistance which can damage fin geometry of FinFET device. This can be minimized by in situ epitaxial growth at the source and drain area [20]. In addition to this, parasitic capacitances are also increased while making a transition from planner device structures to FinFET devices. This increase in capacitances is due to increased overlap of the front and back gates of FinFET device. This can be reduced by varying fin height and fin pitch, i.e., decrease in fin pitch and increase in fin height can minimize the problem of parasitic capacitance in FinFET device [24, 25]. Reliability is better in FinFET devices compared to planner ones due to the reduced transverse field. The threshold voltage becomes unstable for conventional as well as FinFET devices at lower technology nodes. However, this occurs for NMOS devices only, while reliability of PMOS devices remains same for FinFET devices [26]. Another important parameter is integration of FinFET device. Due to the fabrication compatibility of FinFET device with the conventional CMOS device, it is preferred to design digital logic circuits as compared to other modern devices available [27]. Therefore, every parameter mentioned above is crucial for designing of FinFET devices and needs to be addressed properly.

## 5 FinFET Structures of Different Materials

FinFET devices have many advantages which overcome the issues of scalability that exists in MOSFET devices. Single lithographic process flow is used to design FinFET device [28] and hence provides ease in fabrication. Different structures are designed with different materials. These include FinFET-based dual-kk structure, InGaAs-on-Insulator FinFET, double gate-based n-type FinFET using hafnium oxide, MOSFET (Multi-gate), bulk FinFET, SOI-FinFET, selective epitaxial Si growth in FinFET and atomic layer deposition (ALD) in FinFET. In asymmetrical drain extension dual-kk structure, some parameters like trans-conductance, output conductance and cut off frequency are improved by 9.09%, 13.04% and 12.91%, respectively, in comparison with dual-*k* structure [29, 30]. This device shows better tradeoff between on and off currents due to the proper designing of source and drain spaces. Besides this, OFF current is also reduced due to doped extensions of source and drain [24]. Other device structure called double gate-based *n*-FinFET using hafnium oxide has higher $I_{ON}/I_{OFF}$ ratio and lower leakage current. In this devices, combination of gold and hafnium oxide is used as dielectric which has high dielectric constant than aluminum. This device is simulated at 20 nm technology node [31]. In SOI-FinFET, buried oxide separates the fin from substrate, hence provides electrical isolation between adjacent devices. Besides, eliminates the sub-fin leakage current. SOI-FinFET has better subthreshold swing and threshold voltage roll-off as compared to bulk FinFET. In SOI structure, lithography process is used in order to optimize dry etch and trimming conditions of hard mask. Improvement in efficiency occurs as well due to the reduction in resistance in selective epitaxial Si growth. Moreover, other devices like multi-gate MOSFET, scaled CMOS and FinFET have reduced short channel effects, leakage current and chip area. Besides provides high performance as well compared to conventional devices. Other device structure which includes selective epitaxial Si growth in FinFET and atomic layer deposition (ALD) in FinFET provides improvement in outer surface of fin and threshold voltage, respectively [18]. This whole discussion makes us to conclude that these FinFET devices have better electrostatic characteristics and provide wide range of applications. They provide good control on subthreshold leakage current and a device with material InGaAs and GaAs have better crystal quality in comparison with the traditional materials and hence pave the way for research in future.

## 6 Reliability Analysis of FinFET Devices

New device structures, materials and architectures are suggested in integrated circuit designs in the past few decades in order to continue with the scaling challenges in sub-100 nm regime (< 22 nm). Multi-gate devices, e.g., FinFET come into existence [32, 33]. Due to some prominent features like reduced short channel effects, leakage current reduction, compatible fabrication with the conventional devices, high driving capability and good yield [34], it catches the eye of many researchers. But at the same time, many issues of reliability and robustness arise. Reliability issues include process, voltage and temperature (PVT) variations. Variations can be environmental variations (due to temperature and supply voltage) or process variations (systematic and non-systematic). Systematic variations being the variations in circuit characteristics of transistor with same dimensional parameters like length and width of transistors can be managed in manufacturing process during layout analysis. Non-systematic variations (inter-die and intra-die) are the unpredictable ones occur due to technical constraints and lack of manufacturing control [35]. Voltage and temperature variations being the systematic ones are improved from time to time, while the non-systematic ones are hard to identify and control. The process variations being the most crucial in nanometer regime affect the FinFET device as well. The threshold voltage gets affected due to the manufacturing difficulties of small geometric patterns [36] because process variations arise during fabrication process. In addition to this, reliability factors caused by higher electric fields and transistor aging also contribute to process variations. Besides physical factors like geometric and electrical variation in device structures also become the part of process variation. Another important variability parameter of PVT variations is supply voltage variation. This occurs due to noisy power sources and voltage drop in circuits. The direct impact of supply voltage on system performance, dynamic power, logic gate timing and leakage power makes it crucial parameter for FinFET logic circuits [37]. Supply voltage variations also impact propagation delay because gate delay involves the role of saturation current which depends on supply voltage. Therefore, variation in supply voltage can cause circuit degradation and can be responsible for errors in functionality of logic circuits. Besides, results in increase in fabrication cost due to yield loss [38]. In order to improve the control of gate on FinFET device, high-$K$ metal gate stack is used [39]. Metal gate granularity (MGG) being the source of statistical variability affects the FinFET device as well. The diverse work function of metal gates results in higher work function fluctuations [40]. Besides, electrical properties of FinFET devices and variation in power matrices also lead to degradation in performance of logic circuit design [41]. Therefore, supply voltage variation should also be taken into consideration while designing FinFET logic circuits.

Nanoelectronic devices are more prone to variability effect due to the sub-wavelength lithography and unpredictable dopant fluctuation. Temperature variations mainly occur due to switching activity of the device. Temperature in IC's increases due to the latent heat transfer occurred because of increase in power dissipation. This increase in power dissipation occurs due to increase in leakage current. Therefore, there is a need to analyze the variations during early design step. These variations also cause the change in threshold voltage which can result in performance degradation in FinFET logic circuits [42]. In FinFET logic devices, fin engineering is the most crucial part of fabrication. It reduces leakage current and results in ON current improvement in logic circuits [43]. In the literature [44–46], effect of geometrical parameter variation on $I_{ON}$ and $I_{OFF}$ of FinFET transistors at sub-22 nm was studied. Besides, PVT variations impact on power and other performance parameters of standard cells are analyzed. Transistor sizing can be performed to reduce the effect of

PVT variations on FinFET devices. Moreover, only few methods are available in literature to reduce PVT variations in FinFET logic circuits, and lot of work needs to be done in the reliability analysis of FinFET logic circuits.

## 7 Applications of FinFET with Focus on FinFET SRAM Cells

FinFET devices find variety of applications in VLSI domain. Different logic gates are designed at different technology nodes using FinFET devices. FinFET devices have low power dissipation and PDP in comparison with the conventional CMOS devices for same technology nodes [47]. Due to the continuous scaling, it becomes important to design efficient interconnects. The interconnect synthesis using FinFET has efficient performance parameters as compared to conventional ones. Besides, FinFET interconnect synthesis schemes, e.g., threshold voltage scheme through multiple supply voltage (TCMS) in FinFET buffer provides power and area saving by 50.4% and 9.2%, respectively, as compared to dual $V_{DD}$ scheme in interconnect synthesis. Different high performance logic circuits like Schmitt trigger is also designed using FinFET which provides efficient parameters like low power dissipation and large hysteresis window [10]. FinFETs are also used to design short gate latches and independent gate latches. Independent gate latches reduce both switching and leakage power dissipation in comparison with short gate latches [48]. *D*-latches are also designed using FinFET device at 18 nm channel length. These FinFET D-latches consume low power and have less delay compared to CMOS *D*-latches. Hence, these latches can be used in shift registers, power gating circuits and other systems as well [49]. Domino logic circuits suffer from reliability issues due to the technology scaling and dynamic storage of charge. However, domino circuits designed using independent gate (IG) and short gate (SG) FinFETs provide better power delay tradeoffs. IG FinFET domino logic circuits operate at low threshold voltage, and both IG and SG FinFET domino logic circuit provide better noise immunity compared to conventional circuits [50]. FinFET devices are also used to design low power memory cells. The process variations are minimized for lower technology nodes due to less effect of random dopant fluctuations in FinFET device. In addition to this, various SCEs like drain-induced barrier lowering (DIBL), surface scattering, velocity saturation, impact ionization, etc., are minimized in FinFET devices due to the increased gate control as compared to conventional metal oxide semiconductor (MOS) devices. This makes FinFET devices less prone to process variations. Due to these features, FinFET is the promising candidate for SRAM applications [51, 52]. Static random access memory (SRAM) occupies major portion of die area in integrated circuits. They are also used in cache memory in microprocessors. There are various requirements for different applications which are listed below. For battery-operated portable devices, low power dissipation is required in order to provide better battery life. Performance of SRAM cells is measured in terms of static noise margin (SNM). It is used to

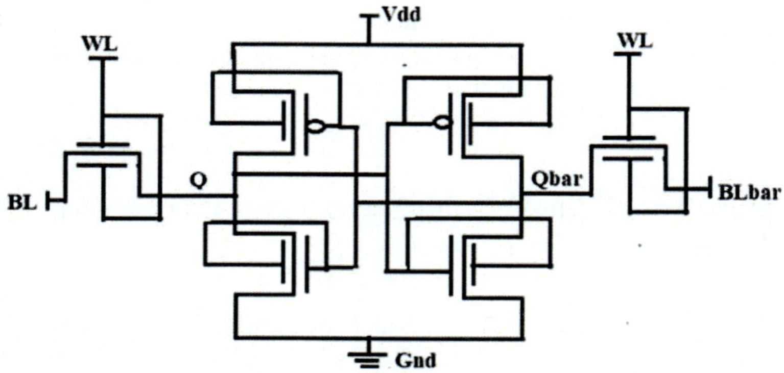


**Fig. 6** 6T short gate FinFET SRAM cell

determine stability of SRAM cell. It can be measured from the butterfly curve by considering the length of the side of largest square formed inside the lobes of butterfly curves. SRAM cells are densely packed to form SRAM arrays and have highest in ICs which make it sensitive to process variations.

Conventional 6T FinFET SRAM cell consists of 2 cross-coupled FinFET inverters and 2 FinFET access transistors as shown in Fig. 6. FinFET SRAM cell operates in 3 modes, i.e., hold, read and write. For hold operation, the word line is set to zero (low voltage). In this case, the SRAM cell retains its data at the output of short gate FinFET inverters. The FinFET access transistors are in off state which results in hold state in FinFET SRAM cell. For read operation, word lines are set to 1 (high voltage) followed by pre-charging of (bit lines) BL and BLbar. The sensing circuits are used to read the state. This state is considered most vulnerable in SRAM cell because it has to maintain cell state when both bit lines are in pre-charge condition. For write operation also word line is kept at high voltage and to write '0' at Qbar, BL is set to high voltage ($V_{dd}$) and BLbar to low voltage (ground) [53].

Many design tradeoffs are taken into consideration while designing SRAM cells. There include area versus yield, read vs. write stability and speed vs. leakage current. As it is obvious, more yield is guaranteed for large area memory arrays. This can be achieved by allowing large design margins, i.e., channel length, channel width, threshold voltage and supply voltage. No doubt increase in size increases noise margins but at the same time increases the SRAM cell area as well. Read voltage in SRAM cell can be determined by voltage division between access transistor and pull down transistor, i.e., more read stability is achieved when access transistor in SRAM cell have less driving strength. In contrast to this, write voltage can be determined by voltage division between access transistor and pull up transistor. In this case, the higher the driving strength of access transistor more is the write stability. Hence, tradeoff exists between read and write stability in SRAM cells. Another tradeoff exists between leakage current and speed. As it is clear, lower the threshold voltage, more is the leakage and higher is the speed. Therefore, high speed SRAM cells have more leakage. SRAM cells are more preferred in comparison to dynamic random access memory (DRAM) and resistive random access memory (ReRAM) due to the reason that SRAM is fast and has the capability to operate at processor speed of 250 MHz and more and ReRAM is not matured yet. In order to maintain the performance at lower technology nodes, non-conventional CMOS structures like ultra-thin body fully depleted SOI and other multi-gate ultra-thin body MOSFETs come into existence. These devices have many drawbacks like improper alignment of front and back gates hence not preferred. FinFET can prove as a best candidate for SRAM design due to the characteristics like improved carrier mobility, reliability of device and gate leakage currents. In addition to this, it shows less propagation delay as compared to conventional planner devices. Besides, it has a better short channel effect. Statistical dopant fluctuations can be minimized due to lower channel doping concentration. Hence, provides low variability of threshold voltage and increase in $I_{ON}/I_{OFF}$ for fixed $V_{DD}$ [54–56].

## 8 Performance Parameters of FinFET SRAM Cell Design

Memory design of majority of systems in VLSI domain chip is performed by SRAM cells. Though CMOS-based SRAM cells have better noise margin, low power dissipation, short access time and better switching speed as well. But as technology node scales below 45 nm, CMOS SRAM cells suffer from short channel effects and process variations. To overcome this, SRAM cells are designed using FinFET devices at lower technology nodes. The stability of SRAM cells is determined by SNM and it can be obtained by the largest side of square formed inside the butterfly curve of read operation. When the threshold voltage is increased, drive current capability decreases which effects the write operation of SRAM cell. However, high drive current even at higher threshold voltage is obtained when FinFET devices are used. Therefore, high SNM and high write capability are obtained when SRAM cell is designed using FinFET. In addition to this, static noise margins show more sensitivity to threshold voltage in n-type MOSFETs but at the same time they show least sensitivity in case of p-type FinFET device. This clearly makes us conclude that FinFET can be a good choice to design SRAM cell. Besides, in order to design robust and stable, FinFET-based SRAM cell, i.e., to improve RNM, upsizing of pull down FinFETs or increase in gate length of access transistors is performed. This results in increase in area and delay, respectively. Hence, affects the write noise margin. This WNM can be improved by using weaker pull up FinFETs and stronger access transistors. Therefore, tradeoff occurs between RNM and WNM in FinFET-based SRAM cell as well. The FinFET-based SRAM cells are advantageous in the way that they have low access time and power dissipation in comparison with the CMOS-based SRAM cell. This is due to lower leakage current and minimum short channel effects in FinFET-based SRAM cell. In these cells, upsizing of FinFET device results in decrease in propagation delay but this occurs at the cost of power dissipation. In order to keep the power delay product constant, tradeoff occurs between the performance indices [54–56]. Designing of FinFET-based SRAM cell is a challenge due to the increase in PVT variations. These variations can result in degradation of performance and parametric failures and due to this cell yield also gets decreased. In case of FinFET device, variation in length and width is not correlated. This is due to the reason that channel length in FinFET is determined by poly and source/drain diffusion steps while as channel width is obtained from field oxide step. In FinFET SRAM cells, fin width, fin thickness and channel doping are the major source of parameter variation hence reliability analysis of FinFET-based SRAM cell is more vulnerable to process variations [53].

## 9 Conclusion

Due to the short channel effects and PVT variations, it becomes challenging to scale down MOS devices continually. Therefore, there should be some novel device which will bridge the gap between planner and non-planner devices. FinFET can prove as the best substitute due to the ease in fabrication, better short channel effects, high driving capability and low power dissipation. This paper clearly presented the scaling challenges of MOS devices and how FinFET overcomes those challenges to a large extent. In addition to this, detailed analysis of PVT variations and various tradeoffs which are important to design FinFET devices is clearly explained. Various methods and techniques are also discussed to reduce the effect of PVT variations. Moreover, special focus is given to FinFET SRAM cell and various design aspects which need to take into consideration are clearly explained. It is a proven fact that FinFET can provide as the best substitute to the conventional devices. But need is to explore it well and lot of work needs to be done for stability and reliability analysis of FinFET logic devices.

## References

1. Patnala M, Yadav A, Williams J, Gopinath A, Nutter B, Ytterdal T, Rizkalla M (2020) Low power-high speed performance of 8T static RAM cell within GaN TFET, FinFET, and GNRFET technologies—a review. Solid-State Electron 163:107665
2. Wong HS, Frank DJ, Solomon PM (1998) Device design considerations for double-gate, ground-plane, and single-gated ultra-thin SOI MOSFET's at the 25 nm channel length generation. In: International electron devices meeting 1998. Technical digest (Cat. No. 98CH36217). IEEE, pp 407–410
3. Suzuki K, Tanaka T, Tosaka Y, Horie H, Arimoto Y (1993) Scaling theory for double-gate SOI MOSFET's. IEEE Trans Electron Devices 40(12):2326–2329
4. Solomon PM, Guarini KW, Zhang Y, Chan K, Jones EC, Cohen GM, Krasnoperova A, Ronay M, Dokumaci O, Hovel HJ, Bucchignano JJ (2003) Two gates are better than one [double-gate MOSFET process]. IEEE Circuits Devices Mag 19(1):48–62
5. Auth C, Allen C, Blattner A, Bergstrom D, Brazier M, Bost M, Buehler M, Chikarmane V, Ghani T, Glassman T, Grover R (2012) A 22 nm high performance and low-power CMOS technology featuring fully-depleted tri-gate transistors, self-aligned contacts and high density MIM capacitors. In: 2012 symposium on VLSI technology (VLSIT). IEEE, pp 131–132

6. Hisamoto D, Kaga T, Kawamoto Y, Takeda E. A fully depleted lean-channel transistor (DELTA)-a novel vertical ultrathin SOI MOSFET. In: International technical digest on electron devices meeting. IEEE, pp 833–836
7. Maszara WP (2001) Integration challenges for double-gate MOSFET technologies. MRS Online Proc Libr 686(1):1–10
8. Hiramoto T (2001) Nano-scale silicon MOSFET: towards non-traditional and quantum devices. In: 2001 IEEE international SOI conference. Proceedings (Cat. No. 01CH37207). IEEE, pp 8–10
9. Hergenrother JM, Oh SH, Nigam T, Monroe D, Klemens FP, Kornblit A (2002) The vertical replacement-gate (VRG) MOSFET. Solid-State Electron 46(7):939–950
10. Mishra P, Muttreja A, Jha NK (2011) FinFET circuit design. In: Nanoelectronic circuit design. Springer, New York, NY, pp 23–54
11. Shehata N, Gaber AR, Naguib A, Selmy AE, Hassan H, Shoeer I, Nabeel R (2015) 3d multi-gate transistors: concept, operation, and fabrication. J Electr Eng 3:1–14
12. Ajayan J, Nirmal D, Tayal S, Bhattacharya S, Arivazhagan L, Fletcher AA, Murugapandiyan P, Ajitha D (2021) Nanosheet field effect transistors-A next generation device to keep Moore's law alive: an intensive study. Microelectron J:105141
13. Gallon C, Fenouillet-Beranger C, Vandooren A, Boeuf F, Monfray S, Payet F, Orain S, Fiori V, Salvetti F, Loubet N, Charbuillet C (2006) Ultra-thin fully depleted SOI devices with thin BOX, ground plane and strained liner booster. In: 2006 IEEE international SOI conference proceedings. IEEE, pp 17–18
14. Choi YK, Asano K, Lindert N, Subramanian V, King TJ, Bokor J, Hu C (1999) Ultra-thin body SOI MOSFET for deep-sub-tenth micron era. In: International electron devices meeting 1999. Technical digest (Cat. No. 99CH36318). IEEE, pp 919–921
15. Yu B, Chang L, Ahmed S, Wang H, Bell S, Yang CY, Tabery C, Ho C, Xiang Q, King TJ, Bokor J (2002) FinFET scaling to 10 nm gate length. In: Digest. International electronic devices meeting. IEEE, pp 251–254
16. Liao K, Cui X, Liao N, Ma K (2013) Leakage power reduction of adiabatic circuits based on FinFET devices. IEICE Trans Electron 96(8):1068–1075
17. Liao N, Cui X, Liao K, Ma K, Wu D, Wei W, Li R, Yu D (2014) Low power adiabatic logic based on FinFETs. Sci China Inf Sci 57(2):1–13
18. Pal RS, Sharma S, Dasgupta S (2017) Recent trend of FinFET devices and its challenges: a review. In: 2017 conference on emerging devices and smart systems (ICEDSS). IEEE, pp 150–154
19. Li Y, Hwang CH (2007) Effect of fin angle on electrical characteristics of nanoscale round-top-gate bulk FinFETs. IEEE Trans Electron Dev 54(12):3426–3429
20. Colinge JP (ed) (2008) FinFETs and other multi-gate transistors, vol 73. Springer, New York
21. Gupta A, Mathur R, Nizamuddin M (2019) Design, simulation and comparative analysis of a novel FinFET based as table multivibrator. AEU-Int J Electron Commun 100:163–171
22. Yeh WK, Zhang W, Chen PY, Yang YL (2018) The impact of fin number on device performance and reliability for multi-fin tri-gate n-and p-type FinFET. IEEE Trans Dev Mater Reliab 18(4):555–560
23. Saini G, Rana AK (2011) Physical scaling limits of FinFET structure: a simulation study. Int J VLSI Des Commun Syst (VLSICS) 2(1):26–35
24. Guillorn M, Chang J, Bryant A, Fuller N, Dokumaci O, Wang X, Newbury J, Babich K, Ott J, Haran B, Yu R (2008) FinFET performance advantage at 22 nm: an AC perspective. In: 2008 symposium on VLSI technology. IEEE, pp 12–13
25. Kang CY, Sohn C, Baek RH, Hobbs C, Kirsch P, Jammy R (2013) Effects of layout and process parameters on device/circuit performance and variability for 10 nm node FinFET technology. In: 2013 symposium on VLSI technology. IEEE, pp T90–T91
26. Ramey S, Ashutosh A, Auth C, Clifford J, Hattendorf M, Hicks J, James R, Rahman A, Sharma V, St Amour A, Wiegand C (2013) Intrinsic transistor reliability improvements from 22 nm tri-gate technology. In: 2013 IEEE international reliability physics symposium (IRPS). IEEE, p 4C-5

27. Hisamoto D, Lee WC, Kedzierski J, Takeuchi H, Asano K, Kuo C, Anderson E, King TJ, Bokor J, Hu C (2000) FinFET-a self-aligned double-gate MOSFET scalable to 20 nm. IEEE Trans Electron Dev 47(12):2320–2325
28. Sairam T, Zhao W, Cao Y (2007) Optimizing FinFET technology for high-speed and low-power design. In: Proceedings of the 17th ACM great lakes symposium on VLSI, pp 73–77
29. Bhoj AN, Joshi RV (2011) Transport-analysis-based 3-D TCAD capacitance extraction for sub-32-nm SRAM structures. IEEE Electron Dev Lett 33(2):158–160
30. Bhoj AN, Joshi RV, Jha NK (2012) Efficient methodologies for 3-D TCAD modeling of emerging devices and circuits. IEEE Trans Comput Aided Des Integr Circuits Syst 32(1):47–58
31. Wu W, Chan M (2007) Analysis of geometry-dependent parasitics in multifin double-gate FinFETs. IEEE Trans Electron Dev 54(4):692–698
32. Pradhan KP, Sahu PK, Ranjan R (2016) Investigation on asymmetric dual-k spacer (ADS) Trigate wavy FinFET: a novel device. In: 2016 3rd international conference on devices, circuits and systems (ICDCS). IEEE, pp 137–140
33. King TJ (2005) FinFETs for nanoscale CMOS digital integrated circuits. In: ICCAD-2005. IEEE/ACM international conference on computer-aided design. IEEE, pp 207–210
34. Mitra S, Brelsford K, Kim YM, Lee HHK, Li Y (2011) Robust system design to overcome CMOS reliability challenges. IEEE J Emerg Sel Top Circuits Syst 1(1):30–41
35. Ezz-Eldin R, El-Moursy MA, Hamed HF (2015) Analysis and design of networks-on-chip under high process variation. Springer International Publishing, Cham
36. Dadgour HF, Endo K, De VK, Banerjee K (2010) Grain-orientation induced work function variation in nanoscale metal-gate transistors—part I: modeling, analysis, and experimental validation. IEEE Trans Electron Dev 57(10):2504–2514
37. Yang Y, Jha NK (2013) Fin Prin: analysis and optimization of FinFET logic circuits under PVT variations. In: 2013 26th international conference on VLSI design and 2013 12th international conference on embedded systems. IEEE, pp 350–355
38. Zimpeck AL, Meinhardt C, Artola L, Hubert G, Kastensmidt FL, Reis RAL (2018) Impact of different transistor arrangements on gate variability. Microelectron Reliab 88:111–115
39. Ortiz RP, Facchetti A, Marks TJ (2010) High-k organic, inorganic, and hybrid dielectrics for low-voltage organic field-effect transistors. Chem Rev 110(1):205–239
40. Dadgour H, De V, Banerjee K (2008) Statistical modeling of metal-gate work-function variability in emerging device technologies and implications for circuit design. In: 2008 IEEE/ACM international conference on computer-aided design. IEEE, pp 270–277
41. Pedram M, Nazarian S (2006) Thermal modeling, analysis, and management in VLSI circuits: principles and methods. Proc IEEE 94(8):1487–1501
42. Almeida RB, Marques CM, Butzen PF, Silva FRG, Reis RA, Meinhardt C (2018) Analysis of 6T SRAM cell in sub-45 nm CMOS and FinFET technologies. Microelectron Reliab 88:196–202
43. Lee C-Y, Jha NK (2013) FinCANON: a PVT-aware integrated delay and power modeling framework for FinFET-based caches and on-chip networks. IEEE Trans Very Large Scale Integr (VLSI) Syst 22(5):1150–1163
44. Meinhardt C, Zimpeck AL, Reis RA (2014) Predictive evaluation of electrical characteristics of sub-22 nm FinFET technologies under device geometry variations. Microelectron Reliab 54(9–10):2319–2324
45. Zimpeck AL, Meinhardt C, Reis RAL (2015) Impact of PVT variability on 20 nm FinFET standard cells. Microelectron Reliab 55(9–10):1379–1383
46. Zimpeck AL, Meinhardt C, Posser G, Reis R (2016) FinFET cells with different transistor sizing techniques against PVT variations. In: 2016 IEEE international symposium on circuits and systems (ISCAS). IEEE, pp 45–48
47. Mushtaq U, Sharma VK (2021) Performance analysis for reliable nanoscaled FinFET logic circuits. Analog Integr Circ Sig Process 107(3):671–682
48. Tawfik SA, Kursun V (2007) Low-power and compact sequential circuits with independent-gate FinFETs. IEEE Trans Electron Devices 55(1):60–70
49. Vallabhuni RR, Yamini G, Vinitha T, Reddy SS (2020) Performance analysis: D-latch modules designed using 18 nm FinFET Technology. In: 2020 international conference on smart electronics and communication (ICOSEC). IEEE, pp 1169–1174

50. Tawfik SA, Kursun V (2007) High speed FinFET domino logic circuits with independent gate-biased double-gate keepers providing dynamically adjusted immunity to noise. In: 2007 international conference on microelectronics. IEEE, pp 175–178
51. Raj B, Saxena AK, Dasgupta S (2011) High performance double gate FinFET SRAM cell design for low power application. Int J VLSI Signal Process Appl 1(1):12–20
52. Mushtaq U, Sharma VK (2019) Design of 6T FinFET SRAM cell at 7 nm. In: 2019 international conference on communication and electronics systems (ICCES). IEEE, pp 104–108
53. Mushtaq U, Sharma VK (2020) Design and analysis of INDEP FinFET SRAM cell at 7-nm technology. Int J Numer Model Electron Netw Dev Fields 33(5):e2730
54. Raja GB, Madheswaran M (2013) Design and performance comparison of 6-T SRAM cell in 32 nm CMOS, FinFET and CNTFET technologies. Int J Comput Appl 70(21):1–6
55. Guo Z, Balasubramanian S, Zlatanovici R, King TJ, Nikolić B (2005) FinFET-based SRAM design. In: Proceedings of the 2005 international symposium on low power electronics and design, p 2
56. Ebrahimi B, Zeinolabedinzadeh S, Afzali-Kusha A (2008) Low standby power and robust FinFET based SRAM design. In: 2008 IEEE computer society annual symposium on VLSI. IEEE, pp 185–190

# Resistive Switching Mechanism in Polymer Embedded Chemically Synthesized Reduced Graphene Oxide

**Nipom Sekhar Das, Avijit Chowdhury, and Asim Roy**

## 1 Introduction

Organic–inorganic materials have drawn growing attention because of their remarkable physical and structural properties. GO and rGO show great interest because of their outstanding charge carrier mobility, unique chemical inertness, and large specific area [1]. Furthermore, these materials are exploited in solar cell [2], supercapacitors [3], memory devices [4], and so forth. Memory properties of GO and its reduced form have extensively studied from many years. The promising candidate for memory devices is one of the families of GO which is the reduced form of GO where some of the oxygen functional groups are removed and restore the C=C bond conjugation which gives better electrical property. Recently, polymeric materials embedded with graphene-based materials have grabbed increasing attention in electronic devices [5]. Al/GO/ITO device has been fabricated by Hong et al. which showed good switching performance with $I_{ON}/I_{OFF}$ $10^3$ [6]. He et al. fabricated the Cu/GO/Pt device that exhibited bipolar resistive switching characteristics with $I_{ON}/I_{OFF}$ 20 and retention $10^4$ s [7]. Gogoi et al. fabricated the device ITO/PMMA/rGO/CuS/PMMA/Al that exhibited bipolar resistive switching performance with $I_{ON}/I_{OFF}$ $10^3$–$10^4$ and $V_{SET} = -0.44 \pm 0.10$ V and $V_{RESET} = 0.50 \pm 0.10$ V and retention $10^4$ s [8]. Khurana et al. fabricated the device Pt/GO/ITO which showed stable bipolar resistive switching nature with $I_{ON}/I_{OFF}$ $10^4$ and $V_{SET} = -1.2$ to 1.8 V and $V_{RESET} = 3$–3.4 V and retention $10^4$ s [9]. ITO/PEDOT + GO/Al device was fabricated by Li et al., and the device showed reversible bistable resistance switching with $I_{ON}/I_{OFF}$ $10^4$ and retention $10^4$ s [10]. In this article, we have reported

N. S. Das · A. Chowdhury · A. Roy (✉)
Department of Physics, National Institute of Technology Silchar, Cachar, Assam 788010, India
e-mail: asim@phy.nits.ac.in

A. Chowdhury
Department of Condensed Matter Physics and Materials Sciences, S.N. Bose National Center for Basic Sciences, Salt Lake, Kolkata, West Bengal 700106, India

 
P. H. J. Chong et al. (eds.), *Emerging Electronics and Automation*, Lecture Notes in Electrical Engineering 937, https://doi.org/10.1007/978-981-19-4300-3_36

the resistive switching characteristics of polyvinyl alcohol (PVA) embedded chemically reduced graphene oxide nanocomposite. Table 1 shows the several non-volatile memory devices that have been already reported.

## 2 Experimental Procedure

### 2.1 Materials and Reagents Used

$NaNO_3$ (Merck), $KMnO_4$ (Merck), Graphite flakes (Alfa Aesar), $H_2SO_4$ (Merck), $H_2O_2$ (Merck), hydrazine hydrate (Merck), DI water, *N*, *N* dimethyl formamide (DMF), polyvinyl alcohol (PVA, Loba Chemie Pvt. Ltd.).

### 2.2 Synthesis of GO

GO was successfully synthesized by conventional Hummer's method where graphite is a starting material which was added with $NaNO_3$ and mixed with $H_2SO_4$ followed by the addition of $KMnO_4$ and kept for overnight stirring [15]. In the next step, temperature of the solution was raised to 96 °C followed by the addition of DI water and hydrogen peroxide and then cleansed with HCl solution and DI water. Finally, graphene oxide powder was obtained after drying in hot air oven.

### 2.3 Synthesis of rGO

GO was reduced by chemical method [16]. 1 mg/mL of GO powder was added to DI water under stirring followed by the addition of 5 ml hydrazine hydrate dropwise and kept under stirring for 24 h at 80 °C. The solution was then rinsed with DI water followed by absolute ethanol. After drying in oven, a black-colored rGO was obtained.

### 2.4 Instrumentation

X-ray diffractometer (XRD) ($Xpert^3$ MRD XL PANALYTICAL), photoluminescence spectroscopy (Fluoromax-4C Spectrofluorometer, Horiba), UV–visible spectroscopy (Agilent Technologies, Model: Cary 60) were used for characterizing the obtained product (Figs. 1 and 2).

**Table 1** Non-volatile memory devices with resistive switching parameters

| Device structure | Process of fabrication | $I_{ON}/I_{OFF}$ | $V_{set}$ (V) | $V_{reset}$ (V) | Retention (s) | Endurance cycles | Explanation | References | Publication year |
|---|---|---|---|---|---|---|---|---|---|
| Ag/PI/GO: PI/PI/ITO | Spin coating | $10^5$ | 3.5 | – | $10^3$ | $10^2$ | The device showed high $I_{on}/I_{off}$ ratio and good endurance property | 11 | 2011 |
| ITO/GO-ZnO/Al | Spin coating | $10^2$ | 2.1 | – 2.0 | $10^4$ | $10^2$ | Good retention and endurance were exhibited by the fabricated device | 12 | 2014 |
| Al/GO/Al | Spin casting | > $10^2$ | – 2.5 | 2.5 | $10^5$ | $10^2$ | Decent retention time was showed by the device along with the bipolar resistive switching phenomenon | 13 | 2010 |
| Al/ZnO/Al | Dual-ion beam sputtering | – | – 6.2 | 6.3 | $10^6$ | 250 | Memristive characteristics and bipolar resistive switching behavior were shown by the device | 14 | 2017 |

**Fig. 1** Schematic diagram of synthesis of rGO

**Fig. 2** Synthesized rGO powder

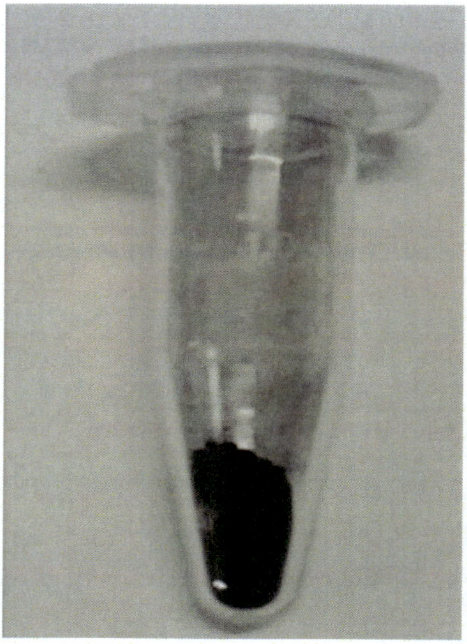

## 2.5 *Fabrication of Device*

The device was fabricated by spin-coating method by the use of programmable spin coater (Apex Instruments Model: Spin NXG-P2). 0.2 mg of rGO was mixed with 99.8 mg of PVA in 3 ml of DMF and ultrasonicated for 30 min. The prepared solution was then spin coated onto the ITO-coated PET substrate at 3000 rotations per minute (rpm) for 30 s and then dried it. Top electrode copper (Cu) was deposited on the rGO-PVA film using thermal evaporation technique (HIND-HIVAC, model: 12A4DM). Fabricated device is schematically displayed in Fig. 3.

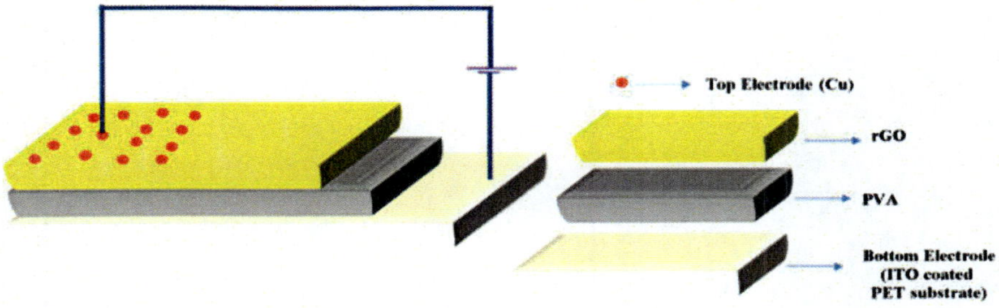


**Fig. 3** Schematic diagram of Cu/rGO-PVA/ITO device

# 3 Results and Discussions

## *3.1 Optical Studies of rGO*

### 3.1.1 UV-visible Absorption Spectroscopy

UV–visible absorption spectrum of rGO was done in the wavelength band of 200–800 nm. An absorption peak occurs at 270 nm due to the transition of $\pi$–$\pi$* which is shifted from the peak of GO because of the removal of some of functional groups on the GO, and the conjugated structure has been restored [17, 18]. It is due to the red shift of the absorption band of GO and partial restoration of the aromatic rings [19, 20]. Optical band gap energy is calculated to be 3.45 eV by Tauc's plot (Fig. 4).

### 3.1.2 Photoluminescence (PL) Spectroscopy

Figure 5 shows the PL spectrum of rGO with $\lambda_{ex}$ ~ 270 nm. rGO exhibits a strong emission peak at 425 nm. The emission peak occurs due to the presence of defect states which may be emanated from the synthesis of GO or from the reduction of GO [21]. The defects arise due to the disordered arrangements of $sp^2$ bonded carbon atoms in rGO.

## *3.2 Structural Studies*

### 3.2.1 X-ray Diffraction (XRD)

XRD of rGO was done in $2\theta$ range of 5–80° by using Cu $K_\alpha$ radiation with $\lambda$ = 1.54 Å. A peak observes at $2\theta = 24.33°$ which is from the (002) plane of rGO [22, 23], and another peak appears at $2\theta = 43.29°$ from the reflection plane (100) [24] as shown in Fig. 6. The interlayer spacing is calculated using Bragg's law, and the

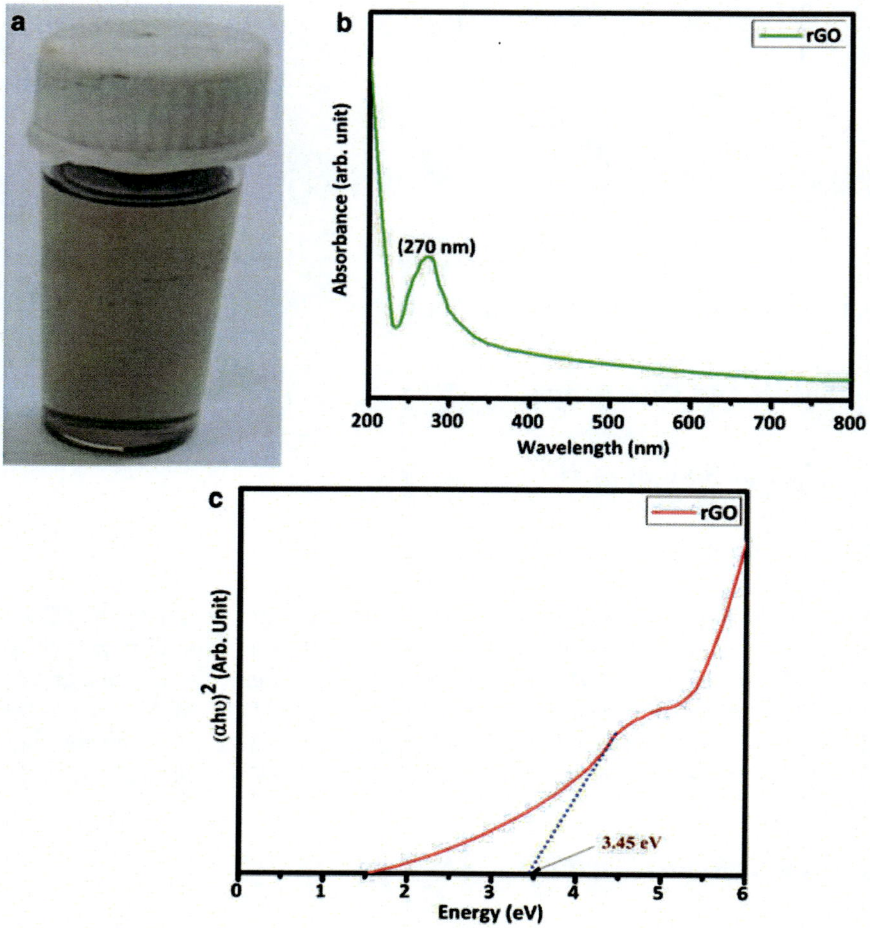


**Fig. 4** **a** Freshly prepared rGO solution in DMF, **b** UV–visible absorption spectra, and **c** Tauc's plot of rGO

value has been estimated to be 3.65Å which is less than the interlayer spacing of GO because of the removal of many oxygen functional groups [25]. Crystallite size is estimated to be 13.91 nm using Scherrer's equation.

### *3.3 Electrical Studies*

Room temperature I-V characteristics of Cu/rGO-PVA/ITO were done using Keithley 4200 Semiconductor Characterization System. The I-V graph is shown in Fig. 7. The fabricated device shows the bipolar resistive switching behavior with $V_{SET} = 1.92$ V,

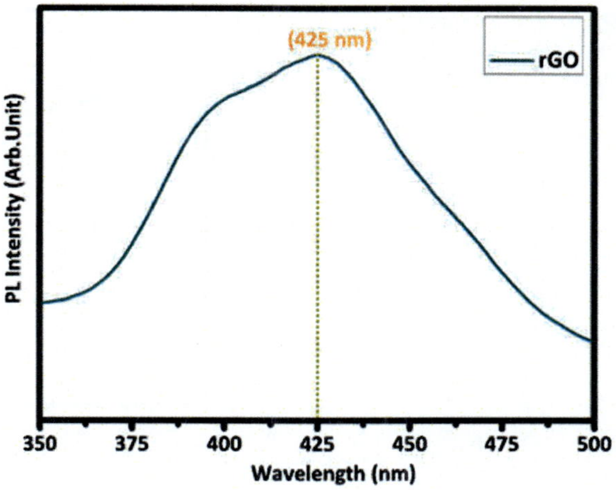


**Fig. 5** Photoluminescence spectra of rGO

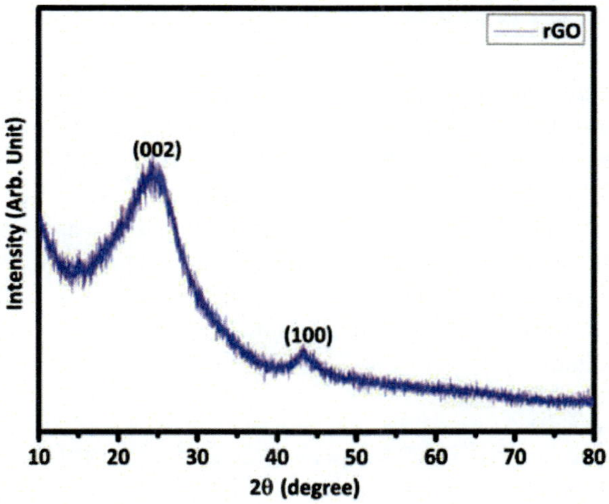


**Fig. 6** XRD pattern of rGO

$V_{RESET} = -1.12$ V, and $I_{ON}/I_{OFF} = 10$. During the analysis, bottom electrode ITO is grounded, and top electrode Cu is biased. After the variation of voltage from 0 to 2 V, the current increases from high resistance state (HRS), i.e., OFF state to low resistance state (LRS), i.e., ON state. The transition from OFF to ON state is basically called the set process. The device is in the LRS state while decreasing from 2 to 0 V. After altering the voltage from 0 to − 2 V, the device initially on the LRS state decreases to HRS state that corresponds to the reset voltage. Set and reset process occurs for the device at different voltage polarities that represents the bipolar resistive switching nature in the device.

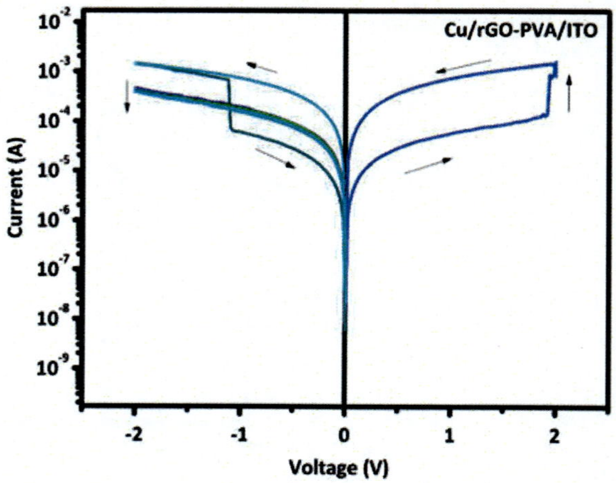


**Fig. 7** Room temperature I-V characteristics of Cu/rGO-PVA/ITO device

## 4 Conclusion

In the present work, we have reported the cost-effective and environmentally friendly chemical synthesis of rGO and are subjected to extensive optical and structural characterizations. The chemically synthesized rGO shows an optical band gap of ~ 3.45 eV along with an interplanar spacing of ~ 3.65Å which is smaller than the graphene oxide and average crystallite size is estimated to be 13.91 nm. The fabricated Cu/rGO-PVA/ITO device exhibits the bipolar resistive switching behavior. The device shows set voltage 1.92 V and reset voltage − 1.12 V and $I_{ON}/I_{OFF} = 10$. These composite materials can be exploited in cost effective electronic devices in near future effectively.

**Acknowledgements** The authors would like to express their sincere gratitude to Central Instrumentation Facility, NIT Silchar for XRD analysis, SMaRT Laboratory, Department of Physics, NIT Silchar for optical characterization.

## References

1. Porro S, Accornero E, Pirri CF, Ricciardi C (2015) Memristive devices based on graphene oxide. Carbon 85:383–396
2. Cheng J, Ma J, Ma Y, Zhou C, Qiang Y, Zhou X, Yang J, Shi H, Xie Y (2018) Highly efficient ZnO-based dye-sensitized solar cells with low-cost Co–Ni/carbon aerogel composites as counter electrodes. New J Chem 42:16329–16334
3. Wang H, Yang Y, Zhou X, Li R, Li Z (2017) $NiCo_2S_4$/tryptophan-functionalized graphene quantum dot nanohybrids for high-performance supercapacitors. New J Chem 41:1110–1118
4. Kao M-C, Chen H-Z, Chen K-H, Shi J-B, Weng J-H, Chen K-P (2020) Resistive switching behavior and optical properties of transparent Pr-doped ZnO based resistive random-access memory. Thin Solid Films 697:137816

5. Thomassin J-M, Trifkovic M, Alkarmo W, Detrembleur C, Jerome C, Makosco C (2014) Poly (methyl methacrylate)/graphene oxide nanocomposites by a precipitation polymerization process and their dielectric and rheological characterization. Macromolecules 47:2149–2155
6. Hong SK, Kim JE, Kim SO, Choi S-Y, Cho BJ (2010) Flexible resistive switching memory device based on graphene oxide. IEEE Electron Dev Lett 31:1005–1007
7. He CL, Zhuge F, Zhou XF, Li M, Zhou GC, Liu YW, Wang JZ, Chen B, Su WJ, Liu JP, Wu YH, Cui P, Li RW (2009) Nonvolatile resistive switching in graphene oxide thin films. Appl Phys Lett 95:232101
8. Gogoi KK, Das S, Maiti S, Chowdhury A (2020) Semiconductor-grafted polymer-embedded reduced graphene-oxide nanohybrid power-efficient nonvolatile resistive memory applications. ACS Appl Nano Mater 3(11):11562–11573
9. Khurana G, Misra P, Katiyar RS (2013) Forming free resistive switching in graphene oxide thin film for thermally stable nonvolatile memory applications. J Appl Phys 114:124508
10. Li Y, Ni X (2016) One-step preparation of graphene oxide-poly(3,4ethylenedioxythiophene) composite films for nonvolatile rewritable memory. RSC Adv 6:16340–16347
11. Wu C, Li F, Zhang Y, Guo T, Chen T (2011) Highly reproducible memory effect of organic multilevel resistive-switch device utilizing graphene oxide sheets/polyimide hybrid nanocomposite. Appl Phys Lett 99:042108
12. Khurana G, Misra P, Kumar N, Katiyar RS (2014) Tunable power switching in non-volatile flexible memory devices based on graphene oxide embedded with ZnO nanorods. J Phys Chem C 118:21357–21364
13. Jeong HY, Kim JY, Kim JW, Hwang JO, Kim J-E, Lee JY, Yoon TH, Cho BJ, Kim SO, Ruoff RS, Choi S-Y (2010) Graphene oxide thin films for flexible nonvolatile memory applications. Nano Lett 10:4381–4386
14. Kumar A, Das M, Garg V, Sengar BS, Htay MT, Kumar S, Kranti A, Mukherjee S (2017) Forming-free high-endurance Al/ZnO/Al memristor fabricated by dual ion beam sputtering. Appl Phys Lett 110:253509
15. Gogoi KK, Das NS, Chowdhury A (2019) Tuning of electrical hysteresis in PMMA/GOs/PMMA multi-stacked devices. Mater Res Express 6:085108
16. Aragaw BA (2020) Reduced graphene oxide -intercalated graphene oxide nano-hybrid for enhanced photoelectrochemical water reduction. J Nanostructure Chem 10:9–18
17. Yang S, Yue W, Huang D, Chen C, Lin H, Yang X (2012) A facile green strategy for rapid reduction of graphene oxide by metallic zinc. RSC Adv 2:8827–8832
18. Zhu C, Guo S, Fang Y, Dong S (2010) Reducing sugar: new functional molecules for the green synthesis of graphene nanosheets. ACS Nano 4:2429–2437
19. Nafiey AA, Addad A, Sieber B, Chastanet G, Barras A, Szunerits S, Boukherroub R (2017) Reduced graphene oxide decorated with $Co_3O_4$ nanoparticles (rGO-$Co_3O_4$) nanocomposite: a reusable catalyst for highly efficient reduction of 4-nitrophenol, and Cr (VI) and dye removal from aqueous solutions. Chem Eng Sci 322:375–384
20. Sunderrajan S, Miranda LR, Pennathur G (2018) Improved stability and catalytic activity of graphene oxide/chitosan hybrid beads loaded with porcine liver esterase. Prep Biochem Biotechnol 48:343–351
21. Chuang C-H, Wang Y-F, Shao Y-C, Yeh Y-C, Wang D-Y, Chen C-W, Chiou JW, Ray SC, Pong WF, Zhang L, Zhu JF, Guo JH (2014) The effect of thermal reduction on the photoluminescence and electronic structures of graphene oxides. Sci Rep 4:4525
22. Soomro SA, Gul IH, Naseer H, Marwat S, Mujahid M (2019) Improved performance of $CuFe_2O_4$/rGO nanohybrid as an anode material for lithium-ion batteries prepared via facile one step method. Curr Nanosci 15:420–429

23. Huang X, Zhang J, Rao W, Sang T, Song B, Wong C (2016) Tunable electromagnetic properties and enhanced microwave absorption ability of flaky graphite/cobalt zinc ferrite composites. J Alloys Compd 662:409–414
24. Latif IA, Merza SH (2016) Fabrication of functionalize reduced graphene oxide and its application in ampicillin detection. Nanosci Nanotechnol 6(2):24–33
25. Jung I, Pelton M, Piner R, Dikin DA, Stankovich S, Watcharotone S, Hausner M, Ruoff RS (2007) Simple approach for high-contrast optical imaging and characterization of graphene-based sheets. Nano Lett 7:3569–3575

# MEMS and VLSI Track—2

# Design and Investigation of PGP SELBOX FinFET with Uniform and Non-uniform Doping Profile at Sub-7 nm Technology Node

Satya Prakash Singh and Md. Waseem Akram

## 1 Introduction

In the ultra-small conventional transistors, it gets very challenging to fabricate ultra-high concentration slope at junctions. Junctionless transistors were introduced and continuously explored after 2009 [1–7]. Junctionless transistors keep high doping in all the regions. Charge carriers move away from channel, when device is off, and bulk conduction starts when device is on. Production of junctionless devices appears easier due to low thermal budget and absence of large concentration gradient. Surface scattering does not play major role due to bulk conduction [8–11]. Although there are various limitations of these devices, it is very tough to get uniform doping in the vertical direction due to technological limitations. Charge carriers are removed more from top of channel than bottom of channel when uniform doping is done. It increases the leakage current. SOI FinFET shows high on-current and minimal parasitic capacitances and enhanced control on short-channel effects. Sub-threshold leakage current flows in the lower part of the channel because lower portion of the channel gets depleted less effectively in SOI-JLT. Gate material must have higher value of gate work function for highly scaled n-channel device for the sake of nice turn-off and subthreshold characteristics [12]. Thus, the channel thickness must be reduced to lower value (10 nm<) to achieve the appropriate control on the channel through gate [13]. The value of on-current will be affected directly due to device channel thickness variation. Command of gate on channel diminishes with movement toward lower part

S. P. Singh · Md. Waseem Akram (✉)
Department of Electronics and Communication Engineering, Jamia Millia Islamia University, New Delhi, India
e-mail: makram1@jmi.ac.in

S. P. Singh
Department of Electronics and Communication Engineering, KIET Group of Institutions, Delhi-NCR, Ghaziabad, India
e-mail: satya.singh@kiet.edu

 
P. H. J. Chong et al. (eds.), *Emerging Electronics and Automation*, Lecture Notes in Electrical Engineering 937, https://doi.org/10.1007/978-981-19-4300-3_37

of the channel. Despite having many advantages of SOI-FinFET, self-heating and low breakdown voltage are its severe limitations. MOSFET with PGP was proposed on SELBOX technology. It shows low short-channel effects suppression as compared to SOI-based devices [14]. The channel is doped uniformly in horizontal direction, and in perpendicular direction, dopants are introduced non-uniformly. To get non-uniformly doped channel, doping profile is implemented according to Gaussian distribution. The short-channel effects give rise to many practical problems, like subthreshold slope degradation, enhancement in leakage effect, surge in drain induced barrier lowering, HCE, dearth of pinch-off, and change in value of threshold voltage as we lower channel length [15]. A novel SELBOX device is presented and simulated through Silvaco Atlas 3-D. The designed structure has all the advantageous factor of the SELBOX structure. It shows reduction in SCEs comprehensively. This design makes more advancement in device scaling plausible. Proposed device is partial ground plane FinFET implemented on SELBOX technology. A narrow opening is provided beneath the channel. It reduces the leakage current [16–18]. The partial ground planes are used beneath the SELBOX. Field lines from drain to source are weakened due to use of massively doped ground planes. This results in the minimization of short channel effects. Various disadvantages of continuous ground planes are eradicated using partial ground planes.

Presented work is organized in three sections. Device structure and its parameters are presented in Sect. 2. Simulation results have been extensively elaborated in Sect. 3. Section 4 includes the conclusion and future scope.

## 2 Device Structure and Its Parameters

Figure 1 presents diagram of simulated devices. SILVACO ATLAS 3D has been used for accurate and reliable result. Effective oxide thickness is taken as 1 nm. Rigorous simulation has been carried out for sub-10 nm channel length. Gate oxide material of high dielectric constant like $Si_3N_4$ is considered. It reduces self-heating due to its high thermal conductivity and hence improves the thermal stability of the device. Small $p+$ region, having high doping ($1.5 \times 10^{19}/cm^3$), has been fabricated to provide partial ground plane. Edge of $p+$ region is aligned with the edge of BOX and opening under the channel. This decreases the binding between source and drain and therefore suppresses the short-channel effects very effectively because field strength between source and drain decreases. The sufferings due to floating body effect in PD-SOI are effectively combated in selective buried oxide and partial ground plane selective buried oxide devices.

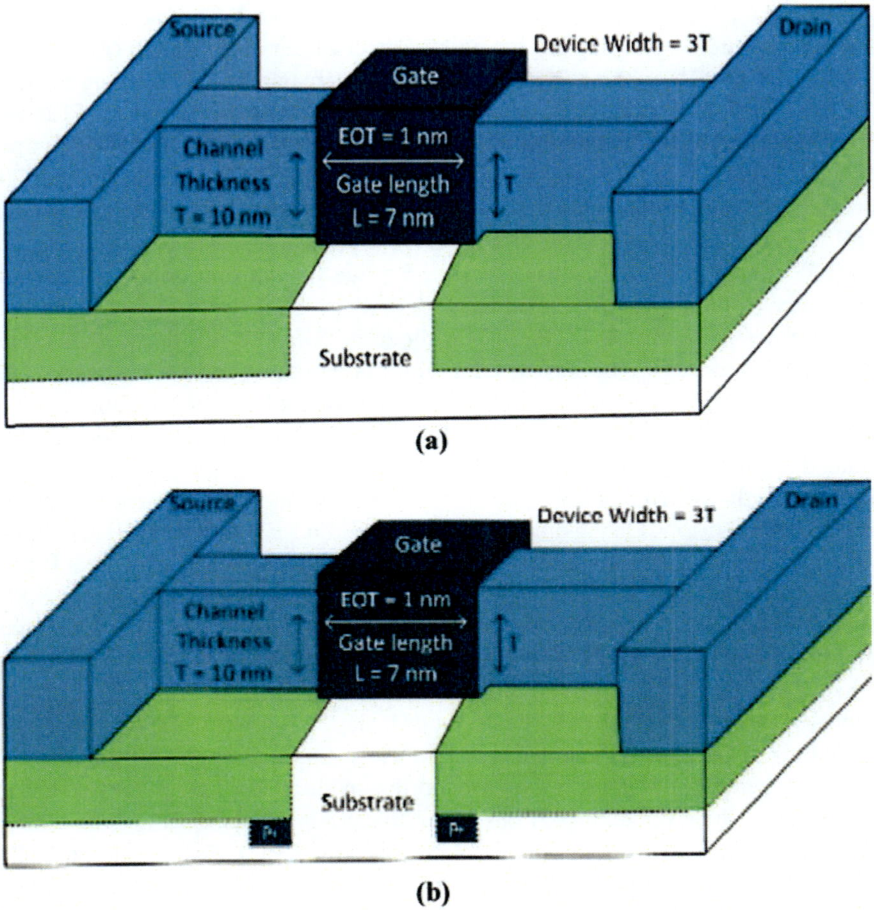


**Fig. 1** **a** Junctionless FinFET based on SELBOX, **b** junctionless FinFET with PGP based on SELBOX

## 3 Simulation Results and Discussions

Performance of different devices, on the basis of various performance parameters, is compared on various channel lengths. 10 nm channel thickness is considered for simulation and comparison. One-dimensional Gaussian doping is incorporated in order to achieve non-uniformly doped channel creation. Gate work function for the *n*-type FinFET is 5 eV. Concentration of carriers in channel is uniform in the horizontal dimension and is equal to $1.5 \times 10^{19}$ cm$^{-3}$ for abovementioned devices. In the SELBOX FinFET, doping concentration of substrate is equal to $5 \times 10^{18}$ cm$^{-3}$ (*p*-type). ni.fermi, bqp.n, cvt, fldmob, consrh, auger, and bgn models are considered in our simulation studies. Ni.fermi includes consequences of statistics of fermi into the calculation. Mobility model, bandgap narrowing model, and band-to-band tunneling model are included due to very low value of channel length (sub-7 nm). PGP SELBOX FinFET with uniform doping has negligible decrement in on current in comparison to SELBOX FinFET without PGP, but it has reasonable drop in off-current. Hence, SELBOX FinFET with PGP gives improved $I_{ON}/I_{OFF}$ ratio by 6% over SELBOX FinFET without PGP for both uniform and non-uniform doping profile. It can be observed from Figs. 2, 3, 4, and 5. Figure 6 shows that SELBOX FinFET with PGP shows higher value of threshold voltage than SELBOX device without PGP. SELBOX FinFET with PGP has SS value of 57.6 mV/decade. On the other hand, SELBOX FinFET without PGP has 90 mV/decade SS value for uniform doping profile, at 7 nm channel length. SELBOX FinFET with PGP has SS value of 72.6 mV/decade. On the other hand, SELBOX FinFET without PGP has 90 mV/decade SS value for non-uniform doping profile, at 7 nm channel length. It can be analyzed from Fig. 7. SELBOX FinFET with PGP attains drain-induced barrier lowering voltage of 118.1 mV/V. On the other hand, SELBOX FinFET without PGP attains drain-induced barrier lowering voltage of 126 mV/V for uniform doping profile, at 7 nm channel length. SELBOX FinFET with PGP attains drain-induced barrier lowering voltage of 43.2 mV/V. On the other hand, SELBOX FinFET without PGP has DIBL value of 52.6 mV/V for non-uniform doping profile, at 7 nm channel length. It can be observed from Fig. 6. In the proposed devices, improvements in DIBL and SS of 5.47 and 36.66% are achieved compared to without the use of PGP technique for uniformly doped channel only. For non-uniformly doped channel, improvement of 17.87 and 19.33% is observed in DIBL and SS values in the proposed structure compared to without the use of PGP technique. It may be observed that non-uniformly doped FinFETs deplete the channel very effectively. Because of non-uniform doping carrier, concentration is lesser in the lower portion of the channel in comparison to the top portion of the channel. It provides lower off-current, by reasonable order which gives improvement in $I_{ON}/I_{OFF}$ ratio in comparison to SELBOX FinFET without PGP. This happens because of the enhanced control of the gate, with non-uniform doping. Above observations can be seen in Fig. 7. The channel below gate is impressively free from carriers when transistor is off because of non-uniformly doped channel. Hence, this makes further scaling down of the device possible.

Because of non-uniform doping, on-current is unaffected. But, it exhibits meticulously very less value of off-current and shows improved $I_{ON}/I_{OFF}$ ratio. Scaling down of the device becomes possible due to open window given beneath gate in SELBOX FinFET. So, leakage current passes through it and so reduces the self-heating of the device. SELBOX technology-based devices show minimal temperature of lattice [16]. Therefore, self-heating effect is extensively minimized in SELBOX Technology FinFET.

Given gap beneath the gate diminishes field binding between source and drain, so punch-through is avoided on smaller value; hence, smaller breakdown voltage issue in SOI FinFET is rectified in SELBOX FinFET. DIBL, SS, and $I_{ON}/I_{OFF}$ degrade as we decrease channel length as gate command on the channel diminishes on lower channel lengths.

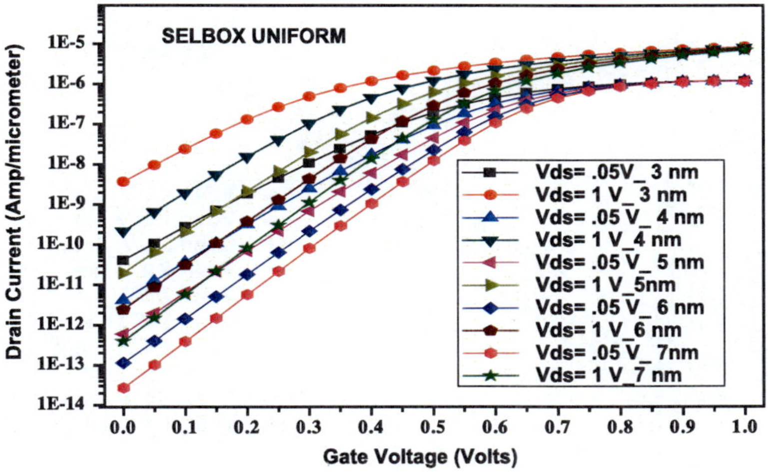


**Fig. 2** $I_{DS}$-$V_{GS}$ graph of uniformly doped SELBOX FinFET with varying channel lengths

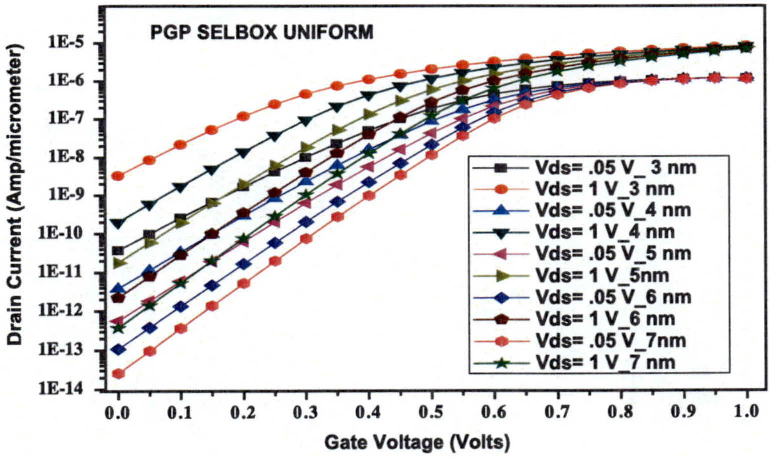


**Fig. 3** $I_{DS}$-$V_{GS}$ graph of uniformly doped PGP SELBOX FinFET with varying channel lengths

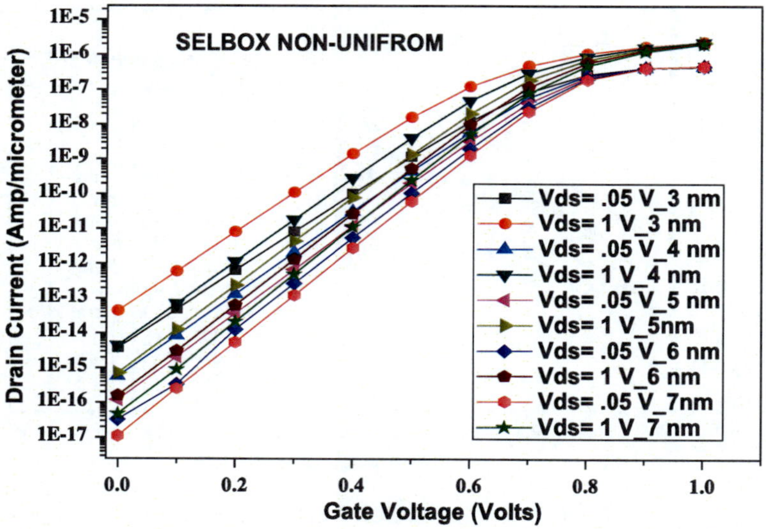


**Fig. 4** $I_{DS}$-$V_{GS}$ graph for SELBOX FinFET without PGP having non-uniform doping for different channel lengths

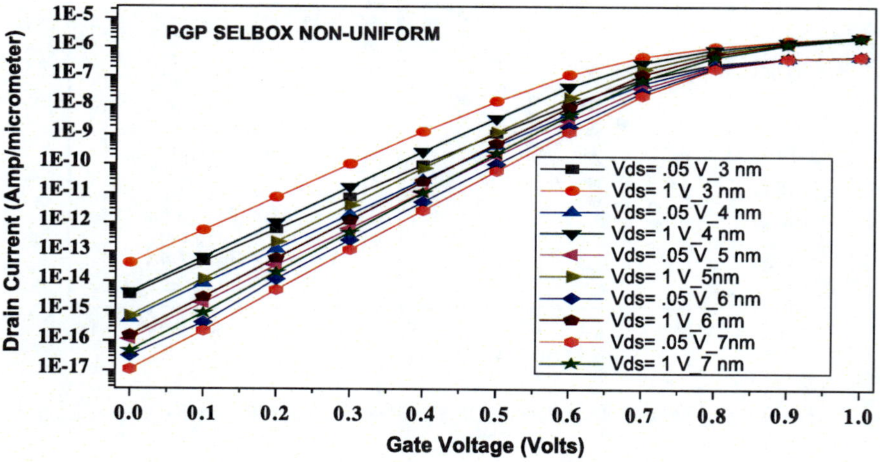


**Fig. 5** $I_{DS}$-$V_{GS}$ graph for PGP SELBOX FinFET having non-uniform doping and different channel lengths

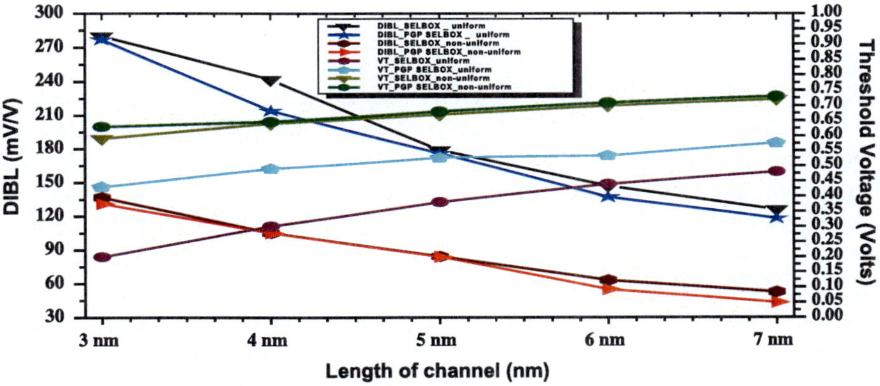


**Fig. 6** Drain-induced barrier lowering and $V_{TH}$ behavior, of SELBOX FinFET having uniform (with and without PGP) and non-uniform doping (with and without PGP) for different channel lengths

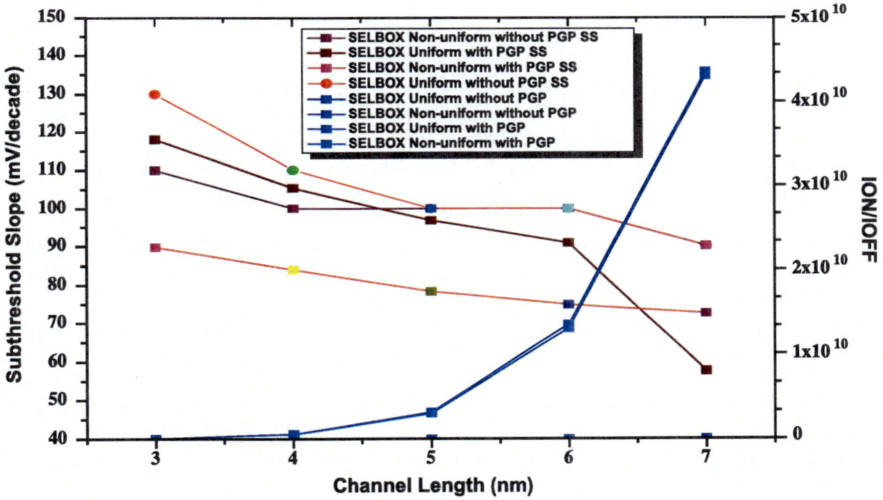


**Fig. 7** SS and $I_{ON}/I_{OFF}$ behavior, of SELBOX FinFET with uniform (with and without PGP) and non-uniform doping (with and without PGP) for different channel lengths

# 4 Conclusion

Uniformly and non-uniformly doped PGP SELBOX FinFET are proposed in presented article, and the results are evaluated with uniformly and non-uniformly doped channel SELBOX FinFET without PGP. Various performance parameters, obtained through rigorous simulation, show that FinFET on SELBOX technology with PGP attains better $I_{ON}/I_{OFF}$ ratio, low leakage current, lower drain-induced barrier lowering, smaller subthreshold slope, and improved SCE behavior than SELBOX FinFET without PGP for uniform as well as non-uniform doping profile. Results of the SELBOX FinFET with PGP improve further as non-uniformity of the channel is increased further; this is due to the fact that control of the gate over the channel improves with this. Improved performance of junctionless SELBOX FinFET with PGP shows that proposed device may play a vital role in future technology node.

# References

1. Colinge J-P, Lee C-W, Afzalian A, Akhavan ND, Yan R, Ferain I, Razavi P, O'Neill B, Blake A, White M, Kelleher A-M, McCarthy B, Murphy R (2010) Nanowire transistors without junctions. Nat Nanotechnol 5(3):225–229
2. Su C-J, Tsai T-I, Liou Y-L, Lin Z-M, Lin H-C, Chao T-S (2011) Gate-all-around junctionless transistors with heavily doped polysilicon nanowire channels. IEEE Electron Dev Lett 32(4):521–523
3. Chen C, Lin J, Chiang M, Kim K (2010) High performance ultra-low power junctionless nanowire FET on SOI substrate in subthreshold logic application. In: Proceeding of the IEEE international SOI conference, pp 1–2
4. Lee C-W, Afzalian A, Akhavan ND, Yan R, Ferain I, Colinge J (2009) Junctionless multigate field-effect transistor. Appl Phys Lett 94(5):053511–053512
5. Colinge JP, Lee C-W, Ferain I, Akhavan ND, Yan R, Razavi R, Yu R, Nazarov AN, Doria RT (2010) Reduced electric field in junctionless transistors. Appl Phys Lett 96(7):073510-1–073510-3
6. Lee C-W, Ferain I, Afzalian A, Yan R, Akhavan ND, Razavi P, Colinge J-P (2010) Performance estimation of junctionless multigate transistors. Solid State Electron 54(2):97–103
7. Gundapaneni S, Ganguly S, Kottantharayil A (2011) Bulk planar junctionless transistor (BPJLT): an attractive device alternative for scaling. IEEE Electron Dev Lett 32(3):261–263
8. Choi S-J, Moon D-I, Kim S, Duarte J, Choi Y-K (2011) Sensitivity of threshold voltage to nanowire width variation in junctionless transistors. IEEE Electron Dev Lett 32(2):125–127
9. Rios R, Cappellani A, Armstrong M, Budrevich A, Gomez H, Pai R, Rahhal-orabi N, Kuhn K (2011) Comparison of junctionless and conventional trigate transistors with Lg down to 26 nm. IEEE Electron Dev Lett 32(9):1170–1172. https://doi.org/10.1109/LED.2011.2158978
10. Sze SM (1983) Physics of semiconductor devices, 2nd edn. Wiley, New York
11. Sallese J-M, Chevillon N, Lallement C, Iniguez B, Pregaldiny F (2011) IEEE Trans Electron Dev 58(8):2628–2637
12. Duarte P, Choi S-J, Moon D-I, Choi Y-K (2011) IEEE Electron Dev Lett 32(6):704–706
13. Zhang G, Shao Z, Zhou K (2008) IEEE Trans Electron Dev 55(3):803–809
14. Loan SA, Qureshi S, Sundar Kumar Iyer S (2010) A novel partial-ground-plane-based MOSFET on selective buried oxide: 2-D simulation study. IEEE Trans Electron Dev 57(3)
15. Suzuki K, Kataoka Y, Nagayama S, Magee CW, Buyuklimanli TH, Nagayama T (2007) IEEE Trans Electron Dev 54(2):262–271

16. Singh SP, Akram MW (2021) Design and performance evaluation of Sub-10 nm Gaussian doped junctionless SOI and SELBOX FinFET. SILICON 13:2125–2133
17. Singh SP, Akram M (2020) Waseem, design and analysis of SOI and SELBOX junctionless FinFET at sub-15 nm technology node. IJEMS 27(5):969–975
18. Nelapati RP, Sivasankaran K (2018) Impact of self-heating effect on the performance of hybrid FinFET. Microelectron J 76:63–68

# Comparison of Memristor-Based SRAM Cells at Subthreshold Voltage

**Zeba Mustaqueem, Abdul Quaiyum Ansari, and Md. Waseem Akram**

## 1 Introduction

The gradual increase in demand for the usage of battery operated and portable devices like mobile phones, wireless sensor-based devices, and medical devices have made power consumption as one of the major factor which is to be kept in mind before considering any design element [1]. As calibrated by the International technology road map for semiconductors (ITRS) in its edition [2] that SRAM occupies maximum of its area when interfaces with CPU and μprocessors and used in low power applications; hence, it is very challenging and desirable to optimize the power consumption and decrease its dissipation to decrease heat on devices and thereby preventing from damaging the circuitry [3]. Working SRAM cell in subthreshold region where applied supply voltage is less than threshold voltage (below 0.7 V) which is the one of the technique adopted to decrease power as power (*P*) is directly proportional to the square of the VDD [4]. In this scaled technology, when basic 6T SRAM cell is simulated, it suffers from read/write degradation as well as leakage currents available thereby occurring failure in the functioning [4]. To overcome this difficulty, many SRAM configurations has been designed at subthreshold so far [4–7].

The SRAM cells are volatile in nature and therefore losses its data when power is removed. It also suffers from leakage problem as well as power dissipation, and so there is intense demand of non-volatile memories which are power efficient as well as having an ability to back up previous information [8]. Introduction of memristors with CMOS technology is fabricated and compatible with less area occupancy which has come in market where researchers are working vastly [9].

Z. Mustaqueem (✉) · A. Q. Ansari
Department of Electrical Engineering, Jamia Millia Islamia University, New Delhi, India
e-mail: zeba.ec@gmail.com

Md. Waseem Akram
Department of Electronics and Communication Engineering, Jamia Millia Islamia, New Delhi, India



P. H. J. Chong et al. (eds.), *Emerging Electronics and Automation*, Lecture Notes in Electrical Engineering 937, https://doi.org/10.1007/978-981-19-4300-3_38

Rest of the paper has memristor-related section, comparison of non-volatile memories, $N$-curve analysis of each of them, and finally concludes it.

## 2 Memristor Behaviour and Properties

Memristor is considered as fourth fundamental passive element after resistor, capacitor, and inductor discovered by Leon [10]. The practical HP Lab memristor model was first discovered by HP labs in 2008 [11] which claimed and proposed a device which resembles ideal memristor shown in Fig. 1 where doped and undoped $TiO_2$ is sandwiched between platinum electrodes. Figure 1 also describes the symbol and polarity of the memristor [12, 13] in which $R_{on}$ (low resistance) and $R_{off}$ (high resistance) condition can be obtained by direction of the current flow. The functional memristor model provides a relation between the memristance and the electric current and the memristor current moves through as explained in Eq. (1).

Relating the charge and flux is called memristive relationship given as in Eqs. (1) and (2):

$$q(t) = \int_{-\infty}^{t} i(\tau)\mathrm{d}\tau \tag{1}$$

$$\varnothing(t) = \int_{-\infty}^{t} v(\tau)\mathrm{d}\tau \tag{2}$$

A generalized expression to understand memristive behaviour is given by:

$$V = R(w_1, w_2, \ldots w_k \ldots w_n)I \tag{3}$$

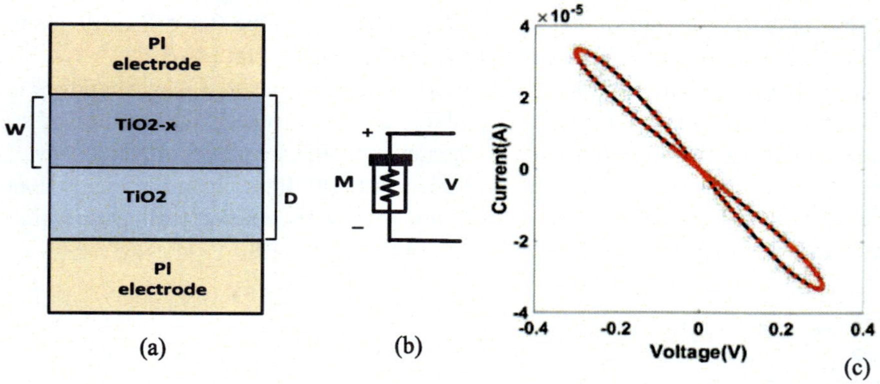


**Fig. 1** **a** Memristive device structure, **b** symbol and polarity of memristor, and **c** hysteresis loop as I-V curve at $V_{SUB} = 0.3$ V

where $w_1$, $w_2 \ldots$ are state variables. Memristor is having unit of resistance itself which is called as memristance.

Equation (3) essentially relates the voltage, current, and the variation in the resistance according to the state of a device at any point of time. It also identifies the charge that passed through the device previously.

The constitutive relation for a memristor is $f(\emptyset, q) = 0$ [13].

The memristor is said to be charge controlled if $\emptyset = \emptyset(q)$. Then the voltage across the device is given by:

$$V = M(q)I \tag{4}$$

where $M(q)$ is termed as memristance given by:

$$M(q) = \frac{\mathrm{d}\emptyset(q)}{\mathrm{d}q} \tag{5}$$

Memristor is said to be flux controlled if $q = q(\emptyset)$ of the flux where current across it is given by [19]:

$$I = W(\emptyset)V \tag{6}$$

$$W(q) = \frac{\mathrm{d}q(\emptyset)}{\mathrm{d}\emptyset} \tag{7}$$

where $W(q)$ is termed as memconductance.

As described earlier, undoped resistance is considered as $R_{\text{off}}$ while the doped is $R_{\text{on}}$ and the total resistance obtained is:

$$R_{\text{total}} = R_{\text{on}}.\frac{W}{D} + R_{\text{off}}\left(1 - \frac{W}{D}\right) \tag{8}$$

where $W$ = doped region and $D$ = length of device. Here, low resistance state (LRS) and high resistance state (HRS) is dependent on doping and direction of the current flow.

Hysteresis loop obtained by analysing the $TiO_2$-based memristor at V-I characteristics is described in Fig. 1c reports the memristive behaviour of device even at lower voltage, i.e., 300 mV and can be used in low voltage SRAM cells.

# 3 Design Methodology

This paper has worked on memristor-based SRAM cell in which subthreshold concept is incorporated with memristor-based SRAM cell. It has taken the memristor model from [12, 13] and used LTSPICE software. It has compared four SRAM configurations 6T2M case1 [14], 6T2M case2 [15], 7T1M [16], and 8T2R [17] at VSUB = 0.4 V with 32 nm technology PTM model [18].

# 4 Non-volatile SRAM Cell

## *4.1 6T2M Case1*

This section has discussed about two architecture of basic 6T2M SRAM cell in which memristors are connected in different fashion shown in Fig. 2a and b. Simulation analysis has been done here for both 6T2M structure at subthreshold voltages ($V_{SUB}$ = 0.3 V and $V_{SUB}$ = 0.4 V) to understand its behaviour at lower supply voltage.

As per the reference [14] mentioned about the 6T2M, Fig shown in 2a is analysed at lower voltage VSUB = 0.3 V, and its performance parameters are calculated. As it is seen by the researchers, working at subthreshold voltage with VDD less than 0.6 V fails the circuit read and write ability [7], and therefore, various other SRAM cell at lower voltage is designed with transistors more than 6T [4–7]. But, due to non-volatile characteristics of memristors and its introduction with SRAM cell, it has become possible to operate for 6T also at low voltage applications. Memristor is compatible with CMOS technology with greater scope of fabrication [9] and with less area usage [19].

A non-volatile SRAM (NVSRAM) cell made up of two two-terminal memory (NVM) devices connected with basic 6T SRAM cell. The non-volatile memory devices (NVM) serve as a storage medium and retain the logic state information. The cell can be completely shut down during standby to minimize all leakage currents because the state information is stored in the non-volatile medium. Memristors are directly built on top of the 6T SRAM cell shown in Fig. 2a. As a result, no area overhead and data transmission requires only a modest amount of power and time.

Simulation results as shown in Fig. 3a is described to estimate the parameters like delay, power, energy, and memristance where it is seen that $V_q$ and $V_{qb}$ is opposite to each other till 12 ns and after that it started following each other which is not as good but the circuit is in working condition even at low voltage. The main disadvantage seen in first 6T2M is power consumption. Since the memristors are connected to power supply directly, flow of current directly dissipated in memristor making one NVM in high resistance state (HRS) and other in low resistance state (LRS). Every parameter is summarized in Table 2 and compared with other existing results in Table 3. Stability is calculated in next section using $N$-curve method.

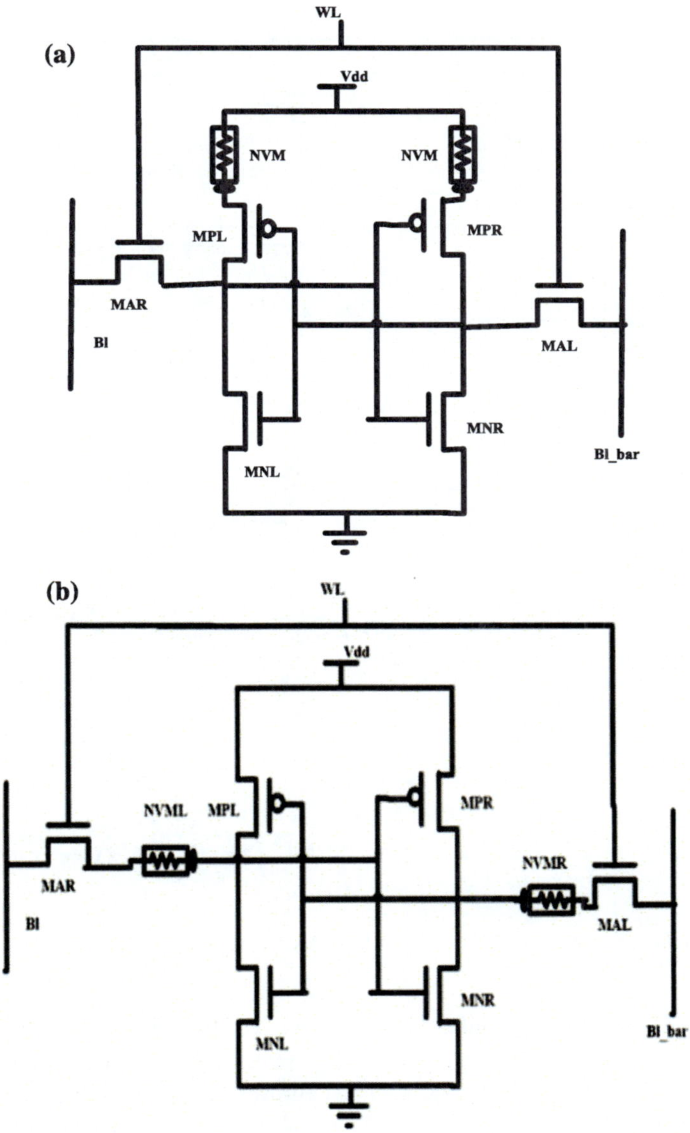


**Fig. 2** **a** Structure of 6T2M case1 SRAM cell and **b** structure of 6T2M case2 SRAM cell

In another architecture of 6T2M shown in Fig. 2b, memristors are connected to the storage nodes instead of connecting to the main logic of SRAM cell. As depicted in Ref. [15], this circuit is been used where RRAM device is connected which $HfO_2$-based resistive memory where as in our work, Fig. 2b has used $TiO_2$-based memristor. Apart from this, this paper is working on subthreshold voltages, and so circuit is analysed and simulated at $V_{SUB} = 0.4$ V and Ref. [15] is simulated

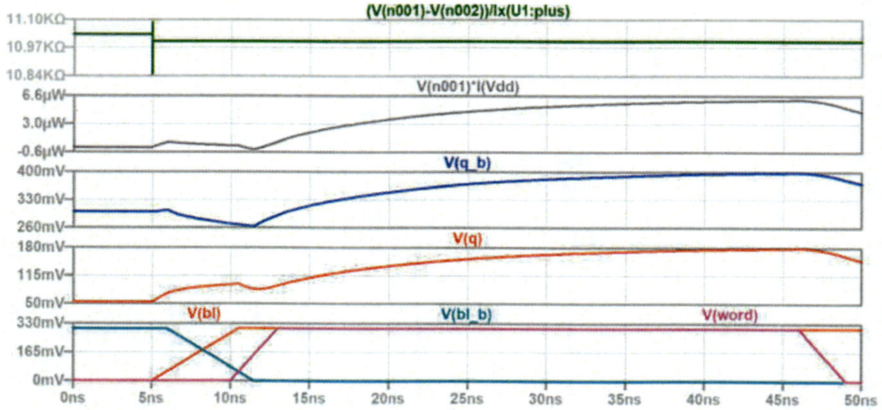


**(a)**

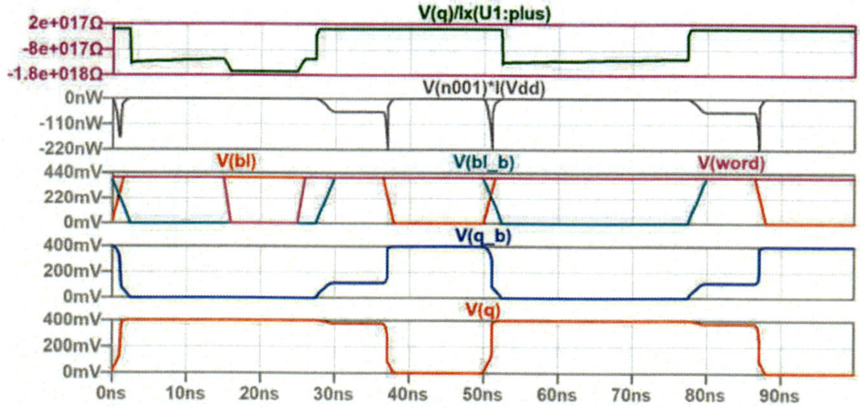


**(b)**

**Fig. 3** **a** Simulation result of 6T2M case1 and **b** simulation result of 6T2M case2

at $V_{DD} = 0.6$ V and has focused only on SET and RESET process but this work has calculated its power, energy, memristance, and stability. Simulation result shown in Fig. 3b has obtained far better results when compared to previous simulated 6T2M SRAM cell. $V_q$ and $V_{qb}$ is also opposite to each other with full range of 400 mV voltage. Power consumption is quite less, and values are discussed in Table 2 later.

## *4.2 7T1M SRAM Cell at Subthreshold*

Figure 4 depicts the 7T 1R NVSRAM design. Only one 1T1R (with a resistive element labelled as NVM) is added to the 6T SRAM core in this circuit (MAR, MAL MPR, MPL, MNR, and MNL). The transistor MWL drives NVM, which is

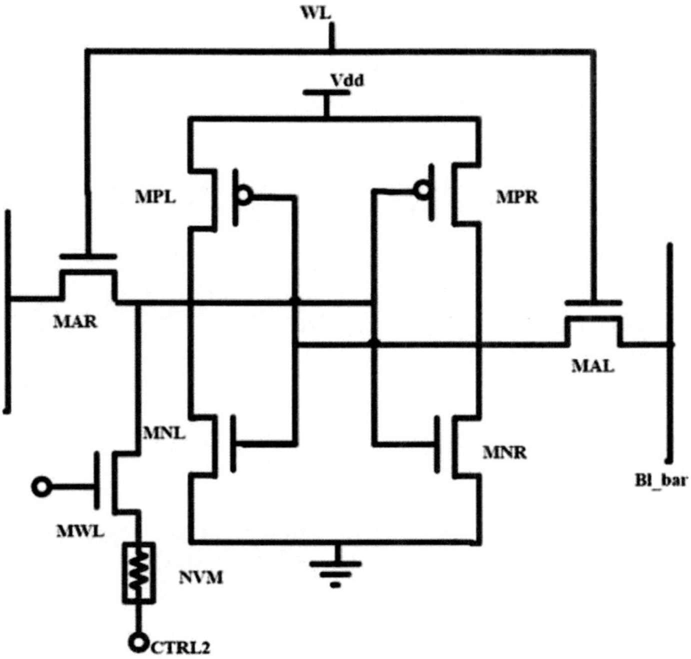


**Fig. 4** Structure of 7T1M SRAM cell

directly attached to the data storage node of the memory core and is used to store the logic information of the SRAM when it is in the "Power down" state. The transistor sizing strategy for creating the 7T1R is based on the cell's core (in this case, a 6T SRAM) and must take into account the correctness of its read/write operation.

The access transistors MAL and MAR are turned on during this store process, and Bitline (BL) "writes" into the storage node *D*. Meanwhile, MWL is turned on and CTRL2 is set to 0 V, allowing NVM to be programmed based on the voltage at node *D*. As a result, if *D* is "1", the positive potential drop on NVM causes the state to switch from HRS to LRS. As a result, both writing data to the 6T SRAM core and programming NVM are finished in one process.

The transistor MWL is switched ON for the restore operation, CTRL2 is set to high, and the power supply is turned ON. In the meantime, $V_{ss}$ is low. When RRAM1 is in LRS, the storage node *D* remains at "1," and DN is discharged via MNR. There is an voltage increase at storage node *D* when NVM is in HRS; however, the quick voltage increase at node DN turns ON transistor MPL. Since CTRL2 is connected through the substantially higher resistance of NVM, storage *D* eventually discharges back to "0". As a result of the asymmetric architecture of this cell, the execution of "Restore" "0" is more difficult than "Restore" "1". The correct execution of the preceding processes is demonstrated further through simulation.

The simulation result is shown in Fig. 5 at subthreshold $V_{\text{SUB}} = 0.3$ V is very challenging because operating at such a low voltage has become very promising

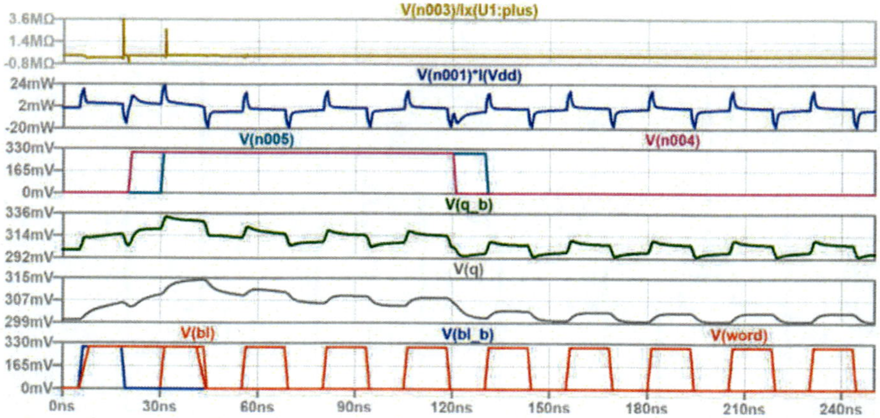


**Fig. 5** Simulation result of 7T1M SRAM cell

these days. As depicted in [16] where this circuit is simulated at greater voltage and has obtained various results with improved WM. 7T1M SRAM cell in this work has obtained the results when simulated at $V_{SUB} = 0.3$ and 0.4 V, and its RSNM/WSNM is also calculated further to understand its performance parameter which is better than existing and more advantageous. As is it seen in Fig. 5, $V_Q$ and $V_{QB}$ obtained is nearly moving within a same pattern with slight difference in voltage. CTRL1 and CTRL2 is denoted by $V$(n005) and $V$(n006), respectively. Then, power waveform is shown whose average is calculated and compared in Table 2, memristance change at this low voltage operation is quite good. Hence energy, power, memristance, and SNM is calculated at subthreshold and compared with existing to understand the benefits of low voltage circuits.

## *4.3 8T2M SRAM Cell at Subthreshold*

There are already various emerging non-volatile memories proposed for achieving functioning in low voltage operation with greater speeds. RRAM. FERAM, MRAM, MTJ, PCM, etc. As depicted in [17], 8T2M circuit is discussed where resistive memory device has been used which is still under development (RRAM) with structure TiN/Ti/HfOx/TiN [17] wherein this paper has evaluated the same circuit 8T2R at subthreshold voltage $V_{SUB} = 0.4$ V using a $TiO_2$-based memristor as discussed in this paper instead of $HfO_2$-based resistive memory device (discussed in paper [17]). In the coming generation memory, RRAM or memristor is a highly promising candidate because of its simpler metal-oxide structure, low voltage application, fast switching activity, less energy and power consumption, and better stability [14–17].

Figures 6 and 7 show the circuit diagram of 8T2M and its simulation waveform.

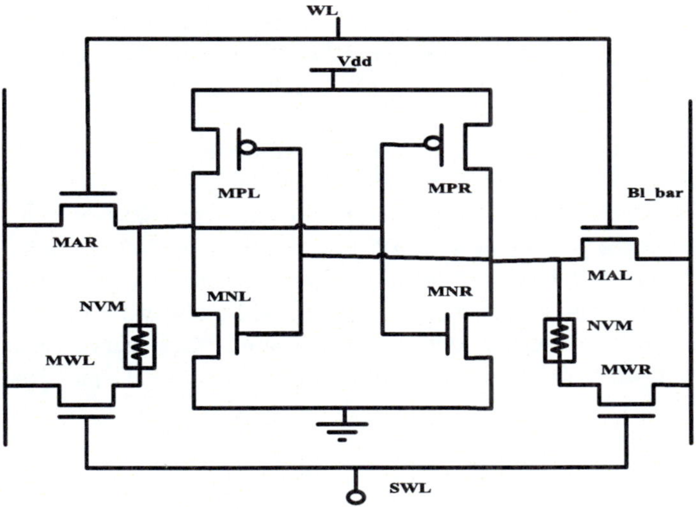


**Fig. 6** Structure of 8T2M SRAM cell

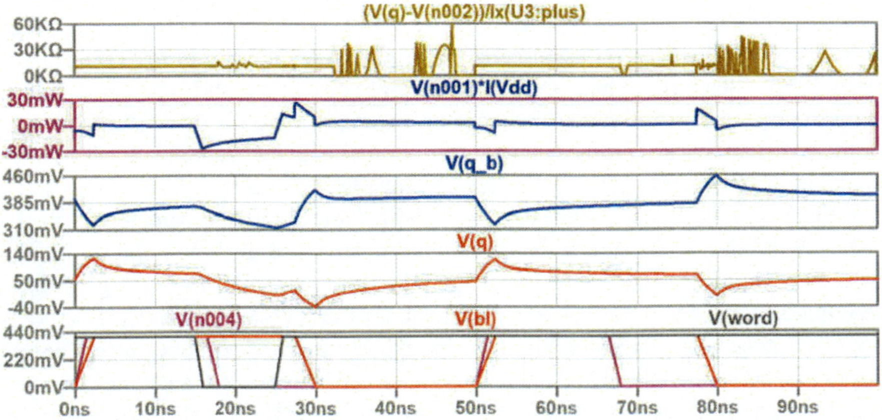


**Fig. 7** Simulation result of 8T2M SRAM cell

All of the benefits of 6T SRAM, including as faster read/write and low voltage operation, are inherited by the 8T2M cell. To enable the storing of complementary backup data while keeping non-volatile features, the resistive devices (memristors) are attached directly to SRAM storage nodes ($Q$ and QB). The 8T2M cell shares bitlines (BL) with the NVM control-line (CL) named as SWL to reduce area overhead, unlike other NVSRAM cells that require an additional control-line (CL) to execute store operations. NVMs can now perform SRAM-mode write-assist functionalities beyond their original storage capacity because of the BL-CL sharing arrangement.

The switch-line (SWL) is kept low during read operations to turn OFF the MWs (MWL and MWR) and prevent the SRAM cell from becoming unstable in the same way as a traditional 6T SRAM, the 8T2M cell also conducts a differential read with storing ability. Both the wordline (WL) and the SWL are switched high during write operations. Through the pass-gates and MWs, the data on the BL/BLB is written into the SRAM cell.

Simulation result shown in figure tells about the performance parameter which is better than the previous 7T1M (asymmetric structure). It shows storage node $V_q$ and $V_{qb}$ apposite to each other but $V_Q$ reaches max 140 mV–min 40 mV which is around 100 mV difference at $V_{SUB} = 400$ mV. $V_{QB}$ reaches max 460 mV–min 310 mV which is 150 mV difference. At this low voltage, storage node behaviour is good. One of the very important parameter to be improved in this work motive is power dissipation, and so low supply voltage concept is used. Memristance obtained is 60 kΩ which is lower than all the cases with energy calculations. Comparison is done later in Table 2.

## 5 *N*-Curve Analysis

The robustness and stability of the SRAM cell is measured using an alternate noise margins technique called as *N*-curve metrics (NCM). This paper has adopted the *N*-curve method in which SRAM cells made with few changes in the circuit. An external DC voltage is applied at the input storage node as $V_{in}$ and is made to swept from 0 to $V_{DD}$ and at the same time, current flowing through it is measured as $I_{in}$. For the NCM analysis, a bitlines and WL are applied with DC voltages (0–$V_{DD}$) and $I_{in}$ versus $V_{in}$ characteristics [20–22] are illustrated as shown in Fig. 8.

In the *N*-curve analysis, basically, four metrics are found and calculated which is SVNM, SINM, WTI, and WTV. This curve is analysed at three points as *A*, *B*, and *C* as described in Fig. 8. Static voltage noise margin (SVNM)—voltage difference obtained between point *A* and *B* tells about the SVNM which describes the maximum noise voltage is tolerable at the internal storage nodes. static current noise margin (SINM)—positive peak current obtained between point *A* and *B* tells about the maximum injected current that is required by the SRAM cell to flip the content. It characterizes the read stability. Write trip current (WTI)—third metrics tells about the amount of current required for writing into the cell. Negative peak obtained between point *B* and *C* is the WTI. Write trip voltage (WTV)—this is voltage difference between *B* and *C* required to flip the internal node content with both bitlines raised to $V_{DD}$ (or dc voltage).

For getting better read and write ability, these metrics are measured in which it is beneficial to achieve maximum values of SVNM and SINM for better read stability and minimum values of WTI and WTV for better write ability. *N*-curve analysis gives information about read and write ability in single simulation process.

After analysis of all four circuits simulated in this paper, with the help of *N*-curve analysis, stability of each circuit is measured by knowing the metrics SVNM, SINM, WTV, and WTI. Figure shows the result of each circuit 6T2M case1, 6T2M case2, 7T1M and 8T2M at subthreshold voltage $V_{SUB} = 0.4$ V. Its calibrated values are

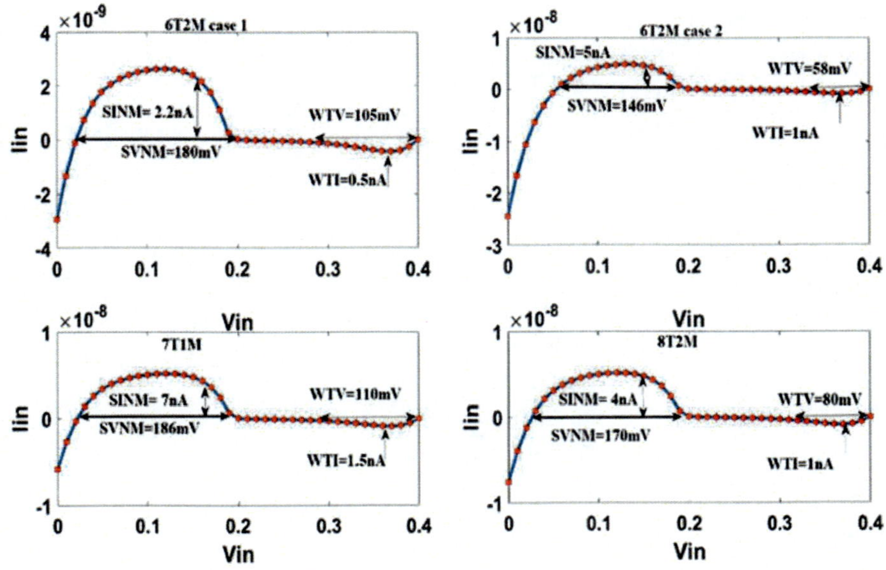


**Fig. 8** *N*-curve analysis of 6T2M case1, 6T2M case2, 7T1M, and 8T2M SRAM cells

together mentioned in Table 1. It is concluded that 7T1M has better read stability with SINM = 7 nA and SVNM = 185 mV, and then 8T2M has good read stability followed by both 6T2M. When talking about write stability. 6T2Mcase2 and 8T2M shows better stability as compared to other two.

It is finally summarized from Tables 1, 2 and 3 that 7T1M has got good read stability and least power dissipation when compared with all four cases simulated at subthreshold voltage $V_{SUB} = 0.4$ V. As in existing 7T1M in [16], only write stability is discussed and improved as read stability doesn't improve at higher supply voltage. Similarly, 6T2M case2 [15] and 8T2M [17] has used ReRAM $HfO_2$ based where this work has used $TiO_2$-based memristor and has achieved good write ability with less energy. Table 3 has described the existing values of the work for the comparison purpose. This paper has not focused on delay as delay decreases with scaling $V_{DD}$, and there is a sacrifice in speed when the power and energy is more significant.

**Table 1** *N*-curve metrics of all four SRAM cell simulated at subthreshold voltages $V_{sub} = 0.4$ V

| SRAM metrics | 6T2Mcase1 | 6T2M case2 | 7T11M | 8T2M |
|---|---|---|---|---|
| SINM (nA) | 2.2 | 5 | 7 | 4 |
| SVNM (mV) | 180 | 146 | 186 | 170 |
| WTV (mV) | 105 | 58 | 110 | 80 |
| WTI (nA) | 0.5 | 1 | 1.5 | 1 |

**Table 2** Parameters obtained at subthreshold voltage

| Parameters | 6T2M case1 | 6T2M case2 | 7T1M | 8T2M |
|---|---|---|---|---|
| Avg. power (nW) | 137.6 | 34.696 | 1.6702 | 48.76 |
| Total energy | 4.878 pJ | 13.769 pJ | 417.54 fJ | 308.25 fJ |
| Memristance | 11 kΩ | 30 MΩ | 4.4 MΩ | 60 kΩ |

**Table 3** Parameters of existing circuit for comparison

| Parameters | 6T2M [15] | 7T1M [16] | 8T2M [17] |
|---|---|---|---|
| Tread (ps) | 35 | 31.40 | 50.6 |
| Twrite (ps) | 27.6 | 38.54 | 26.4 |
| Power (μW) | 5.99 | 3 | 20.54 |
| RSNM (mV) | 3.8 | 172 | 224 |
| WSNM (mV) | 258 | 440 | 337 |

# 6 Conclusion

Low supply voltage ($V_{\text{DDmin}}$) is the main aim in this paper as it is used extensively reducing dynamic power consumption. There are lot of advantages seen when NVSRAM are operable at low $V_{\text{DD}}$ below 0.5 V but unfortunately, circuit suffers from read and write failure in basic 6T, and therefore, all previous memristor-based SRAM are proposed at higher $V_{\text{DDmin}}$ with SRAMs more than 6T. This paper has taken an approach to analyse circuit 6T2M, 7T1M, and 8T2M at low voltage (subthreshold voltage) with results in terms of power, energy, memristance, and stability and made comparative analysis. 7T1M has obtained better read stability as well as least power consumption, 8T2M with less energy consumption and 6T2M case2 has come up with better write ability at this low voltage and can be used in low voltage applications.

# References


1. Calhoun BH, Chandrakasan AP (2006) Static noise margin variation for sub-threshold SRAM in 65-nm CMOS. IEEE J Solid-State Circuits 41(7):1673–1679. https://doi.org/10.1109/JSSC.2006.873215
2. Semiconductor Industry Association (SIA) International technology roadmap for semiconductors 2011 edition [online]. Available http://www.itrs.net/Links/2011ITRS/Home2011.htm
3. Razavipour G, Afzali-Kusha A, Pedram M (2009) Design and analysis of two low-power SRAM cell structures. IEEE Trans Very Large Scale Integr (VLSI) Syst 17(10):1551–1555. https://doi.org/10.1109/TVLSI.2008.2004590
4. Calhoun BH, Chandrakasan AP (2007) A 256-kb 65-nm sub-threshold SRAM design for ultra-low-voltage operation. IEEE J Solid-State Circuits 42(3):680–688. https://doi.org/10.1109/JSSC.2006.891726

5. Calhoun BH, Wang A, Verma N, Chandrakasan A (2006) Sub-threshold design: the challenges of minimizing circuit energy. In: Proceedings of the 2006 international symposium on low power electronics and design. ACM, pp 366–368. https://doi.org/10.1145/1165573.1165661
6. Islam A, Hasan M, Arslan T (2012) Variation resilient subthreshold SRAM cell design technique. Int J Electron 99(9):1223–1237. https://doi.org/10.1080/00207217.2012.669708
7. Ramani AR, Choi K (2011) A novel 9T SRAM design in sub-threshold region. IEEE Int Conf Electro/Inf Technol 2011:1–6. https://doi.org/10.1109/EIT.2011.5978615
8. Seo HJ, Song DK, Lee JH, Ahmed MG, Cho TW (2011, June) A study on the memristor-based non-volatile 4T static RAM cell. In: The 26th international technical conference on circuits/systems, computers and communications
9. Eshraghian K, Cho K, Kavehei O, Kang S, Abbott D, Kang SS (2011) Memristor MOS content addressable memory (MCAM): hybrid architecture for future high performance search engines. IEEE Trans Very Large Scale Integr (VLSI) Syst 19(8):1407–1417. https://doi.org/10.1109/TVLSI.2010.2049867
10. Chua L (1971) Memristor—the missing circuit element. IEEE Trans Circuit Theor 18(5):507–519. https://doi.org/10.1109/TCT.1971.1083337
11. Strukov DB, Snider GS, Stewart DR, Stanley Williams R (2008) The missing memristor found. Nature 453(7191):80–83. https://doi.org/10.1038/nature06932
12. Biolek Z, Biolek D, Biolkova V (2009) SPICE model of memristor with nonlinear dopant drift. Radioengineering 18(2)
13. Biolek D, Di Ventra M, Pershin YV (2013) Reliable SPICE simulations of memristors, memcapacitors and meminductors. Radioengineering 22(4):945–968
14. Wang W, Gibby A, Wang Z, Chen TW, Fujita S, Griffin P, Nishi Y, Wong S (2006) Nonvolatile SRAM cell. IEDM Tech Dig 2006:1–4. https://doi.org/10.1109/IEDM.2006.346730
15. Amara A, Vladimirescu A, Anghel C, Thomas O (2014) Design challenges and solutions for non-volatile SRAMs. In: 2014 IEEE faible tension faible consommation, pp 1–4. https://doi.org/10.1109/FTFC.2014.6828611
16. Wei W, Namba K, Han J, Lombardi F (2014) Design of a nonvolatile 7T1R SRAM cell for instant-on operation. IEEE Trans Nanotechnol 13(5):905–916. https://doi.org/10.1109/TNANO.2014.2329915
17. Chiu P et al (2012) Low store energy, low $V_{\mathrm{DDmin}}$, 8T2R nonvolatile latch and SRAM with vertical-stacked resistive memory (memristor) devices for low power mobile applications. IEEE J Solid-State Circuits 47(6):1483–1496. https://doi.org/10.1109/JSSC.2012.2192661
18. Predictive technology model [online]. Available http://ptm.asu.edu/
19. Louis VJ, Pandey JG (2019) A novel design of SRAM using memristors at 45 nm technology. In: International symposium on VLSI design and test. Springer, Singapore, pp 579–589. https://doi.org/10.1007/978-981-32-9767-8_48
20. Gupta P, Gupta A, Asati A (2015) Leakage immune modified pass transistor based 8T SRAM cell in subthreshold region. Int J Reconfig Comput 2015. https://doi.org/10.1155/2015/749816
21. Grossar E, Stucchi M, Maex K, Dehaene W (2006) Read stability and write-ability analysis of SRAM cells for nanometer technologies. IEEE J Solid-State Circuits 41(11):2577–2588. https://doi.org/10.1109/JSSC.2006.883344
22. Samson M, Srinivas MB (2008) Analyzing N-curve metrics for sub-threshold 65 nm CMOS SRAM. In: 2008 8th IEEE conference on nanotechnology. IEEE, pp 25–28

# Electrochemical Impedance Spectroscopy Analysis of BiMetallic Au–Cu/g-$C_3N_4$ Nanocomposite as a Supercapacitor Electrode Material

**Sebina Yesmin, Rajdeep Dasgupta, and Siddhartha S. Dhar**

## 1 Introduction

The necessity for effective energy storage devices has been increased due to the rush to substitute conventional energy resources due to their fast reduction and carbon emissions. One such technology is the electrochemical energy storage system [1, 2]. Supercapacitors ensure a better specific power and lower specific energy than batteries. However, in a case where high specific energy and specific power are required, both devices fall short [3]. The electrode materials evaluate the efficiency of supercapacitors, and choosing a proper electrode material is one of the most challenging tasks. Many materials, such as carbonaceous material [4–7], metal oxides [8, 9], and polymers [10], are being used as supercapacitors electrode material. Supercapacitors are divided primarily based on their energy storage techniques and electrode materials [3]. Graphitic carbon nitride (g-$C_3N_4$), which has a distinctive arrangement similar as graphite, is one of the most favorable materials for supercapacitors [11]. Different heterostructure composites [12–16] of g-$C_3N_4$ with enhanced electrochemical property have been reported to address the limitations of g-$C_3N_4$ for supercapacitors application. Due to the brilliant tuning properties of g-$C_3N_4$, it can be applied in different fields of applications to improve the overall performances [17–19]. Here, bimetals gold and copper nanoparticles are impregnated over the g-$C_3N_4$ surface to study its performance for energy storage application via different electrochemical analyses.

S. Yesmin (✉) · R. Dasgupta
Department of Electronics and Instrumentation Engineering, National Institute of Technology Silchar, Silchar, Assam, India
e-mail: sebina.nits@gmail.com

S. S. Dhar
Department of Chemistry, National Institute of Technology Silchar, Silchar, Assam, India
e-mail: ssd@che.nits.ac.in



P. H. J. Chong et al. (eds.), *Emerging Electronics and Automation*, Lecture Notes in Electrical Engineering 937, https://doi.org/10.1007/978-981-19-4300-3_39

In the perspective of energy presentations, cyclic voltammetry (CV), galvanostatic charge–discharge (GCD), and electrochemical impedance spectroscopy (EIS) analysis are commonly employed to investigate the efficiency of various electrode materials of supercapacitors. Among all these analyses, EIS is regarded as a solid and low-rate implement for the investigation of charge kinetics due to its non-destructive methodology and ability to take measurements in operational conditions [20–22]. EIS spectra, when used in conjunction with an analogous circuit layout, support in determining the physical basis of the device's various electrical apparatuses. EIS is valuable because it can identify a variety of interfaces (e.g., electrode/electrolyte, electrode/electrode) inside the devices based on their frequency spectrum and the resulting decoupling of various resistive and capacitive circuits parameters [21]. The purpose of this study was to develop an RC circuit and examine the charge kinetics of electrode materials using the EIS analysis to determine the performance of bimetallic nanocomposites grafted over g-$C_3N_4$ electrodes for energy storage applications. This research reveals the various physical processes occurring in Au–Cu/g-$C_3N_4$ composite electrodes with changing weight percent of Au and Cu over g-$C_3N_4$ sheets by calculating different electrical parameters determined from the equivalent circuit.

# 2 Experimental Details

## 2.1 *Synthesis and Electrode Fabrication*

Using melamine as a precursor, a simple approach [23] was used to make pure g-$C_3N_4$. In a muffle furnace at 10 °C/min heating rate, thermal polymerization of 5 gm melamine was carried out at 500 °C for 4 h in a crucible and obtained g-$C_3N_4$ as a light yellow powder. Bimetals gold and copper were decorated over the g-$C_3N_4$ surface with a simple typical reaction [24]. The materials were made into working electrodes by combining the composites, carbon black, and polyvinylidene difluoride in N-Methyl-2-pyridine solvent as 80:10:10 mass ratio by finely grinding it in an agate mortar for 30 min. This slurry was coated on the surface area of 1 $cm^2$ graphite sheet and dried overnight before use as an electrode.

## 2.2 *Electrochemical Measurements*

All electrochemical analyses were performed on a Zahner pro electrochemical instrument. The active material was used as working electrode; Pt wire and Ag/AgCl were acted as counter and reference electrodes, respectively, in the 3-electrode method.

# 3 Results and Discussions

## *3.1 Structural, Morphological, and Thermal Analysis*

Different structural and morphological analyses were performed to investigate the material internal structure and chemical behaviors. The preliminary structural nanostructures were measured with X-ray diffractograms (XRD) using a Bruker, Cu–K source machine. XRD pattern for Au–Cu/g-$C_3N_4$ is shown in Fig. 1a, and the existence of two distinct peaks for pristine g-$C_3N_4$ at 13.2 and 27.5 verified the amorphous nature of g-$C_3N_4$. The gold nanoparticles produced over the g-$C_3N_4$ surface are responsible for the peaks at 38.1, 44.4, 64.5, and 77.9, along with the two classic peaks for g-$C_3N_4$. Owing to the amorphous nature of Cu nanoparticles, they do not show any identifiable peaks on the g-$C_3N_4$ sheets. FT-IR analysis was used to explain the composition and bonding of the produced nanostructures. Bruker 3000 Hyperion microscope was used to take FT-IR spectra. The FT-IR spectra of Au–Cu/g-$C_3N_4$ in Fig. 1b display the same pattern as that of g-$C_3N_4$, with a minor redshift owing to the existence of Au and Cu nanoparticles on the surface of g-$C_3N_4$. A JEOL scanning electron microscope (SEM) and a JEOL high-resolution transmission electron microscope (HRTEM) were used to examine the surface morphology of nanostructures, and Fig. 1c, d shows the FESEM and HRTEM images of Au–Cu/g-$C_3N_4$. The overall structural and morphological characterization showed that Au–Cu/g-$C_3N_4$ nanocomposites had superior performance in supercapacitors electrode application [24].

## *3.2 Electrochemical Impedance Spectroscopy Analysis*

Cyclic voltammetry (CV), galvanostatic charge–discharge (GCD), and electrochemical impedance spectra (EIS) measurements are employed to analyze the electrochemical performances. All the electrochemical experiments were accomplished in 0.5 M $H_2SO_4$ electrolyte. There are various parameters to evaluate the performances of supercapacitor electrode materials, such as specific capacitance, rate capability, cycle stability, energy density, and power density, and all these parameters can be obtained from CV and GCD measurements. CV and GCD measurements were performed in acidic medium electrolytes by considering various scan rates and current densities. Energy density and power density were obtained by fabricating a solid-state asymmetric supercapacitor. Further, the cyclic performance was analyzed to evaluate all synthesized composites cyclic constancy and rate capability. EIS investigations are performed to validate the results obtained from CV and GCD analysis to learn more about the synergistic effect between materials and electrode charge kinetics by realizing the complete influence of different resistances on the electrode materials [20]. In this article, EIS analysis of bimetallic Au–Cu/g-$C_3N_4$ nanocomposite is broadly investigated with Nyquist, Bode, and complex capacitance plots

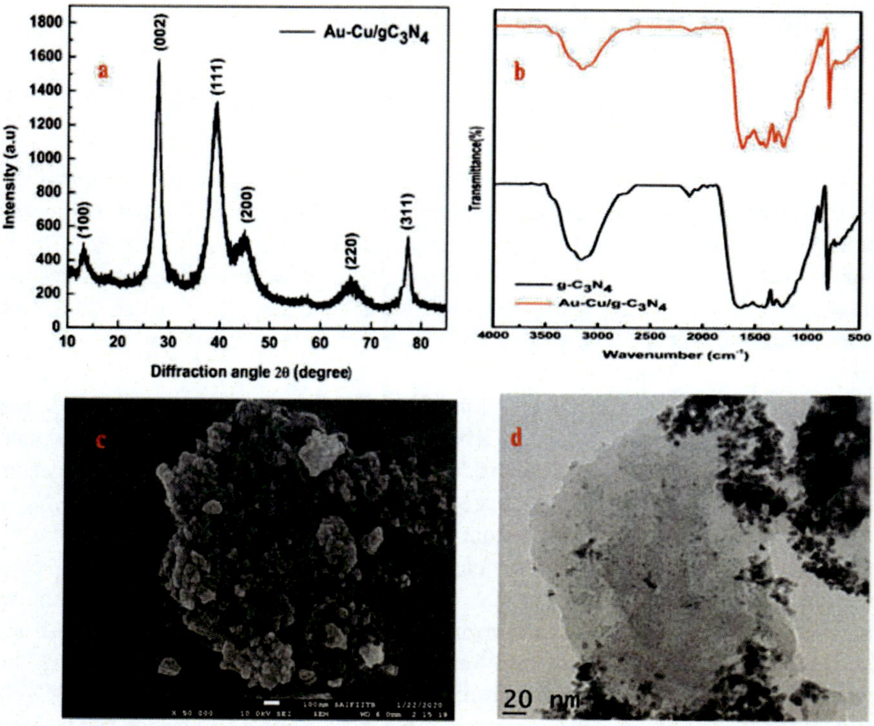


**Fig. 1** **a** XRD, **b** FT-IR, **c** FESEM, and **d** HRTEM images of Au–Cu/g-$C_3N_4$

from 0.1 Hz to 1 MHz frequency ranges. Apart from EIS analysis, some parameters are essential to evaluate the overall performances of supercapacitor electrode material, and those parameters related to Au–Cu/g-$C_3N_4$ nanocomposites were well explained and reported in our previous work [24]. A comparative study of the earlier works in the supercapacitor electrode application is shown in Table 1.

Figure 2a, b shows the Nyquist plots of all synthesized materials and depicts the impedance responses. The diffusion between electrodes and eelctrolytes is generally explained by the Nyquist plot obtained from EIS spectrum [20]. This diffusion results formation of a Warburg impedance which shows a diagonal line with a 45° gradient on a Nyquist plot. The frequency of the potential perturbation determines the impedance. The Warburg impedance is lower because diffusing reactants do not travel as far at high frequencies. The reactants must diffuse even further at low frequencies, resulting in a higher Warburg impedance. The high-frequency region is observed as a semicircle. It reflects the combinational resistance known as ESR of interfacial resistance between electrolyte/electrode interface ($R_{ct}$) and resistance of electrolyte ($R_s$) [21]. The active electrode's capacitance will be enhanced if the semicircle region has a smaller diameter. A lower diameter of the arc region in the Nyquist plot gives less combinational resistance due to the effective electrons transfer and synergistic effect between materials present in the composite. The $R_s$ value of Au–Cu/g-$C_3N_4$ is less

**Table 1** Performance comparison of our results with some other related previous works

| Electrode material | Specific capacitance | Cyclic performance | Refs. |
|---|---|---|---|
| Au–Cu/g-$C_3N_4$ | 506.25 $Fg^{-1}$ at current density 1.25 A/g | 76.1% at 5000 cycles | [24] |
| NFC/g-$C_3N_4$ | 51 $Fg^{-1}$ at current density 1 $mA/cm^2$ | 2000 cycle at 1 $mA/cm^2$ | [25] |
| rGO/AuNPs@ PANI | 212.8 $Fg^{-1}$ at 1 A/g current density | 86.9% at 5000 cycles | [26] |
| Porou carbon @AuNPs | 421.4 $Fg^{-1}$ at current density of 0.5 A/g | 96.4% at 10,000 cycles | [27] |
| NPG/g-$C_3N_4$ | 440 $Fg^{-1}$ at current density 2 A/g | 98% at 10,000 cycles | [28] |
| $MnO_2$–AuNPs | 225 $Fg^{-1}$ at current density of 1 A/g | 90% at 5000 cycles | [29] |
| NP-Au/PPY | 37 $mFcm^{-2}$ at scan rate 10 mV/s | 87% at 1000 cycles | [30] |
| Au@NCNC | 80 $Fg^{-1}$ at current density 1 A/g | 93% at 5000 cycles | [31] |
| Au–$V_2O_5$ | 419 $Fg^{-1}$ at current density 1 A/g | 89% at 5000 cycles | [32] |

than g-$C_3N_4$, as shown in Fig. 2b. The lower $R_s$ value of Au–Cu/g-$C_3N_4$ compared to g-$C_3N_4$ may be credited to Au–Cu/g-$C_3N_4$ having less electrical resistance in acidic electrolytes than g-$C_3N_4$ because Au–Cu/g-$C_3N_4$ is chemically more stable in acidic electrolytes. Continuing to this range is the semicircle (*A-B*), which spans the spectrum from high to low frequencies and is known as charge transfer resistance ($R_{ct}$). The lower value of $R_{ct}$ for Au–Cu/g-$C_3N_4$ than g-$C_3N_4$ is attributable to resistance emerging from ion movement in the electrode pores, as seen in Fig. 2b. The high $R_{ct}$ value could potentially be attributed to increased electrical resistance caused by electrode resistance and electrode-to-electrolyte contact resistance. The morphology of Au–Cu/g-$C_3N_4$ is more suitable for high ionic relocation than g-$C_3N_4$ owing to the higher charge accommodation on the surface of Au–Cu/g-$C_3N_4$.

Equivalent circuits fitted to Nyquist plots can yield a variety of parameters [20–22]. The equivalent circuit in this work is for the mixed kinetic and diffusion control phenomenon. The equivalent circuit for these Nyquist plots and all parameters associated with that equivalent circuit is shown in Fig. 2(b inset). It is clear that Au–Cu/g-$C_3N_4$ reflects a smaller diameter and gives ESR of 1.52 and 2.19 Ω for g-$C_3N_4$. This lowering in ESR value is owing to the effective mass composition of Au and Cu over g-$C_3N_4$ toward synergy phenomenon.

The Bode plot, which represents the magnitude and phase angle versus frequency, can provide valuable information [20]. Figure 2c shows the frequency characteristics of the impedance of different synthesized materials, and it was observed that Au–Cu/g-$C_3N_4$ possesses the most negligible impedance among all samples, which results in more capacitance. In the capacitive characteristic region, the higher the

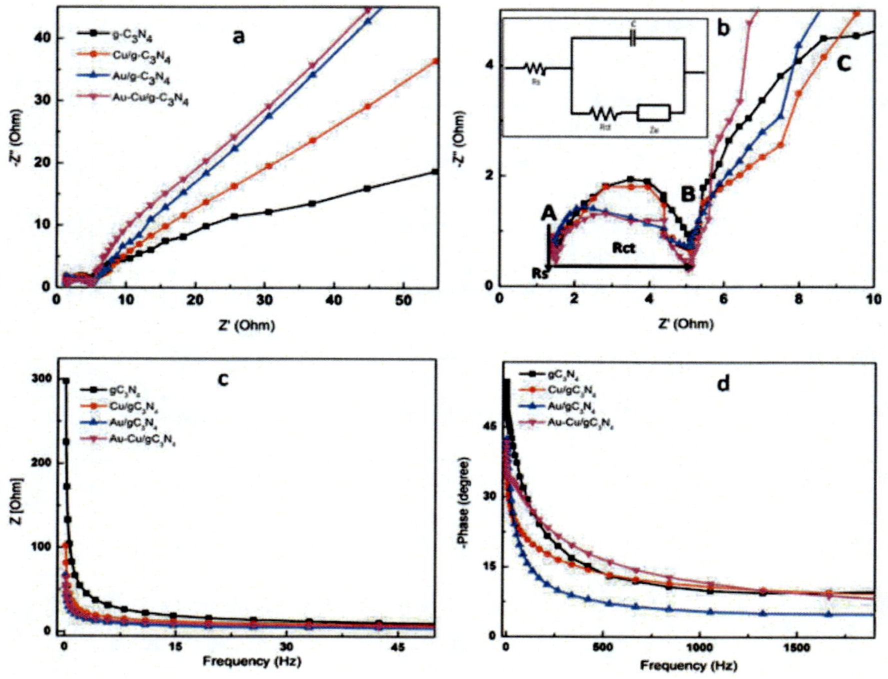


**Fig. 2** Nyquist plots (**a**, **b**), bode plots (**c**, **d**) of all synthesized material, and the equivalent circuit (**b** inset)

capacitance, the lower impedance. The resistive behavior of all samples is the same in the high-frequency range, but the curves begin to diverge slightly in the low-frequency section, as seen in Fig. 2d. Phase angle values found as 54.7, 41.8, 42.4, and 41.3 for g-$C_3N_4$, Cu/g-$C_3N_4$, Au/g-$C_3N_4$, and Au–Cu/g-$C_3N_4$, respectively, demonstrating that all samples had capacitive behavior. Complex capacitance response, which is $C(\omega)$, of all synthesized materials is analyzed based on EIS measurement. Real and imaginary parts of complex capacitance are $C'(\omega)$ and $C''(\omega)$, respectively, and these capacitances are evaluated using these equations [33]

$$C(\omega) = C'(\omega) - jC''(\omega), \omega = 2\pi f \tag{1}$$

$$\mathrm{C}'(\omega) = \frac{-Z''(\omega)}{\omega|Z(\omega)|^2} \tag{2}$$

$$\mathrm{C}''(\omega) = \frac{Z'(\omega)}{\omega|Z(\omega)|^2} \tag{3}$$

where $Z(\omega)$ is the complex impedance, $|Z(\omega)|$ is the impedance modulus, $Z'(\omega)$, $Z''(\omega)$ are real and imaginary parts of $Z(\omega)$, respectively. Figure 3a, b shows the

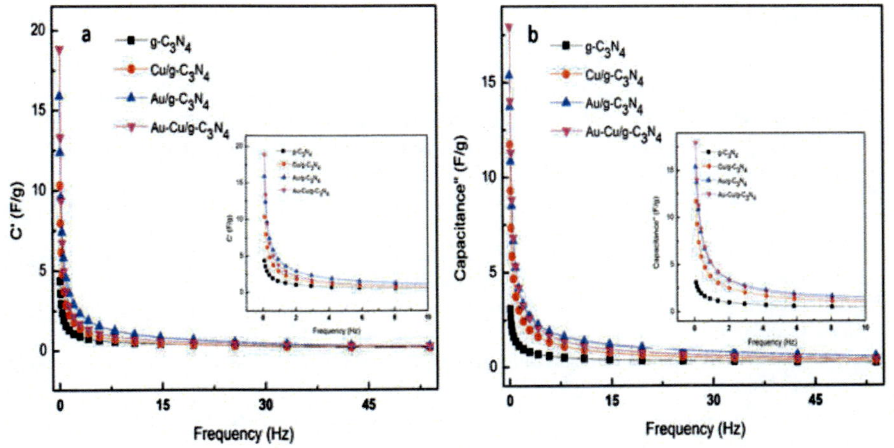


**Fig. 3** **a, b** Complex capacitance analysis of all synthesized material

complex capacitance analysis of all synthesized materials, and it can be seen that the capacitance remains reasonably stable at high frequencies. Real-part capacitance $C'(\omega)$ represents the electrode capacitance and drops dramatically with frequency, while imaginary part capacitance $C''(\omega)$ reflects irreversible energy dissipation and depicts a relaxation process during ion transport.

## 4 Conclusion

A new nanocomposite based on bimetals over graphitic carbon nitride for supercapacitor electrode material has been reported, and various methods were carried out for electrochemical impedance spectroscopy evaluation. The key feature of this electrochemical method was to investigate the dependency of fundamental electrical parameters on the applied frequency spectrum to understand the flow of electrons between the various interfaces by modeling an equivalent circuit. Hence, accurate modeling of equivalent circuits from the EIS data is regarded as one of the most critical tasks to reveal information about the capacitances and resistances of the designed system. In this hybrid electrode material, the mass ratio of components is critical for achieving the optimal synergistic effect, and a lower ESR was obtained Au–Cu/g-$C_3N_4$ (1.52 Ω) than the base material g-$C_3N_4$ (2.19 Ω). As a result, Au–Cu/g-$C_3N_4$ provided overall improved electrochemical properties compared to g-$C_3N_4$ sheets, Au/g-$C_3N_4$, and Cu/g-$C_3N_4$, and our research on bimetallic nanocomposites offers a fresh look at energy storage applications.

**Acknowledgements** The authors would like to acknowledge the SAIF IIT Bombay, STIC Coachin, CIF IIT Guwahati, CIF, and Energy Laboratory, NIT Silchar.

**Conflicts of Interest** There are no conflicts to declare.

## References


1. Gur TM (2018) Review of electrical energy storage technologies, materials and systems: challenges and prospects for large-scale grid storage. Energy Environ Sci 11:2696–2767
2. Svasta P, Negroiu R, Vasile A (2017) Supercapacitors—an alternative electrical energy storage device. Int Symp Electr Electron Eng (ISEEE) 5:1–5
3. Iro ZS, Subramani C, Dash SS (2016) A brief review on electrode materials for supercapacitor. Int J Electrochem Sci 11:10628–10643
4. Yang Z, Ren J, Zhang Z, Chen X, Guan G, Qiu L, Zhang Y, Peng H (2015) Recent advancement of nanostructured carbon for energy applications. Chem Rev 115(11):5159–5223
5. Zhang LL, Zhao XS (2009) Carbon-based materials as supercapacitor electrodes. Chem Soc Rev 38:2520–2531
6. Frackowiak E (2007) Carbon materials for supercapacitor application. Phys Chem Chem Phys 9:1774–1785
7. Su DS, Schlcgl R (2010) Nanostructured carbon and carbon nanocomposites for electrochemical energy storage applications. Chemsuschem 3:136–168
8. Chan PY, Majid SR (2017) Metal oxide-based electrode materials for supercapacitor applications. Adv Mater Appl Micro Nanoscale 13–30
9. Qu B, Hu L, Chen Y, Li C, Li Q, Wang Y, Wei W, Chen L, Wang T (2013) Rational design of Au–NiO hierarchical structures with enhanced rate performance for supercapacitors. J Mater Chem A 1:7023–7026
10. Snook GA, Kao P, Best AS (2011) Conducting-polymer-based supercapacitor devices and electrodes. J Power Sources 196:1–12
11. Luo Y, Yan Y, Zheng S, Xue H, Pang H (2019) Graphitic carbon nitride-based materials for electrochemical energy storage. J Mater Chem A 7:901
12. Xu L, Xia J, Xu H, Yin S, Wang K, Huang L, Wang L, Li H (2014) Reactable ionic liquid assisted solvothermal synthesis of graphite-like $C_3N_4$ hybridized a-$Fe_2O_3$ hollow microspheres with enhanced supercapacitive performance. J Power Sour 245:866–874
13. Liu L, Wang J, Wang C, Wang G (2016) Facile synthesis of graphitic carbon nitride/nanostructured α-$Fe_2O_3$ composites and their excellent electrochemical performance for supercapacitor and enzyme-free glucose detection applications. Appl Surf Sci 390:303–310
14. Wu YZ, Chen M, Yan XH, Ren J, Dai Y, Wang JJ, Pan JM, Wang YP, Cheng XN (2017) Hydrothermal synthesis of $Fe_3O_4$ nanorods/graphitic $C_3N_4$ composite with enhanced supercapacitive performance. Mater Lett 198:114–117
15. Guan B, Shan QY, Chen H, Xue D, Chen K, Zhang YX (2016) Morphology dependent supercapacitance of nanostructured $NiCO_2O_4$ on graphitic carbon nitride. Electrochim Acta 200:239–246
16. Zhao Y, Xu L, Huang S, Bao J, Qiu J, Lian J, Xu L, Huang Y, Xu Y, Li H (2017) Facile preparation of $TiO_2/C_3N_4$ hybrid materials with enhanced capacitive properties for high-performance supercapacitors. J Alloy Compd 702:178–185
17. Hussain I, Kumar M (2017) Design and performance analysis of a 3–2 compressor by using improved architecture. J Active Passive Electron Dev 12:3–4
18. Yesmin S, Chetia D, Basumatary T, Singh HK (2018) Fiber optic sensor for detection of chlorine level in water. In: Advances in communication, devices and networking. Springer, Singapore, pp 429–437
19. Hussain I, Kumar M (2017) A fast and reduced complexity Wallace tree multiplier. J Active Passive Electron Dev 12(1–2):63–71
20. Bredar ARC, Chown AL, Burton AR, Farnum BH (2020) Electrochemical impedance spectroscopy of metal oxide electrodes for energy applications. ACS Appl Energy Mater 3(1):66–98
21. Sunil V, Pal B, Misnon II, Jose R (2021) Characterization of supercapacitive charge storage device using electrochemical impedance spectroscopy. Mater Today Proc 46(4):1588–1594

22. Zhang GQ, Zhao YQ, Tao F, Li HL (2006) Electrochemical characteristics and impedance spectroscopy studies of nano-cobalt silicate hydroxide for supercapacitor. J Power Sour 161(1):723–729
23. Bhuyan B, Devi M, Bora D, Dhar SS, Newar R (2018) Design of a photoactive bimetallic Cu–Au@g-$C_3N_4$ catalyst for visible light-driven hydroxylation of the benzene reaction through C–H activation. Eur J Inorg Chem 34:3849–3858
24. Yesmin S, Devi M, Hussain I, Dasgupta R, Dhar SS (2021) In-situ grafting of Au and Cu nanoparticles over graphitic carbon nitride sheets and unveiling its superior supercapacitive performance as a hybrid composite electrode material. J Energy Storage 44(A):103308
25. Li F, Gu X, Zhang K, Nguyen TT, Guo M (2019) Fabrication of freestanding NFC/g-$C_3N_4$ composite film as supercapacitor electrode via vacuum-induced self-assembly. Vacuum 160:54–59
26. Ciplaka Z, Yildizb A, Yildiza N (2020) Green preparation of ternary reduced graphene oxide-au@polyaniline nanocomposite for supercapacitor application. J Energy Storage 32:101846
27. Albashir AIM, Shang W, Hadi MK, Zhang J, Zhang T, Ran F (2021) Straightforward solution polymerization synthesis of porous carbon@gold nanoparticles electrode for high-performance supercapacitor. J Energy Storage 33:102041
28. Chen AY, Zhang TT, Qiu YJ, Wang D, Wang P, Li HJ, Li Y, Yang JH, Wang XY, Xie XF (2019) Construction of nanoporous gold/g-$C_3N_4$ heterostructure for electrochemical supercapacitor. Electrochim Acta 294:260–267
29. Chen YL, Chen PC, Chen TL, Leeb CY, Chiu HT (2013) Nanosized $MnO_2$ spines on Au stems for high-performance flexible supercapacitor electrodes. J Mater Chem A 1:13301–13307
30. Lu P, Ohlckers P, Chen XY (2016) On-chip supercapacitor electrode based on polypyrrole deposited into nanoporous Au scaffold. J Phys Conf Ser 773:012054
31. Anjana PM, Bindhu MR, Rakhi RB (2019) Green synthesized gold nanoparticle dispersed porous carbon composites for electrochemical energy storage. Mater Sci Energy Technol 2:389–395
32. Rudra S, Nayak AK, Chakraborty R, Maji PK, Pradhan M (2018) Synthesis of Au–$V_2O_5$ composite nanowire through shape transformation of vanadium(III) metal complex for high performance solid-state supercapacitor. Inorg Chem Front 5:1836–1843
33. Yan D, Li Y, Liu Y, Zhuo R, Geng B, Zhiguo W, Wang J, Ren P, Yan P (2015) Design and influence of mass ratio on supercapacitive properties of ternary electrode material reduced graphene oxide@$MnO_2$@ poly(3,4-ethylenedioxythiophene)-poly(styrene sulfonate). Electrochim Acta 169:317–325